Lecture Notes in Artificial Intelligence 10384

Subseries of Lecture Notes in Computer Science

More information about this series at http://www.springer.com/series/1244

Michael Mangan · Mark Cutkosky
Anna Mura · Paul F.M.J. Verschure
Tony Prescott · Nathan Lepora (Eds.)

Biomimetic and Biohybrid Systems

6th International Conference, Living Machines 2017
Stanford, CA, USA, July 26–28, 2017
Proceedings

 Springer

Editors
Michael Mangan (iD)
University of Lincoln
Lincoln
UK

Mark Cutkosky (iD)
Stanford University
Stanford, CA
USA

Anna Mura
Universitat Pompeu Fabra
Barcelona
Spain

Paul F.M.J. Verschure
Universitat Pompeu Fabra
Barcelona
Spain

Tony Prescott (iD)
University of Sheffield
Sheffield
UK

Nathan Lepora
Bristol University
Bristol
UK

ISSN 0302-9743   ISSN 1611-3349   (electronic)
Lecture Notes in Artificial Intelligence
ISBN 978-3-319-63536-1   ISBN 978-3-319-63537-8   (eBook)
DOI 10.1007/978-3-319-63537-8

Library of Congress Control Number: 2017946685

LNCS Sublibrary: SL7 – Artificial Intelligence

Printed on acid-free paper

This Springer imprint is published by Springer Nature
The registered company is Springer International Publishing AG
The registered company address is: Gewerbestrasse 11, 6330 Cham, Switzerland

neuroscientist and cyberneticist Valentino Braitenberg, in his seminal book *Vehicles: Experiments in Synthetic Psychology*. Yet emergent systems by their very nature are notoriously hard to predict particularly when scaling to large integrated systems such as neural networks or collective robots. In this regard, modern computer simulation and embodied hypothesis in robots are now offering previously unfeasible insights into the functioning of many aspects of bio-inspired systems from morphological and neural computation, to sensing and control. Advances in each of these areas were presented in detail at the conference.

In addition, Living Machines 2017 provided an opportunity for researchers working on the next generation of tool-sets to present their work. Such approaches included novel materials for soft robots, next-generation biohybrid 3D printing methods, and bioinspired power systems. All of these offered a glimpse of the tools that may underpin future breakthroughs in the field.

With this installment of the Living Machines conference taking place in the USA for the first time we are reminded of the wonder upon which European explorers considered the boundless possibilities when discovering the Americas with Amerigo Vespucci coining the term "the New World" in 1503. This term appears newly appropriate in the context of the technological possibilities we face as demonstrated by the recent realization of fully automated cars, mind-controlled prosthetics, and AI systems capable of beating the best human Go players. As the relationship between man and artificial systems becomes increasingly intertwined, an enhanced understanding of the ethical, societal, and economic impacts of automated systems will be required. Such philosophical discussions have always been welcomed at Living Machines conferences and will only grow in their importance.

The main conference, during July 25–28, took the form of a three-day single-track oral and poster presentation program that included five plenary lectures from leading international researchers in biomimetic and biohybrid systems: Kwabena Boahen (Standford University, USA) on neuromorphic computing; Robert J. Full (Berkeley University, USA) on comparative biomechanics and physiology; Koh H. Hosoda (Osaka University, Japan) on the use of robots to explain adaptive intelligence of biological systems; Rebecca Kramer (Yale University, USA) on bioinspired manufacturing, materials and robotics; and Cecilia Laschi (Scuola Superiore Sant'Anna, Pisa, Italy) on soft robotics, humanoid robotics, and neurodevelopmental engineering. There were also 22 regular talks and two poster sessions featuring approximately 40 posters. Session themes included: advances in soft robotics; 3D-printed bio-machines; robots and society; biomimetic vision and control; utility and limits of deep learning for biorobotics; collective and emergent behaviors in animals and robots; and bioinspired flight.

The conference was complemented with a day of workshops on July 24, covering three prominent themes related to biomimetic and bioinsired systems: Bioinspired Aerial Vehicles (Alexis Lussier Desbiens, Mark Cutkosky and David Lentink); Evo Devo of Living Machines (Tony Prescott and Leah Krubitzer), and Control Architectures for Living Machines (Paul Verschure and John Doyle).

We would like to thank our hosts at Stanford University. Since its opening in 1891, Stanford University has grown to be one of the world's leading teaching and research universities. It is famed for its ability to take research from the laboratory to the market

Preface

These proceedings contain the papers presented at Living Machines: The 6th International Conference on Biomimetic and Biohybrid Systems, held at Stanford University, USA, during July 25–28, 2017. The international conferences in the Living Machines series are targeted at the intersection of research on novel life-like technologies inspired by the scientific investigation of biological systems, biomimetics, and research that seeks to interface biological and artificial systems to create biohybrid systems. The conference aim is to highlight the most exciting international research in both of these fields united by the theme of "Living Machines."

The Living Machines conference series was first organized by the Convergent Science Network (CSN) of biomimetic and biohybrid systems to provide a focal point for the gathering of world-leading researchers and the presentation and discussion of cutting-edge research in this rapidly emerging field. The modern definition of biomimetics is the development of novel technologies through the distillation of principles from the study of biological systems. The investigation of biomimetic systems can serve two complementary goals. First, a suitably designed and configured biomimetic artifact can be used to test theories about the natural system of interest. Second, biomimetic technologies can provide useful, elegant, and efficient solutions to unsolved challenges in science and engineering. Biohybrid systems are formed by combining at least one biological component — an existing living system — and at least one artificial, newly engineered component. By passing information in one or both directions, such a system forms a new hybrid bio-artificial entity.

Although one may consider this approach to be modern, the underlying principles are centuries old. For example, Leonardo Da Vinci took inspiration from the elegance and functionality he observed in the wings of birds when designing his famous flying machines in 15th-century Italy. Two centuries later, René Descartes, taking inspiration from the automatons he observed in Paris, postulated that the bodies of animals could be considered as nothing more than sophisticated machines. In contrast to Descartes, the mind is now commonly considered to be similarly machine-like allowing investigation through experimentation and replication. The logic of abstracting from nature for engineering purposes is clear in the light of evolutionary theory, which affirms that nature has had millions of years during which to hone biological systems to function robustly in their competitive and complex world.

One reason for the recent expansion and progress of the field of biomimetics is the availability of the necessary tools. For example, the observation of seemingly complex behaviors emerging from the interaction of animals with limited processing capabilities and their environment has been reported many times in nature. Social insects, with their ability to perform complex tasks such as nest building through simple stigmery methods being a foremost example. Building machines with similarly robust functionality governed by computationally cheap methods represents a defining goal for many researchers in this field. Such systems were famously explored by the

Organization

Conference Chairs

Tony J. Prescott   University of Sheffield, UK
Paul F.M.J. Verschure   Universitat Pompeu Fabra and Catalan Institution
                                   for Research and Advanced Studies, Barcelona,
                                   Spain

Program Chair

Michael Mangan   University of Lincoln, UK

Local Organizer

Mark Cutkosky   Stanford University, USA

Communications

Anna Mura   Universitat Pompeu Fabra, Spain
Nathan Lepora   University of Bristol, UK

Conference Website

Anna Mura   Universitat Pompeu Fabra, Spain

Workshop Organizers

Mark Cutkosky   Stanford University, USA
Alexis Lussier Desbiens   Université de Sherbrooke, Canada
John Doyle   Caltech, USA
Leah Krubitzer   University of California, Davis, USA
David Lentink   Stanford University, USA
Tony Prescott   University of Sheffield, UK
Paul Verschure   UPF, IBEC, ICREA, Barcelona, Spain

Program Committee

Andrew Adamatzky   UWE, Bristol, UK
Robert Allen   University of Southampton, UK
Farshad Arvin   University of Manchester, UK
Tareq Assaf   Bristol Robotics Laboratory, UK
Joseph Ayers   Northeastern University, USA

Contents

Straight Swimming Algorithm Used by a Design of Biomimetic Robotic Fish

M.O. Afolayan[✉] [iD]

Mechanical Engineering Department, Ahmadu Bello University, Zaria, Nigeria
tunde_afolayan@yahoo.com

Abstract. *Teleost* species of fish mostly move with their peduncle. They are the fastest moving underwater creature. In this work, focus is specifically on the algorithm that was used for propelling a design of a robotic fish based on Mackerel in a straight swimming motion. The approach used is fundamentally called built in motion pattern algorithm as against follow the leader approach and mathematically generated serpentine motion used in hyper-redundant robot motion control strategies. The design requires just 3 actuators (RC servomotors) that are controlled using Microchip PIC18F4520 microcontroller. The 3 PWM controlling the motors are dynamically adjusted to be at fixed phase to each other at all times. This design was able to produce a travelling wave which propels the robot forward. Though not a very flexible implementation in terms of dynamically reprogramming the robot firmware while executing code, like the other methods that involve onboard mathematical position generation, it can however save battery life. This method works perfectly because of the unique design of the robot hardware. A field test yield 1/3 of the speed (3.66 km/h) of a life Mackerel.

Keywords: Biomimicry · Robotic fish · Swimming

1 Introduction

Fish are known for their fulgurating acceleration inside water. "It is well known that the tuna swims with high speed and high efficiency, the pike accelerates in a flash and the eel swims skillfully into narrow holes" – [1, 2]. Such astonishing swimming ability has inspired several researchers [3–9] to imitate fishes in an attempt to improve the performance of aquatic man-made systems.

Furthermore, a robotic fish as a form of Hyper-Redundant Robots (HRR) [10] has its own merits; such as ability to function after losing mobility in one or more sections, stability on all terrain because of low center of gravity, terrain-ability which is the ability to traverse rough terrain, small size that can penetrate small crevices and are capable of being amphibious by sealing the whole body, the same body motion on land is then used for swimming in water as exemplified by ACM-R5 robot [11, 12]. Unlike conventional rotary propeller used in ships and several underwater vehicles designs, the undulating movement of the body or part of it provides the main propulsion of robotic fishes. The observation on the real fish shows that this kind of propulsion is more noiseless, effective,

© Springer International Publishing AG 2017
M. Mangan et al. (Eds.): Living Machines 2017, LNAI 10384, pp. 1–12, 2017.
DOI: 10.1007/978-3-319-63537-8_1

and maneuverable than the propeller-based propulsion [13]. For *teleost* species of fish like mackerel, the peduncle is more dominant as their propulsive system.

HRR have their own limitations also, such as low speed since the whole body is used for motion, [14], poor thermal control because of low surface to volume ratio, [15]. Also, in hyper-redundant robot joints design, it is a common knowledge that the challenge still exist on how to control the multiple degree-of-freedom joints (DOF) to produce usable motion. A hyper-redundant body can take a very large number of possible shapes without constraints [15–21]. Another well known challenge is selecting or incorporating a compact actuator that will have enough strength and tenacity to carry the weight of other links (or part) and still be fast enough while not generating too much heat [11, 22].

On theoretical side, a number of researchers [9, 13, 23–26] have tried to simulate or modeled fish various swimming modes with different success using various assumptions. It is worth to mention it now that most mathematical models are still in development. Some existing models are resistive models [27], 2D wave plate theory [28], wake theories for oscillating foil propulsion [29], kinematic model [30].

Motion generation in HRR commonly follows either of these three methodologies, [14]; (1) Serpenoid curve (Fig. 1), which was a result of the study on live snakes [12] from which it was discovered that snake motion does not follow sine wave as thought earlier. The parametric equation for the Serpenoid curve is shown in Eqs. (1) and (2). The s is a function of time, x and y are the Cartesian co-ordinates of the snake segment, l is the snake length, J is the segment Jth. (2) Follow the leader approach is another motion generation methodology wherein the head segment (or the tail if reversing) is controlled. The information is passed to the next segment till the last segment is reached. The desired motion is mathematically generated. (3) The third method is the built in motion pattern; in this case, the microcontroller (or microprocessor) will have to adjust each segment according to the predefined motion stored in its memory, it is less mathematically involving but not very flexible.

$$x(s) = sJ_o(\alpha) + \frac{4l}{\pi} \sum_{m=1} \frac{(-1)^m}{2m} J_{2m}(\alpha)\sin\left(m\pi\frac{s}{l}\right) \tag{1}$$

$$y(s) = \frac{4l}{\pi} \sum_{m=1}^{\infty} (-1)^{m-1}\frac{J_{2m-1}(\alpha)}{2m-1}\sin\left(\frac{2m-1}{2}\pi\frac{s}{l}\right) \tag{2}$$

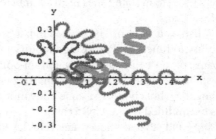

Fig. 1. Serpenoid curves [5]

1.1   Aim of This Work

This work is about a simple algorithm that produces a traveling wave (Fig. 2) suitable for *teleost* species of fish based robot.

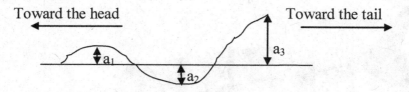

Fig. 2. Teleost fish swimming pattern – tail amplitude increases toward the tail fin. Amplitude $a_1 < a_2 < a_{23}$

1.2   Justification for the Work

Teleost species of fish are the fastest moving creature object inside water and imitating them will be a step further into designing an efficient and fast underwater system, since they are also efficient swimmers. Examples of Teleost fishes are (speed, length): Herring (6 km/h, 0.3 m), Pike (6 km/h, 0.5 m), Carp (6 km/h, 0.8 m), Cod (8 km/h, 1.2 m), Mackerel (11 km/h, 0.5 m), Salmon (45 km/h, 1 m), Bonito (60 km/h, 0.9 m), Black Tuna (80 km/h, 3 m) and Sword Fish (96 km/h, 4 m). Bonito has the highest speed to length ratio (V/L) of 18.6 and the lowest value is by Cod at 1.6 (www.nmri.go.jp/eng/khirata/fish/general/speed/-speede.htm).

2   The Design of the Robotic Fish

2.1   Mechanical Description

Figure 3 shows the 3D CAD model of the robotic fish tail. The rubber joint (A) (stripes of rubber) is sandwiched between pairs of rigid support segments (1) to (6). The support (6) is attached to oval support (B) having six pass through holes (C) for the cables support. The servomotor (D) is attached to the oval support (B) having pass through holes for the cables. The cables are connected to the servomotor lever (E) by tying. The servomotor lever oscillates at angle (F). To get a serpentine motion, the microcontroller uses it's built in pattern generator to control the sequence of turning of the servo motors (D) and hence the segments (1), (3) and (5) they are connected to. It sends the angular displacement information to the servo motors in such a manner that its lever (E) will oscillate at ± angle (F). On both sides of each segment (1) to (6) are located quarter pulleys (H) over which the nylon cable (G) passes before hooking to those segments. Only one cable is shown for clarity. The nylon cables (G) are attached to the servomotor lever (E). Its support passes through the pass-through holes (C). The other segment simplifies the design as they act to restore the joints to their static states and in getting the desired serpentine shape without complicated design – just like nature has simplified its designs by appropriate use of material. Furthermore this approach simplifies the number of motors required and hence the control scheme.

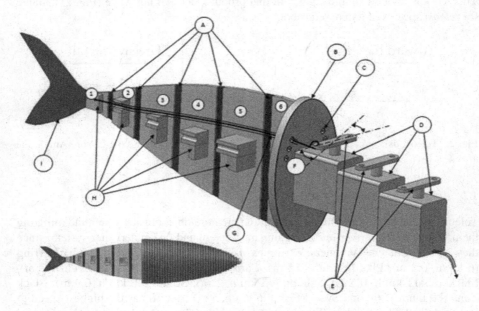

Fig. 3. 3D CAD model of the robotic fish tail. The inset is the complete robotic fish

For swimming to take place, the microcontroller sends angle data to each servomotor using pulse-width-modulated scheme. The servo motors then turn to that angle which are out of phase with the next servomotor. The servomotor then pulls the cable which in turn pulls the segment it is attached to. In this manner, the tail generates a travelling wave that has its origin at the segment (6) and ends at segment (1). The amplitude increases from segment (6) to segment (1).

2.2   Materials Selected

The following materials were selected for constructing the robotic fish tail:

1. The revolute joints are made from 1.5 mm thick vulcanized rubber – from motor car inner tube (14 in. or 35.56 cm) made by King Rubber Tire Company of China;
2. The joint supporting material is 3.175 mm (1/8 in.) thick seasoned plywood - this is equivalent to an insect cuticle;
3. The quarter pulleys is made from 19.05 mm (¾ in.) thick plywood.
4. 4 min setting epoxy glue was used for assembling of the parts.
5. 0.5 mm diameter nylon cables - connect the servomotor lever to the joints.
6. 2.5 mm diameter unplasticized PVC tubing - for supporting the nylon cables.
7. Three RC servomotors (Futaba 3003) acts as the actuators.
8. A PIC18F4520 microcontroller for managing the robot subsystems.

9. A Mackerel (394 mm) fish was selected as the biological model for the robot (that is, it was copied, with the tail design of a fast moving mackerel).

2.3 Control Algorithm

The specification for the robot used for this work is that it should use minimal number of hardware. Therefore, one microcontroller (PIC18F4520) was used for managing the three actuators in addition to generating ultrasonic signals, one for distant measurement and another for object detections and avoidance algorithm and other peripheral functions.

The three RC servomotors was implemented and found to be adequate in controlling the robotic fish motion.

Plain Swimming Mode. The flow chart (Fig. 4) shows a software based PWM signal generator with three rigidly coupled PWM that are always out of phase by a fixed angle. The outputs have continuously varying duty cycles that are never the same at any given point in time. Furthermore, the three PWM outputs have the same period. A deliberate delay is incorporated into the design so that the servomotors can follow the changing pattern in real time. This is necessary because of the inertia of the servomotors. This implementation also allows the microcontroller to perform other functions during the idling time.

The PWM signal generator uses timer based interrupt method mixed with instruction time based method. Timer0 interrupt (INT0) was used as the trigger. The period is fixed at 20 ms. The pulse length could be taken from a table of pre-calculated values (built in motion pattern) or range of sequential values hardcoded into the firmware. The values could also be mathematically generated. The value generation is performed during the idle time.

As soon as there is an interrupt, critical register are saved and the Timer0 is prepared for the next interruption. The Ports are all set high, the difference (lag) between the start of the first and the last is calculated as follows:

$$\text{If Port 0 start time} = T_{port0}$$

$$\text{Then Port X start time} = T_{port0} + T_{port1} + T_{port2} + \ldots . + T_{portx-1}$$

$$= T_{port0} + (x - 1) * T_{CY}$$

where T_{CY} is the length of an Instruction Cycle

$$T_{CY} = (1/32\,\text{Mhz}) * 4 = 0.000000125\,\text{s or }125\,\text{ns} \quad \text{(The clock was set to run at 32 Mhz)}$$

$$\text{Hence the lag for Port x} = (x - 1) * T_{CY} = (x - 1) * 125\,\text{ns}$$

For this scenario, the lag between the first port and last port will be

$$\text{Lag} = (3 - 1) * 125\,\text{ns} = 250\,\text{ns}$$

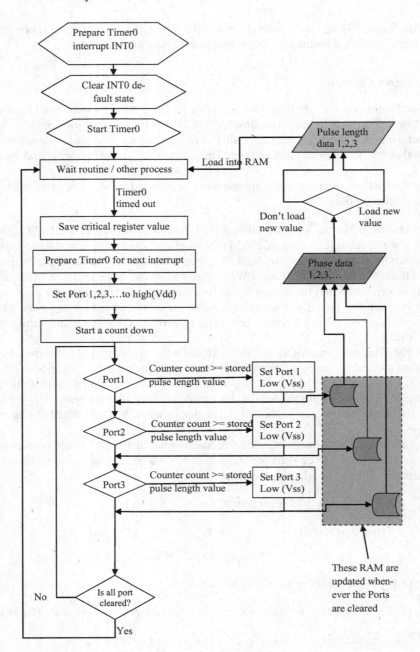

Fig. 4. Flow chart of the PWM generator

Fortunately, this lag will be fixed as the PWM signals have the same reference, i.e. they have the same period that is fired up by a single timer (Timer0). The period for any channel will therefore remain constant at ≈20 ms as designed. The count down process check and compared the RAM value of the pulse against the current count, if greater or equal, the corresponding port is set low. The routine is exited as soon as all the ports are set low.

Swimming Speed. The swimming speed is adjusted using the flowchart of Fig. 5. The PWM generators refer to the Ports outputs of Fig. 4. The delay causes the introduced dead band earlier discussed. This is for compensating the inertial due to the motor shaft. The longer the delay, the slower the rate of changes of the servomotor levers and consequently the slower the tail beating rate (frequency).

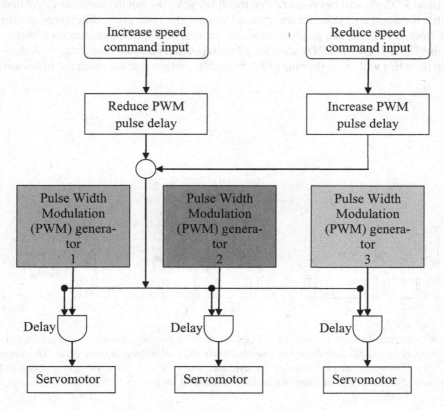

Fig. 5. Tail oscillation speed control routine

3 Experimental Verification

The experiment conducted was first aimed at verifying that the microcontroller outputs generates the PWM concurrently and at specified duty cycles for each Port using Logic

Analyzer. Also, the output interval was checked using an Oscilloscope as a complimentary measure. Secondly, the output Ports were connected to the servomotors and
tested to ensure that traveling wave pattern is generated (Fig. 2) - according to Lighthill
[31], teleost species of fish swim using that pattern. This was done by extracting still
images from the video taken during the experiment using Microsoft Movie Maker. The
camera used was Sony Cyber-shot digital camera (model DSC-S730) at a resolution of
3 megapixels, F-stop of f/2.8, exposure time of 1/40 s, ISO speed of 100.

4   Results and Discussion

The Logic Analyzer output (Fig. 6) shows the output of the programmed Ports (RD0,
RD1 and RD2). It will be observed that the duty cycles are not the same, as calculated.
At that instant, the servomotor lever would all have the same phase differences in their
turn from zero reference position as shown in Fig. 7. Furthermore, the oscilloscope
output (Fig. 8) of the pin RD1 shows a 20 ms repetition rate as a confirmation. A deviation from this will cause jittering of the assembly – which was not observed in the test.

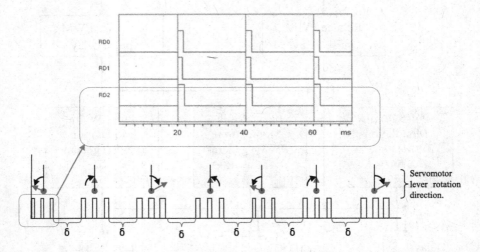

Fig. 6. The output shows varying duty cycles, all occurring simultaneously at ≈20 ms interval.
The RD0, RD1 and RD2 are the microcontroller ports physical address. δ is the delay. The larger
the δ, the slower the rate of change of the lever and hence the slower the swimming speed and
vise versa. Servomotor lever rotation direction depends on the pulse width.

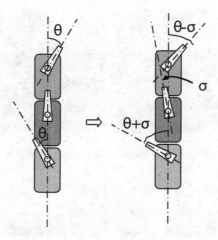

Fig. 7. The RD0 will be at $0°$ (Blue servomotor), while RD1 will be at $\theta°$ (Orange Servomotor) and RD2 is at $+\theta°$ (Green servomotor). To maintain constant phase difference, if RD0 rotate by $\sigma°$, others will have to adjust either by addition (RD1) or by subtraction (RD2) (Color figure online)

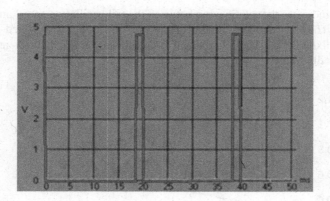

Fig. 8. Oscilloscope plot of one channel showing repeated pulses at 20 ms

An amplitude variation in the tail fin as the PWM generator was pulsed is observable in Fig. 9. Amplitude decreases from A to C as desired. This is comparable to Fig. 2. Figure 10 shows the final assembly with the amplitude of the generated traveling wave increasing significantly from the middle to the peduncle.

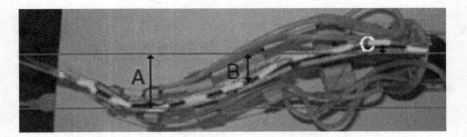

Fig. 9. Plain swimming motion showing increasing amplitude towards the tail

Fig. 10. Five sequence of still picture of the robot showing swimming mode. Frame interval is 0.5 s. The tail covering is obscuring the perfect serpentine motion achieved during the test.

Figure 11 indicates the relationship between the tail oscillation speed and the robot linear speed. The maximum excursion of the tail from the lateral based on the last segment was fixed at 90°. The relationship is not linear, though it shows an increased speed with oscillation.

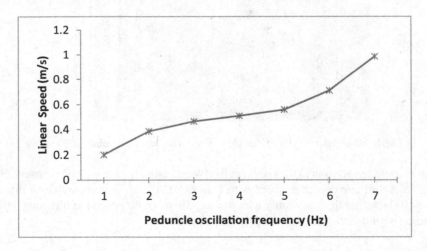

Fig. 11. Plot of tail (peduncle) against the fish linear speed

5   Recommendations

The design is promising, but there is an obvious limitation, the actuator used. It is hereby recommended that a different actuating methods be employed that can move the tail faster so as to gain higher speed. The author is aware that vortices influence the swimming efficiency, it should be used advantageously also especially in the way and manner the body motion capture and extract useful propulsive energy from it. In our scenario, the fixed phase difference may need to be adjusted based on the vortices pattern sensed so as to present a better cut-in angle to it. It also implies that the purely built in motion pattern will not suffice, some onboard calculations has to be done by the microcontroller so as to dynamically adjust to unforeseen situations presented by the environment.

References

1. NMRI: National Maritime Research Institute (2000). http://www.nmri.go.jp/eng/khirata/fish/. Accessed 4 July 2013
2. Jindong, L., Huosheng, H.: A methodology of modelling fish-like swim patterns for robotic fish. In: Proceedings of the 2007 IEEE International Conference on Mechatronics and Automation, 5–8 August 2007, Harbin, China, pp. 1316–1321 (2007)
3. Streitlien, K., Triantafyllou, G.S., Triantafyllou, M.S.: Efficient foil propulsion through vortex control. AIAA J. **34**, 2315–2319 (1996)
4. Anderson, J.M.: Vorticity control for efficient propulsion. Ph.D. dissertation, Massachusetts Inst. Technol./Woods Hole Oceanographic Inst. Joint Program, Woods Hole, MA (1996)
5. Guo, S., Fukuda, T., Kato, N., Oguro, K.: Development of underwater microrobot using ICPF actuator. In: Proceedings of the 1998 IEEE International Conference on Robotics & Automation, pp. 1829–1834 (1998)
6. Kato, N.: Control performance in the horizontal plane of a fish robot with mechanical pectoral fins. IEEE J. Ocean. Eng. **25**, 121–129 (2000)
7. Liang, J.: Researchful development of underwater Robofish II- development of a small experimental robofish. Robot **24**(3), 234–238 (2002)
8. Yu, J.Z., Chen, E.K., Wang, S., Tan, M.: research evolution and analysis of biomimetic robot fish. Control Theor. Appl. **20**(4), 1–10 (2002)
9. Jindong, L., Huosheng, H.: Building a 3D simulator for autonomous navigation of robotic fishes. In: Proceedings of 2004 IEEE/RSJ International Conference on Intelligent Robots and Systems, 28 September–2 October 2004, Sendai, Japan, pp. 613– 618 (2004)
10. Yamada, H., Chigisaki, S., Mori, M., Takita, K., Ogami, K., Hirose, S.: Development of amphibious snake-like robot ACM-R5, ISR2005. In: Proceedings of ISR 2005, p. 133 (2005)
11. William, C.: Snake-like robot can crawl on land or swim (2013). http://www.gorobotics.net/The-News/Latest-News/Snake%11Like-Snake-Like-Robot-Can-Crawl-on-Land-or-Swim/. Accessed 1 June 2013
12. Yamada, H., Hirose, S.: Development of practical 3-dimensional active cord mechanism ACM-R4. J. Rob. Mech. **18**(3), 305–311 (2006)
13. Jindong, L., Huosheng, H.: Building a simulation environment for optimising control parameters of an autonomous robotic fish. In: Proceedings of the 9th Chinese Automation & Computing Society Conference in the UK, Luton, England, 20 September 2003, pp. 317–322 (2003)
14. Kevin, J.D.: Limbless locomotion: learning to crawl with a snake robot. A Ph.D. thesis at the Robotics Institute Carnegie Mellon University, 5000 Forbes Avenue, Pittsburg, PA 15213 (1997)
15. Shugen, M.A., Watanabe, M.: Time-optimal control of kinematically redundant manipulators with limit heat characteristics of actuators. Adv. Rob. **16**(8), 735–749 (2002)

16. Masayuki, A., Yoshinori, T., Hirose, S.: Development of "Souryu-VI" and "Souryu-V:" serially connected crawler vehicles for in-rubble searching operations. J. Field Rob. **25**, 31–65 (2008). doi:10.1002/rob.20229
17. Choset, H., Lee, J.Y.: Sensor-based construction of a retract-like structure for a planar rod robot. IEEE Trans. Rob. **17**, 435–449 (2001)
18. Kier, W.M., Smith, K.K.: Tongues, tentacle and trunks: the biomechanics of movement in muscular-hydrostat. Zool. J. Linn. Soc. **83**, 307–324 (1985)
19. Skierczynski, B.A., Wilson, R.J.A., Kristian Jr., W.B., Skalak, R.: A model of the hydrostatic skeleton of the Leech. J. Theor. Biol. **181**, 329–342 (1996)
20. Wilbur, C., Vorus, W., Cao, Y., Currie, S.: A lamprey-based undulatory vehicle. In: Ayers, J., Davis, J., Rudolph, A. (eds.) Neurotechnology for Bio-mimetic Robots, pp. 285–296. MIT Press, Cambridge (2002)
21. Crespi, A., Badertscher, A., Guignard, A., Ijspeert, A.J.: Amphibot I: an amphibious snake-like robot. Rob. Auton. Syst. **50**, 163–175 (2005)
22. Robinson, G., Davies, J.B.C.: Continuum robots - a state of the art. In: IEEE Conference on Robotics and Automation, pp. 2849–2854 (1999)
23. Gwenaël, A.: Control of a free-swimming fish using fuzzy logic. Int. J. Virtual Real. **6**(3), 23–28 (2007)
24. Morgansen, K.A., Triplett, B.I., Klein, D.J.: Geometric methods for modeling and control of free-swimming fin-actuated underwater vehicles. IEEE Trans. Rob. **23**(6), 1184–1199 (2007)
25. Mbemmo, E., Chen, Z., Shatara, S., Tan, X.: Modeling of biomimetic robotic fish propelled by an ionic polymer-metal composite actuator. IEEE/ASME Trans. Mech. **15**(3), 448–452 (2010)
26. Korkmaz, D., Ozmen, K.G., Akpolat, Z.H.: Robust forward speed control of a robotic fish. In: 6th International Advanced Technologies Symposium (IATS 2011), 16–18 May 2011, Elazig, Turkey, pp. 33–38 (2011)
27. Taylor, G.: Analysis of the swimming of long narrow animals. Proc. R. Soc. Lond. A **214**, 158–183 (1952)
28. Wu, T.Y.: Swimming of a waving plate. J. Fluid Mech. **10**, 321–344 (1961)
29. Anderson, J.M., Streitlien, K., Barrett, D.S., Triantafyllou, M.S.: Oscillating foils of high propulsive efficiency. J. Fluid Mech. **360**, 41–72 (1998)
30. Jindong, L., Huosheng, H.: Mimicry of sharp turning behaviours in a robotic fish. In: Proceedings of the 2005 IEEE. International Conference on Robotics and Automation, Barcelona, Spain, April 2005, pp. 3329–3334 (2005)
31. Lighthill, M.J.: Note on the swimming of slender fish. J. Fluid Mech. **9**, 305–317 (1960). Abstract and introduction only

Animal and Robotic Locomotion on Wet Granular Media

Hosain Bagheri, Vishwarath Taduru, Sachin Panchal, Shawn White,
and Hamidreza Marvi[✉]

Arizona State University, Tempe, AZ 85281, USA
hmarvi@asu.edu
http://birth.asu.edu

Abstract. Most of the terrestrial environments are covered with some
type of flowing ground; however, inadequate understanding of moving
bodies interacting with complex granular substrates has hindered the
development of terrestrial/all-terrain robots. Although there has been
recent performance of experimental and computational studies of dry
granular media, wet granular media remain largely unexplored. In particular, this encompasses animal locomotion analysis, robotic system performance, and the physics of granular media at different saturation levels.
Given that the presence of liquid in granular media alters its properties
significantly, it is advantageous to evaluate the locomotion of animals
inhabiting semi-aquatic and tropical environments to learn more about
effective locomotion strategies on such terrains. Lizards are versatile and
highly agile animals. Therefore, this study evaluated the brown basilisk,
which is a lizard species from such habitats that are known for their
performance on wet granular media. An extensive locomotion study was
performed on this species. The animal experiments showed that on higher
saturation levels, velocity of the animal was increased due to an increase
in the stride length. A basilisk-inspired robot was then developed to further study the locomotion on wet granular media and it was observed
that the robot can also achieve higher velocities at increased saturation levels. This work can pave the way for developing robotic systems
which can explore complex environments for scientific discovery, planetary exploration, or search-and-rescue missions.

Keywords: Wet granular media · Bipedal/quadrupedal locomotion ·
Basilisk lizard · Bio-inspired robot

1 Introduction

Robots are poised to enter our everyday lives, transforming how humans interact
with the physical world. In the near future, multi-functional robots will advance
scientific discovery across disciplines (e.g. archaeology, biology and geology),
planetary exploration, aid first responders in search-and-rescue missions and even
traverse hazardous environments to mankind [1]. Unfortunately, typical robotic

© Springer International Publishing AG 2017
M. Mangan et al. (Eds.): Living Machines 2017, LNAI 10384, pp. 13–24, 2017.
DOI: 10.1007/978-3-319-63537-8_2

systems are still restricted to maneuvering upon hard planar surfaces. Difficulties in improving terrestrial robots arise due to the inadequate understanding of moving bodies interacting with complex natural substrates (i.e., sand, dirt, mud, and rocks) [2–6]. Soft granular media and friable terrains have already seen to pose complications in regards to wheel sinkage, entrapment, and eventual immobilization of NASA Mars rovers, Opportunity [7–9]. Rover wheels have gotten trapped in sand, preventing it from maneuvering out without digging itself deeper into the trap, almost ending the mission [10]. Therefore, robots must have the ability to effectively traverse on complex, diverse, and extreme terrain topographies. Future robots will require legs to reach and operate on such terrains for cross-discipline applications. While legged robots are certainly not the only solution, they have been observed to perform better on granular media compared to wheeled or limbless robots [11,12]. Therefore, such legged robots will need to address granular media interaction.

Interaction of foreign bodies with granular media, like that of sand and mud, introduces a highly nonlinear behavior [3–6,13]. Complications arise when attempting to formulate mathematical models for these interactions. This is due to the fact that individual particles hold an inhomogeneous force distribution, causing the ground reaction forces for interaction to be quite difficult to predict [2,14–16]. Furthermore, it is infeasible to measure the influence of grain inertial forces [14,17,18] upon the moving body as it accelerates over the medium. This factor is of significance when evaluating moving bodies at high speeds. When it comes to the mechanics of locomotion, there have only been few studies on dry granular media [3,6,12,13,19–23], whereas wet granular media is largely unexplored [24–28]. The properties of granular media significantly change in the presence of liquids, due to inter-grain cohesion [29–32].

Few robots, such as Big Dog [24], AmphiHex [26,27], SeaDog [28], and [25] have shown potential to move on wet granular media. However, the mechanics of animal locomotion and physics of the environment have not been adequately explored. Thus, the performance of these robots on wet granular media cannot be compared efficaciously to that of their biological counterparts. There is a need for carefully designed studies on animal species that are highly agile on such terrain to seek potential solutions for robotics locomotion on wet granular media.

In comparison to man-made devices, organisms such as snakes, lizards, and insects can effectively move on nearly all natural environments. Scientists and engineers have sought to systematically ascertain the biological principles of their movement and implement them within robotics [33]. This "bioinspired robotics" approach [34] has proven to be fruitful for designing laboratory robots with advanced capabilities (gaits, morphologies, control schemes) including rapid running [24,33], slithering [35], flying [36], and "swimming" in sand [12]. By using biologically inspired robots as the basis of the organisms' "physical models," scientific insights have revealed the principles governing movement in biological systems and low-dimensional dynamical systems. However, minimal studies have transmitted biological principles into conventional field-ready mobile devices [24,37], capable of operating and interacting in and with natural terrains.

Lizards are known to be versatile animals which can walk on all four legs, run on two legs, swim, climb, and burrow in, on, and through diversified terrains [38–42]. Lizard species which live in habitats covered with wet granular media, such as rain-forests, utilize different interfacial structures and locomotion patterns than other terrestrial animals for effective locomotion on muddy terrain. Therefore, our focus is on species which inhabit semiaquatic environments; ones that live along flowing stream beds, use water as a mean of retreat, and interact with vast saturation levels of granular media. During our preliminary studies, one reptile stood out amongst the others-the brown basilisk lizard (*Basiliscus vittatus*). This specimen which has gained recognition for running across water, was observed to leap into bipedal gait when running on granular media. While previous locomotion studies have been performed on basilisk, they have mainly been on water [40,43,44] and dry hard surfaces (i.e. sandpaper) [45]. Furthermore, previous studies on bipedal motion of other lizards [46,47] have focused on the development of mechanical modeling and running in water [41]. While, to the best of our knowledge, no bipedal behavior study has been performed on wet granular media. This research provides an insight to the animal's morphology and locomotion transformation as the medium transition from dry to wet.

This work aims to simplify and optimize elegant designs found in highly agile and versatile animals, such as lizards, toward developing bioinspired robotic systems for diversified terrains. Through our animal studies, our main objective is to identify the optimal morphology and gait parameters for mobility on complex deformable terrain, specifically granular media at different saturation levels. A bioinspired robot has been designed and fabricated that addresses one of the deficiencies existing in "all-terrain" robots, namely the inability to traverse on complex granular terrains. More precisely, transitioning from quadrupedal to bipedal gait (and vice versa) within a few steps and effectively traversing bipedal on granular media at different saturation levels.

2   Methods

2.1   Experimental Setup

A Fluidized Bed sized at $2 \times 1\,m^2$ was built to perform dry and saturated granular experimentations (Fig. 1a). The setup consists of a porous and honeycomb sheet separating the sand from air-duct system, which is positioned underneath it. The air-duct system controls compactness (volume fraction) of the sand. In the fluidized bed, above a critical air flow rate, the granular material behaves like a fluid. By gradually decreasing the air flow to zero, the media reaches a loosely packed state with a flat surface, removing previous disturbances to the medium. Air fluidization only occurs prior to each test run and not during experiments. After fluidization and once the media has settled, the bed can slowly be inclined to an angle from 0° to 45°. For these series of experiments the bed was kept at an inclination angle of 0°.

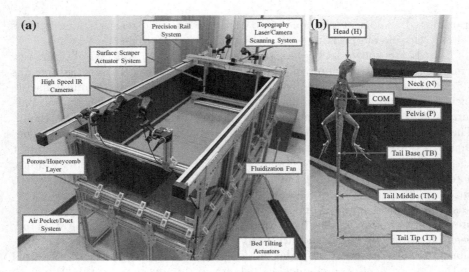

Fig. 1. (a) Air-fluidized bed setup. (b) A basilisk lizard with IR markers.

For saturated granular media experiments, fluidization was infeasible and thus not performed. Instead, saturation was performed manually by evenly pouring water over the bed and mixing the medium thoroughly with an industrial hand mixer. A soil moisture sensor (GS1 manufactured by Decagon Devices) was used to measure the saturation level (i.e., volumetric water content) of the medium prior to experimentation. Nine distributed points throughout the bed were sampled (with the sensor fully embedded under the soil). For saturation level, the average was recorded as the saturation level of the bed. This was performed before and after each experiment to ensure the saturation level remains constant throughout the experiment.

Three saturation levels were used for experimentation: 0%, 15%, and 30% water content. With the experimental focus being on the change in saturation level, the volume fraction of the medium was not varied and ensured to be compact. To preserve a relatively smooth and compact surface, an actuated scraper was grazed back and forth across the bed. The nine distributed locations throughout the bed were measured for soil height using a pair of digital calipers to obtain the relative height of the soil.

Six infrared (IR) OptiTrack motion capture cameras were mounted on the bed and orientated to obtain the maximum image overlap. With a horizontal field of view of 56° and a frame rate of 120 fps, the cameras recorded the 3D locomotion of the animal and robot through the attached IR reflective markers on the desired locations of analysis. Prior to each experiment, the cameras were calibrated and a reference coordinate system was established. Utilizing the Motive software developed by OptiTrack, post capture analysis of each marker's position with respect to time was evaluated. The speed was calculated as the measured distance traveled of a marker per unit time. Penetration depth was obtained based on the z-displacement of markers. Aside from the IR motion capture cameras, a Canon DLSR (EOS 70D) camera along with a GoPro (Hero5) were used to capture visual observations from the top and side view, respectively.

The sand used for the dry and saturated experiments was commercially available playground sand (washed, sterilized, and dried) that was sieved with a sieve shaker to particle size range of 250–600 μm. A total volume of about 102 L of sand was sieved and displaced into the bed.

2.2   Animal Studies

Three juvenile Brown Basilisk (*Basiliscus vittatus*) were used for the animal experiment, with SVL = 8.37 ± 0.43 cm and mass = 18.6 ± 3.6 g (Fig. 1b).

All of the animal experiments were reviewed and approved by ASU Institutional Animal Care and Use Committee (IACUC) under IACUC Protocol #: 16-1504R. Arizona State University DACT (Department of Animal Care Technologies) were delegated with their housing and care. All personnel handling the animals were approved and trained by ASU IACUC. When the basilisk lizards were brought to the lab for experimentation, they were placed into an incubator at 30 °C between trials to preserve their desired environmental temperature condition and agility.

3 mm hemispherical adhesive backing reflective (facial) markers (0.04 g) were placed at each main joint of the lizard and throughout its body for motion capture analysis. Minute amount of nail polish was applied to the base of the markers for better adhesion. The markers were left on to gradually fall off on their own. A total of 17 markers were placed upon each basilisk lizard (Fig. 1b): 3 were placed on each hind limb (Knee, Ankle, and Foot), 2 were placed on each fore limb (Knee and Foot), 3 were placed from the snout to vent (Head, Neck, COM, Pelvis), and 3 were evenly distributed along the tail (Tail Base, Tail Middle, Tail Tip). Only 7 markers from the snout to the tip of the tail were used for these series of locomotion analysis. More specifically, the following parameters were evaluated: Pelvis velocity, stride frequency, and stride length (measured in coordinates of absolute space).

Each lizard was placed in one end of the Fluidized Bed and its locomotion on closely packed sand at different saturation levels (0%, 15%, 30%) was recorded using the IR motion capturing system. Each trial consisted of observing the lizard running along a straight path without any interruptions. Three trials were completed per basilisk lizard per saturation level. 40 min in the incubator was allocated for their rest between trials. The order of experimentation with the lizards were randomized for each saturation level evaluated.

2.3   Robotic Studies

The dimensions of the 3 basilisk lizards were taken from the Motive software and averaged. In particular, the length of fore limbs (3.05 ± 0.25 cm), hind limbs (6.63 ± 0.44 cm), and tail (17 ± 2.89 cm) were measured. The basilisk-inspired robot, Basiliskbot (Fig. 2) was designed based on the averaged dimensions of

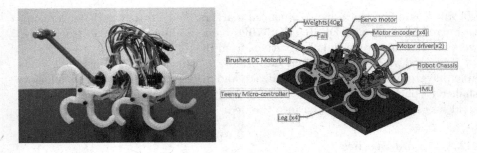

Fig. 2. The Basiliskbot

the animal. The four-spoked legs were 3D printed with Acrylonitrile-Butadiene Styrene (ABS) with a 30% fill and thickness of 8.5 mm, while the body was laser cut using 4.3 mm acrylic. Mass of the robot (without the tail) was 273 g with the following components on board: four 50:1 micro DC geared motors, four magnetic encoders, four aluminum motor mounts, two dual motor drivers, a micro servo, a Teensy 3.2 micro-controller, and an inertial measurement unit (IMU). Total mass of the tail was 45 g (a 40 g mass was attached to the end of a 5 g tail with the length of 12.5 cm).

Both bipedal and quadrupedal gaits were implemented for the robot and tested on closely packed dry and wet sand (0%, 15%, and 30% saturation levels). In addition, a PID controller was used to control the tail angle during the bipedal gait and ensure body stability. Thus, the robot was able to push off the ground and stand up on its hind limbs in a couple of steps and then to perform a stable bipedal gait.

Body velocity was evaluated at five different motor frequencies per saturation level, where three trials were performed per frequency, giving a total of 45 test runs for each gait. Only successful runs were considered, in which the robot was observed to move on a relatively straight path. The experimental setup, procedure, and method were kept the same as that of animal experiments. The DSLR Canon and GoPro cameras were used for top and side view observations, respectively.

3   Results and Discussion

3.1   Animal Experiments

Here we report our experimental results for basilisk lizards and a basiliskbot moving on dry and wet sand. In particular, we look at the effect of saturation level on stride length and stride frequency, and thus the body velocity. The animal's pelvis velocity was averaged between the trials and evaluated per saturation level (Fig. 3a, b). A significant difference ($P < 0.05$ and $P < 0.01$, respectively) in pelvis velocity was observed when the saturation increased from 0% to 15%

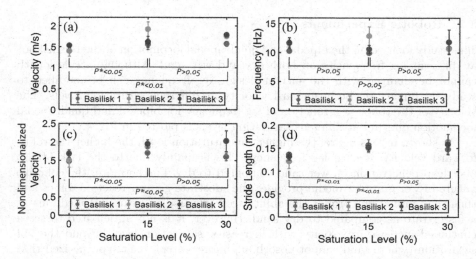

Fig. 3. (a), (b) Pelvis velocity, (c) stride frequency, and (d) stride length of the basilisk lizards performing bipedal running at different saturation levels of sand. SPSS (ANOVA and T-Test) was used with $P* < 0.05$ and $P* < 0.01$ indicating statistically and statistically highly significance. Sample size: $n = 3$

saturation and from 0% to 30%, but minimal difference was observed from 15% to 30% saturation.

The animal's body velocity increased as the saturation increased. However, no significant difference was observed when evaluating the stride frequency (Hz) from 0% to 15%, 15% to 30%, or 0% to 30% (Fig. 3c). Therefore, the average stride length (about the Pelvis marker) was naturally the next point of evaluation. Significant difference ($P < 0.01$) was observed in stride length when the saturation was increased from 0% to 15% and from 0% to 30%, but minimal change when going from 15% to 30% (Fig. 3d).

This indicates that the significant change in the animal's body velocity was due to the change in stride length at different levels of saturation. As the saturation level increased, the animal's ability to make longer stride lengths increased and thus increased its body velocity. The animal was anecdotally observed to struggle with sinkage when running across dry sand and thus took shorter strides to stay afloat and stabilize. However, saturated sand decreased sinkage and gave the animal confidence to take wider strides of motion. Stride length increased as saturation increased from 0% to 30% with an average stride length of 13.18 ± 1.83 cm to 15.68 ± 2.20 cm. Velocity increased as saturation increased from 0% to 30% with average absolute velocity of 1.44 ± 0.14 m/s to 1.70 ± 0.13 m/s. Interestingly enough, it has been cited [41] that basilisk lizards on water experience an average velocity of 1.3 ± 0.1 m/s, regardless of their size, which is not too far from its performance on 0% saturation.

3.2   Robotic Experiments

This study focused on the bipedal and quadrupedal locomotion of the Basiliskbot at different leg frequencies on both dry and wet sand. Although stride length cannot be changed with the current design, leg frequency can be modified to study its impact on body forward velocity at different saturation levels. The velocity of the robot as a function of leg frequency for bipedal and quadrupedal locomotion on sand at different saturation levels is plotted in Fig. 4.

As shown in this figure, the higher the saturation level, the higher the robot forward velocity at each leg frequency. This is mainly due to the lack of slip and deep penetration on wet sand compared to dry. Furthermore, the cohesive nature of wet granular media provides the substrate greater reaction forces and thus the (animal and) robot the ability to transverse on saturated mediums at a faster rate as compared to dry sand. However, it is not anticipated to see a continued increase in velocity with increasing saturation level beyond the full saturation (where sand cannot absorb any more water). It is hypothesized that the additional water on the surface would decrease the ground reaction force and decrease the drag force and thus could slow down the robot.

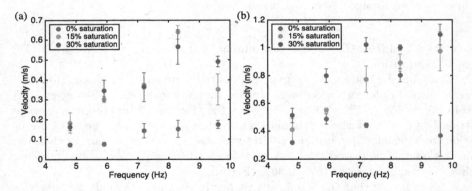

Fig. 4. Velocity of the basiliskbot versus leg frequency for (a) bipedal and (b) quadrupedal locomotion at different saturation levels of sand.

In addition, it is observed that bipedal velocities are smaller than those of quadrupedal. This can perhaps be attributed to multiple factors. First, in bipedal mode hind legs exert a higher normal pressure to the sand and thus penetrate further. As a result, they need to spend more energy and time to come out of sand. This is observed more on the dry sand. Moreover, the controller was sometimes forcing the tail to touch the ground to ensure stability of bipedal gait. This would reduce the bipedal velocity by introducing additional drag force. The tail was also removed from the robot when performing quadrupedal gait to avoid pitch instability due to the presence of a passive tail. This would correspond to 14% reduction in the total mass of the robot.

Finally, it was observed that high frequencies can sometimes result in decreased forward velocities. Reduced velocity at high frequency quadrupedal locomotion on dry sand has been reported by Li *et al.* [19]. In this study, the above mentioned phenomena was also observed for bipedal locomotion on wet sand which needs to be further explored.

4   Conclusion

An experimental study of animal and robot locomotion on wet and dry granular media is presented in this paper. In particular, the effect of sand saturation level on stride length, stride frequency, and body velocity of a basilisk lizard performing bipedal locomotion has been explored. It was found that the animal velocity increased at higher saturation levels due to an increase in stride length (the change in stride frequency at different saturation levels was not statistically significant). Next, the Basiliskbot was developed to explore maximal speeds on dry and wet granular media. In particular, the effect of stride frequency on body velocity at different saturation levels of sand was studied for both quadrupedal and bipedal gaits. It was found that increased saturation level will result in increased body speed at each leg frequency for both bipedal and quadrupedal gaits. In addition, the robot was able to achieve higher speeds using quadrupedal gait, although the effect of tail mass and touching the ground needs to be further explored. Finally, increased leg frequency may sometimes result in decreased forward velocity depending on the gait and the saturation level. Through the series of performed animal and robotic experiments, we have addressed significant contributing factors (e.g. morphology, gait, and adaptation) to legged locomotion on granular media and possible optimal solutions for traversing complex deformable terrain. Using the Basiliskbot, we would like to further study optimal locomotion patterns on complex unstructured environments and develop effective robotic systems for locomotion on such terrains.

Acknowledgements. The authors would like to thank ASU Institutional Animal Care and Use Committee (IACUC) for animal husbandry (IACUC Protocol #: 16-1504R), Professor Dale DeNardo for greatly valuable suggestions on the choice of animal and animal experiments, Professor Heather Emady and Spandana Vajrala for fruitful discussions on studying wet granular media, Carolyn Harvey for her contributions to the setup development, Daniel Lee, Isaac Charcos, and John Millard for helping with animal data collection/analysis, and Arizona State University for funding.

References

1. Murphy, R.R., Tadokoro, S., Kleiner, A.: Disaster Robotics. Springer Handbook of Robotics. Springer, Cham (2016)
2. Aguilar, J., Zhang, T., Qian, F., Kingsbury, M., McInroe, B., Mazouchova, N., Li, C., Maladen, R., Gong, C., Travers, M., et al.: A review on locomotion robophysics: the study of movement at the intersection of robotics, soft matter and dynamical systems. Rep. Prog. Phys. **79**(11), 110001 (2016)

3. Mazouchova, N., Gravish, N., Savu, A., Goldman, D.I.: Utilization of granular solidification during terrestrial locomotion of hatchling sea turtles. Biol. Lett. **6**(3), 398–401 (2010)
4. Sharpe, S.S., Kuckuk, R., Goldman, D.I.: Controlled preparation of wet granular media reveals limits to lizard burial ability. Phys. Biol. **12**(4), 046009 (2015)
5. Richefeu, V., El Youssoufi, M.S., Azéma, E., Radjai, F.: Force transmission in dry and wet granular media. Powder Technol. **190**(1), 258–263 (2009)
6. Li, C., Zhang, T., Goldman, D.I.: A terradynamics of legged locomotion on granular media. Science **339**(6126), 1408–1412 (2013)
7. Reina, G., Ojeda, L., Milella, A., Borenstein, J.: Wheel slippage and sinkage detection for planetary rovers. IEEE/ASME Trans. Mech. **11**(2), 185–195 (2006)
8. Ghotbi, B., González, F., Kövecses, J., Angeles, J.: Mobility evaluation of wheeled robots on soft terrain: effect of internal force distribution. Mech. Mach. Theor. **100**, 259–282 (2016)
9. Zhou, F., Arvidson, R.E., Bennett, K., Trease, B., Lindemann, R., Bellutta, P., Iagnemma, K., Senatore, C.: Simulations of Mars rover traverses. J. Field Robot. **31**(1), 141–160 (2014)
10. Heverly, M., Matthews, J., Lin, J., Fuller, D., Maimone, M., Biesiadecki, J., Leichty, J.: Traverse performance characterization for the mars science laboratory rover. J. Field Robot. **30**(6), 835–846 (2013)
11. Li, C.: Biological, robotic, and physics studies to discover principles of legged locomotion on granular media. Georgia Institute of Technology (2011)
12. Maladen, R.D., Ding, Y., Li, C., Goldman, D.I.: Undulatory swimming in sand: subsurface locomotion of the sandfish lizard. Science **325**(5938), 314–318 (2009)
13. Marvi, H., Gong, C., Gravish, N., Astley, H., Travers, M., Hatton, R.L., Mendelson, J.R., Choset, H., Hu, D.L., Goldman, D.I.: Sidewinding with minimal slip: snake and robot ascent of sandy slopes. Science **346**(6206), 224–229 (2014)
14. Goldman, D.I., Umbanhowar, P.: Scaling and dynamics of sphere and disk impact into granular media. Phys. Rev. E **77**(2), 021308 (2008)
15. Gravish, N., Franklin, S.V., Hu, D.L., Goldman, D.I.: Entangled granular media. Phys. Rev. Lett. **108**(20), 208001 (2012)
16. Hubicki, C.M., et al.: Tractable terrain-aware motion planning on granular media: an impulsive jumping study. In: 2016 IEEE/RSJ International Conference on Intelligent Robots and Systems (IROS). IEEE (2016)
17. Brzinski, T., Mayor, P., Durian, D.: Depth-dependent resistance of granular media to vertical penetration. Phys. Rev. Lett. **111**(16), 168002 (2013)
18. Katsuragi, H., Durian, D.J.: Unified force law for granular impact cratering. Nat. Phys. **3**(6), 420–423 (2007)
19. Li, C., Umbanhowar, P.B., Komsuoglu, H., Koditschek, D.E., Goldman, D.I.: Sensitive dependence of the motion of a legged robot on granular media. Proc. Natl. Acad. Sci. **106**(9), 3029–3034 (2009)
20. Li, C., Umbanhowar, P.B., Komsuoglu, H., Goldman, D.I.: The effect of limb kinematics on the speed of a legged robot on granular media. Exp. Mech. **50**(9), 1383–1393 (2010)
21. Maladen, R.D., Ding, Y., Umbanhowar, P.B., Kamor, A., Goldman, D.I.: Mechanical models of sandfish locomotion reveal principles of high performance subsurface sand-swimming. J. Roy. Soc. Interface **8**(62), 1332–1345 (2011)
22. Mcinroe, B., Goldman, D.: Construction of a mudskipper-inspired robot to study crutching locomotion on flowable ground. Integr. Comp. Biol. **54**, E316–E316 (2014)

23. Lejeune, T.M., Willems, P.A., Heglund, N.C.: Mechanics and energetics of human locomotion on sand. J. Exp. Biol. **201**(13), 2071–2080 (1998)
24. Raibert, M., Blankespoor, K., Nelson, G., Playter, R.: BigDog, the rough-terrain quadruped robot. In: Proceedings of the 17th World Congress, pp. 10822–10825 (2008)
25. Asif, U., Iqbal, J.: On the improvement of multi-legged locomotion over difficult terrains using a balance stabilization method. Int. J. Adv. Robot. Syst. **9**(1) (2012). doi:10.5772/7789
26. Ren, X., Liang, X., Kong, Z., Xu, M., Xu, R., Zhang, S.: An experimental study on the locomotion performance of elliptic-curve leg in muddy terrain. In: Proceedings of IEEE/ASME International Conference on Advanced Intelligent Mechatronics (AIM), pp. 518–523 (2013)
27. Xu, L., Liang, X., Xu, M., Liu, B., Zhang, S.: Interplay of theory and experiment in analysis of the advantage of the novel semi-elliptical leg moving on loose soil. In: Proceedings of IEEE/ASME International Conference on Advanced Intelligent Mechatronics (AIM), pp. 26–31 (2013)
28. Klein, M., Boxerbaum, A.S., Quinn, R.D., Harkins, R., Vaidyanathan, R.: SeaDog: a rugged mobile robot for surf-zone applications. In: Proceedings of 4th IEEE RAS and EMBS International Conference on Biomedical Robotics and Biomechatronics (BioRob), pp. 1335–1340 (2012)
29. Mitarai, N., Nori, F.: Wet granular materials. Adv. Phys. **55**(1–2), 1–45 (2006)
30. Tegzes, P., Vicsek, T., Schiffer, P.: Avalanche dynamics in wet granular materials. Phys. Rev. Lett. **89**(9), 094301 (2002)
31. Albert, R., Albert, I., Hornbaker, D., Schiffer, P., Barabási, A.L.: Maximum angle of stability in wet and dry spherical granular media. Phys. Rev. E **56**(6), R6271 (1997)
32. Richefeu, V., El Youssoufi, M.S., Radjai, F.: Shear strength properties of wet granular materials. Phys. Rev. E **73**(5), 051304 (2006)
33. Cutkosky, M.R., Kim, S.: Design and fabrication of multi-material structures for bioinspired robots. Philos. Trans. Roy. Soc. Lond. A Math. Phys. Eng. Sci. **367**(1894), 1799–1813 (2009)
34. Bhushan, B.: Biomimetics: lessons from nature-an overview (2009)
35. Tesch, M., Lipkin, K., Brown, I., Hatton, R., Peck, A., Rembisz, J., Choset, H.: Parameterized and scripted gaits for modular snake robots. Adv. Robot. **23**(9), 1131–1158 (2009)
36. Ma, K.Y., Chirarattananon, P., Fuller, S.B., Wood, R.J.: Controlled flight of a biologically inspired, insect-scale robot. Science **340**(6132), 603–607 (2013)
37. Holmes, P., Full, R.J., Koditschek, D., Guckenheimer, J.: The dynamics of legged locomotion: models, analyses, and challenges. SIAM Rev. **48**(2), 207–304 (2006)
38. Li, C., Hsieh, S.T., Goldman, D.I.: Multi-functional foot use during running in the zebra-tailed lizard (callisaurus draconoides). J. Exp. Biol. **215**(18), 3293–3308 (2012)
39. Irschick, D.J., Jayne, B.C.: Effects of incline on speed, acceleration, body posture and hindlimb kinematics in two species of lizard callisaurus draconoides and uma scoparia. J. Exp. Biol. **201**(2), 273–287 (1998)
40. Glasheen, J., McMahon, T.: Size-dependence of water-running ability in basilisk lizards (basiliscus basiliscus). J. Exp. Biol. **199**(12), 2611–2618 (1996)
41. Hsieh, S.T.: Three-dimensional hindlimb kinematics of water running in the plumed basilisk lizard (basiliscus plumifrons). J. Exp. Biol. **206**(23), 4363–4377 (2003)

42. Irschick, D.J., Jayne, B.C.: Comparative three-dimensional kinematics of the hindlimb for high-speed bipedal and quadrupedal locomotion of lizards. J. Exp. Biol. **202**(9), 1047–1065 (1999)
43. Hsieh, S.T., Lauder, G.V.: Running on water: three-dimensional force generation by basilisk lizards. Proc. Natl. Acad. Sci. U.S.A. **101**(48), 16784–16788 (2004)
44. Bush, J.W., Hu, D.L.: Walking on water: biolocomotion at the interface. Annu. Rev. Fluid Mech. **38**, 339–369 (2006)
45. Snyder, R.C.: Bipedal locomotion of the lizard basiliscus basiliscus. Copeia **1949**(2), 129–137 (1949)
46. Aerts, P., Van Damme, R., D'Août, K., Van Hooydonck, B.: Bipedalism in lizards: whole-body modelling reveals a possible spandrel. Philos. Trans. Roy. Soc. Lond. B Biol. Sci. **358**(1437), 1525–1533 (2003)
47. Park, H.S., Sitti, M.: Compliant footpad design analysis for a bio-inspired quadruped amphibious robot. In: IEEE/RSJ International Conference on Intelligent Robots and Systems, IROS 2009, pp. 645–651. IEEE (2009)

Tunable Normal and Shear Force Discrimination by a Plant-Inspired Tactile Sensor for Soft Robotics

Afroditi Astreinidi Blandin[1,2(✉)], Massimo Totaro[1], Irene Bernardeschi[1], and Lucia Beccai[1(✉)]

[1] Center for Micro-BioRobotics, Istituto Italiano di Tecnologia, Pontedera, Italy
{afroditi.astreinidi,lucia.beccai}@iit.it
[2] The BioRobotics Institute, Scuola Superiore Sant'Anna, Pontedera, Italy

Abstract. In plants, particular biomechanical protruding structures, tactile bleps, are thought to be specialized tactile sensory organs and sensitive to shear force. In this work, we present a 2D finite element analysis of a simplified plant-inspired capacitive tactile sensor. These preliminary results show that the variation of geometrical and material parameters permits to tune the sensitivity to normal and shear force and, with particular configurations, to discriminate between the two forces with a simple electrical layout and no signal processing.

Keywords: Bio-inspired tactile sensor · Tactile blep · Shear force discrimination · Soft robotics

1 Introduction

The mechanosensing filter of biological structures is found to be not only composed of a traditional neural filter, but also of a mechanical filter [1]. This acknowledgement increases the significance of mechanical properties of the sensory organs, like geometry and material properties, and of the surrounding medium.

While thoroughly studied in many living beings, tactile sensing mechanisms have still not been unveiled for plants, despite being of primary importance for survival. Plant mechanical perception ability, i.e. 'mechanoperception', is a key characteristic of all its cells, which deform because of external and internal mechanical forces [2]. In some plants, protruding hemispherical structures have been identified and studied to a certain extent, like the tactile bleps of *Bryonia dioica* Jacq. tendrils (Figs. 1A, B and 2A), which are supposed to act as specialized sensory organs for mechanoperception despite lack of evidence [3]. In particular, they are assumed to be more sensitive to shear than normal stimulation. Regarding the dome shape, some artificial sensors take advantage from it. One of them is a tactile sensor inspired by cucumber tendrils papillae, based on a working principle of light intensity variation using fiber optics [4]. The shape of the tactile blep was taken as an inspiration to create a tactile sensor capable of sensing in all orthogonal directions, based on a piezoelectric working principle, but without exploiting its material properties [5]. In recent years, soft three-axial force sensors with remarkable perform- ance have been proposed, exploiting capacitive [6], resistive [7] or piezoelectric [8]

© Springer International Publishing AG 2017
M. Mangan et al. (Eds.): Living Machines 2017, LNAI 10384, pp. 25–34, 2017.
DOI: 10.1007/978-3-319-63537-8_3

transduction mechanisms. However, the shear force detection is obtained by a specific electrode layout that, in combination with the processing of different signals, too often complicates the overall sensing system.

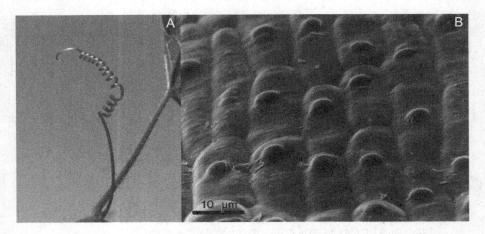

Fig. 1. A. *Bryonia dioica* Jacq. tendril (A. Moro, Dept. of Life Sciences, Univ. of Trieste CC BY-SA 4.0). B. Scanning-electron-microscope image of protruding tactile bleps on *B.dioica* Jacq. tendril (adapted with permission from [3]).

Starting from those two observations, in this work we study the principle of the biomechanical filter in mechanoperception, applied to a bio-inspired bilayer dome-shaped structure, exploiting a capacitive transduction principle. We therefore investigate the relevance of its biomechanical aspects, in particular the geometry (shape and proportions) and the mechanical properties of materials, by means of finite element method (FEM) simulations. The motivation of this work is to create a novel artificial tactile sensor capable of discriminating between normal and shear force that is entirely soft and smart because of its mechanical structure, in order to permit its integration in soft robots without complex signal processing and without affecting the mechanical characteristics of the robot body.

2 Plant-Inspired Design of Simplified Tactile Blep

Inspired by the natural structure, different aspects are taken into consideration for the design of an extremely simplified tactile blep having a bilayer structure (Fig. 2B):

1. The mechanical properties of **materials** used are thought to play a decisive role in the behavior of the artificial sensor. Increasing their elasticity allows to build a flexible structure which will undergo increased deformation and consequently increase sensor sensitivity.
2. A striking feature of the biological structure is the large diversity of the mechanical properties of its building components, ranging from a fluid inner cytoplasm to a rigid

cell wall [3]. The integration of materials with different mechanical properties in various **proportions** is therefore another aspect considered to be critical.

3. The dome **shape** suggests that the displacement of the dome lower region will be larger when the dome is subjected to shear force than to a normal force. This leads to designing a capacitive based artificial sensor, with the ground electrode put at the bottom side of the device base, and the sensing electrode put circularly around the lower surface of the inner dome (Fig. 2B). With this design, since capacitance is proportional to the distance between electrodes, a larger capacitance change is expected under shear force.

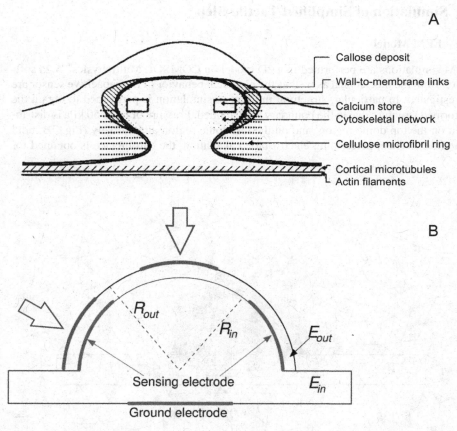

Fig. 2. A. Model of natural tactile blep architecture (adapted with permission from [3]). B. Artificial dome-shaped bilayer structure. The soft inner part (base and inner hemisphere with radius R_{in}) has Young's modulus E_{in}, and a more rigid outer part (shell with radius R_{out}) has Young's modulus E_{out}. The sensing electrode is positioned around the lower surface of the inner dome, whereas the ground electrode is positioned under the base. The open arrows indicate the direction of the applied loads, distributed either on top or side regions (in blue) of the blep. (Color figure online)

In Fig. 2B a planar section of the **artificial sensor** is depicted. It represents an extremely simplified tactile blep, exploiting its dome shape. Using a bilayer structure permits to vary the proportion of diverse composition materials (through the inner layer thickness determined by the inner radius R_{in}) and the material in each layer (through the mechanical properties, and in particular the inner Young's modulus E_{in}). The inner part, composed by the inner hemisphere and the base, is softer than the outer shell, as in the tactile blep. The material of the base is selected to be the same as for the inner dome, since such a configuration is more feasible from a technological point of view. The boundary case of the monolayer structure (i.e. a dome made of homogeneous material) is also studied.

3   Simulation of Simplified Tactile Blep

3.1   FEM Model

FEM simulations are performed in a 2D model on COMSOL Multiphysics® 5.2a software. Both the mechanical as well as the electrical behavior of the capacitive sensor are investigated. In particular, structural mechanics simulation is performed to obtain the deformed structure due to the boundary load applied. Pressure of up to 50 kPa is distributed on the top dome region, and on the left dome region alternatively (Fig. 2B, bold blue lines). Then, performing electrostatic calculation, the capacitance is obtained for

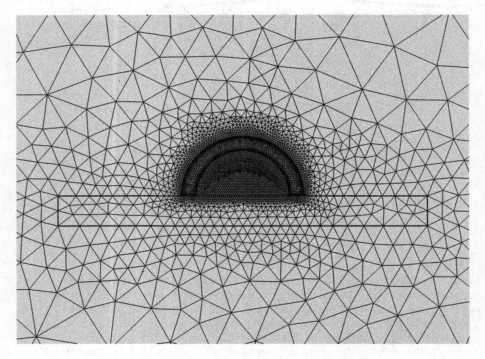

Fig. 3. FEM mesh of artificial dome-shaped bilayer model

each pressure step. The two physics are coupled by means of the deformed mesh interface, used in electrostatic calculations. It modifies the original mesh by applying the displacements obtained in the structural mechanics simulation.

In Fig. 3, the mesh used in FEM structure simulations is shown. For the interfaces between different materials, boundary layers mesh elements are used, considering eight layers for each interface. For the remaining structure a free triangular mesh is considered. Variations in the capacitance values were observed for maximum element size up to $R_{out}/20$, while the differences between the results for $R_{out}/30$ and $R_{out}/50$ are below 0.1%. Therefore, a maximum element size of $R_{out}/30$ is chosen for all simulations. Finally, in the electrostatic FEM section an outer square with $10R_{out}$ side length is introduced in order to calculate the electric field in the air region around the device and thus include fringing field effects. At the boundary of this box, the floating potential condition was set.

Polydimethylsiloxane (PDMS), a silicone elastomeric material widely used in soft robotics [9], is employed for the design of simulated devices. In this first study, (i) the dimensions of the sensor are at millimeter scale, since this allows us to experiment typical tactile sensor sizes while avoiding unnecessary large size deviations from the natural tactile blep, and (ii) the Young's moduli range is restricted with respect to the biological model considering the values of the soft elastomer (PDMS) available. The radius of the outer shell $R_{out} = 1$ mm and its Young's modulus $E_{out} = 5$ MPa are held constant, as well as the thickness (0.5 mm), the length (6 mm) of the base and the ground electrode size (300 μm). A parametric study is performed, varying R_{in} from 0.25 to 1 mm (in the latter case a monolayer structure is obtained) and with E_{in} spanning from 0.4 MPa to 4 MPa.

3.2 Results

From a mechanical point of view, having a softer inner material, along with the augmentation of proportion of the inner hemisphere, leads to greater deformations when subjected to side force (Fig. 4A and C) than when subjected to normal force (Fig. 4B and D). In particular, the maximum displacement of 85 μm occurs under side pressure of 50 kPa for $R_{in} = 0.75$ mm and $E_{in,1} = 0.4$ MPa (Fig. 4C), while the displacement caused by top pressure for the same structure is 72 μm (Fig. 4D). For a structure with $R_{in} = 0.25$ mm and $E_{in,2} = 4$ MPa, the displacements are of 11 μm and 7 μm, for side (Fig. 4A) and top (Fig. 4B) externally applied forces, respectively.

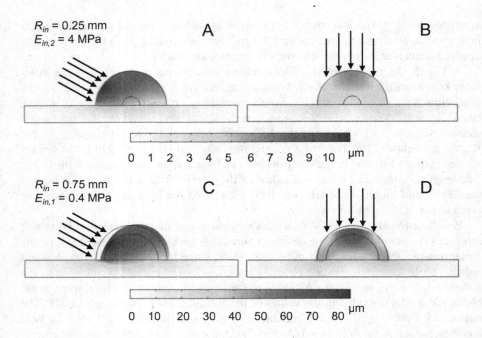

Fig. 4. FEM images showing displacement of the structure for two case studies. **Top**: $R_{in} = 0.25$ mm and $E_{in,2} = 4$ MPa for a side (A) and normal (B) pressure of 50 kPa, with displacement maximum values of 11 µm and 7 µm, respectively. **Bottom**: $R_{in} = 0.75$ mm and $E_{in,1} = 0.4$ MPa for side (C) and normal (D) pressure of 50 kPa, with displacement maximum values of 85 µm and 72 µm, respectively.

From an electrostatic point of view, relative capacitance variations for pressures up to 50 kPa are evaluated for structures with $R_{in} = 0.25$ mm (Fig. 5A), $R_{in} = 0.50$ mm (Fig. 5B), $R_{in} = 0.75$ mm (Fig. 6A) and $R_{in} = 1$ mm (Fig. 6B). Four different configurations are considered for each structure with constant R_{in} (Table 1):

Table 1. Configurations for each structure with constant R_{in} including graph codes used in Figs. 5 and 6

	$E_{in,1}$ [MPa]	$E_{in,2}$ [MPa]
Side applied pressure | Solid red line (——)	0.4 (×)	4 (◊)
Top applied pressure | Dashed black line (– – –)	0.4 (×)	4 (◊)

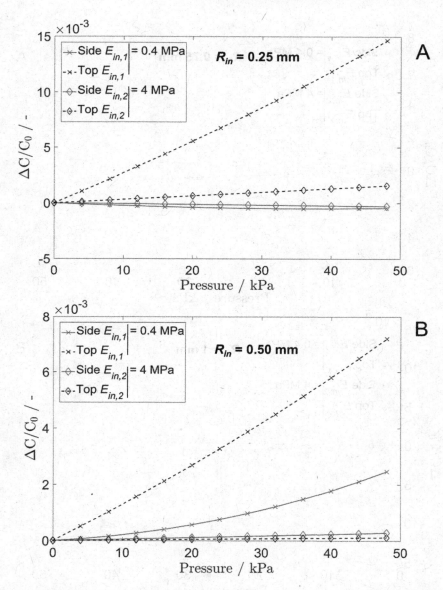

Fig. 5. FEM results of the dome-shaped bilayer structure. Pressure is applied either to the side region (solid red line) or to the top region (dashed black line) of the structure, with an inner and base material of $E_{in,1}$ = 0.4 MPa (cross marker) or $E_{in,2}$ = 4 MPa (diamond marker) and E_{out} = 5 MPa. The radius of the inner hemisphere (R_{in}) is 0.25 mm (A) and 0.50 mm (B). (Color figure online)

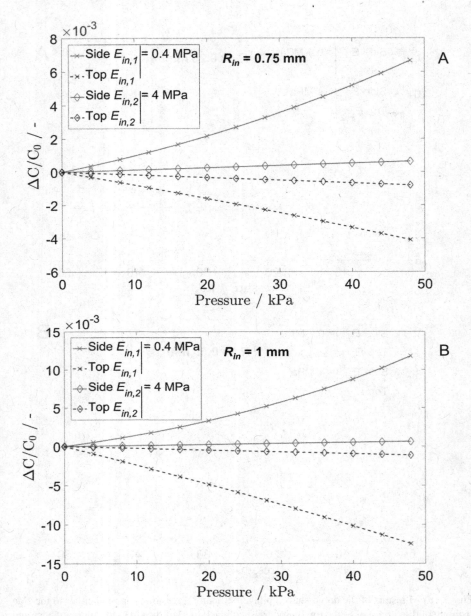

Fig. 6. FEM results of the dome-shaped bilayer structure. Pressure is applied either to the side region (solid red line) or to the top region (dashed black line) of the structure, with an inner and base material of $E_{in,1} = 0.4$ MPa (cross marker) or $E_{in,2} = 4$ MPa (diamond marker) and $E_{out} = 5$ MPa. The radius of the inner hemisphere (R_{in}) is 0.75 mm (A). For the boundary case $R_{in} = 1$ mm, the structure becomes monolayer (B). (Color figure online)

Regarding the three design principles, following statements are made based on these results:

1. The mechanical properties of the **materials** play a decisive role. The sensor sensitivity ($S = \Delta C/C_0$) is higher for softer inner and base material ($E_{in,1} = 0.4$ MPa), due to larger deformations occurring.
2. With the dome **shape** (constant R_{in} and E_{in}), there is a difference in sensitivity to side and top force in all cases. However, the shape alone does not lead to similar side force sensitivity with respect to top force response. In the cases of the bilayer structures with $R_{in} = 0.25$ mm and $R_{in} = 0.50$ mm (Fig. 5A and B, respectively) the top force sensitivity is much higher than for side force, which displays a negligible or slightly negative response.
3. Oppositely, with $R_{in} = 0.75$ mm (Fig. 6A) and $R_{in} = 1$ mm (monolayer structure, Fig. 6B) the behavior is different. In these cases, the side response is positive, while the top response is negative. Therefore, varying the proportion of the inner hemisphere leads to salient results, since the sensitivity to side force increases by increasing R_{in}. This reveals the significance of the **proportions** of the diverse building components. It should be noted that the fabrication of the monolayer structure implies the addition of a thin outer shell, in order to insulate the electrode on the side.

. Noticeably, by varying inner dimensions and material composition of a quite simple device, a tunable sensitivity for normal and shear force would be achievable. According to the application, the device can be designed to be more sensitive to normal or shear components. In specific configurations, normal and shear stimuli could be discriminated with a single signal by looking at the sign of capacitance variation.

The change in the device behavior implies that the adopted capacitive mechanism is much more complex than what initially hypothesized, and the simple consideration on electrode displacement is not sufficient in order to explain all cases studied. Therefore, future investigations of the electrostatic behavior of such a device are needed. In particular, the influence of ground electrode dimension (kept constant in this work) should be investigated, since fringing fields should play a key role in the different configurations.

4   Conclusion

The FEM simulation results suggest that, by exploiting a simplified bilayer dome-shaped structure inspired by the tactile blep of *Bryonia dioica* Jacq. tendrils, a tunable discrimination between normal and shear force should be possible, in particular by varying inner dimensions and material composition. The influence of the diversity of building components will be further investigated, as well as more complex geometries, inspired by the study of the natural structure. This preliminary study encourages the conception of novel capacitive-based soft tactile sensor prototypes, with a dome-shaped structure and simple electrode layout. They could be integrated in challenging soft robotics applications or in scenarios needing tunable normal/shear sensitivity. For instance, an array of sensors with opposite sensitivities in particular locations could discriminate normal and shear stimuli without any additional computational effort.

References

1. Sane, S.P., McHenry, M.J.: The biomechanics of sensory organs. Integr. Comp. Biol. **49**, i8–i23 (2009)
2. Monshausen, G.B., Haswell, E.S.: A force of nature: molecular mechanisms of mechanoperception in plants. J. Exp. Bot. **64**, 4663–4680 (2013)
3. Engelberth, J., Wanner, G., Groth, B., Weiler, E.: Functional anatomy of the mechanoreceptor cells in tendrils of Bryonia dioica Jacq. Planta **196**, 539–550 (1995)
4. Sareh, S., Jiang, A., Faragasso, A., Noh, Y., Nanayakkara, T., Dasgupta, P., Seneviratne, L.D., Wurdemann, H.A., Althoefer, K.: Bio-inspired tactile sensor sleeve for surgical soft manipulators. In: 2014 IEEE International Conference on Robotics and Automation (ICRA), pp. 1454–1459. IEEE (2014)
5. Taya, M., Wang, J., Xu, C., Kuga, Y.: Tactile sensors (2010). http://www.google.com/patents/US7823467
6. Viry, L., Levi, A., Totaro, M., Mondini, A., Mattoli, V., Mazzolai, B., Beccai, L.: Flexible three-axial force sensor for soft and highly sensitive artificial touch. Adv. Mater. **26**, 2659–2664 (2014)
7. Park, J., Lee, Y., Hong, J., Lee, Y., Ha, M., Jung, Y., Lim, H., Kim, S.Y., Ko, H.: Tactile-direction-sensitive and stretchable electronic skins based on human-skin-inspired interlocked microstructures. ACS Nano **8**, 12020–12029 (2014)
8. Yu, P., Liu, W., Gu, C., Cheng, X., Fu, X.: Flexible piezoelectric tactile sensor array for dynamic three-axis force measurement. Sensors **16**, 819 (2016)
9. Case, J.C., White, E.L., Kramer, R.K.: Soft material characterization for robotic applications. Soft Robot. **2**, 80–87 (2015)

Pop! Observing and Modeling the Legless Self-righting Jumping Mechanism of Click Beetles

Ophelia Bolmin[1]([✉]), Chengfang Duan[1], Luis Urrutia[1], Ahmad M. Abdulla[1],
Alexander M. Hazel[2], Marianne Alleyne[2], Alison C. Dunn[1], and Aimy Wissa[1]

[1] Department of Mechanical Science and Engineering,
University of Illinois at Urbana-Champaign, Urbana, IL 61801, USA
obolmin2@illinois.edu
[2] Department of Entomology, University of Illinois at Urbana-Champaign, Urbana,
IL 61801, USA
http://bamlab.mechse.illinois.edu/

Abstract. Click beetles (Coleoptera: Elateridae) have evolved a jumping mechanism to right themselves when on their dorsal side, without using their legs or any other appendages. This paper describes and analyzes the stages of the click beetle jump using high-speed video recordings and scanning electron micrographs of four beetle species, namely *Alaus oculatus*, *Ampedus nigricollis*, *Ampedus linteus* and *Melanotus* spp. The body of the click beetle is considered as two masses linked by a hinge. Dynamic and kinematic models of the jump stages are developed. The models were used to calculate the hinge stiffness and the elastic energy stored in the body during the jump. The modeling results show agreement with the experimental values. The derived models provide a framework that will be used for the design of a click beetle inspired self-righting robot.

Keywords: Legless jumping · Self-righting robot · Click beetle inspired robot

1 Introduction

1.1 Motivation

Robots inspired by animals have been designed to perform specific tasks autonomously in remote places. Current conventional autonomous robots are either wheeled or multi-pedal [1–5]. Some of these robots encounter difficulties when traversing unknown and unstructured environments due to limited robot motion gaits. Their design focuses on improved stability and balance to prevent falls. However, when navigating uneven terrains, falling is inevitable. One of nature's solutions to overcome obstacles and recover from falls can be found in insects, such as the click beetle (Coleoptera: Elateridae).

Click beetles (Fig. 1) have evolved a unique self-righting jumping mechanism that does not depend on legs or other appendages. From an initially inverted

© Springer International Publishing AG 2017
M. Mangan et al. (Eds.): Living Machines 2017, LNAI 10384, pp. 35–47, 2017.
DOI: 10.1007/978-3-319-63537-8_4

Fig. 1. From left to right: *Ampedus nigricollis* (Photograph by Katja Schulz, distributed under a CC-BY 2.0 license), *Ampedus linteus* (Photograph by WonGun Kim, used with permission), *Melanotus* spp. (Photograph by Judy Gallagher, distributed under a CC-BY 2.0 license) and *Alaus oculatus* (Photograph by Henry Hartley, distributed under a CC-BY 3.0 license). Live specimens of the first three species were studied as part of this work as well as dry specimens of *Al. oculatus* and *Melanotus* spp.

position, they jack-knife their body to catapult themselves into the air and land back on their feet. The objective of the work presented here is to understand the jumping mechanism of the click beetles and use its governing principles to advance self-righting techniques for robotic applications.

1.2   Background on Click Beetles

About 10,000 species in 400 genera of click beetles can be found worldwide [6]. While many insects jump to escape predators, click beetles have evolved a unique jumping mechanism to right themselves. The click beetle's jump is usually vertical [7] and they therefore land within 50 cm of their take-off location. Such jumps would make a very poor escape method. When feeling threatened, click beetles rather play dead or run away [8]. Thus the jump of click beetles is primarily used for self-righting.

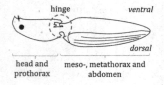

Fig. 2. The beetle body can be divided into two masses and a hinge: the head and the prothorax make up one mass (M1) and the mesothorax, metathorax and abdomen make up second mass (M2). The two masses are linked by a hinge.

In 1971, Evans described in detail click beetle morphology and the mechanics of the jump [10]. Subsequent work by Ribak further explained the morphological constraints and jump characteristics [7–9]. Dissections of click beetles and environmental scanning electron microscopy (ESEM) micrographs allowed for a better understanding of the beetle's exoskeletal structure [10]. Click beetles perform the jump from an inverted position by arching their body while storing energy in enlarged muscles in their prothorax, cuticle (exoskeleton), tendons and other bio-mechanical springs, until suddenly releasing this stored energy. The basic body morphology of a click beetle can be divided into three parts: two subunits or masses and the hinge, as shown in Fig. 2. The first mass includes the head and prothorax, the second mass includes the mesothorax, metathorax and abdomen, the hinge is comprised of a peghold and a mesosternal lip.

These previous studies explained how the beetle is able to jump without legs but many of the details are still unclear. These details are important to accurately model the click beetle's jump and emulate the jump in a novel robotic design. For instance, where exactly is the energy for the jump stored? What are the contributions of the muscles and the exoskeleton to this energy storage? What are the friction forces involved in the peghold - mesosternal lip system? How do the total mass and mass ratio between the two body subunits influence the take-off and aerial stages of the jump?

The remainder of the paper is organized as follows: the stages of the jump are detailed in Sect. 2. Dynamic and kinematic models are derived based on observing the jump of live beetles and are presented in Sect. 3. Results are presented and discussed in Sect. 4.

2   The Click Beetle Jump

2.1   Morphological Measurements

Morphological and kinematic measurements of 3 different click beetle species were obtained. *Melanotus* spp. individuals are 11–13 mm length and weigh 0.039–0.053 g (n = 7). The *Ampedus nigricollis* beetle we studied is 10.6 mm long and weighed 0.039 g (n = 1). The *Ampedus linteus* was the smallest beetle we studied, it weighed 0.029 g and was 9.3 mm in length (n = 1). The mass ratio between the head+prothorax and mesothorax+metathorax+abdomen varies between the different species. The smaller the beetle, the larger the mass ratio. The smaller *Am. linteus* has a mass ratio of 0.38 whereas the larger mass ratios for *Am. nigricollis* and *Melanotus* spp. beetles were 0.34 and 0.29–0.36, respectively. The rear heavier part (mesothorax+metathorax+abdomen), is the primary source of mass variation among species. The separate weights of the head+prothorax and mesothorax+metathorax +abdomen masses of *Am. linteus*, *Am. nigricollis* and *Melanotus* spp. show that all beetles have similar head+prothorax unit mass (0.008–0.0012 g), and smaller beetles have a lighter mesothorax+metathorax+abdomen (0.029–0.037 g). The muscular mass in the prothorax is therefore presumed to be similar for all beetles filmed using high-speed cameras. For the remainder of the paper, we made the assumption that all the filmed beetles are able to transfer the same amount of energy from the muscles to the cuticle and other biomechanical springs. Biomechanical springs are defined as all the body parts that can store elastic energy.

2.2   Stages of the Click Beetle Jump

The click beetle jump can be divided into three consecutive stages (Fig. 3). During the *pre-jump stage* or energy storage stage (Fig. 3-1), the beetle arches its body and stores energy. This position is held by friction for a varying amount of time, ranging from half a second to 2–3 s. During the *take-off stage* (Fig. 3-2), the beetle starts releasing energy while still in contact with the floor and generates the necessary impulse to propel its body into the air. Both the pre-jump

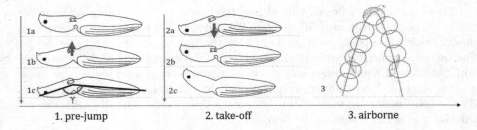

Fig. 3. The observed jump stages of click beetles

and take-off stages are characterized by translation of the beetle's hinge and the rotation of the two masses relative to the hinge (Fig. 3-1b–2a). The head of the beetle leaves the ground and is followed by the body. The beetle then somersaults in the air during the *airborne stage* (Fig. 3-3).

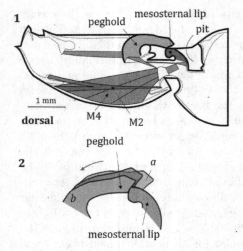

Fig. 4. 1: Anatomy of the hinge, based on Evans morphological sketch [10]. The beetle is shown resting on its dorsal side. The M4 and M2 muscles located in the prothorax store the mechanical energy prior to the jump. 2: During the pre-jump stage, the peghold slides over the mesosternal lip and comes to rest as its friction plate drops into a semi-spherical ledge on the mesosternal lip (position *b*).

Pre-jump Stage: The hinge is composed of a peghold and a mesosternal lip, which comprise an interface with topography that allows them to be held in place or slip. The anatomy of the hinge is sketched in Fig. 4-1. Removal or significant damage to these parts prevents the ability to jump. This was proven by Evans through removing the ventral part of the hinge of living specimens [10]. Once part of the peghold (Fig. 4) was removed, the beetle could not maintain the arched, energy storage position (Fig. 3-1c), and therefore could not perform the jump [10].

The peghold and mesosternal lip of *Melanotus* spp. and *Al. occulatus* were photographed using a Phonom Pro environmental scanning electron microscope (ESEM) (Fig. 5). Variations in shapes between species were observed (Fig. 5-1). During the pre-jump stage, the peghold's friction plate, shown in Fig. 5-1a, slides over the mesosternal lip. The peghold

slides from position *a* to position *b* as shown in Fig. 4-2. At the end of the stage, the peghold comes to rest as the friction plate drops into a semi-spherical ledge on the mesosternal lip, as shown in Fig. 5-2b.

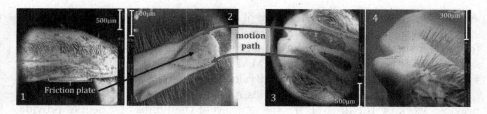

Fig. 5. ESEM micro-graphs of the side (1) and dorsal (2) views of the peghold and ventral (3) and side (4) views of the mesosternal lip of *Alaus oculatus*.

The resting position is held by friction while the energy is transferred from the M4 and M2 muscles (Fig. 4) to the cuticle and other biomechanical springs [10,11]. Once the energy required to perform the jump is stored, the arched overall body position position is released: the peghold slips back over the mesosternal lip and inserts into the cavity (i.e. the peghold moves from position *b* to position *a* as shown in Fig. 4-2). We postulate that the mechanical slip over the ledge is the source of the audible "click" as the beetle launches. Given that the hinge's motion is constrained by the junction of the prothorax and mesothorax sections, both peghold and mesosternal lip deform elastically to allow the slip of the friction plate over the lip, as shown by the different deformed states of the peghold shown in Fig. 4-2. Two rows of hairs, aligned parallel to the mesosternal cavity, guide the peghold into the cavity after the energy is released from the muscles and into the body's biomechanical springs (Fig. 5-4).

Takeoff Stage: The take-off stage is defined by the motion of the click-beetle from the first instance of the energy release (the "click", when the peg frees itself from its resting position on the edge of the lip) to the moment when the beetle leaves the ground. The motion of *Am. linteus* (n = 1), *Am. nigricollis* (n = 1) and *Melanotus* spp. (n = 9) was recorded using a Photron Fastcam SA-Z at a rate of 20,000 frames per second. The frames of interest are presented in

Fig. 6. Frames of high speed video footage of the take-off stage. (a) shows the initial resting position and (e) shows the frame when the abdomen leaves the ground.

Fig. 6. The downwards motion of the hinge is characteristic of this stage and causes the rotation of the two masses around the hinge. Angular accelerations and velocities were derived from high-speed footage of three different species (see Sect. 2.1). The acceleration is considered to be constant during the take-off stage as a first assumption.

Airborne Stage: The airborne stage of *Am. nigricollis* (n = 1) and *Melanotus* spp. (n = 6) was recorded using a Photron FastCam-X1280PCI at a rate of 2,000 frames per second. White dots of paint were placed on the specimens to reconstruct their airborne trajectory using MATLAB [15]. The jumps of seven beetles were analyzed and take-off angles and velocities were extracted.

The values of the take-off velocities and take-off angle are very similar to previous work by Ribak *et al.* [7]: the analysis of the jumps of 7 beetles provide take-off angles ranging from 82° to 89°, and take-off velocities of 1.4 to 1.9 m/s (Fig. 7). Maximum height and distance covered between the take-off and landing locations were derived from the videos. Previous work on click beetles has shown that the insects only had a 50% change of landing on their feet [7].

The beetles studied also landed on their feet or on their dorsal side randomly, and if needed, would jump again until landing on their feet, ventral side down. However, we observed several individuals which would open their wings before landing. The beetles either flapped their wings or used them to glide attempting to stabilize their body in the air. The wing deployment behavior was also random, as several individuals from the same species would perform consecutive jumps with or without deploying their wings.

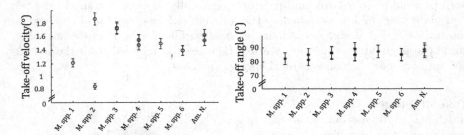

Fig. 7. Take-off velocity (left) and angle (right) of *Melanotus* spp.(M.spp.) and *Am. nigricollis* (Am.n.)

3   Modeling Approach

3.1   Energetics

An energetic approach was taken to model how click beetles store and release energy in order to jump. We assume that all energy is stored during the pre-jump stage. The muscles in the prothorax (Fig. 4) contract, extending the ventral length and pulling the hinge (center of hinge rotation) away from the substrate and rotating the two masses around the hinge. Gronenberg *et al.* suggested that the accelerations and velocities observed during the energy release stage cannot be created by muscle contraction alone, because muscles have inherently slow contraction velocity. Small animals and insects have other elastic biomechanical springs, such as the cuticle, tendons and rubbery proteins [11]. For the remainder of this paper, we will assume that the beetles' combined biomechanical springs are responsible for the energy release during the take-off and airborne stages. The amount of energy released during the jump can be derived using an energy balance approach.

Energy Balances During the Jump: During the pre-jump stage, we assumed the following:

- The work of the muscles is used to position the body in the stored-energy state, and "spring load" the hinge to its launch-ready configuration. The body's upwards motion is due to the hinge rotation and the deformation of the peghold (snap-fit mechanism, see Sect. 3.2).
- The remainder of the muscle energy is transfered onto the beetle's cuticle, tendons and other biomechanical springs, referred to as $E_{cuticle}$.
- Friction with the ground and inertial effects can be neglected.
- There are no observed bulk body deformations; all deformations are local to the mechanisms.

Based on the aforementioned assumptions, the energy balance during the pre-jump stage is expressed by:

$$E_{muscles} = E_{motion} + E_{snap-fit} + E_{biom,spring} \qquad (1)$$

Where $E_{bio-spring}$ is defined as the energy stored in the beetle's body:

$$E_{bio-spring} = E_{cuticle} + E_{proteins} + E_{deflectionSF} + \dots \qquad (2)$$

Given that the hinge's motion is constrained by the junction of the prothorax and mesothorax sections, both peghold and mesosternal lip *deform elastically to allow the slip of the friction plate over the lip*, as shown by the different deformed states of the peghold shown in Fig. 4. The strain energy generated by this deformation is referred in Eq. (2) as $E_{deflectionSF}$ (see Fig. 4-2).

The peghold/lip interaction will be modeled as a mechanical "snap-fit" mechanism, which is described in detail in Sect. 3.2. This purely mechanical model assumes that the energy needed to both load and release the mechanism is equal hence $E_{snap-fit}$ is similar during the pre-jump and take-off stages. The energy released during the downwards impulse of the hinge is converted to the airborne motion, referred to as $E_{take-off}$. Therefore the energy balance for the take-off stage is expressed by:

$$E_{biom,spring} = E_{snap-fit} + E_{take-off} \qquad (3)$$

During the airborne stage, the center of gravity of the beetle has an overall ballistic motion. The difference in masses between the two main body sections create angular momentum, which causes somersaulting maneuvers. The energy of the somersaulting motion is referred to as $E_{rotation}$. The head and prothorax of the beetle oscillate around the hinge. We neglect head and prothorax oscillations in this paper. The energy equilibrium during the airborne stage is expressed by:

$$E_{take-off} = E_{ballistics} + E_{rotation} \tag{4}$$

Considering the click beetle's body as a *biomechanical spring*, we can interpret $E_{muscles}$, $E_{cuticle}$ and $E_{snap-fit}$ introduced in Eq. 1 through Eq. 4 as the energies developed and transfered between the body's *subsystems*, key constituents of the jumping mechanism. These energies are derived from high-speed video recordings and friction experimental data (see Sect. 3.1). Muscles *actuate* the system, creating the motion and the potential energy storage in the cuticle, which acts like a *spring*. The snap-fit mechanism constitutes the *trigger* to energy release. We obtain the equivalences presented in Eqs. 5–7.

$$E_{muscle} = E_{actuator} \tag{5}$$

$$E_{cuticle} = E_{spring} \tag{6}$$

$$E_{snap-fit} = E_{trigger} \tag{7}$$

3.2   Pre-jump Model

The hinge mechanism, composed of the peghold and the mesosternal lip (Fig. 4-2) is modeled as a *snap-fit mechanism*, as shown in Fig. 8. Snap-fits geometrically combine the deflection of a cantilevered beam with a friction wedge at the tip of the cantilevered beam to create a discontinuous motion (Fig. 8). As the friction wedge slips over a lip, the elastic deformation of the beam is recovered quickly as it snaps back to a neutral position. In the click beetle, the peghold is the cantilever beam, and the friction plate is the wedge. The energy needed to load and release this trigger (i.e. to reach the resting position in Fig. 4) is defined as $E_{snap-fit}$.

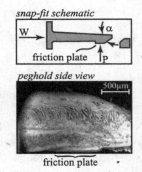

Fig. 8. The peghold (top) is modeled as a snap-fit mechanism (bottom) [12]

The inputs to this model are the effective peghold cross-sectional area, the deflection distance, the angle of the frictional pad with respect to the axis of slip, the effective Young's modulus of the peghold, and the friction coefficient of the peghold/lip interface. We assume that the peghold can be approximated as a hollow cylinder, as an internal cavity has been observed (images not shown). This geometry is measured directly from the ESEM images, and is summarized in Sect. 4. The equation for the deflection, δ, of a fixed-free beam of modulus E, length L, and area moment of inertia I due to load P (Eq. 8) is combined with the friction coefficient, μ, and the angle of the friction wedge, α, to provide the insertion force W of the snap-fit (Eq. 9). This insertion force over some effective slip distance and angular translation is considered to contribute substantially to the hinge stiffness, K (see Sect. 4.1), and therefore the launch energy, $E_{snap-fit}$.

$$\delta = \frac{PL^3}{3EI} \tag{8}$$

$$W = P\frac{\mu + tan(\alpha)}{1 - \mu * tan(\alpha)} \tag{9}$$

A micro-mechanical experiment on the bending of the hinge was constructed to measure the force required to load/release the snap-fit mechanism. Dry specimens of *Melanotus* spp. were mounted in foam from the posterior tip of the abdomen up to about 1 mm short of the thorax-abdomen junction (i.e. the hinge). The hinge began in a neutral position, and was not loaded during drying. A nylon thread was affixed to the dorsal side of the head and a force transducer with a sensitivity on the order of single microNewtons and a maximum force of approximately 200 mN. A linear stage translated the mounted body at v = 0.25 mm/s, causing the hinge to rotate and moving the peghold from position a to position b as shown in Fig. 4-2. When it is loaded and the peghold snaps over the lip, a force drop is observed, and the magnitude of the force drop is reported in Sect. 4.

3.3 Take-off Model

The energy release stage is initiated when the beetle is resting on the substrate. When the muscle tension is released, the peghold slips over the mesosternal lip with low frictional resistance, launching its mass into the air through leverage against the ground. This motion is modeled by a crank-slider mechanism as shown in Fig. 9.

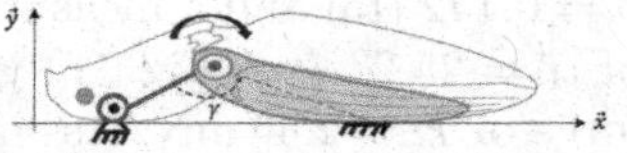

Fig. 9. Crank-slider mechanism modeling the dynamics of the take-off stage

We assume that the friction with the ground can be neglected and that there are no body deformations. The acceleration is assumed to be constant, and we consider as an input to the crank-slider model the peak acceleration (measured when the body leaves the ground). The two links of the crank-slider are connected by an ideal pin joint, which is actuated by a torque (T). We assume as a first approximation that $T = K * \gamma$, where γ is the body's angle and K is the stiffness of the modeled spring, which is composed of the stiffness of the snap-fit mechanism and the beetle body's biomechanical springs [11]. The equation of motion of the crank-slider is solved for the stiffness coefficient K given a certain mass ratio. Initial angles, velocities and accelerations are derived from the high-speed video recordings of the beetle's motion (see Fig. 6).

3.4   Airborne Model

During the airborne phase, we neglected the oscillations of the head and prothorax around the hinge.
The click beetle is modeled as two point masses connected with a rigid massless link, as shown in Fig. 10.
The dynamics of the system are derived and simulated using MATLAB [15]. Given an initial velocity, take-off angle and angular momentum of the two
masses, the overall trajectory of the beetle in the air,
described by the landing location, maximum height

Fig. 10. Two-mass rigid-
body model of the beetle's
body during the airborne
stage.

of the center of mass (which has ballistics motion), and the number of rotations
of the two masses around the center of mass is predicted. The model inputs were
derived from the high speed recording of the complete beetle jumps.

4   Results

4.1   Pre-jump Model Results

The force drop as the peghold/lip mechanism snapped into the loaded position
was measured to be $W = 1.6$ mN. There are multiple force drops in the measurement (Fig. 12), but video timing was used to confirm that the force drop
corresponded to the "snap" of the beetle hinge.

Implementing Eq. 9 with a friction coefficient of
$\mu = 0.142$ [13] and a measured friction wedge angle
of $\alpha = 25.2°$ (see Fig. 11) gives a beam deflection
force of $P = 2.44$ mN. Assuming a hollow cylindrical peg with outer diameter $\sim$1 mm, wall thickness
$\sim$200 µm, length L = 4 mm, and Young's modulus
of $E \sim 7$ GPa [14], the deflection of the beam δ
is 0.25 µm (by Eq. 8). This value is likely too low,
further work is needed to refine the specific snap-fit
model properties.

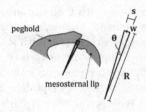

Fig. 11. Geometry and force
vector used to calculate the
stiffness of the peghold/lip
mechanism.

The force drop is used to estimate the stiffness
contribution of this mechanism to the hinge stiffness K. Rough estimates of the slip distance across
the mesosternal lip width of $s \sim 50$ µm (see Fig. 11) gives a launch torque of
80 nNm. This torque acting over an impulse angular displacement of $\theta \sim 0.1$
radians at a distance away from the center of rotation of $R \sim 0.5$ mm gives a
stiffness of $K = \frac{T}{\theta} = 8 * 10^{-7}$ Nmrad^{-1}.

4.2 Take-off Model Results

The equation of motion of the crank-slider mechanism (Fig. 9) was derived and solved for the torque T. The hinge stiffness K was then calculated assuming that the torque necessary to create the downwards motion of the hinge is linearly proportional to γ. Analysis of the high-speed recording of the take-off stage (Fig. 6) provided angular velocities and accelerations, as well as the body's angle in the energy storage position. Figure 13 shows the calculated results for the stiffness coefficient resulting of the jumps of five different species.

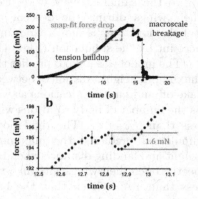

The value of hinge stiffness calculated from the force drop and geometry of the peghold/lip interface gives a value that falls within the range of stiffness values for *Am. nigricollis* (n = 1) and *Melanotus* spp. (n = 1) calculated from the crank slider

Fig. 12. The force needed to actuate the hinge of a *Melanotus* spp. beetle showing the force drop during the snap-fit and subsequent breakage as the motion was pushed past biological limits

dynamic model. Additional future data on the geometry of the interface and body motion will allow for higher confidence in this value. Also further morphological measurements will allow for comparisons between different click beetle species.

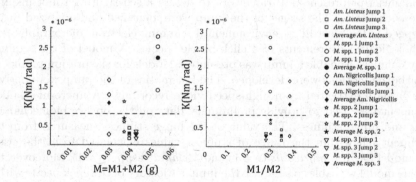

Fig. 13. Coefficient of stiffness K calculated in terms of the mass ratio (left) and total mass (right) of the tested 3 beetles species, namely *Ampedus linteus* (Am.L.), *Ampedus nigricollis* (Am.N.) and *Melanotus* spp. (M.spp.)

4.3 Airborne Model Results

The equation of motion of the two mass-rigid body model (Fig. 10) was derived and a MATLAB [15] simulation was developed to replicate the airborne motion of the beetle. The jumps of *Am. nigricollis* (n = 1) and

Melanotus spp. (n = 6) were analyzed. Only jumps that were perpendicular to the camera's view were used for further analysis. The beetles opened their wings during the descent stage in approximately half of the jumps, a behavior that is not taken into account by the simulation (Fig. 14).

The model was able to predict the maximum height, distance between take-off and landing locations as well as the number of body rotations with less than 5% error. The rigid body simulation predicted the maximum height and landing distance when the beetles open their wings with an error less than 10%. Given that the ballistic motion model was able to predict the jump kinematics of the click beetle even when they deployed their wings during the descent supports the hypothesis that click beetles open their wings to stabilize themselves, and not to cover distance.

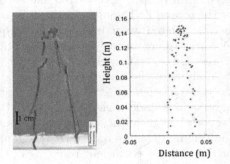

Fig. 14. *Left:* The airborne trajectory reconstructed from high speed video recordings. *Right:* A MATLAB [15] simulation of the airborne trajectory calculated using the two-mass rigid body model for a jump of a *Melanotus* spp. individual.

5   Conclusions and Future Work

This manuscript presented a framework to design a self-righting robot inspired by the click beetle. The stages of the jump were identified and analyzed, using high-speed video recordings, environmental scanning electron micrographs, and morphological measurements of 4 click beetle species. A model of the required energy balance to enable a jump was presented. Models for the pre-jump, take-off and airborne motions were developed. The kinematics of the jump were derived based on data extracted from high-speed video recordings. A micro-mechanical experiment was used to provide preliminary data on the body's biomechanical spring and hinge stiffness. The value of the hinge stiffness measured from the experiment fell within the range of stiffness values calculated from the crank slider model for *Am. nigricollis* (n = 1) and *Melanotus* spp. (n = 1). Moreover, the airborne model was able to simulate the jump trajectory of the click beetle within 5%. Future work will include collecting morphological data from more individuals and from other species of beetles. More experiments are planned to confirm the energy exchange assumptions and to measure any body deflections during the pre-jump and take-off stages. Moreover, the pre-jump, take-off and airborne models will be validated through physical prototypes. Various head+prothorax and mesothorax+metathorax+abdomen units masses will be tested in order to evaluate the influence of the mass ratio and the total mass on the jump.

Acknowledgements. This research was supported by the Department of Mechanical Science and Engineering and the Department of Entomology at the University of

Illinois at Urbana-Champaign (UIUC). Ophelia Bolmin was fully supported by Fulbright and the Monahan Fundation and the Aerospace Engineering department at the UIUC and ENSTA Bretagne. We thank Joshua Gibson and Andy Suarez from the Suarez Laboratory (UIUC) for their experimental resources and expertise. We also appreciate Prof. Shelby Hutchens (UIUC) for allowing the use of the ESEM.

References

1. Seok, S., Wang, A., Chuah, M.Y., Hyun, D.J., Lee, J., Otten, D.M., Lang, J.H., Kim, S.: Design principles for energy-efficient legged locomotion and implementation on the MIT Cheetah robot. IEEE/ASME Trans. Mech. **20**(3), 1117–1129 (2015). doi:10.1109/TMECH.2014.2339013
2. Kovac, M., Fuchs, M., Guignard, A., Zufferey, J.-C., Floreano, D.: A miniature 7g jumpins robot. In: IEEE International Conference on Robotics and Automation (ICRA 2008), pp. 373–378 (2008). doi:10.1109/ROBOT.2008.4543236
3. Nguyen, Q.-V., Park, H.C.: Design and demonstration of a locust-like jumping mechanism for small-scale robots. J. Bionic Eng. **9**, 217–281 (2012). doi:10.1016/S1672-6529(11)60121-2
4. Lia, F., Liua, W., Fua, X., Bonsignorib, G., Scarfoglierob, U., Stefaninib, C., Dariob, P.: Jumping like an insect: design and dynamic optimization of a jumping mini robot based on bio-mimetic inspiration. Mechatronics **22**(2), 167–176 (2012). doi:10.1007/978-1-4419-9985-6_11
5. Quinn, R.D., Offi, J.T., Kingsley, D.A., Ritzmann, R.E.: Improved mobility through abstracted biological principles. In: IEEE/RSJ International Conference on Intelligent Robots and Systems (2002). doi:10.1109/IRDS.2002.1041670
6. Slipinski, S.A., Leschen, R.A.B., Lawrence, J.F.: Order coleoptera linnaeus, 1758. In: Zhang, Z.-Q. (ed.) Animal Biodiversity: An Outline of Higher-Level Classification and Survey of Taxonomic Richness, vol. 3148, pp. 203–208. Zootaxa (2011)
7. Ribak, G., Weihs, D.: Jumping without using legs: the jump of the click-beetles (Elateridae) is morphologically constrained. PLoS One **6** (2011). doi:10.1371/journal.pone.0020871
8. Ribak, G., Reingold, S., Weihs, D.: The effect of natural substrates on jump height in click-beetles. Funct. Ecol. **26**, 493–499 (2012). doi:10.1111/j.1365-2435.2011.01943.x
9. Ribak, G., Mordechay, O., Weihs, D.: Why are there no long distance jumpers among click-beetles (Elateridae)? Bioinspir. Biomimin. **8**, 036004 (2013). doi:10.1088/1748-3182/8/3/036004
10. Evans, M.E.G.: The jump of the click-beetle (Coleoptera, Elateridae) - a preliminary study. J. Zool. **167**, 3179–36 (1972). doi:10.1111/j.1469-7998.1972.tb03115.x
11. Gronenberg, W.: Fast actions in small animals: springs and click mechanisms. J. Comput. Physiol. A **178**, 727–734 (1996). doi:10.1007/BF00225821
12. Bayer Material Science LLC: Snap-fit joints for plastics - a design guide, Pittsburg (2013)
13. Dai, Z., Min, Y., Gorb, S.N.: Frictional characteristics of the beetle head-joint material. Wear **260**, 168–174 (2006). doi:10.1016/j.wear.2004.12.042
14. Vincent, J.F.V., Wegst, U.G.K.: Design and mechanical properties of insect cuticle. Arthropod. Struct. Dev. **33**, 187–199 (2004). doi:10.1016/j.asd.2004.05.006
15. MATLAB and Statistics Toolbox Release 2016b. The MathWorks Inc., Natick (2016)

Bioinspired Magnetic Navigation Using Magnetic Signatures as Waypoints

Brian K. Taylor[1]([✉]) and Grant Huang[2]

[1] Air Force Research Laboratory, Eglin AFB, USA
brian.taylor.48@us.af.mil
[2] University of Florida, Shalimar, USA
grant.huang@ufl.edu

Abstract. Diverse taxa use Earth's magnetic field in conjunction with other sensor modes to accomplish navigation tasks that range from local homing to long-distance migration across continents and ocean basins. However, despite extensive research, animal magnetoreception remains a poorly understood, and active research area. Concurrently, Earth's magnetic field offers a signal that engineered systems can leverage for navigation and localization in environments where man-made systems such as GPS are either unavailable or unreliable. Using a proxy for Earth's magnetic field, and inspired by migratory animal behavior, this work implements behavioral strategies to navigate through a series of magnetic waypoints. The strategies are able to navigate through a closed set of points, in some cases running through several "laps". Successful trials were observed in both a range of environmental parameters, and varying levels of sensor noise. The study explores several of these parameter combinations in simulation, and presents preliminary results from a version of the strategy implemented on a mobile robot platform. Alongside success, limitations of the simulated and hardware algorithms are discussed. The results illustrate the feasibility of either an animal, or engineered platform to use a set of waypoints based on the magnetic field to navigate. Additionally, the work presents an engineering/quantitative biology approach that can garner insight into animal behavior while simultaneously illuminating paths of development for engineered algorithms and systems.

1 Introduction

Earth's magnetic field offers a useful source of information for both engineered [1,2], and biological systems (e.g., [3–5]). The magnetic field provides a 3-dimensional omnipresent vector signal that can be used to gain insight into one's absolute and/or relative position (e.g., [1]), and orientation. Extending from the crust up through the atmosphere, the Earth's magnetic field can be described by its inclination angle (i.e., angle from the local horizontal), direction

© Springer International Publishing AG 2017
M. Mangan et al. (Eds.): Living Machines 2017, LNAI 10384, pp. 48–60, 2017.
DOI: 10.1007/978-3-319-63537-8_5

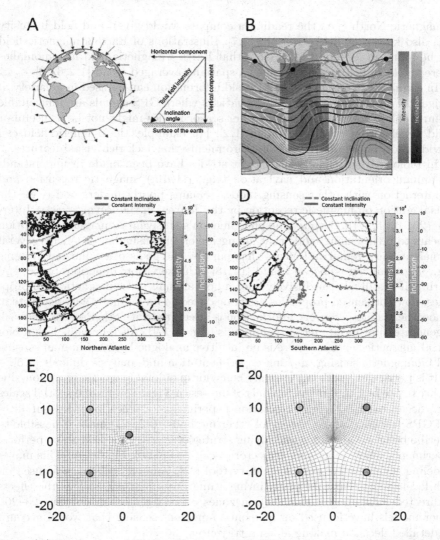

Fig. 1. (A) Illustration of the Earth's magnetic field that shows the definitions of magnetic inclination, and intensity (taken from [8,9]). (B) Illustration of lines of constant inclination (black and orange lines), and lines of constant intensity (blue, white, and green lines). Specific combinations of intensity and inclination are relatively distinct and rare across the world (e.g., the combination of parameters marked by either the black dots or orange triangles only occurs 4 times, and only occurs in the northern hemisphere). (C) and (D) More detailed illustrations of lines of constant inclination and intensity in the northern (C - left) and southern (D - right) Atlantic Ocean. In these regions a given combination of intensity and inclination appears to be either rare, or unique. The spatial scale of latitude and longitude lines is 0.25°. (E) and (F) Composite contour maps that are used as proxies for magnetic field properties in this study. E shows the map for a low value of Γ, while F shows a high value of Γ. The encircled shaded regions denote waypoint locations that are used in this study. Axes are in meters. (Color figure online)

of magnetic North (i.e., the reading a compass would give), and field intensity (i.e., also known as field strength) [6,7]. Illustrations of Earth's magnetic field can be found in [8], and Fig. 1. Note that Earth's magnetic field has anomalies where the local field does not follow a spatially averaged trend [1,2,6,10].

In engineered systems, magnetic field information can be used in a variety of navigation applications, including situations where GPS signals are unavailable or unreliable, and scenarios where other sensor modes might not be as useful or could benefit from a form of aiding [1,2]. For example, the magnetic field can provide additional information in environments that lack rich visual features.

In biological systems, though great strides have been made in understanding animal orientation and navigation (e.g., [11–16]), magnetic reception and its interaction with other sensing modes remains poorly understood (e.g., the special edition on magnetic reception by the Journal of the Royal Society: Interface (2010), [5,17]). Despite this phenomenon's elusiveness, magnetic reception appears to be a sensing mode that, alongside other sensing modes, helps several animals achieve remarkable feats that parallel the goals of engineered systems (e.g., migration across ocean basins [12,18]). Unlike vision, olfaction, touch, and hearing, it is still not conclusively known how animals sense magnetic field properties, how the sensed properties are transduced to an animal's nervous system, or how the magnetic sensory information is integrated with other sensor feedback to generate appropriate motor commands (e.g. [4,17]). Also, humans lack a biological magnetic sense, so there is no intuitive analogue between human sensing and biomagnetic sensing, making experimentation and analysis difficult [4,5].

It is particularly difficult to study behavior in animal magnetic reception, due in part to the fact that many animals of interest navigate over large spatial scales (e.g., ocean basins). While orientation experiments, studies involving satellites and GPS tracking, and a variety of other methods [19? –21] make it possible to describe behavior and infer underlying strategies [8,22], it is difficult to perform experiments where migratory trajectories can be analyzed in detail. This makes modeling an attractive complementary tool [23]. Some models have focused on high-level events that take place during a migratory journey, such as the effects of directed swimming within certain zones of a larger ocean current [23–26]. Other models have focused on moment-to-moment sensing that would account for detailed decision making along a trajectory [36–38].

Building on these studies, and more broadly, on work that has used engineering tools and mathematical models to advance biological understanding and/or engineered technnologies [27–34], this work uses a moment-to-moment approach to explore the feasibility of animals using rare and/or unique combinations of magnetic properties (i.e., signatures) as a type of waypoint to help them navigate. This idea has been posed in a conceptual manner [8], and has shown promise in studies involving turtles and salmon [18,24–26,35]. Specifically, building on example navigation concepts presented by [36,37], this work challenges a strategy to migrate to a series of geometric locations by moving towards magnetic signatures associated with these locations. A software simulation is used to execute several closed loops around a series of goal locations in a variety of environmental and system conditions. Additionally, to investigate how this approach might work in a real-world setting, an algorithm developed in [39] was tasked with moving to a goal, and then back to the starting location. The findings provide insight into how an animal might navigate using magnetic signatures as waypoints, and offer concepts that an engineered system might use for navigation.

2   Methods

For both the software and hardware experiments, a simulated magnetic environment was used to serve as a proxy for a real-world magnetic field. While either real-world or modeled magnetic field data could be used (e.g., [40]), the character and geometry of Earth's magnetic field varies over the Earth's surface (e.g., lines of constant intensity an inclination are orthogonal in some locations, but not others). This makes it difficult to construct and evaluate an algorithm's strengths and weaknesses in a controlled way (e.g., an algorithm may succeed or fail because of its own mechanics, the environment, or some interplay of the two). In contrast, by constructing a proxy for the field that contains real-world characteristics, we can better understand the performance of an algorithm prior to testing it on real-world data. To this end, we first describe our simulated magnetic enviroment that is used for both the software and hardware experiments. We then describe the software and hardware navigation strategies that are employed, and the experiments that are performed.

2.1   Simulated Magnetic Environment

For both the simulated and hardware experiments, a simulated static magnetic environment was employed. The environment uses two separate contour maps that serve as proxies for the isolines of two independently sensed magnetic field properties (e.g., lines of constant inclination and intensity), a technique used in previous theoretical biology studies [36,37]. In this study, the proxy magnetic contour maps were generated by using the streamlines (Eq. 1) and velocity potential contours (Eq. 2) of an aerodynnamic lifting cylinder [41].

$$\gamma = (V_\infty y)\left(1 - \frac{R^2}{x^2 + y^2}\right) + \frac{\Gamma}{2\pi}\ln\left(\frac{\sqrt{x^2 + y^2}}{R}\right) \tag{1}$$

$$\beta = (V_\infty x)\left(1 + \frac{R^2}{x^2 + y^2}\right) - \frac{\Gamma}{2\pi}\arctan\left(\frac{y}{x}\right) \tag{2}$$

Here, γ and β represent streamlines (lines that trend towards horizontal in Fig. 1E and F) and velocity potential lines (lines that trend towards vertical in Fig. 1E and F), respectively. x and y are spatial coordinates. R, Γ, and V_∞ are the cylinder radius, flow velocity, and flow circulation, respectively. Note from Eqs. 1 and 2 that there is a singularity at the origin (i.e., $x = y = 0$), and a jump discontinuity along either the x axis or the y axis depending on whether a standard, or 4 quadrant inverse tangent is used. Also, these contours are orthogonal to each other at all points. R, Γ, and V_∞ change the shape and curvature of the contours. Notably, increasing the value of Γ relative to V_∞ creates a more curvilinear (and still orthogonal) map. Illustrations of these contour maps for two different values of Γ are shown in Fig. 1E and F.

These features allow the proxy field to vary smoothly over large spatial scales while still providing discontinuities, potential ambiguity, and a local anomaly that does not conform to the field's general pattern of variation, characteristics that are seen in Earth's magnetic field. This provides an environment that is simple to implement and modify for testing, yet retains real world elements that can cause the agent navigational difficulty and allow for probing questions to be addressed.

2.2 Software Agent

For the software experiments, a simulated agent is tasked with moving to a series of magnetic waypoints. We base our agent's strategy on animal navigation concepts from [36,37], which showed how a mismatch between an animal's perception of its relative location to a goal and its actual relative location to a goal based on sensor feedback causes a systematic error in its motion. Properly followed, the motion eventually leads to the goal in a nonlinear trajectory that depends on the mismatch between the sensory environment's geometry, and the animal's "cognitive" representation of this geometry.

In the present strategy, the agent measures the magnetic properties at its current location (γ^*, β^*), which may each be corrupted by zero mean normally distributed sensor noise (σ, ν) (Eq. 3). The current location's properties, and the magnetic properties of the goal location (γ_g, β_g) are used to form a magnetically based vector that points to the goal location. For convenience we decompose this vector into a magnitude (d), and unit vector $(\hat{d})$ (Eq. 4). Based on these measurements, the agent's position is updated from one timestep to the next by creating a velocity vector in the direction of $\hat{d}$ (Eq. 5).

$$\begin{bmatrix} \gamma^* \\ \beta^* \end{bmatrix} = \begin{bmatrix} \gamma + \sigma \\ \beta + \nu \end{bmatrix} \tag{3}$$

$$\vec{d} = \begin{bmatrix} \gamma_g - \gamma^* \\ \beta_g - \beta^* \end{bmatrix}$$

$$d = |\vec{d}|, \hat{d} = \frac{\vec{d}}{d} \tag{4}$$

$$\vec{x}_{k+1} = \vec{x}_k + s\hat{d}\Delta t \tag{5}$$

$\vec{x}_k$ and $\vec{x}_{k+1}$ are the agent's (x, y) positions at the current (i.e., k^{th}) and next (i.e., $k + 1^{th}$) time steps. s is the agent's speed, and Δt is the time step size. Notice that the agent does not have an explicit representation of the full magnetic environment, but instead, only notional ideas about where and how to move based on what is currently sensed.

2.3   Simulation Experiments

The software simulation was set up to travel to either 3 points or 4. The 3 point scenario has 2 points that are far from the singularity at the origin, and one that is close to it (Fig. 1E). The 4 point scenario has all of its points far from the singularity at the origin (Fig. 1F). In both scenarios, the agent was run for 20 migrations to determine whether the implemented strategy could travel to a specified set of points multiple times. Successfully reaching a given waypoint was defined as moving within 1 m of that waypoint as defined by an outside observer. The agent had to successfully reach each waypoint before being allowed to search for the next one.

The vehicle was run with Γ set to either zero or 20. The measurement noise was set to either a standard deviation of zero or 2. The frequency of measurements (f) was set to either 0.5 Hz, or 0.1 Hz. The vehicle's speed (s) was 1 m/s, and its position was updated using Forward Euler steps with a step size of $\Delta t = 0.1\,s$. To verify result repeatability in cases with nonzero noise, we ran 100 trials for each parameter combination, and recorded the success rate (q), and mean sinuosity (ζ) of all trials as defined by [42] (i.e., actual path length divided by straight line distance to the goal).

2.4   Hardware Experiments

For the robotic experiments, a TurtleBot mobile robotic platform (Fig. 2A) running the Robot Operating System (ROS) is tasked with moving to a goal waypoint, and then moving back to the starting location using an algortihm developed in [39]. Briefly, the robot computes cost functions based on the Euclidian distance between the sensed magnetic properties at the current location (γ_k, β_k), and the goal's magnetic properties (γ_g, β_g) (Eqs. 6 and 7). This is done for both the current and previous timesteps. The difference of these cost functions (Eq. 8), and their size relative to one another are used to determine translation and rotation commands that are sent to the robot. The robot moves forward at a specified speed for a period of time, or rotates at a specified angular velocity for a period of time (Fig. 2B).

$$cost_k = \sqrt{(\gamma_k - \gamma_g)^2 + (\beta_k - \beta_g)^2} \tag{6}$$

$$cost_{k-1} = \sqrt{(\gamma_{k-1} - \gamma_g)^2 + (\beta_{k-1} - \beta_g)^2} \tag{7}$$

$$\text{diffcost} = |cost_k - cost_{k-1}| \tag{8}$$

The robot was started from a variety of starting locations and orientations relative to the goal location. The question posed here was whether or not the robot was able to reach both the goal, and then the starting location again. Success was defined when $cost_k \leq 0.6$. Ten trials were performed from each initial condition. For the hardware experiment, Γ was set to zero.

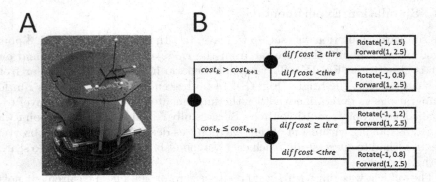

Fig. 2. (A) Picture of a TurtleBot mobile robot platform. (B) Illustration of the logic that is used to determine how the robot moves based on the sensor values that it receives. In the *Forward* and *Rotate* commands, the first argument is the speed or angular velocity, while the second argument is how long the command should be executed for. In this study, *thre* is set to 0.01.

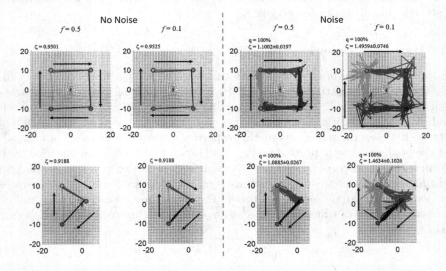

Fig. 3. Illustration of the software simulation moving through 20 migrations of four points (top row), and three points (bottom row) for a relatively rectilinear environment (i.e., $\Gamma = 0$). In this and all other plots, arrows indicate the notional direction of motion. The first and third columns show the agent taking measurements at 0.5 Hz, while the second and fourth columns show the agent taking measurements at 0.1 Hz. Plots to the left of the dashed line have no measurement noise, while plots to the right of the dashed line have a measurement noise standard deviation of 2. The color of the trajectory corresponds to the goal being sought out at that point in time (e.g., for 4 points, the light yellow part of the trajectory is trying to reach the shaded yellow point in the second quadrant). ζ denotes the sinuosity, and q denotes the success rate in cases where 100 trials were performed. (Color figure online)

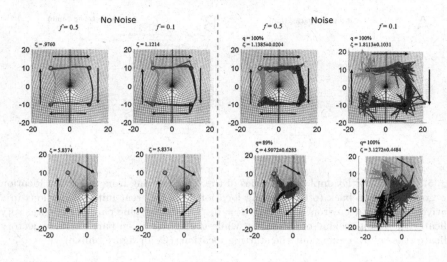

Fig. 4. Illustration of the software simulation moving through 20 migrations of four points (top row), and three points (bottom row) for a curvilinear environment (i.e., $\Gamma = 20$). Magenta parts of the trajectory illustrate where the agent was unable to successfully find a waypoint (Color figure online)

3   Results

3.1   Software Simulation

Figure 3 illustrates the software simulation running through 20 migrations with $\Gamma = 0$. For all parameter combinations, the agent was able to successfully able to migrate through all waypoints (i.e., $q = 100$) with sinuosities between 1 and 1.5. Sinuosities slightly less than 1 are due to the agent trying to move to within 1 m of a specified goal location. This means that the straight line distance between two subsequent migration points can be slightly greater than the path taken by the agent. A smaller success radius would push the minimum sinuosity closer to 1.

Figure 4 illustrates the software simulation running through 20 migrations with $\Gamma = 20$. Here, while the agent is able to succeed in all cases for the 4 points that are far from the singularity, it is only able to succeed in the three point case when sensor noise is present. In cases with sensor noise, the agent was able to succeed 100% of the time in four point migration experiment with sinuosities between 1 and 2. In contrast, in the three point experiment, the agent succeeded 89% of the time with a large measurement frequency ($\zeta \sim 4.9$) and 100% of the time with a small measurement frequency ($\zeta \sim 3$). The sinuosities in these two cases are statistically significantly different (Student's T Test, $P \leq 0.05$).

3.2   Hardware Simulation

Figure 5 shows sample results from the hardware. For all starting locations (Northwest, Northeast, Southwest, and Southeast of the goal) and orientations

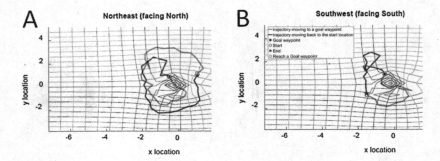

Fig. 5. (A) and (B) Example trajectories of the robot moving from a starting location to a goal, and then back to the starting location from different initial conditions in a relatively rectilinear environement (i.e., $\Gamma = 0$). The darker blue part of the trajectory indicates the robot seeking out the goal, while the lighter green part of the trajectory indicates the robot seeking out the starting location. (Color figure online)

(either facing North or South), the robot was able to successfully move towards the goal, and then back to the initial location for each of 10 trials. Note that the displayed trajectories are average trajectories over 10 trials.

4   Conclusions and Discussion

The results show one plausible way of migrating through a series of magnetic waypoints under a number of conditions, and in a variety of environments. This is illustrated in a software simulation, and confirmed in a hardware platform. The software simulation demonstrates that it is possible to navigate around a closed set of waypoints several times over, though the measurement frequency and noise play a role in how direct the path is. For example, with no noise, the smaller measurement frequency's behavior over 20 migrations is nearly identical to that of the larger measurement frequency. However, when noise is introduced, the larger measurement frequency trajectories appear to become more inefficient due to consistently overshooting the goal and having to backtrack.

The no-noise four-point trajectories of Fig. 4 illustrate the initial orientation error phenomenon theoretically described by [36]. Notice that the trajectories each reach their destination, but do so in a curvilinear manner due to the mismatch between the agent's notional ideas of what constitute North/South and East/West movement, vs. what the map actually looks like. This is an interesting observation, as is the work of [36] because it demonstrates that an animal may successfully navigate without having a full map or representation of the actual environment. This also demonstrates that a map need not have a human-centered conceptualization or implementation to be useful [8].

It is interesting that in the $\Gamma = 20$ case, the the three point migration only succeeds when noise is present, and improves somewhat when the measurement frequency is lower. This is likely due to the fact that the agent has to move near the singularity for the given migration task. Moving near the singularity causes ambiguity, likely causing the agent to get stuck in an area where the magnetic signature is correct but the geographic location is wrong. With sensor noise, even if the agent hits this point, the injected noise is likely to push it out of this situation, thereby increasing the chance that it will reach the correct location. This is aided by the fact that with a lower measurement frequency, if the agent takes a measurement in a problematic area, it then has time to move away from a problematic area before taking a new measurement, as opposed to taking measurements more rapidly and getting stuck in an effective local minimum. Interestingly, if this particular migration route is run in reverse, the agent succeeds even with no noise (data not shown). The reasons for this phenomenon will be explored more in a future study.

One might notice that the robot's trajectory is more circuitous than the software simulation. There are several reasons for this. First, the strategy used by the robot is related to, but not the same as that used in the simulation. The simulation presents a kinematic agent that simply translates towards a desired location, and does not take into account any potential effects or constraints of orientation and translation on motion. This is done both for simplicity, and to gain a general perspective on how different system and environmental navigation conditions affect performance. In contrast, the mobile robot's motion is tied to its orientation, and there are constraints on how the robot is able to move. Therefore, any algorithm implemented on the robot must take these factors into account. Additionally, because of how the proxy environment is scaled relative to the size limitations of the robot's working environment, it spends a larger portion of its time near the singularity at the origin (compare the spatial scales of the simulation experiments (e.g., Fig. 3) with that of the hardware experiments (Fig. 5)). Also, the robot must contend with real-world control, actuation, and dynamics (i.e., ground contact) that are not simulated in the software agent.

In the future, we hope to expand this work by running multiple trials for each noise case, expanding the set of parameters that is varied, and performing statistical analyses on the resulting success rates and sinuosities. We would also like to incorporate other sensor modes to explore how this kind of strategy could be used in a multimodal sensory context. In particular, multimodal sensing may be useful in allowing the agent to detect when it has reached the goal on its own by using magnetic cues in it's midcourse phase, and then using other sensory cues in its terminal/endgame phase (e.g., [43,44]). Particularly in the software simulation, an additional sensor mode or modes capable of declaring success or failure could take the place of an outside observer. In addition to these steps, we would like to implement this type of strategy on different robotic platforms (e.g., fixed wing aircraft vs. skid-steered ground vehicle vs. a ground vehicle that steers its wheels to turn), and standardize the strategy as much as possible across each platform. This will allow for direct comparison to the simulated results, and garner better insight into how well the concept works in real-world systems. We also wish to adapt the formulation here for comparison to modeling and experimental studies on animals such as salmon [26] and turtles [19,24].

Results from the present study hold promise because they mirror key features of previous animal studies [24,26]. In particular, these studies point to the feasibility of using sparse intermittent sensing, in some cases in specific locations, to move from a starting location to a goal. The present study demonstrates this in the 3-point $\Gamma = 20$ environment with an improved sinuosity at a lower measurement frequency.

Acknowledgments. We thank Dr. Kevin Brink (AFRL/RWWI) for the use of his robotics laboratory for the hardware portion of this study.

References

1. Shockley, J.A., Raquet, J.F.: Navigation of ground vehicles using magnetic field variations. Navigation **61**, 237–252 (2014)
2. Canciani, A., Raquet, J.F.: Absolute positioning using the earth's magnetic anomaly field. Navigation **63**, 111–126 (2016)
3. Johnsen, S., Lohmann, K.J.: The physics and neurobiology of magnetoreception. Nat. Rev. Neurosci. **6**, 703–712 (2005)
4. Johnsen, S., Lohmann, K.J.: Magnetoreception in animals. Phys. Today **61**, 29–35 (2008)
5. Wiltschko, R., Wiltschko, W.: Magnetic Orientation in Animals. Springer, Heidelberg (1995)
6. Knecht, D.J., Shuman, B.M.: The geomagnetic field. In: Jursa, A. (ed.) Handbook of Geophysics and the Space Environment. Air Force Geophysics Laboratory (1985)
7. Wajnberg, E., Acosta-Avalos, D., Alves, O.C., de Oliveria, J.F., Srygley, R.B., Esquivel, D.M.S.: Magnetoreception in eusocial insects: an update. J. R. Soc. Interface **7**, S207–S225 (2010)
8. Lohmann, K.J., Lohmann, C.M.F., Putman, N.F.: Magnetic maps in animals: nature's GPS. J. Exp. Biol. **210**, 3697–3705 (2007)
9. Taylor, B.K., Johnsen, S., Lohmann, K.J.: Detection of magnetic field properties using distributed sensing: a computational neuroscience approach. Bioinspiration & Biomimetics **12**(3), 036013 (2017)
10. Walker, M.: On a wing and a vector: a model for magnetic navigation by homing pigeons. J. Theor. Biol. **192**, 341–349 (1998)
11. Stoddard, P.K., Mardsen, E.J., Williams, T.C.: Computer simulation of autumnal bird migration over the western north Atlantic. Animal Behav. **31**, 173–180 (1983)
12. Lohmann, K.J., Hester, J.T., Lohmann, C.M.F.: Long-distance navigation in sea turtles. Ethol. Ecol. Evol. **11**, 1–23 (1999)
13. Goyret, J., Markwell, P.M., Raguso, R.A.: The effect of decoupling olfactory and visual stimuli on the foraging behavior of Manduca sexta. J. Exp. Biol. **210**, 1398–1405 (2007)
14. Willis, M.A., Avondet, J.L., Finnell, A.S.: Effects of altering flow and odor information on plume tracking behavior in walking cockroaches, peripleneta americana (L.). J. Exp. Biol. **211**, 2317–2326 (2008)
15. Reppert, S.M., Gegear, R.J., Merlin, C.: Navigational mechanisms of migrating monarch butterflies. Trends Neruosci. **33**, 399–406 (2008)
16. Baker, T.C., Kuenen, L.P.S.: Pheromone source location by flying moths: a supplementary non-anemotactic mechanism. Science **216**, 424–427 (1982)

17. Jensen, K.K.: Light-dependent orientation responses in animals can be explained by a model of compass cue integration. J. Theor. Biol. **262**, 129–141 (2010)
18. Lohmann, K.J., Putman, N.F., Lohmann, C.M.F.: The magnetic map of hatchling loggerhead sea turtles. Curr. Opin. Neurobiol. **22**, 336–342 (2012)
19. Luschi, P., et al.: The navigational feats of green sea turtles migrating from Ascension Island investigated by satellite telemetry. Proc. R. Soc. Lond. B Biol. Sci. **265**(1412), 2279–2284 (1998)
20. Luschi, P., et al.: Marine turtles use geomagnetic cues during open-sea homing. Curr. Biol. **17**(2), 126–133 (2007)
21. Brothers, J.R., Lohmann, K.J.: Evidence for geomagnetic imprinting and magnetic navigation in the natal homing of sea turtles. Curr. Biol. **25**, 392–396 (2015)
22. Mouritsen, H.: Spatiotemporal orientation strategies of long-distance migrants. In: Berthold, P., Gwinner, E., Sonnenschein, E. (eds.) Avian Migration, pp. 493–513. Springer, Berlin (2003)
23. Painter, K.J., Hillen, T.: Individual and continuum models for homing in flowing environments. J. Royal Soc. Interface **12** (2015)
24. Putman, N.F., Verley, P., Shay, T.J., Lohmann, K.J.: Simulating transoceanic migrations of young loggerhead sea turtles: merging magnetic navigation behavior with an ocean circulation model. J. Exp. Biol. **215**, 1863–1870 (2012)
25. Putman, N.F., Verley, P., Endres, C.S., Lohmann, K.J.: Magnetic navigation behavior and the oceanic ecology of young loggerhead sea turtles. J. Exp. Biol. **218**, 1044–1050 (2015)
26. Putman, N.F.: Inherited magnetic maps in salmon and the role of geomagnetic change. Integrat. Comparat. Biol. **55**, 396–405 (2015)
27. Servedio et al.: Not just a theory - the utility of mathematical models in evolutionary biology. PLoS Biol. **12** (2014). doi:10.1371/journal.pbio.1002017
28. Rutkowski, A.J.: A Biologically-inspired sensor fusion approach to tracking a wind-borne odor in three dimensions. PhD Dissertation, Case Western Reserve University (2008)
29. Rutter, B.L., Taylor, B.K., Bender, J.A., Blümel, M.A., Lewinger, W.A., Ritzmann, R.E., Quinn, R.D.: Descending commands to an insect leg controller network cause smooth behavioral transitions. In: International Conference on Intelligent Robots and Systems, pp. 215–220 (2011)
30. Rutter, B.L., Taylor, B.K., Bender, J.A., Blümel, M.A., Lewinger, W.A., Ritzmann, R.E., Quinn, R.D.: Sensory coupled action switching modules (SCASM) for modeling and synthesis of biologically inspired coordination. In: International Conference on Climbing and Walking Robots (2011)
31. Webb, B.: What does robotics offer animal behaviour? Anim. Behav. **60**, 545–558 (2000)
32. Talley, J.L.: A comparison of flying Manduca sexta and walking Periplaneta americana male tracking behavior to female sex pheromones in different flow environments. Ph.D Dissertation, Case Western Reserve University (2010)
33. Grasso, F.W., Consi, T.R., Mountain, D.C., Atema, J.: Biomimetic robot lobster performs chemoo-orientation in turbulence using an pair of spatially separated sensors: progress and challenges. Robot. Auton. Syst. **30**, 115–131 (2000)
34. Grasso, F.W., Atema, J.: Integration of flow and chemical sensing for guidance of autonomous marine robots in turbulent flows. Environ. Fluid Mech. **2**, 95–114 (2002)
35. Putman, N.F., Endres, C.S., Lohmann, C.M.F., Lohmann, K.J.: Longitude perception and bicoordinate magnetic maps in sea turtles. Curr. Biol. **21**, 463–466 (2011)

36. Postlethwaite, C.M., Walker, M.M.: A gemoetric model for initial orientation errors in pigeon navigation. J. Theor. Biol. **269**, 273–279 (2011)
37. Benhamou, S.: Bicoordinate navigation based on non-orthogonal gradient fields. J. Theor. Biol. **225**(2), 235–239 (2003)
38. Benhamou, S.: How to reliably estimate the tortuosity of an animal's path: straightness, sinuosity, or fractal dimension? J. Theor. Biol. **229**(2), 209–220 (2004)
39. Huang, G., Taylor, B.K., Brink, K.M., Miller, M.M.: Engineered and bioinspired magnetic navigation, Institute of Navigation Pacific Positioning, Navigation, and Timing Meeting (2017)
40. https://www.ndgc.noaa.gov/geomag/geomag.shtml
41. Anderson, J.D.: Fundamentals of Aerodynamics, 3rd edn. McGraw-Hill, New York (2001)
42. Dey, S.: Fluvial Hydrodynamics. Springer, Berlin (2014)
43. Endres, C.S., Putman, N.F., Ernst, D.A., Kurth, J.A., Lohmann, C.M.F., Lohmann, K.J.: Modeling dual use of geomagnetic and chemical cues in island-finding. Frontiers Behav. Neurosci. **10** (2001)
44. Bingman, V.P., Cheng, K.: Mechanisms of animal global navigation: comparative perspectives and enduring challenges. Ethol. Ecol. Evol. **17**, 295–318 (2001)

Jellyfish Inspired Soft Robot Prototype Which Uses Circumferential Contraction for Jet Propulsion

George Bridges, Moritz Raach[✉], and Martin F. Stoelen

School of Computing, Electronics and Mathematics, University of Plymouth, Plymouth, England
{George.Bridges,Moritz.Raach}@students.plymouth.ac.uk,
Martin.Stoelen@plymouth.ac.uk

Abstract. Several robotic jellyfish have been designed over the years, yet none have properly mimicked the very efficient method of propulsion that jellyfish use. Using circumferential contraction, water is pushed out the bottom of the bell creating upwards thrust. Jellyfish use this basic movement along with more complex features to move around the seas. In this paper, we attempt to mimic this circumferential contraction using hydraulically actuated silicone bellows that expand and contract a bell made of flexible silicone skin. 3D printed polylactic acid (PLA) was used to make the structure of the robot, and hinges and jubilee clips were used to fasten it together in order to maintain exchangeability of parts. The jellyfish expands and contracts using a pump with a simple on-off control which switches dependent on the internal pressure of the hydraulic system. This very simple control mechanism is similar to real jellyfish, and much like jellyfish, our design attempts to use both passive and active movements to maximize thrust.

Keywords: Jellyfish · Jet-propulsion · Bio-inspired robot · Bio-mimicry · Silicone · Soft robotics · Passive energy recapture

1 Introduction

Jellyfish, although not the most advanced or agile swimmers, are considered to be one of the most energy efficient swimmers on the planet [1]. This is largely due to their reliance on jet propulsion for locomotion, which is a simple and effective method of moving through water. Propulsion is achieved by contracting their umbrella-shaped bell, forcing a jet of water out of the opening and propelling the jellyfish in the opposite direction. Upon relaxing, the bell expands to its original size and therefore refills with water.

High efficiency and relative simplicity make jellyfish an ideal target for mimicry in robotics. Many designs have used traditional engineering methods to attempt to mimic movement of jellyfish using rigid materials and joints, such as the JetPro [2]. JetPro uses a mechanical iris mechanism to perform the contraction of the bell, which, although proved to be effective, requires a very complex array of moving parts. The FESTO AquaJelly [3] is another robot which on the surface appears to mimic the movement of a jellyfish. However, it instead propels itself using a system of tentacles with fins at the

© Springer International Publishing AG 2017
M. Mangan et al. (Eds.): Living Machines 2017, LNAI 10384, pp. 61–72, 2017.
DOI: 10.1007/978-3-319-63537-8_6

bottom as opposed to the contraction of a bell. This method of movement is actually more similar to fish than the jet propulsion used by most jellyfish.

We propose that in order to appropriately mimic a jellyfish whilst maintaining simplicity, a soft robotics approach would be more appropriate. Soft robotics can not only be used to reduce the number of moving parts but also to create a more adaptable and dynamic system [4]. Several approaches have been used previously in order to incorporate the use of soft robotic actuators to mimic jellyfish locomotion. One such robot used a dielectric elastomer as an actuator [5]. Whereas in a real jellyfish the bell contracts in order to reduce internal volume, the robot instead uses a rigid bell with an internal chamber made of the dielectric elastomer. When a voltage is applied to this elastomer it expands and as such its volume increases, this decreases the available volume within the bell and therefore ejects a jet of water through the opening in the bell. The robot provides an excellent example of how soft actuators can provide a massive reduction in complexity and moving parts whilst maintaining effectiveness. However, despite similar principles being used as in real jellyfish this design could not be considered a real mimic.

One soft robotics approach which more appropriately mimicked the actuation of a jellyfish used an array of springs to perform the bell contraction [6]. Each spring could be contracted individually using a servo motor and winch, meaning that the pattern of contraction could be altered to mimic that of a jellyfish. The result was a soft robot with almost no rigid structure which was proven to be capable of replicating jellyfish propulsion cycles, although its propulsion capability was not tested. The necessity for motors and winches resulted in a complex array of moving parts, and buckling of the springs also proved to be a problem.

Another key aspect of jellyfish propulsion that contributes to their efficiency is a method of passive energy recapture that enables them to recover energy which would otherwise be wasted. It would be expected that when the jellyfish relaxes its muscles, thrust would be generated in opposition to the desired movement due to the refilling of the bell with water, however, this is not the case. Instead, the passive movement of the bell margin (the bottom of the bell which is not directly actuated by the muscles) creates vortex rings beneath the bell which provide additional thrust in the direction of movement. This thrust is simply generated by relaxing the muscles used for bell contractions, as such it does not use any extra energy. This free 'boost' has been shown to allow jellyfish to move up to 30% further for each contraction cycle [1].

Again, soft robotics would seem like an ideal approach for the replication of this process. Soft materials could be used to create elements of a design which move passively like the bell margin [4]. Similarly, elastic materials could also be perfect for the storage and release of energy in efficient ways, such as the contraction and relaxation cycle of the jellyfish.

Other underwater actuation methods include the use of real biological tissue [7] and using transverse and longitudinal actuators along an artificial octopus arm [8]. However, the aim was to develop a robot which could mimic jellyfish, whilst maintaining simplicity, and also being capable of efficiently and effectively travelling through water by replicating the jet propulsion and passive energy recapture methods.

2   Design

The overall design of the robot jellyfish was to reflect that of a real jellyfish, the majority of its body would take the form of a flexible umbrella-shaped bell. The electronics to drive and control the robot would be housed in an air-tight dome positioned on top of the bell. An engineering drawing of the design can be seen in Fig. 1.

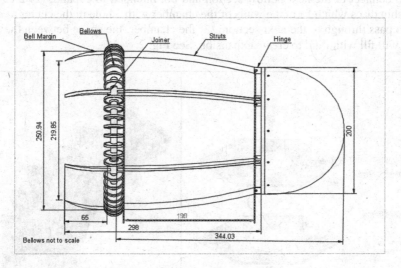

Fig. 1. Engineering drawing of overall concept

Actuation is achieved through the use of hydraulically powered bellows which sit around the opening at the bottom of the bell. When pressurized the bellows will expand, causing the opening of the bell to enlarge and therefore the volume of the bell to increase. When the bellows are depressurized they contract to their original length, therefore reducing the volume of the bell and as such forcing the contents out through the opening.

The overall shape of the bell is maintained by 6 3D printed PLA struts which are connected to the dome via a simple hinge which allows each strut to rotate outwards. The struts increase the diameter from the top to the bottom of the jellyfish by 50 mm. This increases the volume under the bell significantly while also mimicking the shape of a real jellyfish. The skin was securely fastened to the struts using 3D printed PLA clamps.

Pre-made hydraulic push fit T-junctions were fastened to the struts via a 3D printed hinge allowing two sections of bellows to be connected to the struts and the hydraulic system. Jubilee clips were used to tighten the bellows around the joiners, whilst maintaining interchangeability. 6 mm PVC piping was used to connect each of the joiners together into one hydraulic system along with the pump. As this design is intended as a proof of concept it is assumed that some of the hydraulic system will be external to the robot and as such is not included in the design.

2.1 Bellow Design and Manufacture

In order for the bellows to create the desired effect, they should consist of air pockets sandwiched between two flexible elastic sides. As the bellows are filled, the chambers expand causing an extension along the length of the bellow. Each bellow therefore consisted of 3 main parts; (1) the spacer, a centimeter-long tube like section which was used to connect to the next bellow section and not intended to expand; (2) 2 circular sides, thin discs which form the walls of the chamber with a hole in the center to allow fluid to pass through to the next section; (3) the chamber, the space between the sides which will fill with fluid to create expansion. See Fig. 2.

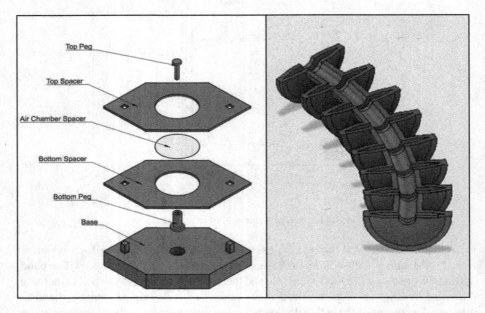

Fig. 2. CAD model showing breakdown of parts of the bellow mold (left) and a cross-sectional view of the resulting bellows (right).

The primary obstacle with using bellows for actuation is finding a material which is in keeping with the soft robot ethos of the project whilst also being strong enough to be capable of holding pressure and maintaining the shape and structure of the robot. The material is also required to be elastic enough such that expansion is maximized for a given pressure.

The chosen material for the creation of the bellows was a substance which has been nicknamed Oogoo, it is a mixture of 3 parts 100% silicone caulk to 1 part corn starch. The silicone, therefore, makes up the bulk of the substance and creates a very strong elastomer which has a very high tensile strength and very low Young's modulus. The addition of corn starch has several benefits; it results in a more rigid end-material; it makes the silicone easier to manipulate and mold; it substantially decreases the curing time of silicone thus speeding up the process of manufacture [9].

A mold was designed such that thin layers of Oogoo could be spread on top of one another in order to build up each bellow section. Each mold had 5 3D printed parts which were designed such that they could be easily disassembled to remove the cured bellow from the mold; the design of the mold can be seen in Fig. 2. The chamber was created by sandwiching a small disc of polythene between the two circular sides. Polythene was used because Oogoo does not adhere to it and therefore a pocket remains between the sides once the substance has cured. It is also soft enough that it would not obstruct the flow of fluid through the bellows.

Dimensions of the bellows were chosen based on the practicalities of working with the Oogoo substance, as such each side of each bellow section was designed to have a diameter of 30 mm and a thickness of 1.5 mm. This ensured that minimal pressure would be required inside the chamber to force the sides to expand. A cross sectional view of the resulting bellow design can also be seen in Fig. 2.

2.2 Bellow Testing

In order to fully understand the effectiveness and appropriateness of the silicone actuators it was necessary to understand a number of properties of their behavior:

- Extension – How length changes relative to the input pressure, and the maximum extension before failure. Required to determine the change in volume of the bell.
- Force – How much force the bellows exert relative to pressure. Required to determine the necessary characteristics of the material to make the skin.
- Pressure – What is the maximum pressure the average bellow can hold before failure.

The testing of both force and pressure were completed simultaneously. 7 bellow sections were joined together to form a bellow which would approximate those between the struts. One end of the bellow was sealed closed so that it could be pressurized, the other end was connected to a hand pump such that it could be inflated with air. The sealed end of the bellow was placed onto a force gauge, and the bellow was placed inside of a rigid pipe so that it could not buckle and all force was directed to the gauge. The bellow and pipe were clamped such that they could not move and a pressure gauge was connected to the system to measure the internal pressure of the bellow. The pressure inside the bellow was increased until the bellow failed, this was repeated three times. The results of the testing can be seen in Fig. 3.

A similar setup was used for the testing of extension relative to pressure. The connected end of the bellow was fastened to a rule such that its length could be measured. The spacers between each bellow section were also loosely tied to the rule such that they could slide, this ensured that all extension was directed along the length of the rule. The pressure inside the bellow was slowly increased and regular readings of the length were taken from the rule. The test ended when the bellow began to show signs of failure. This was done on three bellows and the results can be seen in Fig. 4.

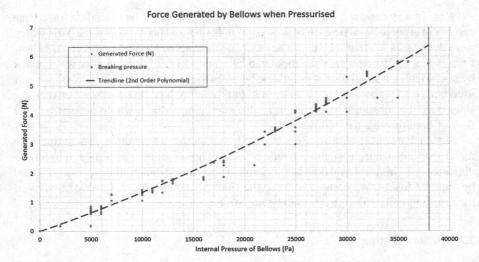

Fig. 3. Graph showing the force generated by the bellows as they are pressurized and the pressure at which they fail

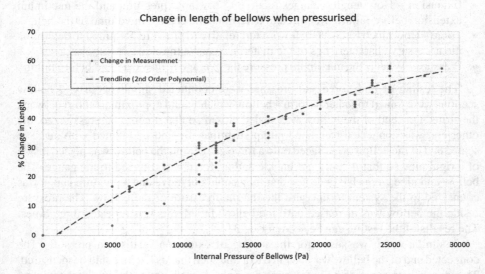

Fig. 4. Graph showing the percentage change in length of the bellows relative to their internal pressure

2.3 Skin Design and Manufacture

Due to the low force exerted by the bellows, it was necessary for the skin to be formed of a material with a low Young's Modulus in order to maximize the expansion. The material was also required to be nonporous so that water must be forced out of the bell.

Thin sheets of silicone were therefore the chosen material for the skin. Sheets were produced by depositing 100% silicone caulk between two parallel 0.5 mm high rails. The rails sat on a flat surface covered in a layer of polypropylene backed tape, this prevented the silicone from sticking to the surface. The silicone was then screeded between the rails and left to cure. The sheets produced ranged between 0.38 mm and 0.89 mm thick, with an average thickness of 0.60 mm.

2.4 Bell Margin Design and Manufacture

In order to mimic the passive energy recapture process used by some jellyfish it was necessary to design a bell margin which behaved similarly to that of a real jellyfish. The required behavior is for the bell margin to dramatically flick inwards at the point at which contraction stops and expansion begins; this motion forms the vortex ring which then follows the bell margin as it expands.

The proposed design to replicate this passive motion was for there to be a flexible 'spine' attached to the bottom of each strut below the actuator ring. This spine would support the skin which falls beneath the bellow in order to maintain its shape. These spines and excess skin would form the bell margin. The flexibility of the spines is key for the movement of the margins; they need to be flexible enough that the passive motion can occur, but also rigid enough to maintain the structure of the margin and provide enough force to create the desired vortex rings in the water. The material and structure of the spines is therefore very important.

It was decided that the spines would take the form of an inwardly curved 50 mm long strut which is tapered along its length such that its movement through water would cause it to deform more towards its bottom than at its top. The purpose of the curve is to mimic the shape of the jellyfish. It was decided that the spine would be 3D printed in NinjaFlex (NinjaTek, Manheim, PA, USA), a flexible 3D printing filament. This would allow its flexibility to be altered by adjusting the percentage infill of the component.

2.5 Bell Margin Testing

10 spines were printed with infills which varied from 0% to 100% and their properties were examined. It was found that spines with very low infills (0%–15%) would buckle far too easily, meaning that they could not maintain the structure of the margin. Spines with high infills (40% upwards) were very rigid and a substantial amount of force was required for them to flex, meaning that they were unlikely to provide the desired passive motion. It was therefore chosen to print the spines with a 20% infill. During informal testing this appeared to approximate the correct motion, as seen in Fig. 5.

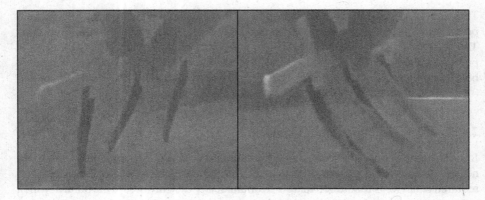

Fig. 5. Frames from the recorded slow-motion footage of the bell margin 20% spine. The central spine is the spine of interest. Left – Frame showing the angle of the spine shortly before the change from contraction to expansion. Right – Frame showing the angle of the spine shortly after the change from contraction to expansion.

3   Testing

At the time of testing, a complete prototype with on-board electronics and control systems was not implemented. However, an external compressor was attached to the system in order to pressurize and depressurize the bellows using air. The assembled system at the time of testing can be seen in Fig. 6. The aim of the testing was to verify that the mechanical design worked, and gain information about performance of the system. Expansion rates were not tested due to dependency on the pump and lack of detailed information on bell margin motion. Goals of the testing were as follows:

- Verify that the bellow ring is a practical and effective actuator for circumferential contraction and expansion
- Verify that the bell is capable of expanding and contracting when submerged
- Gain information about contraction rates

All of the testing was completed with the jellyfish sub-merged in a 145 l clear plastic tank filled with water. An image of the robot inside the testing environment can be seen in Fig. 6. Pressurization was achieved with a small 12 V DC air compressor capable of inflating and deflating. The compressor was powered by a 7.2 V Ni-Cd battery, and attached to a switch such that its polarity could be reversed. The compressor was linked to the jellyfish using 4 mm pipe. An electronic pressure gauge was also linked to the system such that the internal pressure of the bellows could be monitored.

Several cycles of expansion and contraction were completed. In order to prevent damage to the bellows, the expansion phase was allowed to continue until an internal pressure of between 25000 and 27000 Pascals was reached. At this point, the polarity of the compressor was switched and the contraction phase began. The contraction phase was allowed to continue until the bellows stopped contracting further. Another cycle was completed, in which, once the expansion phase completed, the compressor was

Fig. 6. Images of the assembled robot inside the water tank for testing.

disconnected from the system. This meant that the contraction phase here was completely passive and generated only from the elasticity of the bellows and skins.

4 Results

Upon pressurization, the circumference of the bell opening increased evenly, resulting in an increase in diameter of 130 mm (52%) at a pressure of 27000 Pascals. Upon contraction, the bellows successfully returned to their original size. Through both phases the bellows maintained their circular shape which is an important factor when considering the direction of the thrust created from jet propulsion.

Bellow failure was very uncommon unless the system was over-pressurized. Similarly, the bellows were capable of withstanding a considerable amount of handling, movement and contact without being damaged.

The system as a whole behaved as expected, the force and extension generated through the pressurization of the bellows was capable of stretching the skin. This meant that the volume of the bell would increase from approximately 0.0065 m3 to approximately 0.0138 m3 (113% change) during the expansion phase. The contraction phase was successful for both passive and active contraction, both therefore resulting in the displacement of 0.0074 m3 of water.

The contraction rates differed between passive and active. The average active contraction time was 2.2 s compared to 2.7 s for a passive contraction. Contraction did not occur linearly in either case however accurate measurements were only recorded for the passive contractions. Analysis of videos of the passive contractions revealed that the rate of contraction slowed as it approached the end of the contraction period.

4.1 Extrapolation of Results

From the results, it is possible to extrapolate how much thrust will be generated by a passive contraction of the bell. Equation 1 was used to estimate the thrust generated throughout the contraction phase [10].

$$T = \left(\frac{\rho_{water}}{S_{bell}} \right) \left(\frac{\Delta v}{\Delta t} \right)^2 \tag{1}$$

Where ρ_{water} is the density of water, S_{bell} is the area of the bell opening, Δv is the change in volume and Δt is the change in time.

The measured dimensions for the passive contraction phase were inserted into this equation using MATLAB in order to simulate the change in thrust over time. The results can be seen in Fig. 7.

5 Discussion

The results verify the overall design of the robot and indicate that it could be a promising platform for further testing and development. The Oogoo bellows prove themselves to be an effective form of actuation for this purpose, as well as being reliable and easily replaceable in the case of failure. The results show promise for effective jet propulsion, with contractions being able to regularly eject large masses of water. This is

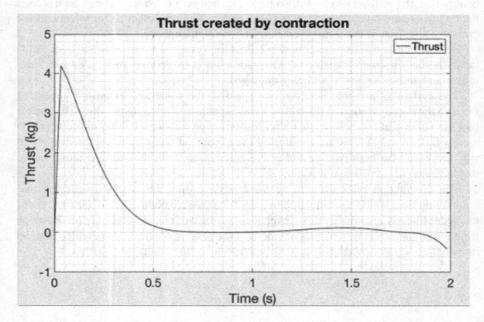

Fig. 7. Graph showing extrapolated thrust variation over time throughout a single contraction of the bell.

demonstrated in Fig. 7, which shows a very similar thrust profile to that which would be expected from the measured speeds of real jellyfish [1, 10].

Interestingly the elasticity of the skins and bellow provide enough energy to fully actuate a contraction. This means that the design could be controlled similarly to that of a real jellyfish, in that it only needs to use energy for one half of the propulsion cycle. This means that the thrust generated by contraction is actuation 'free', improving the efficiency of the design. Results also proved, however, that active contraction would generate faster contraction rates and therefore more thrust if necessary, providing the future possibility of dynamic control based on thrust requirements.

Further testing on the jellyfish will include a full functionality test of the passive movement created by the bell margin. In order to check if the vortex is being created in a way to produce thrust, the water flow and pressure will have to be studied closely around the bell margin for each expansion and contraction.

The whole system will need to be tested in a large body of water. This testing would include buoyancy tests, motion tests, and speed tests. All of which would be required to fully verify the effectiveness the design.

Future development of this prototype should include alterations to achieve a self-contained robot, namely, an airtight housing for the electronic systems, this would considerably increase the complexity of the design and material choice. If self-containment was achieved and energy demands proven to be low enough, then future developments could strive to achieve a self-sustaining system, perhaps using a microbial fuel-cell, as used in Row-Bot [11].

6 Conclusion

The robot successfully demonstrates how a soft robotics approach to biomimicry can be very effective. The soft design of the jellyfish resulted in a robust robot prototype which, on the surface, successfully mimicked the movement of a jellyfish. The results and thrust calculations show promise for the robot being able to move through water using jet propulsion alone, as well as showing a thrust profile similar to that of real jellyfish, indicating effective mimicry. On top of this, the robot may achieve a method of reducing energy demands using elastic energy stored in the actuators. The design also attempts to mimic the vortex ring generation of real jellyfish, if testing reveals this to be effective, the robot would be capable of generating thrust in both the expansion and contraction phase of its movement. This would dramatically increase the efficiency of the design.

The design goal of maintaining simplicity is also achieved in terms of required electronics and sensing. The only necessary sensor for locomotion is a pressure sensor which can be used to indirectly measure bell diameter, similarly all actuation for forward motion could be controlled through the use of a single pump. The design therefore shows a great deal of potential in providing a simple, efficient and effective method of underwater locomotion based on the natural design of jellyfish.

Acknowledgments. This work was completed as a student project for the ROCO504 Advanced Robot Design module at the University of Plymouth, UK, in autumn 2016.

References

1. Gemmell, B.J., Costello, J.H.: Passive energy recapture in jellyfish contributes to propulsive advantage over other metazoans. PNAS (2013). http://www.pnas.org/content/110/44/17904.full. Accessed 5 Dec 2016
2. Marut, K., Stewart, C., Michael, T., Villanueva, A., Priya, S.: A jellyfish-inspired jet propulsion robot actuated by an iris mechanism. Smart Mat. Struct. **22**(9) (2013)
3. Festo AG and Co. KG: AquaJelly (2015). www.festo.com/cms/en_corp/9772_10378.htm. Accessed 12 Dec 2016
4. Rus, D., Tolley, M.T.: Design, fabrication and control of soft robots. Nature **521**, 467–475 (2015)
5. Godaba, H., Li, J., Wang, Y., Zhu, J.: A soft jellyfish robot driven by a Dielectric Elastomer Actuator. IEEE Robot. Autom. Lett. **1**(2), 624–634 (2016)
6. Nir, S.: A jellyfish-like robot for mimicking jet propulsion. In: IEEE 27th Convention of Electrical and Electronics Engineers in Israel (2012)
7. Nawroth, J., Lee, H., Feinberg, A., Ripplinger, C., McCain, M., Grosberg, A., Dabiri, J., Parker, K.: A tissue-engineered jellyfish with biomimetic propulsion. Nat. Biotechnol. **30**(8), 792–797 (2012)
8. Mazzolai, B., Margheri, L., Cianchetti, M., Dario, P., Laschi, C.: Soft-robotic arm inspired by the octopus: II. From artificial requirements to innovative technological solutions. Bioinspiration & Biomimetics **7**(2) (2012)
9. Instructables. How To Make Your Own Sugru Substitute (2010). http://www.instructables.com/id/How-To-Make-Your-Own-Sugru-Substitute/. Accessed 10 Nov 2016
10. McHenry, M.J., Jed, J.: The ontogenetic scaling of hydrodynamics and swimming performance in jellyfish (Aurelia aurita). J. Exp. Biol. **206**, 4125–4137 (2003)
11. Philamore, H., Stinchcombe, A., Ieropoulos, I.: Row-bot: an energetically autonomous artificial water boatman. In: IEEE Intelligent Robots and Systems (IROS), pp. 3888–3893 (2015)

You Made Him Be Alive: Children's Perceptions of Animacy in a Humanoid Robot

David Cameron[1]([✉]), Samuel Fernando[2], Emily C. Collins[1], Abigail Millings[1], Michael Szollosy[1], Roger Moore[2], Amanda Sharkey[2], and Tony Prescott[1]

[1] Department of Psychology, University of Sheffield, Sheffield, UK
{d.s.cameron,e.c.collins,a.millings,m.szollosy,
t.j.prescott}@sheffield.ac.uk
[2] Department of Computer Science, University of Sheffield, Sheffield, UK
{s.fernando,r.k.moore,a.sharkey}@sheffield.ac.uk

Abstract. Social robots are becoming more sophisticated; in many cases they offer complex, autonomous interactions, responsive behaviors, and biomimetic appearances. These features may have significant impact on how people perceive and engage with robots; young children may be particularly influenced due to their developing ideas of agency. Young children are considered to hold naive beliefs of animacy and a tendency to mis-categorise moving objects as being alive but, with development, children can demonstrate a biological understanding of animacy. We experimentally explore the impact of children's age and a humanoid's movement on children's perceptions of its animacy.

Our humanoid's behavior varied in apparent autonomy, from motionless, to manually operated, to covertly operated. Across conditions, younger children rated the robot as being significantly more person-like than older children did. We further found an interaction effect: younger children classified the robot as significantly more machine-like if they observed direct operation in contrast observing the motionless or apparently autonomous robot. Our findings replicate field results, supporting the modal model of the developmental trajectory for children's understanding of animacy. We outline a program of research to both deepen the theoretical understanding of children's animacy beliefs and develop robotic characters appropriate across key stages of child development.

Keywords: Human-robot interaction · Humanoid · Animacy · Psychology

1 Introduction

With the increased use of social robots as companions and tutors for children, it becomes essential for effective progress in human-robot interaction (HRI) to understand how children perceive and evaluate robots. Advances in robots' responsive interaction-capabilities, autonomy, and biomimetic morphology may blur boundaries between object and agent in a child's perspective [1].

© Springer International Publishing AG 2017
M. Mangan et al. (Eds.): Living Machines 2017, LNAI 10384, pp. 73–85, 2017.
DOI: 10.1007/978-3-319-63537-8_7

Research already explores high-level social issues concerning children's perceptions of robots, such as its 'role' as co-learner or teacher [2] and its adherence to social norms (e.g., as an (in)attentive listener [3]). However, much of the fundamentals concerning children's perceptions of social robots remain to be explored. In this paper, we explore factors that can shape children's perceptions of social robots as being machines, agents, or agent-machine hybrids.

A theoretical understanding of the perceptions children have of social robots can not only inspire better HRI design, but also address questions in developmental psychology. Opfer and Gelman highlight children's understanding of the object-agent distinction as a research area that biomimetic robotics could benefit, '*As technology improves and robots become increasingly animal-like in appearance and capacities, it will be intriguing to examine if and/or how children interpret such entities.*' [4]. If advanced social robots genuinely have the potential to span boundaries between object and agent [1], these can be used as research tools to explore development in children's understanding of animacy.

Children show a developmental pattern in their understanding of animacy. In early-childhood, children show naive theories of animacy [4,5]. They may attribute life to simple objects [6] and describe an object's behaviour (e.g., a ball rolling down a hill) in terms of it's *intentions* (e.g., the ball wanted to go home). The naive theory of animacy is considered to be supplanted by a theory of biology in late-childhood [4,5] and so animacy is less likely to be mis-attributed. However, even among adults, users may describe robots as if they are animate and make judgments based on their beliefs of a robot's animacy [7]. Robots may be an ideal tool to explore children's development in their understanding of animacy from early- through to late-childhood, as, unlike simple objects, they can cross boundaries that challenge even adults' theories of animacy.

1.1 Developmental Understanding of Animacy

Children's understanding of animacy as a reasoned process is presented by Piaget as an extension of their understanding of causality [6]. It is observed that young children often make category errors in regarding beings as animate or inanimate, typically favoring animate explanations for an object's behavior [4].

Piaget offers explanation for these errors in terms of children's accessible reasoning. The understanding of cause-and-affect for a child is intertwined with an understanding of their own intentions [6] (i.e. my actions occur because I first intend them to; thus objects perform actions because they intend to). However, this perspective has been criticized on account of children as young as four being able to articulate mechanical cause-and-effect with no reference to intention [8].

In contrast to Piaget's model, a substantial body of research tracks recognition of animacy in terms of infants' social development [9]. Despite an infant's limited movement and communication, his or her world is nevertheless a social one and, so, animate beings are uniquely important. Carey's model [9] for children's understanding of animacy suggests that children hold two innate processes for determining animate beings: face-detection and recognition of reactive, or autonomous, goal-directed movement. These processes are argued to lay the foundation for early-childhood category mistakes made with animacy [5].

Newborns have been shown to attend to schematic face-like stimuli but not the same schematic stimuli in a scrambled arrangement [10]. This is argued to arise from newborns needing carers for survival and that many significant changes in newborns' environment arise from a carer's actions. These changes coincide with the carers' face in the newborns' vision: the presence of a face *is* the presence of agency [9]. In infancy, further signs of recognizing agency from faces emerge, with infants following another person's gaze [11,12]. Remarkably, this behaviour is apparent even with infants observing a robot's 'gaze' [13].

Like Piaget, Carey argues that young children's understanding of animacy is influenced by movement [9]. However, Carey narrows her model to argue that animacy is inferred from *autonomous* and *goal-directed* movement to manipulate, or respond to, the environment [9]. A wealth of research, using abstracted agents as stimuli, supports this position: infants habituate towards animations of one shape 'chasing' another until the roles of the shapes change [14,15]; children prefer animated shapes that 'help' other shapes in apparent movement goals over shapes that hinder the same [16,17]. Adults' descriptions of similar animations are rich with intentions, emotions, and even invented relationships between shapes [18], yet adults do not make the same animacy mistakes as children.

With development, children can articulate their naive theories of animacy and demonstrate a transitioning through stages of what constitutes an animate being [4]. Piaget presented a series of stages that children progress through: from random judgments, via progressively more refined understandings autonomous movement, to adult concepts of life [6]. Carey posits that autonomy is a factor in children's decision making but not the *sole* factor [5]. Rather, young children attempt to apply their limited biological knowledge [5], leading to the categorical errors not seen from older children. From seven onward, children appear to rapidly develop a deeper biological understanding and application of this knowledge to address the question: What is alive? [19]. Bio-inspired robotics can be used to explore contrasting models of children's development in understanding animacy. By shaping robots' appearances and behaviors in a program of experimental studies we may determine particular and combined influences on childhood perceptions of animacy in robots.

1.2   Animacy in HRI

Current HRI research offers mixed indications concerning children's perceptions of robots as animate beings [20,21]. Studies point towards aspects of Carey's model of conceptual change [5] but offer contrasting outcomes. For example, in one study, children aged three to five present inconsistent beliefs regarding a robot dog's agency and biology [22], while in another children of this age evaluate a robotic dog as comparable to a stuffed dog toy in terms of animacy, biology, and mental states [23]. Studies further indicate children articulate beliefs that biomimetic robots can be between alive and inanimate [24–26].

It is this boundary-spanning nature of biomimetic robots that presents HRI research as a unique opportunity to study children's understanding of animacy, through use of embodied characters and agents. Key factors in Carey's model, such as autonomy in movement, can be isolated and tested in physical agents. We identify three common, key factors that could influence children's perceptions of animacy in robots: morphology, responsiveness, and autonomous movement.

Morphology. A robot's appearance can substantially influence users' immediate impressions of its capabilities and interaction potential (e.g., [7]). Users may expect more sophisticated interaction potential from a life-like humanoid than that of a robotic table. Indeed, biomimicry in particular, both in terms of the physical structure and movement, can promote engagement [27], and suggest animacy [22].

Young children show low reliability when classifying a humanoid as a machine or living thing; however, they did not show reliability difficulties to the same extent with a picture of a girl, nor of a camera [28]. This suggests that children struggle with discrete classification when it comes to humanoids. More recent work identifies that children classify a humanoid robot as a person-machine hybrid [25] following a short, interactive game with the robot. Their classifications are somewhat stable across repeated interactions with the robot: neither familiarity with the interaction scenario nor differences in the robot's behavior substantially influenced participants' ratings.

Responsiveness. A robot's responsiveness is considered to impact on users' perceptions of animacy. Johnson and colleagues demonstrated that infants will direct their visual attention to follow apparent attention-cues of a robot that is responsive to their babbling or movement but not one that exhibits the same behavioral signals independent of the infants' own behaviors [13]. The authors conclude that infants are attributing some agency to the robot: a capacity for attending, and responding, to the environment.

Older children are also seen to shape their social behavior and visual attention, during interaction with a social robot, based on the robot's responsiveness [3]. In a question-and-answer scenario, children tended to engage in human-human-like social behaviors with a robot that showed socially-appropriate attention cues but not with one that showed attention cues outside social norms. It appears that responsiveness *meaningful to the user* could contribute to individuals perceptions of animacy.

Autonomous Movement. A robot's autonomy in movement can shape user perceptions of animacy, particularly if the movement can be interpreted as goal-directed. Somanader and colleagues report that children aged four or five attribute fewer living properties to a robot with a mechanical appearance, if its complex, goal-directed behavior is revealed to be controlled by a human [21]. These findings are mirrored in adult HRI, where goal-directed movement supports perceptions of animacy, unless the user is *directing* the same robot towards a goal [29].

While studies manipulating the apparent origin of a robot's movement suggest autonomy has an important role in perceptions of animacy, further work offers a less clear picture. Children playing with a robot- and a stuffed-toy-dog make no meaningful distinctions between the animacy of the two objects, despite the robot presenting autonomous movement [22]. Inconsistencies, such as these, across the literature warrant deeper exploration; yet a systematic approach to exploring the factors behind children's understanding of animacy through robotics still remains to be undertaken.

1.3   Testing Children's Perceptions of Animacy

The Expressive Agents for Symbiotic Education and Learning (EASEL) project [30, 31] explores children's interactions with, and perceptions of, a humaniod synthetic tutoring assistant, using the Robokind Zeno R25 platform ([33], Fig. 1). The current study draws from EASEL research exploring children's representations - through interview, questionnaire, and open-ended HRI - of the humaniod as being something between animate and inanimate [25, 26] to provide children the opportunity to clearly articulate the potentially boundary-spanning nature that a humanoid robot can appear to have [1].

Fig. 1. The Robokind Zeno R25 platform (humanoid figure approximately 60 cm tall)

The results from Cameron et al.s' study indicate a difference in animacy perceptions between children aged six-and-under and those aged over-six [25]. As anticipated by Carey's model of conceptual change [5], younger children regarded the robot as more animate than older children. Results further indicate that children, irrespective of their age, tended to report Zeno as being between person and machine; their ratings of Zeno were largely stable across repeated interaction.

We propose that the stability of ratings across conditions (the presence or absence of life-like facial expressions) may be due to a ceiling effect, arising from the interaction scenario used [25]. The children interacted with a responsive and autonomous humanoid: such a scenario would present all animacy cues identified by Carey [9]. Further cues, such as facial expression, may not suggest 'more' animacy. To address this potential ceiling effect on children's perceptions, we create a minimal model for an HRI scenario, in which only autonomy is manipulated. We anticipate: (1) while younger children will report the humanoid to be more animate than older children will, (2) only younger children will be influenced by the autonomy of the robot. Younger children will report an autonomous robot as being more animate than one they observe being operated.

2   Method

2.1   Design

We employed a between subjects design. Participants were randomly allocated to one of three conditions for the interaction: (1) the robot remained motionless (*Still*), (2) the robot was directly activated by the experimenter (*Operated*), and (3) the robot was covertly activated by the experimenter (*Autonomous*). In the Operated condition, the experimenter pressed a button on the robot's chest screen to initiate the robot's synthetic facial-expressions; in the Autonomous condition the same facial-expressions were covertly activated remotely via laptop.

This design isolates autonomous movement as the variable of interest. While children can observe that the robot is responsive to the experimenter's actions in the Operated condition, earlier work indicates that only robot responsiveness *meaningful to the observer* impacts on beliefs of animacy [3,13].

2.2   Measure

To assess perceptions animacy we used a one-item likert-scale measure, used in prior HRI research on animacy [25,32], contrasting the humanoid as a *person* and as a *machine*. This measure uses age-appropriate terms, drawn from qualitative HRI research on children's self-generated questions to Zeno [26][1].

2.3   Procedure

The experiment took place in publicly-accessible spaces. The animacy measure was presented as the first of a series of questions about the robot's appearance (further questions were used for an unrelated study on robotic expressions). For Operated and Autonomous conditions, participants completed the measure after observing the robot demonstrate two types of facial expressions (Happiness and Sadness). For the Still condition, participants completed the measures before they observed any movement from the robot. Interactions lasted approximately five minutes. Brief information about the experiment was provided to parents/carers; informed consent for participation was obtained from parents/carers prior to participation; and parents/carers were available to offer reassurance, if needed. Ethical approval for this study was obtained prior to any data collection.

---

[1] Children's questions fall into two distinct umbrella categories: the robot's experiences as a *person* and its capabilities as a *machine*.

2.4 Participants

The study took place across two university public-engagement events. Children aged four and over visiting the events were invited to participate in playing a game with Zeno, titled "Guess the robot's expressions". In total, 72 children took part in the study (32 female and 40 male; M age = 6.71, SD = 1.91; 35 aged six-and-under and 37 aged over-six, evenly divided across conditions). Power analysis indicates that the sample size of 72 participants across the three conditions would be sufficient to detect medium-small size effects (eta-squared = .05) with 95% power using an ANOVA between means with alpha at .05.

3 Results

A preliminary check was run to ensure even distribution of participants across age groups to conditions. There was no significant difference between conditions for participants' ages $F(1, 69) = 1.64$, $p = .20$. This outcome was not affected when ages were categorised across the theory-driven bounds of: six-and-under and over-six (35 six-and-under, 37 over-six; χ^2 (2, N = 72) = 1.30, $p = .52$).

A regression of age against animacy rating indicated that older children view the robot as being significantly less like a person ($\beta = -.30$, $t = 2.68$, $p < .01$). When ages are categorized as being six-and-under and over-six, we observed significant main effects for age ($F(1, 66) = 4.48$, $p < .05$) on children's reports of animacy. Older children viewed the robot as being significantly more like a machine (M = −.42, SD = 1.16) than younger children did (M = .14, SD = 1.06)[2]. This is a medium sized effect (Cohen's $d = .50$).

There was no significant main effect for condition ($F(2, 66) = 2.25$, $p = .11$) on children's reports of animacy. There were no differences observed between the Still (M = .10, SD = 1.07), Operated (M = −.56, SD = 1.07), or Autonomous conditions (M = .03, SD = 1.07).

There was a significant interaction effect between age and condition on children's reports ($F(2, 66) = 3.48$, $p = .04$; see Fig. 2). This interaction is a medium-large sized effect (partial eta^2 = .10). Older children's (aged over-six) responses were no different across conditions: Zeno appeared to be more machine-like across the Still (M = −.67, SD = 1.04) Operated (M = −.43, SD = 1.06) and Autonomous (M = −.17, SD = 1.06) conditions. In contrast, younger children (aged six-and-under) showed a marked difference across conditions: those who saw the Still and Autonomous robot reported it as being more person-like (M = .88, SD = 1.05; M = .24, SD = 1.07 respectively), whereas those who saw the Operated robot reported it as being more machine-like (−.70, SD = 1.04).

Simple effects tests indicate that there is a significant difference for younger children between the Still and Operated conditions ($p < .01$) and a further difference between the Autonomous and Operated conditions ($p = .03$); there was no significant difference between the Still and Autonomous conditions ($p = .16$). There were no differences across conditions for older children.

---

[2] Positive scores indicate more person-like; negative scores indicate more machine-like, 0 indicates an even mix of person and machine.

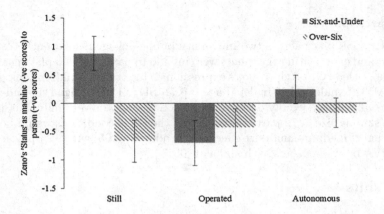

Fig. 2. Mean ratings of Zeno as machine-like (−ve) or person-like (+ve); SE bars shown.

4   Discussion

Our findings further indicate a sizable influence of age and autonomy on whether children regard humanoid robots as being person- or machine-like. *Hypothesis 1* is supported through the difference in responses between older and younger children. Younger children classify the humanoid robot, with life-like facial features, as being significantly more person-like than older children do. Younger children's responses are influenced by the robot's autonomy, supporting *hypothesis 2*. Direct robot-operation by an individual impacts on young children's reports of animacy; the robot is classified as significantly more machine-like than an Autonomous or even Still robot. In contrast, older children are not influenced by the autonomy, reporting the robot as slightly more machine-like across conditions.

The primary finding lends support to Carey's model of children applying their developing ideas of biology to determine animacy [5,19] and challenges Piaget's model [6]. Younger and older children show a distinct difference in their responses, even in the Still condition; Piaget's model would suggest that younger children would rate this as less-animate (more machine-like on our scale), however this is not the case. Children may be drawing from other cues, such as the robot's biomimetic features (Fig. 1, [33]), to evaluate the robot.

Nevertheless, autonomy of movement is seen to have an impact on judgments of animacy for younger, but not older, children. Again, Carey's model would suggest this difference as younger children attempt to draw together naive cues of animacy in lieu of a well-rounded working model of biology [5]. The isolated finding of autonomous movement's impact also supports Piaget's model [6]. Autonomous movement may suggest animacy to more of the younger children than externally driven movement does (the former being a more advanced stage than the latter), whereas older children, in final stage of understanding animacy are uninfluenced. However, in the context of the complete study, support is limited.

In comparison to the earlier study, this replication produced a reduced main effect size. Younger children report the robot as being closer to a machine-person hybrid overall than children of the same age do in earlier work (Fig. 3). This is as anticipated for two reasons: (1) the current study replicates with autonomy only and does not include relevant responsiveness from the robot, which is indicated to be an animacy cue [13], (2) the current study includes a condition (Operated) explicitly designed to reduce young children's perceptions of animacy. It is also worth noting that older participant's responses are extremely similar across both studies and across conditions within the current study.

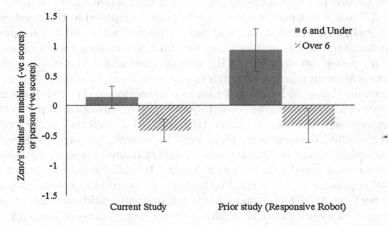

Fig. 3. Mean ratings of children's perceptions of Zeno. Main effects of age shown for current study and prior research [25]. SE bars shown

Remarkably, the Still condition in the current study shows the closest resemblance to prior findings. Morphology alone replicates the distinction between ages in a study using morphological, responsiveness, and autonomy cues. The Autonomous condition in the current study shows the same trend as prior findings [25] but, again, as a much smaller effect. Both these outcomes may be explained through a model of perceptual coherence.

In essence, perceptual coherence refers to the degree that user percepts of a robot align (i.e. physical features, movements, communication, apparent 'intelligence', and so on). The uncanny-valley phenomena [34], particularly the 'eeriness and repulsion' may arise from incoherent perceptual cues [35]. It is possible that low coherence limits children's perceptions of animacy.

In prior work, morphology, movement, and expression had a coherent *context* of playing a game [25]; movements and expressions are life-like, not just in their form [33], but also in their context. However, efforts in the current study to create minimally animate conditions mean expressions lose context, potentially creating incoherence in perceptual cues; young children may regard the robot as less animate because life-like movement occurs without context. In contrast, children may be using the coherent morphology cues of the Still robot, absent of possibly incongruent movement. This explanation presents a further possible factor for determining animacy in robots (beyond morphology, responsiveness, and autonomy) of *coherency* as an avenue for explorative research.

Future Directions

The current research indicates that children can use specific criteria to determine animacy of humanoid robots and that this is contingent on children's ages. The existent literature and current findings indicate that a systematic program of study is needed to address the individual and interactive effects of factors influencing children's views of animacy in robots. As suggested in the developmental literature [4], HRI literature [1], and current results, advanced social robots may present as interesting boundary cases for children's understanding of animacy.

In this paper, we draw from HRI and developmental literature to highlight key factors that may influence children's perceptions of animacy. Each of these can be manipulated, either in isolation in conjunction, to (1) develop a multidimensional model of animacy cues in robotics, (2) explore optimal parameter sets for effective user- and context-driven HRI, and (3) deepen the theoretical understanding of child development. First steps to achieve this are the development of a research program testing across and within factors for animacy cues.

Further work could explore within factors. Use of alternate robots of varying 'life-like' appearance such as the Nao humanoid [36], or bio-mimetic mammalian-robots such as MIRO [37], may further illuminate children's use of physical appearance to identify animacy. Alternatively, perceptions of animacy may be further explored by manipulating the extent or nature of autonomous motion: reaching for a ball may better suggest goal-directed motion than expressive moments which may better suggest affective states. Autonomy may be further manipulated by shaping the apparent control children observe others to have over a robot's actions. Varying operation from direct control, to wireless, to remote location operation may influence apparent autonomy and animacy.

Changes in perceptions of animacy may be explored across a developmental trajectory, exploring children's progress with age through Carey's or Piaget's stages [5,6], or in an interventionist manner, examining if perceptions can change through alteration of animacy cues. Education on a robot's behavior, physical mechanisms, and computation, delivered by a researcher or even the robot itself, could also help shape perceptions of animacy.

Finally, it is important to acknowledge the value of extended interviews or multiple questions [22]; despite their time-consuming nature, they may draw out nuances in children's reasoning. However, longer, more-involved questionnaires may fatigue young participants without any justifiable improvement in data quality. A continued effort in developing unobtrusive or brief measures, built on the systematic categorization of children's spontaneous interactions and remarks towards or about a robot, may be beneficial (e.g., [26]). Children's spontaneous, brief comments could reflect their understanding of social robots as boundary-spanning [1], and research may benefit from efforts made to capture these.

Acknowledgments. This work was supported by European Union Seventh Framework Programme (FP7-ICT-2013-10) under grant agreement no. 611971.

References

1. Sharkey, A., Sharkey, N.: Children, the elderly, and interactive robots. IEEE Robot. Autom. Mag. **18**, 32–38 (2011)
2. Blancas, M., Vouloutsi, V., Grechuta, K., Verschure, P.F.M.J.: Effects of the robot's role on human-robot interaction in an educational scenario. In: Wilson, S.P., Verschure, P.F.M.J., Mura, A., Prescott, T.J. (eds.) Living Machines 2015. LNCS(LNAI), vol. 9222, pp. 391–402. Springer, Cham (2015). doi:10.1007/978-3-319-22979-9_39
3. Cameron, D., Fernando, S., Collins, E., Millings, A., Moore, R., Sharkey, A., Prescott, T.: Impact of robot responsiveness and adult involvement on children's social behaviours in human-robot interaction. In: Proceedings of the AISB Convention (2016)
4. Opfer, J.E., Gelman, S.A.: Development of the animate-inanimate distinction. Wiley-Blackwell Handbook Childhood Cogn. Dev. **2**, 213–238 (2011)
5. Carey, S.: Conceptual change in childhood (1985)
6. Piaget, J.: The Child's Conception of the World. Rowman & Littlefield, Totowa (1951)
7. Bartneck, C., Kanda, T., Mubin, O., Al Mahmud, A.: Does the design of a robot influence its animacy and perceived intelligence? Int. J. Soc. Robot. **1**, 195–204 (2009)
8. Bullock, M., Gelman, R., Baillargeon, R.: The development of causal reasoning. In: The Developmental Psychology of Time, pp. 209–254 (1982)
9. Carey, S.: The Origin of Concepts. Oxford University Press, Oxford (2011)
10. Goren, C.C., Sarty, M., Wu, P.Y.: Visual following and pattern discrimination of face-like stimuli by newborn infants. Pediatrics **56**(4), 544–549 (1975)
11. Hood, B.M., Willen, J.D., Driver, J.: Adult's eyes trigger shifts of visual attention in human infants. Psychol. Sci. **9**, 131–134 (1998)
12. Driver, J., Davis, G., Ricciardelli, P., Kidd, P., Maxwell, E., Baron-Cohen, S.: Gaze perception triggers reflexive visuospatial orienting. Vis. Cogn. **6**, 509–540 (1999)
13. Johnson, S., Slaughter, V., Carey, S.: Whose gaze will infants follow? the elicitation of gaze-following in 12-month-olds. Dev. Sci. **1**, 233–238 (1998)
14. Rochat, P., Striano, T., Morgan, R.: Who is doing what to whom? young infants' developing sense of social causality in animated displays. Perception **33**, 355–369 (2004)
15. Schlottmann, A., Surian, L.: Do 9-month-olds perceive causation-at-a-distance? Perception **28**(9), 1105–1113 (1999)
16. Hamlin, J.K., Wynn, K., Bloom, P.: Social evaluation by preverbal infants. Nature **450**(7169), 557–559 (2007)
17. Hamlin, J.K., Wynn, K.: Young infants prefer prosocial to antisocial others. Cogn. Dev. **26**, 30–39 (2011)
18. Heider, F., Simmel, M.: An experimental study of apparent behavior. Am. J. Psychol. **57**, 243–259 (1944)
19. Carey, S.: Sources of conceptual change. In: Conceptual Development: Piagets Legacy, pp. 293–326 (1999)

20. Kahn, P., Reichert, A., Gary, H., Kanda, T., Ishiguro, H., Shen, S., Ruckert, J., Gill, B.: The new ontological category hypothesis in human-robot interaction. In: 6th International Conference on Human-Robot Interaction, pp. 159–160. IEEE (2011)

21. Somanader, M.C., Saylor, M.M., Levin, D.T.: Remote control and children's understanding of robots. J. Exp. Child Psychol. **109**, 239–247 (2011)

22. Okita, S.Y., Schwartz, D.L.: Young children's understanding of animacy and entertainment robots. Int. J. Hum. Robot. **3**, 393–412 (2006)

23. Kahn, P.H., Friedman, B., Perez-Granados, D.R., Freier, N.G.: Robotic pets in the lives of preschool children. Interact. Stud. **7**(3), 405–436 (2006)

24. Melson, G.F., Kahn, P.H., Beck, A., Friedman, B., Roberts, T., Garrett, E., Gill, B.T.: Children's behavior toward and understanding of robotic and living dogs. J. Appl. Dev. Psychol. **30**(2), 92–102 (2009)

25. Cameron, D., Fernando, S., Millings, A., Moore, R., Sharkey, A., Prescott, T.: Children's age influences their perceptions of a humanoid robot as being like a person or machine. In: Wilson, S.P., Verschure, P.F.M.J., Mura, A., Prescott, T.J. (eds.) Living Machines 2015. LNCS(LNAI), vol. 9222, pp. 348–353. Springer, Cham (2015). doi:10.1007/978-3-319-22979-9_34

26. Cameron, D., Fernando, S., Cowles-Naja, E., Perkins, A., Millings, A., Collins, E., Szollosy, M., Moore, R., Sharkey, A., Prescott, T.: Age differences in animacy beliefs observed in childrens questions for a humanoid. In: Proceedings of the Conference Towards Autonomous Robotic Systems (in Press)

27. Collins, E.C., Prescott, T.J., Mitchinson, B.: Saying it with light: a pilot study of affective communication using the MIRO robot. In: Wilson, S.P., Verschure, P.F.M.J., Mura, A., Prescott, T.J. (eds.) Living Machines 2015. LNCS(LNAI), vol. 9222, pp. 243–255. Springer, Cham (2015). doi:10.1007/978-3-319-22979-9_25

28. Saylor, M.M., Somanader, M., Levin, D.T., Kawamura, K.: How do young children deal with hybrids of living and non-living things: the case of humanoid robots. Brit. J. Dev. Psychol. **28**, 835–851 (2010)

29. Fukuda, H., Ueda, K.: Interaction with a moving object affects ones perception of its animacy. Int. J. Soc. Robot. **2**, 187–193 (2010)

30. Vouloutsi, V., et al.: Towards a synthetic tutor assistant: the EASEL project and its architecture. In: Lepora, N.F.F., Mura, A., Mangan, M., Verschure, P.F.M.J., Desmulliez, M., Prescott, T.J.J. (eds.) Living Machines 2016. LNCS, vol. 9793, pp. 353–364. Springer, Cham (2016). doi:10.1007/978-3-319-42417-0_32

31. Reidsma, D., et al.: The EASEL project: towards educational human-robot symbiotic interaction. In: Lepora, N.F.F., Mura, A., Mangan, M., Verschure, P.F.M.J., Desmulliez, M., Prescott, T.J.J. (eds.) Living Machines 2016. LNCS, vol. 9793, pp. 297–306. Springer, Cham (2016). doi:10.1007/978-3-319-42417-0_27

32. Cameron, D., Fernando, S., Millings, A., Szollosy, M., Collins, E., Moore, R., Sharkey, A., Prescott, T.: Congratulations, it's a boy! bench-marking children's perceptions of the robokind zeno-R25. In: Alboul, L., Damian, D., Aitken, J.M.M. (eds.) TAROS 2016. LNCS, vol. 9716, pp. 33–39. Springer, Cham (2016). doi:10.1007/978-3-319-40379-3_4

33. Hanson, D., Baurmann, S., Riccio, T., Margolin, R., Dockins, T., Tavares, M., Carpenter, K.: Zeno: a cognitive character. AI Mag., 9–11 (2009)

34. Mori, M., MacDorman, K.F., Kageki, N.: The uncanny valley [from the field]. IEEE Robot. Autom. Mag. **19**, 98–100 (2012)

35. Moore, R.K.: A bayesian explanation of the uncanny valleyeffect and related psychological phenomena. Sci. Rep. **2**, 864 (2012)

36. Gouaillier, D., Hugel, V., Blazevic, P., Kilner, C., Monceaux, J., Lafourcade, P., Marnier, B., Serre, J., Maisonnier, B.: Mechatronic design of nao humanoid. In: International Conference on Robotics and Automation, pp. 769–774. IEEE (2009)

37. Mitchinson, B., Prescott, T.J.: MIRO: a robot "Mammal" with a biomimetic brain-based control system. In: Lepora, N.F.F., Mura, A., Mangan, M., Verschure, P.F.M.J., Desmulliez, M., Prescott, T.J.J. (eds.) Living Machines 2016. LNCS, vol. 9793, pp. 179–191. Springer, Cham (2016). doi:10.1007/978-3-319-42417-0_17

Analysing the Limitations of Deep Learning for Developmental Robotics

Daniel Camilleri[(✉)] and Tony Prescott

Psychology Department, University of Sheffield, Western Bank, Sheffield, UK
d.camilleri@sheffield.ac.uk
http://www.sheffield.ac.uk

Abstract. Deep learning is a powerful approach to machine learning however its inherent disadvantages leave much to be desired in the pursuit of the perfect learning machine. This paper outlines the multiple disadvantages of deep learning and offers a view into the implications to solving these problems and how this would affect the state of the art not only in developmental learning but also in real world applications.

Keywords: Deep learning · Artificial narrow intelligence · Transfer learning · Probabilistic models

1 Introduction

Deep learning approaches currently dominate the scene of (AI) and soon robotics. This approach to learning has been consistently yielding incredible results in multiple fields of AI. From object recognition [1] and object segmentation [2] to speech recognition [3] and even human-like speech synthesis with applications like WaveNet [4] and Deep Speech [5]. Yet, behind the illusion that more depth, more data and more computational power can solve any task, one can start to see the limitations that this approach to learning entails. As we shall see, depth carries with it a troublesome companion: obscurity in the understanding of *what* has been learned. Deep Learning has proven to be a sufficient and easy solution for limited, ad-hoc problems. However, the more complex a task becomes and the more flexibility an environment requires, the more obsolete feed-forward systems become in dealing with this challenge.

In this paper, we discuss the various disadvantages that deep learning methods exhibit. We start with the low transparency and interpretability of this approach and how these models require incredible amounts of data to train. Subsequently we see how the models created are very specialised and narrow and furthermore lack the ability to perform transfer learning between tasks. Thus the aim of this paper is to present a case for how different approaches are required to overcome these drawbacks in order to create machines that are capable of learning from experiences in a more neuromimetic manner.

© Springer International Publishing AG 2017
M. Mangan et al. (Eds.): Living Machines 2017, LNAI 10384, pp. 86–94, 2017.
DOI: 10.1007/978-3-319-63537-8_8

2 Disadvantages of Deep Learning

2.1 Low Transparency and Interpretability

The information and experiences encountered by a robot can be stored in multiple forms of which two are the most commonplace: either in the form of Formal Symbolic Models or in the form of Sub-Symbolic Models. A Formal Symbolic model makes use of meaningful representations to express information in a compact manner. Instances of Formal Symbolic models include Hidden Markov Models and Bayesian Graphs. They are defined by graphs, where each node in the graph represents a whole. This makes the encoding of information in this way a tedious manual process which would need to define the features that are to be extracted from the input sensory streams. But it also makes the model explanatory in nature because given a series of activations within the graphs making up these models, one can trace these activations and thus discern how and why the model responded in the way it did.

On the other hand, Sub-Symbolic model approaches like Deep Learning represent information as parts that when combined in different ways make different wholes. Taking the example of a spoken word, a Symbolic Model would represent the whole waveform that describes the word while a Sub-Symbolic Model stores different features in different nodes such as the different power spectra making up the sound. This approach is powerful in that it can be trained automatically and will discover which patterns in the data are relevant to the required task and which are to be ignored leading to a really good understanding of the data at a very low level. However this understanding cannot be conveyed to the user or even to an expert because of the cryptic inner workings of a deep model. Thus once trained and optimised, sub-symbolic models are, for all intents and purposes, a black box. This results in two major disadvantages with the use of Sub-Symbolic models for robot learning.

The first is that the correct operation of the model cannot be formally verified. Dataset-centred verification is the only method of checking for correct operation under normal conditions and the operation of the model in abnormal conditions is covered in so far as abnormal conditions are covered within the training dataset. Therein lies the problem because the construction of an all-encompassing dataset is impractical and in reality unachievable.

To compound the issue, it has also been demonstrated that the current state of the art can be fooled into consistently making wrong decisions by what are called 'adversarial' inputs which is the second major drawback in this category. Adversarial inputs are inputs to a model which have been modified by a very small percentage in key areas, leading to a consistently wrong classification [6, 7]. In a real world application this is a serious vulnerability especially because the changes that create these adversarial inputs are small enough to be imperceptible by humans.

This weakness combined with the lack of transparency presents an obstacle in the application of these models to certain real world applications which are legally required to demonstrate a high level of safety across all operating conditions like in the case of fully autonomous cars and drones [8] and in general any fully automated physical body including robots.

As an extension of this argument, the use of non-transparent models raises an issue whenever decisions taken by these models carry with them a level of required accountability. In the end, the users of these models are legally accountable for actions taken based on the prediction of these models and as such when queried about their decisions they are required to fully justify said decision. The absence of interpretability in the current state of the art renders their use, in financial situations like loan approval or healthcare applications like cancer diagnosis, problematic because they lack the explanatory power that humans have when reaching a decision.

There is also a human interaction component to the requirement of transparency and that is trust in the model's performance which ties in with accountability. The motivation for building more and more intelligent algorithms and models is to automate and improve and therefore we hold these models to a higher standard than we would a fellow human performing the same task. This, therefore, also demonstrates the requirement for transparency in trusting that a model is performing the correct decision.

Thus the ideal algorithm would exhibit the automatic pattern extraction properties and the classification accuracy of a sub-symbolic algorithm while still retaining the high level representations of a symbolic model. In this way, the model can be used to explain the 'reasoning' behind a decision in the manner that a symbolic algorithm is capable of.

2.2 Large Data Requirements

Deep learning, even in its application to narrow intelligence as we shall see later on, calls for the optimisation of millions of parameters via training. Consequently this leads to the requirement for large amounts of data in order to train a well-performing deep model. Furthermore, the data in most cases is required to be labelled which introduces an additional hindrance in the training of deep models although unsupervised approaches are also available.

These requirements mean that several problems which could benefit from the pattern classification power of deep learning, do not contain enough data to leverage this benefit. Data for particular applications like object segmentation or speech recognition are easy to collect and there are plenty of resources to obtain this data. However, certain complex applications such as emotion recognition require information from multiple sources such as ECG, galvanic skin response and vision which is not easy to collect. Furthermore, due to the complexity of the inputs for the model, this further increases the amount of data required to train a good performing model. The same obstacle applies in industry with small businesses which do not generate enough data to train a meaningful model. Thus, these businesses cannot glean its benefits like larger businesses do. Consequently the collection of clean and abundant data is currently the biggest hurdle in the design of an AI system for both industry and academia.

Finally, since their learning is based on statistical processes, this type of learning cannot be used to remember one off events, like for example the event of purchasing your first car, something that humans do so well with the use of episodic memory. A plausible solution to this problem would be an architecture that becomes dynamically more complex as experiences are encountered by a robot and is then compressed when enough observations have been collected.

This would allow a robot to learn specific concepts quickly but also provide a means for performing generalisation when enough statistical data is collected. This generalisation would compress the complex experiential nodes into a simpler network that nonetheless preserves or even improves the classification accuracy of the uncompressed state.

2.3   Artificial Narrow Intelligence

Algorithms that learn have existed for more than 30 years in one form or another. Of interest is the fact that these algorithms have barely changed in that time, and the term learning is often used as a synonym of performance optimization. However, having robots present in highly variable environments like our homes, conference venues, or the streets, that can act robustly enough to be moved from an environment to another requires them to have the necessary mechanisms to acquire a vast array of diverse knowledge that includes grounding between concepts in different types of knowledge.

The current state of the art is however very good for (ANI) applications: applications which have a very specific and narrow but deep knowledge of a certain task or process such as speech generation with WaveNet [4] or playing Go with AlphaGo [9]. Both of these applications are examples of models which have a very specific task (narrow area of knowledge) at which they have an incredible, sometimes super-human, performance because their knowledge of the area is very deep. However, what makes these models narrow, is that neither of these models can be used for anything other than the task that they are trained on. AlphaGo cannot be trained to play chess while still being capable of playing Go and WaveNet cannot be used to generate animal sounds while remaining capable of generating human sounds.

This means that ANI lacks the ability of generalising between different but similar applications and also lacks the ability of combining multiple separate sensory streams to perform multi-modal sensory integration. These properties are however quite important for the development of a learning robot because when combined they provide a manner of grounding one sensory experience within another and provided the basis for the creation of models that have less deep but much broader knowledge.

2.4   Transfer Learning

Transfer learning is the application of knowledge learned during the execution of a primary task to the execution of a new but similar secondary task. An example of transfer learning would be applying techniques learned from one game to a similar but different game which makes it a crucial skill for a robot to have in order to speed up learning across different tasks. However, due to the gradient-based nature of the algorithm, Deep Learning techniques are not by themselves capable of transfer learning. This has been demonstrated in [10,11] where a network which is trained on a primary task and then trained on a similar secondary task can learn the second task much faster but at the expense of forgetting the first. This is called catastrophic forgetting.

Catastrophic forgetting is one of the primary research areas in the field of Deep Learning and a recent approach by Deep Mind [12] demonstrates the challenges inherent in achieving this and a possible solution which combines genetic algorithms with Deep Networks. This approach freezes the configuration of certain dominant pathways that are present within the network for the primary task in order to retain their functionality while allowing the rest of the network to learn normally. This is done in such a manner with the use of genetic algorithms such that it allows the learning of the second task as a function of the first, keeping the most fit pathways between different iterations of the genetic algorithm.

This has been the first demonstration of transfer learning within deep learning but due to its requirement of genetic algorithms is a trial and error driven approach to transfer learning requiring thousands of hours of training in order to achieve transfer learning between two tasks.

The difficulty of transfer learning derives mainly from the backpropagation algorithm which is central to all deep learning techniques and this is because the backpropagation algorithm essentially changes the representation of each node within the network in order to find the best features for the task at hand. This issue can be solved by having a fixed representation at each node but instead dynamically growing the network when new representations are required to handle a new problem.

Furthermore, the fixed representation of nodes within the network opens a new possibility for greatly accelerated learning when performed across multiple robots. Due to the fixed representations, a network learned in robot A can be matched and aligned with that learned by robot B and subsequently fused in order to combine the knowledge and experiences of the two into a single network. This can then be redistributed to A and B thus effectively making their learning rate significantly higher.

2.5 Probabilistic Models

Another issue with deep learning is that the models created are not fully probabilistic models. They can provide a probability vector of length n that describes how probable the current classification is with respect to the n classes that the model has been trained on. However this is not sufficient in a real world environment especially in the case of a developmental approach to learning. In these cases one must be able to also distinguish whether the current classification is something known or unknown to the model.

This capability is crucial for a learning machine because it defines the extent of knowledge for said machine. Given the capability of recognising known and unknown provides the machine with the capability for active learning where it can ask a human about the unknown object and thus expand its knowledge. Furthermore, a fully probabilistic model with the property of fixed node representations implies that the transitions of each node are assigned a probability that has been learnt from experience. This probabilistic representation can therefore be used to carry out predictions much like the associative action of System I thinking in the human brain [13]. This makes for a powerful system that conceivably brings us a step closer to implementing human intelligence in a machine.

2.6 Generative Models

Generative models are models that operate in two directions. They can go from a high dimensional input to a low dimensional classification or vice versa generate a typical high dimensional output given one of the classifications it has been trained on. The brain is a generative model due to the evidence we have for episodic replay where FMRI studies have shown the different parts of the brain involved in sensory processing light up when recollecting a memory [14]. Furthermore, using the human brain as inspiration, its generative connections allow it to perform what is called top-down control: where an activation in a high level network biases the functionality of a lower level network. A classic example of this is guided search where the features of an object that is being actively searched become more salient in the visual field [15].

Deep learning methods have been applied to the field of generative models and the current state of the art are deep convolutional generative adversarial networks [16]. These GAN operate by having two networks connected together in a cyclic form where the first network takes a high dimensional input such as an image and learns to classify that image. The input of the second network is then connected to the output of the first and its role is to take the classification vector received and generate from it a high dimensional output. This cycle can then be repeated several times in order to train the generative model. The issue with deep learning in this case lies in the complexity of the system which results in unstable training [16] when creating generative model.

3 Impact of Solving the Challenges

An approach that is capable of solving these challenges would create a great upheaval in the domain of AI and robotics much like the mainstream adoption of deep learning has but on a bigger scale due to the flexibility of a system without these disadvantages. The impact, would be incredibly far reaching and would mark the start of a new era of interactive learning machines. From fully autonomous transport, improved finance, advancements in the efficiency and accuracy of healthcare and last but not least the realisation of autonomous, embodied learning robots.

First and foremost, solving the issue of interpretability would mean that machine learning techniques can now be deployed in safety critical areas. This is not possible with current deep learning techniques but highly interpretable models provide the opportunity for formal validation and verification of operation under all conditions. Examples of safety critical areas include all forms of automated transport: autonomous cars, autonomous drones and even autonomous spacecraft where the risk is high but the option for failure is extremely low requiring reliable systems of guidance. The same applies for the use of robots in an industrial setting where robots work alongside humans.

Second, with the use of techniques that do not require extraordinary amounts of data one would see an increased adoption of these techniques. Furthermore when combined with interpretability this provides a method that is capable of analysing and extracting structure from small datasets that can be conveyed to the user. This is especially useful in the case of data analysis in small businesses in order to improve their product, an advantage which is currently reserved for only big companies with lots of data. This can even be extended to create a technique which is capable of storing information in a manner similar to the human brain with multiple hierarchies representing sensory, episodic and semantic memory. This provides an architecture that is capable of performing one shot learning on scarce unrepeatable data but at the same time capable of generalised learning.

Additionally, overcoming ANI and bypassing the issues with transfer learning in deep learning applications, one would create an algorithm that is capable of applying knowledge learned in the solution of one problem to other problems via the process of analogy. Solving the problem of ANI would allow for the creation of robots capable of general developmental learning much like infants do in their early years by supporting multi-sensory integration and hence the grounding of concepts within different senses. But it is only with a solution for transfer learning that these robots can move past the early stages of human development and become intelligent machines.

Finally, with the use of fully probabilistic models that can recognise between activations that are known and unknown one can create machines that actively learn about their environment and also machines that are capable of interactive learning, where an unknown stimulus can be demonstrated to a human with the use of a generative property and then the human can provide a label to the stimulus thus driving the semi-supervised learning in the robot.

4   Conclusion

The approach of deep learning has provided a significant boost to the performance of models in different areas but contain the above problems which limit its incorporation in industry in safety critical applications and especially in robotics. At the same time, AI is becoming increasingly dominant and present in every aspect of our lives and its proliferation will continue to increase together with its power at solving increasingly difficult problems. Yet in order to have more interactive human-robot interactions, the robots need to be able to communicate in a human-understandable way and provide reasons for their actions. This is the focus of our future work which will look into a machine learning approach that performs as well as deep networks but at the same time does not have the outlined issues.

References

1. Krizhevsky, A., Sutskever, I., Hinton, G.E.: Imagenet classification with deep convolutional neural networks. In: Advances in Neural Information Processing Systems, pp. 1097–1105 (2012)
2. Girshick, R., Donahue, J., Darrell, T., Malik, J.: Rich feature hierarchies for accurate object detection and semantic segmentation. In: Proceedings of the IEEE Conference on Computer Vision and Pattern Reognition, pp. 580–587 (2014)
3. Graves, A., Mohamed, A.r., Hinton, G.: Speech recognition with deep recurrent neural networks. In: 2013 IEEE International Conference on Acoustics, Speech and Signal Processing (icassp), pp. 6645–6649. IEEE (2013)
4. van den Oord, A., Dieleman, S., Zen, H., Simonyan, K., Vinyals, O., Graves, A., Kalchbrenner, N., Senior, A., Kavukcuoglu, K.: Wavenet: A generative model for raw audio. CoRR abs/1609.03499 (2016)
5. Arik, S.O., Chrzanowski, M., Coates, A., Diamos, G., Gibiansky, A., Kang, Y., Li, X., Miller, J., Raiman, J., Sengupta, S., et al.: Deep voice: Real-time neural text-to-speech. arXiv preprint (2017). arXiv:1702.07825
6. Papernot, N., McDaniel, P., Jha, S., Fredrikson, M., Celik, Z.B., Swami, A.: The limitations of deep learning in adversarial settings. In: 2016 IEEE European Symposium on Security and Privacy (EuroS P), pp. 372–387 (2016)
7. Nguyen, A., Yosinski, J., Clune, J.: Deep neural networks are easily fooled: High confidence predictions for unrecognizable images. In: 2015 IEEE Conference on Computer Vision and Pattern Recognition (CVPR), pp. 427–436 (2015)
8. Koopman, P., Wagner, M.: Challenges in autonomous vehicle testing and validation. SAE Int. J. Transp. Saf. 4(2016—-01-0128), 15–24 (2016)
9. Silver, D., Huang, A., Maddison, C.J., Guez, A., Sifre, L., Van Den Driessche, G., Schrittwieser, J., Antonoglou, I., Panneershelvam, V., Lanctot, M., et al.: Mastering the game of go with deep neural networks and tree search. Nature 529(7587), 484–489 (2016)
10. McCloskey, M., Cohen, N.J.: Catastrophic interference in connectionist networks: The sequential learning problem. Psychol. Learn. Motiv. 24, 109–165 (1989)
11. Goodfellow, I.J., Mirza, M., Xiao, D., Courville, A., Bengio, Y.: An empirical investigation of catastrophic forgetting in gradient-based neural networks. arXiv preprint (2013). arXiv:1312.6211
12. Fernando, C., Banarse, D., Blundell, C., Zwols, Y., Ha, D., Rusu, A.A., Pritzel, A., Wierstra, D.: Pathnet: Evolution channels gradient descent in super neural networks. arXiv preprint (2017). arXiv:1701.08734
13. Kahneman, D.: Thinking, Fast and Slow. Macmillan, New York (2011)

14. Wheeler, M.E., Petersen, S.E., Buckner, R.L.: Memory's echo: Vivid remembering reactivates sensory-specific cortex. Proc. Nat. Acad. Sci. **97**(20), 11125–11129 (2000)
15. Wolfe, J.M.: Guided search 2.0 a revised model of visual search. Psychon. Bull. Rev. **1**(2), 202–238 (1994)
16. Radford, A., Metz, L., Chintala, S.: Unsupervised representation learning with deep convolutional generative adversarial networks. arXiv preprint (2015). arXiv:1511.06434

Reducing Training Environments in Evolutionary Robotics Through Ecological Modularity

Collin Cappelle[✉], Anton Bernatskiy, and Josh Bongard

The University of Vermont, Burlington, VT 05405, USA
collin.cappelle@uvm.edu

Abstract. Due to the large number of evaluations required, evolutionary robotics experiments are generally conducted in simulated environments. One way to increase the generality of a robot's behavior is to evolve it in multiple environments. These environment spaces can be defined by the number of free parameters (f) and the number of variations each free parameter can take (n). Each environment space then has n^f individual environments. For a robot to be fit in the environment space it must perform well in each of the n^f environments. Thus the number of environments grows exponentially as n and f are increased. To mitigate the problem of having to evolve a robot in each environment in the space we introduce the concept of *ecological modularity*. Ecological modularity is here defined as the robot's modularity with respect to free parameters in its environment space. We show that if a robot is modular along m of the free parameters in its environment space, it only needs to be evolved in n^{f-m+1} environments to be fit in all of the n^f environments. This work thus presents a heretofore unknown relationship between the modularity of an agent and its ability to generalize evolved behaviors in new environments.

1 Introduction

One of the major challenges to evolutionary robotics in particular, and evolutionary computation in general, is the relatively slow rate of convergence toward acceptable solutions due to these algorithms' stochastic elements. This challenge is exacerbated when robust behavior is desired: In such cases robots must be evolved in multiple environments until the robots exhibit the desired behavior in all of them. However, because of catastrophic forgetting [8], it is not usually possible to evolve robots in one environment, discard that environment, continue evolving them in a different environment, and have them retain their ability to succeed in the first environment. Thus, robots must be trained in some set of static environments, or gradually exposed to a growing set of training environments over evolutionary time. [14] pointed out convergence time explodes in such multiple-environment contexts because of the combinatorics of parametrically-defined environments. Typically, a set of training environments is generated before evolution commences by defining a number of free parameters f, which

© Springer International Publishing AG 2017
M. Mangan et al. (Eds.): Living Machines 2017, LNAI 10384, pp. 95–106, 2017.
DOI: 10.1007/978-3-319-63537-8_9

represent aspects of the environment that change from one to another. For each of these free parameters, there are n possible settings. For example, given an object in the environment, a free parameter could be the starting position of that object. If the object may have different sizes as well as starting positions in a given environment from taken for the total set of possible environments, then $f = 2$. If there are two possible sizes, and two different starting positions, then $n = 2$. [14] showed that if we wish evolved robots to succeed in all environments defined for a given f and n, then the robots will have to be evolved in n^f environments.

1.1   Robustness

Much work has been done to increase the robustness of evolved behavior in robots. For instance, Jakobi [10] investigated the introduction of noise to guard against evolutionary exploitation of any inaccuracies in the simulator used to the evolve the behaviors. Lehman [12] demonstrated experiments in which explicit selection pressure was exerted on robots to respond to their sensor input, thus ensuring that evolved robots would behave differently when placed in different environments where they could sense the changes. Bongard [2] demonstrated that robots with ancestors that changed their body plans during their lifetimes tended to be more robust than robots with fixed-morphology ancestors, because the former lineages tended to experience wider ranges of sensorimotor experiences than the latter lineages. However, these and similar works did not investigate the role that modularity might play in the evolution of robust behavior. One exception is the work of Ellefsen *et al.* [6], in which an evolutionary cost is placed on the synapses of disembodied neural networks trained to compute logical functions. They had previously shown that such connection cost tends to lead to the evolution of modular networks [5], and, in [6], this neural modularity enabled evolved networks to rapidly adapt to new environments without losing their ability to succeed in the original environments.

1.2   Modularity

Like robust behavior, the ubiquity of modularity in evolved systems has spawned an active literature. Work in this area can be divided into investigations into the evolution of modularity in disembodied systems and embodied systems, such as robots.

Wagner [17] forwarded a theoretical argument that modularity evolves when systems experience combinations of directional and stabilizing selection on different parts of their phenotypes. This was subsequently verified by experiments using non-embodied data structures [13], neural networks [5,11], and models of gene networks [7].

Investigations into the evolution of modularity in embodied systems begin with Gruau [9], who employed an indirect genotype to phenotype mapping that allowed for the construction of neural modules in a robot. Yamashita et al. [18] demonstrated robots capable of learning independent motor primitives and then combining them in novel sequences. In [1,3], Bongard *et al.* showed how to evolve structurally modular neural controllers for autonomous robots.

However, none of these approaches investigate the relationship between both morphological and neurological modularity as a way to increase generalization. That is, how the structure of the robot's morphology and controller may interact with the environment to reduce the minimum number of environments robots must be evolved in to generalize across the entire environment space.

1.3 Morphological and Neurological Modularity

Modularity has shown to be important in evolution of networks and robots because it helps the agent avoid catastrophic forgetting when presented with a new environment [6]. Catastrophic forgetting is a problem when, in order to learn a new task, an agent must forget what it previously learned [8].

However, most of the modularity research in robotics has focused on modularity with respect to the controller of the robot. Most often the controller is a neural network so network metrics are used. Most notably the Q-metric has been used to define modularity in networks [15]. Q is a metric which measures the fraction of edges which fall between within a group subtracted by the expected fraction of edges within that group given a random network with the same degree distribution. However, Q disregards many aspects of the morphology and control of robots which may be important in determining if the robot is made up of actual useful modules.

More recent work has defined both neural and morphological modularity in terms of the sensor-motor feedback loop [4]. It was shown that the number of necessary training environments for robots that are morphologically and neurologically modular in this manner grows less rapidly than the number of necessary training environments in non-modular cases when the number of free parameters, f, was held constant and the number of variations, n, was increased.

In this work, we build upon this research by holding n constant and increasing f. Also, we continue to use those definitions of neurological and morphological modularity and expand them by considering the robot's interactions with its environment space a property which we here term 'ecological modularity'.

1.4 Ecological Modularity

We define the following terms and variables to be used throughout the paper:

- **F** - The set of free parameters in the system with cardinality f. Free parameters are the dimensions of the environment space which change.
- **n** - number of variations of each free parameter in F. Because we are only considering free parameters that vary, $n \geq 2$. For simplicity, all free parameters are assumed to have the same number of variations.

- **Discrete Environment Space** - The set, E, comprised of all the possible combinations of free parameter variations. These are environment spaces which can be discretized and organized into an f-dimensional hypercube with n^f hypervoxels each corresponding to one individual environment. Therefore there are a total of n^f environments in E. Each environment can therefore be defined as a f-tuple consisting of the variations of each free parameter.
- **Orthogonal Environments** - Orthogonal environments are those in which none of the variations of the free parameters are equal. Thus given two environments e_1 and e_2, $\pi_{e_1}^{(j)} \neq \pi_{e_2}^{(j)}$ for all j in F. For example,

$$\left(\pi_1^{(1)}, \pi_1^{(2)}, \ldots, \pi_1^{(f)} \right) \perp \left(\pi_2^{(1)}, \pi_2^{(2)}, \ldots, \pi_2^{(f)} \right)$$

because $\pi_1^{(j)} \neq \pi_2^{(j)}$ for each j.

- **Orthogonal Environments along a Subset of Free Parameters** - Let $D \subset F$. Then two environments, e_1, e_2, are orthogonal along D if for each $d \in D$, $\pi_{e_1}^{(d)} \neq \pi_{e_2}^{(d)}$.
- **Modularity along U Free Parameters** - Let U be a subset of F with cardinality u. Let $O_U \subset E$ represent a subset of orthogonal environments along U. A robot is said to be modular along U if, when the robot achieves sufficient fitness in all u-environments in O_U, the robot will maintain its fitness in the remaining environments where the variations along the $F \setminus U$ free parameters remain fixed. We note $1 \leq u \leq f$ for every robot and environment space.
- **Ecological Modularity** - Let M be a subset of F with cardinality m such that M is the maximal subset of F a robot is modular with respect to. That is the robot is modular with respect to every free parameter in M but none of the free parameters in $F \setminus M$. Then ecological modularity is defined to be the degree to which the robot is modular with respect to its environment, m. Robots with $m = f$ are said to be fully ecologically modular, robots with $1 > m > f$ are said to be partially ecologically modular, and robots with $m = 1$ are said to be ecologically non-modular.

Using the definitions above, we claim the total number of environments necessary for a robot to be evolved in is $n^{(f-m+1)}$. Meaning when we have a robot which is fully ecologically modular ($m = f$) we only need to evolve the robot in n mutually orthogonal environments, the easiest example of which is the 'grand diagonal' of the hypercube representation of the environment space. When the robot is ecologically non-modular ($m = 1$) we need to evolve the robot in all n^f environments. The term $f - m + 1$ represents the number of free parameters the robot is not modular with respect to.

2   Methods

In this section we describe the structure of the environment spaces, robot design, evolutionary algorithm, and experimental design.

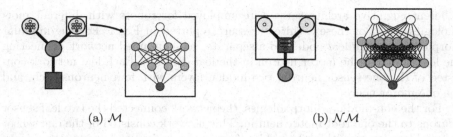

Fig. 1. The modular (1a) and non-modular (1b) robots' morphology and control structures. The morphology consisted of fixed hinges (red squares), free hinges (large blue circles), and sensor nodes (large beige circles). The networks are presented blown up for each robot. They consisted of sensor (white circles), hidden (yellow circles), and motor (blue circles) neurons. Motor neuron output controls the hinge joint at the base of the node the motor neuron is in. Connections between layers were feed-forward and feed-back. There are also recurrent connections for each hidden and motor neuron not depicted. (Color figure online)

2.1 Robot Design

Robot Morphology. The robots were designed with a branching, hierarchical morphology. A tree structure was chosen because it is symmetric, can easily be made modular/non-modular by fixing different branch hinges, and is easily expandable. Each tree consisted of one root node and two leaf nodes. The root node was connected to a point in space by a hinge joint. The leaf nodes were connected to the root node by hinge joints. Sensors were distance sensors placed in the leaf nodes of the robot. When the robot was pointing at an object they returned the distance to that object. When the robot was not pointing at anything, the sensor values returned a default value of ten.

Each robot was composed of three cylinders, one root node and two leaf nodes, of length one. The base of each leaf node was attached to the tip of the root node. Robots were initially positioned such that the leaves were horizontally rotated $+45°$ and $-45°$ with respect to the root node. In this paper we explored two variations of robot morphology.

First is the modular morphology, $\mathcal{M}$. In the modular morphology, the root node of the robot is fixed while the leaf nodes of the robot are free to move. Each leaf could rotate horizontally $[-45°, +45°]$ with respect to its starting position.

Second is the non-modular morphology, $\mathcal{NM}$. In the non-modular morphologies, the root of the robot is free to move while its leaf nodes are fixed. The root could rotate horizontally $[-120°, +120°]$.

Robots were simulated using Open Dynamics engine.

Robot Controllers. Robots were controlled by artificial neural networks. All networks were layered networks with both feed-forward and feed-back synapses as well as recurrent connections on both the hidden neurons and motor neurons.

Different cognitive architectures were employed for robots with different morphologies. Each of these architectures are reported in Fig. 1. For the modular morphologies, each leaf node had a separate, self-contained network connecting the leaf sensor to the motor neuron in the leaf (Fig. 1a). Each leaf network consisted of the one sensor neuron, two hidden layers with four neurons each, and the one motor neuron.

For the non-modular morphologies, the network connected the two leaf sensor neurons to the one root motor neuron. This network consisted of the one sensor neuron, two layers with eight hidden neurons each, and the one motor neuron (Fig. 1b).

Sensor neurons could take values between $[0, 10]$. Hidden and motor neurons could take values between $[-1, 1]$. Sensors could take any real valued number. Neurons in the network were updated at each time step in the simulation. The value of each neuron was determined by

$$y_i^{(t)} = \tanh\left(y_i^{(t-1)} + \sum_{j \in J} w_{ji} y_j^{(t-1)} \right) \tag{1}$$

where y_i^t denotes the i neuron's new value at time step t. $y_i^{(t-1)}$ denotes that neurons value in the previous time step. w_{ji} denotes the weight of the synapse connecting neuron j to neuron i.

2.2   Environmental Setup

Environments consisted of two clusters of cylinders set up on the left and right of the robot such that on the first time step of simulation, the robot pointed at the center of each cluster as shown in Fig. 2. Cylinders were organized on a line segment perpendicular to the direction of the leaf nodes. Clusters were placed such that the robot was initially pointing at their center. The diameter of each cylinder was equal to the length of the line segment divided by the number of cylinders in the cluster. A small constant value of $\varepsilon = .1$ was then added to the

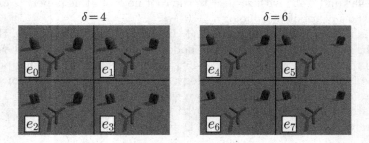

Fig. 2. The starting point of the robots in simulation for each environment. The environment space $E_2 = \{e_0, e_1, e_2, e_3\}$ is shown by the four left environments in the figure and $E_3 = \{e_0, e_1, \ldots, e_7\}$ is shown by all eight environments which make up the figure. The δ variable defines the initial distance of both clusters from the robot.

diameters so there were no gaps between cylinders in the cluster. Thus, each environment consisted of three free parameters:

- c_L: The number of cylinders in the left cluster. $c_L \in \{1, 2\}$.
- c_R: The number of cylinders in the right cluster. $c_R \in \{1, 2\}$.
- δ: The distance, in simulator units, the center point of each cluster is from the tip of its corresponding sides leaf node of the robot. $\delta \in \{4, 6\}$.

From the variables described above, we can categorize each environment as a 3-tuple (δ, c_L, c_R).

We can generate environment spaces by restricting which parameters are free and which are fixed. In this manner we generate two different environment spaces we are interested in:

- $E_2 = (\delta = 4, c_L = *, c_R = *)$
- $E_3 = (\delta = *, c_L = *, c_R = *)$

where $*$ indicates that parameter is free to vary. From this we see E_2 is a 2×2 environment space with four total environments and E_3 is a $2 \times 2 \times 2$ environment space with eight total environments.

We can then enumerate individual environments by the corresponding tuples parameter values. For example, we let $e_{(0,0,0)}$ represent an environment that consists of the first variation of each parameter, namely $e_{(0,0,0)} = e_0 = (\delta = 4, c_L = 1, c_R = 1)$. Thus $e_{(1,1,1)} = e_7 = (\delta = 6, c_L = 2, c_r = 2)$ and so on for each environment. All of the environments considered in this work are presented in Fig. 2.

2.3 Physical Implementation

The robot was also made in out of Legos as shown in Fig. 3. While the physical implementation can move and respond in the same manner as the simulation, it is still in development so no evolution was performed using the physical robot.

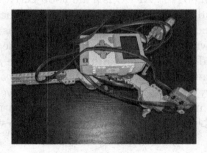

Fig. 3. Physical robot made out of Legos. Can represent the $\mathcal{M}$ or $\mathcal{NM}$ robot by fixing/freeing motors.

2.4   Evolutionary Setup

The goal of the robot was to point towards clusters containing an even number of cylinders and away from clusters containing an odd number of cylinders. This was implemented using a simple counting method detailed in Eq. (4).

The fitness scores of each sensor for each time step, $(s_L(t), s_R(t))$, were then summed and normalized with respect to the environment so the overall fitness was in $[0, 1]$ for each environment in the space (Eq. 3).

The fitness scores of each individual environment were then sorted from lowest to highest (worst to best) and a weighted average was performed meaning the overall fitness of the entire environment space also in the range $[0, 1]$ (Eq. 2). Weighting was performed by the geometric sequence $w_i = 1/(2i)$ for $i = \{1, 2, \ldots, \|O\| - 1\}$ where O is subset of the environment space considered. In order to make the weights sum to one, the last weight was set equal to the second to last weight. The other weighting schemes considered were a mean average and simply taking the worst individual environment fitness as the fitness for the whole environment set. Both converged more slowly than method we use.

$$\text{Overall Fitness} = \sum_{i \in \|O\|} w_i \text{ fit}(e_{(i)}) \tag{2}$$

$$\text{fit}(e_i) = \frac{1}{\text{normalize}(e_i)} \sum_{t=T/2}^{T} s_L(t) + s_R(t) \tag{3}$$

$$s_{\{L,R\}}(t) = \begin{cases} 1 & \text{if the sensor is pointing at} \\ & \text{an even cluster at time } t \\ 0 & \text{if the sensor is not pointing at} \\ & \text{an object at time } t \\ -1 & \text{if the sensor is pointing at} \\ & \text{an odd cluster at time } t \end{cases} \tag{4}$$

Evolution was performed using Age Fitness Pareto Optimization (AFPO) with a population size of 50 [16]. AFPO is a multi-objective optimization method using a genome's age and fitness as objectives. Mutations occurred by way changing synapse values in the neural network. If a synapse was chosen for mutation, a new weight was drawn from a random Gaussian value with mean equal to the previous weight and standard deviation equal to the absolute value of the previous weight. This mutation operator is employed because it allows weights near zero to mutate very slightly, while large-magnitude weights can be mutated in a single step over a much broader range. A mutation rate was chosen such that the expected number of synapses mutated each step was one.

2.5   Experimental Setup

Robots were evolved in a subset, O, of the total environment space, E. O was designated as the training set. When the best robot in the population achieved a certain fitness threshold for each environment in O, evolution was halted and the best robot was then tested in the remaining unseen environments, $E \setminus O$. We chose a fitness value of 0.9 as the threshold. We performed 30 trials for each experiment.

3   Results

The first environment space explored was E_2, the 2×2 environment space where only c_L and c_R were varied. The training set of the robots was $O_{2,2} = \{e_0, e_3\}$. In E_2, $O_{2,2}$ represents the grand diagonal of the space. Figure 4a shows that $\mathcal{M}$ was able to achieve sufficient fitness in the entirety of E_2 when the robot achieved sufficient fitness in $O_{2,2}$. Figure 4b shows that $\mathcal{NM}$ was not able to achieve sufficient fitness in all environments of E_2. The robot was not above the fitness threshold in any unseen environment for any trial.

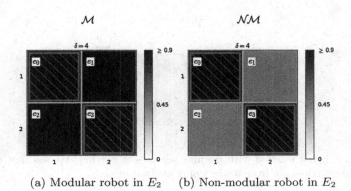

(a) Modular robot in E_2 (b) Non-modular robot in E_2

Fig. 4. Average fitness scores for $\mathcal{M}$ (4a) and $\mathcal{NM}$ (4b) robots in E_2 with training set $O_{2,2} = \{e_0, e_3\}$. $O_{2,2}$ is represented by the blue outlines around the environments. (Color figure online)

The second environment space we explored was E_3, the $2 \times 2 \times 2$ environment space where c_L, c_R and δ were free parameters. For this environment space, the training set was $O_{3,4} = \{e_0, e_3, e_4, e_7\}$. This training set represents diagonal sub-spaces of environments for each value of δ. The $\mathcal{M}$ robot was able to be sufficiently fit in the whole space while only being evolved in the environments in $O_{3,4}$ (Fig. 5a). The $\mathcal{NM}$ robot was not able to achieve sufficient fitness in the rest of E_3 after achieving sufficient fitness in $O_{3,4}$ (Fig. 5b). We also evolved the $\mathcal{M}$ robot in E_3 using a different training subset. For this experiment, $O_{3,2} = \{e_0, e_7\}$ which is the grand diagonal of E_3. Results presented in Fig. 6 show the $\mathcal{M}$ was not able to be sufficiently fit.

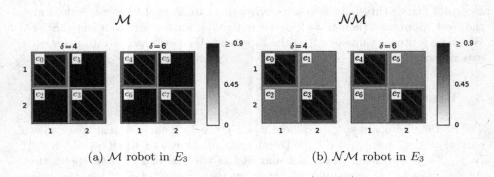

(a) $\mathcal{M}$ robot in E_3 (b) $\mathcal{NM}$ robot in E_3

Fig. 5. Average fitness scores for $\mathcal{M}$ (a) and $\mathcal{NM}$ (b) robots in E_3 with training set $O_{3,4} = \{e_0, e_3, e_4, e_7\}$. $O_{3,4}$ is represented by the blue outlines around the environments. (Color figure online)

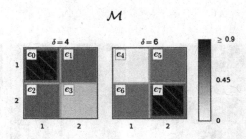

Fig. 6. The $\mathcal{M}$ robot evolved with training set $O_{3,2} = \{e_0, e_7\}$. We see that the robot does not achieve adequate fitness when only evolved in $O_{3,2}$.

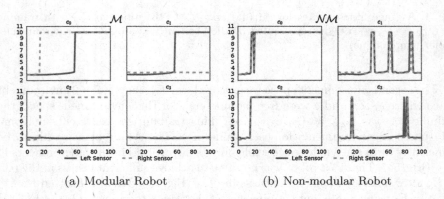

(a) Modular Robot   (b) Non-modular Robot

Fig. 7. Sensor values of the left and right distance sensors for randomly chosen $\mathcal{M}$ and $\mathcal{NM}$ robots evolved in training set $O_{2,2} = \{e_0, e_3\}$. We see that the modular robot can move one leaf node without affecting the sensor value of the other arm. In contrast the non-modular robot cannot.

4   Discussion and Conclusion

When a robot with ecological modularity is presented with a new environment, it is able to break down that environment along M, the free parameters it is modular with respect to. The robot then can recognize this new environment as a combination of percepts it has seen before and act accordingly. This means the robot only needs to be evolved in a subset of the whole environment space. Specifically, the minimal size of this subset is n^{f-m+1}.

As the ecological modularity in the robot increases, it is able to break down more free parameters in the environment space. This is shown by the fact the M robot can achieve fitness at or above 0.9 in environments it has not seen before while the NM robot cannot as seen in Figs. 4 and 5.

We claim that in both the E_2 and E_3 environment spaces the M robot has $m = 2$ while the NM robot has $m = 1$. This is because the robot can break down its environment because it is able to move its sensors independently (Fig. 7a). The NM robot cannot break down its environment into left and right percepts because it is morphologically and neurologically non-modular (Fig. 7b). When it senses a new percept on the right it fundamentally changes how it views its environment even if the percept on the left remains constant. Therefore, in the E_2 environment space $m = 2$ for M and $m = 1$ for NM.

Neither the M nor NM robots are modular with respect to δ. This is shown by the fact that they cannot be simply trained along the diagonal of E_3 to be sufficiently fit in the whole space (Figs. 5 and 6). Therefore, in the E_3 environment space $m = 2$ for M and $m = 1$ for NM. We hypothesize that if a robot was able to categorize the clusters independently of distance then, in E_3, the robot would have $m = 3$ and only $n^{3-3+1} = n$ environments would be necessary.

In this paper we introduced the concept of ecological modularity and showed that robots which are ecological modular can be sufficiently fit in an entire environment space even though they are only evolved in a subset of its environments. Robots that are not morphologically modular cannot move without changing their entire perception of their environment and thus cannot break down their environment into familiar percepts. Similarly, ecologically non-modular robots cannot view the varying environments in terms of unfamiliar combination of familiar percepts because they cannot sense their world in a manner which breaks down the environment into individual percepts.

In future work we would like to investigate whether ecological modularity can be discovered by evolution instead of from human construction of these robots.

Acknowledgments. We like to acknowledge financial support from NSF awards PECASE-0953837 and INSPIRE-1344227 as well as the Army Research Office contract W911NF-16-1-0304. We also acknowledge computation provided by the Vermont Advanced Computing Core.

References

1. Bongard, J., Bernatskiy, A., Livingston, K., Livingston, N., Long, J., Smith, M.: Evolving robot morphology facilitates the evolution of neural modularity and evolvability. In: Proceedings of the 2015 Genetic and Evolutionary Computation Conference, p. 129136. ACM, Madrid (2015)
2. Bongard, J.: Morphological change in machines accelerates the evolution of robust behavior. Proc. Nat. Acad. Sci. **108**(4), 1234–1239 (2011)
3. Bongard, J.C.: Spontaneous evolution of structural modularity in robot neural network controllers. In: Proceedings of the 2011 Genetic and Evolutionary Computation Conference, pp. 251–258. ACM, Dublin (2011)
4. Cappelle, C.K., Bernatskiy, A., Livingston, K., Livingston, N., Bongard, J.: Morphological modularity can enable the evolution of robot behavior to scale linearly with the number of environmental features. Front. Rob. AI **3**, 59 (2016)
5. Clune, J., Mouret, J.B., Lipson, H.: The evolutionary origins of modularity. Proc. R. Soc. B Biol. Sci. **280**(1755), 20122863 (2013)
6. Ellefsen, K.O., Mouret, J.B., Clune, J.: Neural modularity helps organisms evolve to learn new skills without forgetting old skills. PLoS Comput. Biol. **11**(4), e1004128 (2015)
7. Espinosa-Soto, C., Wagner, A.: Specialization can drive the evolution of modularity. PLoS Comput. Biol. **6**(3), e1000719 (2010)
8. French, R.M.: Catastrophic forgetting in connectionist networks. Trends Cogn. Sci. **3**(4), 128–135 (1999)
9. Gruau, F.: Automatic definition of modular neural networks. Adapt. Behav. **3**, 151–183 (1994)
10. Jakobi, N., Husbands, P., Harvey, I.: Noise and the reality gap: the use of simulation in evolutionary robotics. In: Morán, F., Moreno, A., Merelo, J.J., Chacón, P. (eds.) ECAL 1995. LNCS, vol. 929, pp. 704–720. Springer, Heidelberg (1995). doi:10.1007/3-540-59496-5_337
11. Kashtan, N., Alon, U.: Spontaneous evolution of modularity and network motifs. PNAS **102**(39), 13773 (2005)
12. Lehman, J., Risi, S., D'Ambrosio, D., Stanley, K.O.: Encouraging reactivity to create robust machines. Adapt. Behav. **21**, 484–500 (2013)
13. Lipson, H., Pollack, J.B., Suh, N.P., Wainwright, P.: On the origin of modular variation. Evolution **56**(8), 1549–1556 (2002)
14. Matarić, M., Cliff, D.: Challenges in evolving controllers for physical robots. Rob. Auton. Syst. **19**(1), 67–83 (1996)
15. Newman, M.E., Girvan, M.: Finding and evaluating community structure in networks. Phys. Rev. E **69**(2), 026113 (2004)
16. Schmidt, M., Lipson, H.: Age-fitness pareto optimization. In: Riolo, R., McConaghy, T., Vladislavleva, E. (eds.) Genetic Programming Theory and Practice VIII. Genetic and Evolutionary Computation, vol. 8, pp. 129–146. Springer, New York (2011). doi:10.1007/978-1-4419-7747-2_8
17. Wagner, G., Pavlicev, M., Cheverud, J.: The road to modularity. Nat. Rev. Genetics **8**(12), 921–931 (2007)
18. Yamashita, Y., Tani, J.: Emergence of functional hierarchy in a multiple timescale neural network model: a humanoid robot experiment. PLoS Comput. Biol. **4**(11), e1000220 (2008)

Automated Calibration of a Biomimetic Space-Dependent Model for Zebrafish and Robot Collective Behaviour in a Structured Environment

Leo Cazenille[1,2(✉)], Yohann Chemtob[1], Frank Bonnet[3], Alexey Gribovskiy[3], Francesco Mondada[3], Nicolas Bredeche[2], and José Halloy[1]

[1] Univ Paris Diderot, Sorbonne Paris Cité, LIED, UMR 8236, 75013 Paris, France
leo.cazenille@univ-paris-diderot.fr
[2] Sorbonne Universités, UPMC Univ Paris 06, CNRS, ISIR, 75005 Paris, France
[3] Robotic Systems Laboratory, School of Engineering,
Ecole Polytechnique Fédérale de Lausanne,
ME B3 30, Station 9, 1015 Lausanne, Switzerland

Abstract. Bio-hybrid systems made of robots and animals can be useful tools both for biology and robotics. To socially integrate robots into animal groups the robots should behave in a biomimetic manner with close loop interactions between robots and animals. Behavioural zebrafish experiments show that their individual behaviours depend on social interactions producing collective behaviour and depend on their position in the environment. Based on those observations we build a multilevel model to describe the zebrafish collective behaviours in a structured environment. Here, we present this new model segmented in spatial zones that each corresponds to different behavioural patterns. We automatically fit the model parameters for each zone to experimental data using a multi-objective evolutionary algorithm. We then evaluate how the resulting calibrated model compares to the experimental data. The model is used to drive the behaviour of a robot that has to integrate socially in a group of zebrafish. We show experimentally that a biomimetic multilevel and context-dependent model allows good social integration of fish and robots in a structured environment.

Keywords: Collective behaviour · Model fitting · Evolutionary algorithms · Decision-making · Multilevel model · Zebrafish

1 Introduction

Robotics stands now as a convenient tool to study the animal behaviour. In recent ethological and animal behavioural studies, robots are used to induce specific and controlled stimuli and assess the response of the animals under scrutiny. This allows to test various hypothesises on the nature of the signals used by the animals for social interactions [16,21].

© Springer International Publishing AG 2017
M. Mangan et al. (Eds.): Living Machines 2017, LNAI 10384, pp. 107–118, 2017.
DOI: 10.1007/978-3-319-63537-8_10

Autonomous robots interacting in real-time with animals [19] makes it possible to create social interactions between both of them. This has already been demonstrated by several authors for studying the behaviours of sheepdogs [26], cows [12] or drosophila [27] to cite a few. In this paper, we focus on zebrafish (*Danio rerio*), and we describe a biomimetic model that can be implemented in a robotic lure and validated its acceptance by four zebrafish in a structured environment.

The main difficulty is to make the robotic lure behave in such a way that it is accepted by the animals as social companion, just as any other interacting fish would be. Beyond the scope of this paper, this is a first step to enable the modulation (though action) of the collective behaviours of the observed zebrafish [15].

Different approaches have been proposed to control the movement of fish-lures [9]. Most of them do not involve a closed loop of social interaction with the fish. This is often the case for lures fixed to a robotic arm that performs repeated movements, but also for studies with autonomous fish-lures. Closing the loop of social interactions requires a real-time tracking, or perception, of the agents (fish and robot), and a decision-making algorithm to control the robot behaviours. In most of the experiments reported in the literature, the robots driven with closed-loop control are programmed to follow the centroid of the fish group, to ensures that the robot will join and follow the group of fish. However, this type of controller implies that the robot is more a passive follower than a real group-member making its own decisions. The embodiment of bio-inspired models can lead to a better social integration of the artificial agents in animal groups and can allow the robots to influence the collective decision of the mixed group by giving specific preferences to the robot by tuning parameter values of the model [9,15].

We present a method to calibrate automatically a new behavioural zebrafish model by evolutionary parameters optimisation. This multilevel model describes collective behaviour in a structured environment in agreement with experimental observations. This model makes important extensions to our previous model for collective behaviour in a homogeneous environment [10]. The model takes into account a simple structured environment composed of two rooms and the fact that the fish adapt their behaviour to the zones where they are while performing collective behaviour. For such multilevel and spatially dependent social behaviour model it is an issue to calibrate the model because it involves trade-offs between social tendencies (aggregation, group formation), and response to the environment (wall-following, zone occupation). We use an evolutionary algorithm (NSGA-II [13]) to optimise the parameters of this model so that the exhibited collective dynamics correspond to those observed in biological experiments. Then, we validate experimentally this model by implementing it as the controller of robots that are integrated in small fish groups.

2   Materials and Methods

2.1   Experimental Set-Up

We use the experimental set-up described in [3,9,11,24], with the arena presented in [9,24]. This set-up (Fig. 1A) consists of a white plexiglass arena (Fig. 1C) of

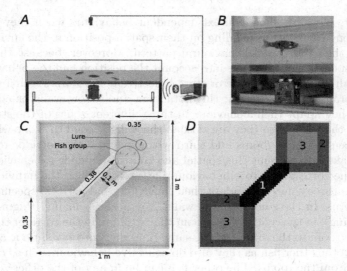

Fig. 1. Panel A: Experimental set-up used during the experiments [2–5,9]. **Panel B**: FishBot [2,4,5]: the robot used for mimicking fish motion patterns, with the biomimetic lure used during the reference experiments. This robot was developed by the EPFL for the ASSISI project [23]. **Panel C**: Experimental arena composed of a tank containing two square rooms (350×350 mm at floor level) connected by a corridor (380×100 mm at floor level). The fish tend to swim from one room to the other, either in small groups, or individually. This set-up is used to study the zebrafish collective dynamics. **Panel D**: Positions of the three different zones corresponding to different types of behaviours: in the corridor (zone 1), in the center of each room (zone 2), and near of the walls of each room (zone 3).

$1000 \times 1000 \times 100$ mm, that is composed of two rooms linked by a corridor. To validate experimentally our calibrated model, we use a robot developed by the EPFL [2–5] for the ASSISI project [23]. This robot is powered by two conductive plates under the aquarium. An overhead camera captures frames that are then processed for tracking and control purposes (see Fig. 1A).

All trials have a duration of 15 min. We tracked the positions of the agents by using the idTracker software [22]. Using this software, we obtain the positions $P(x, y, t)$ of all agents at each time step $\Delta t = 1/15$ s for all experiments, and build the trajectories of each agent. The experiments performed in this study were conducted under the authorisation of the Buffon Ethical Committee (registered to the French National Ethical Committee for Animal Experiments #40) after submission to the French state ethical board for animal experiments.

2.2   Behavioural Model

Most of the fish collective behaviour models do not take into account the environment i.e. the walls or the structure of the tanks because they only focus on the social interactions [18,25].

However, zebrafish show context-dependent behaviours when they are in a structured environment. Depending on their spatial position in the environment they adapt their individual behavioural pattern. Moreover, because they are a gregarious species they also take into account the position and the behaviours of the other fish and can aggregate or start collective behaviours. As many animal species, zebrafish display strong thigmotactism and follow walls or edges. We show that they adapt their behaviour in three different zones of the structured set-up: first the zone when they are close to the walls, second the zone when they are in the centre of the rooms and third when they use the corridor to change room. We take into account this spatial and context-dependent behaviours.

Each zone corresponds to a behavioural attractor. When the individuals are in one of the three zones they adapt their behaviour and perform specific behavioural patterns. In the zone near the walls they perform mainly thigmotactism (wall following), in the centre of the room they explore, in the corridor they transit from one room to the other. At the same time they also take into account the behaviour of the other fish as they also do collective behaviour such as collective departures from the rooms. The other fish can be in any of the other zones and thus can also induce behavioural attractor switching of their companions.

We extend the biomimetic hybrid model [9,10] using microscopic and macroscopic information [7,8]. This new model (described in Fig. 2) takes into account zones that correspond to different behavioural attractors and thus allows context-dependent behaviours. The individual can switch from one behavioural attractor to the other and at the same time perform collective behaviour. Our model describes individual choices close to action selection and collective behaviours at the same time. It is a step towards modelling action selection in the context of collective behaviours.

We present a multi-level and multi-agent biomimetic model, inspired from [9, 10] that describes the individual and collective behaviours of fish. As in [10], this model makes the link between fish visual perception (of congeners and walls) and motor response (*i.e.*: trajectories of the agents). However, it is also capable of expressing a variability in agents behaviours when they occupy specific zones of the arena (behavioural attractors). Figure 3B lists the model parameters.

In this model, the agents update their position vector X_i with a velocity vector V_i:

$$X_i(t + \delta t) = X_i(t) + V_i(t)\delta t \tag{1}$$

$$V_i(t + \delta t) = v_i(t + \delta t)\Theta_i(t + \delta t) \tag{2}$$

The model computes a circular probability distribution function (PDF) [10] corresponding to the probability of the agent to move in a specific direction (Θ_i). This PDF is as a mixture of von Mises distributions, an equivalent to the Gaussian distribution in circular probability. The computation of this PDF involves the calculation of two other PDF functions: the first one describing agent behaviour when no stimuli is present, and the second one characterising agent behaviour when conspecifics are perceived by the agent.

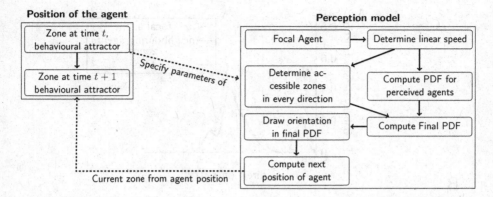

Fig. 2. Multilevel model used to describe fish behaviour. The agents display different behavioural attractors depending on the zone where they are situated. Thus, according to the agent spatial position, the physical features of the zone drive them towards a specific behavioural attractor. A behavioural attractor corresponds to a set of behavioural patterns adapted to the zone where they are located. It can correspond to different parameters sets for the same behaviour kind.

The PDF capturing agent behaviour when no stimuli is present is given by:

$$f_{0,z_j}(\theta) = \frac{\exp(\kappa_{0,z_j}\cos(\theta))}{2\pi I_0(\kappa_{0,z_j})} \tag{3}$$

for an agent situated in zone z_j, and with I_0 the modified Bessel function of first kind of order zero. When the agent is situated in a zone close to a wall (zones 1 and 2 of Fig. 1D), we implement a wall-following behaviour, by increasing the probabilities of moving towards either side of the closest wall. This is achieved by using the following PDF:

$$f_{0,z_j,w}(\theta) = \frac{1}{2}\sum_{k=1}^{2}\frac{\exp(\kappa_{0,z_j}\cos(\theta - \mu_{w_k}))}{2\pi I_0(\kappa_{0,z_j})} \tag{4}$$

with μ_{w_k} the two possible directions along the considered wall.

Examples of agents trajectories are found in Fig. 5B. The probability of the focal fish to orient towards a perceived fish is given by a von Mises distribution clustered around the fish position:

$$f_{F,z_j}(\theta) = \sum_{i=1}^{n}\frac{A_{f_i}}{A_{T_f}}\frac{\exp(\kappa_{f,z_j}\cos(\theta - \mu_{f_i}))}{2\pi I_0(\kappa_{f,z_j})} \tag{5}$$

with μ_{f_i} the direction towards the perceived agent, $A_{f_T} = \sum_{i=1}^{n_f} A_{f_i}$ the sum of the solid angles A_{f_i} captured by each agent and n_f the number of perceived agents.

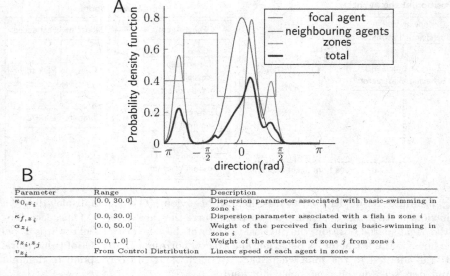

B

Parameter	Range	Description
κ_{0,z_i}	[0.0, 30.0]	Dispersion parameter associated with basic-swimming in zone i
κ_{f,z_i}	[0.0, 30.0]	Dispersion parameter associated with a fish in zone i
α_{z_i}	[0.0, 50.0]	Weight of the perceived fish during basic-swimming in zone i
γ_{z_i,z_j}	[0.0, 1.0]	Weight of the attraction of zone j from zone i
v_{z_i}	From Control Distribution	Linear speed of each agent in zone i

Fig. 3. Panel A Computation of the PDFs functions used by the model. One function corresponds to the focal fish; another corresponds to the perceived neighbouring agents. The final PDF is a weighted sum of these functions, with a normalisation factor γ_{z_1,z_2} corresponding to the affinity between the zones z_1 (origin) and z_2 (destination). The direction taken by an agent is drawn randomly from the resulting PDF by inverse transform sampling. **Panel B** Table of model parameters for each agent. The zone z_i corresponds to the zone where the agent is situated at time t, and z_j to the zone where the agent would be at time $t + 1$. The linear speed distributions of the agents are the same as the ones observed in the Control experiments, and they are not optimised. The other parameters in the table are optimised.

The final PDF $f(\theta)$ is computed as follow:

$$f_{z_j,z_k}(\theta) = \gamma_{z_j,z_k} \frac{f_{0,z_j}(\theta) + \alpha_{z_j} A_{T_f} f_{F,z_j}(\theta)}{1 + \alpha_{z_j} A_{T_f}} \tag{6}$$

The parameter γ_{z_1,z_2}, used as a multiplicative term of the final PDF, modulates the attraction of agents towards target zones. Figure 3A describes how the final PDF is computed and how it is used to determine the agents next positions.

Unreachable areas of the PDF (e.g. the walls) are attributed a probability of 0. Then, we numerically compute the cumulative distribution function (CDF) corresponding to this custom PDF $f(\theta)$ by performing a cumulative trapezoidal numerical integration of the PDF in the interval $[-\pi, \pi]$. Finally, the model draws a random direction Θ_i in this distribution by inverse transform sampling. The position of the fish is then updated according to this direction and his velocity with Eqs. 1 and 2.

3   Results

We consider four cases. We define the **Control** results as obtained from biological experiments with five zebrafish in the experimental set-up described in Sect. 2.1. The **Sim-MonoObj** and **Sim-MultiObj** results are defined to correspond to the model in simulation with five agents, calibrated respectively using mono-objective or multi-objective optimisation. The **Biohybrid** results are obtained from experiments with four zebrafish and one robot driven by the model using the best optimised parameters.

3.1   Optimisation of Model Parameters

We define a similarity measure (ranging from 0.0 to 1.0) to compare two experiments (e_1 and e_2), and define it as:

$$S(e_1, e_2) = \sqrt[3]{I(O_{e_1}, O_{e_2})I(T_{e_1}, T_{e_2})I(D_{e_1}, D_{e_2})} \tag{7}$$

with O_e the distribution of zones occupation, T_e the transition probabilities from zone e to the others, and D_e the distribution of inter-individual distances of all agents in zone e. The similarity measure $S(e_1, e_2)$ corresponds to the geometric mean of these three features. The function $I(P, Q)$ is defined as such:

$$I(P, Q) = 1 - H(P, Q) \tag{8}$$

The $H(P, Q)$ function is the Hellinger distance between two histograms [14]. It is defined as:

$$H(P, Q) = \frac{1}{\sqrt{2}}\sqrt{\sum_{i=1}^{d}(\sqrt{p_i} - \sqrt{q_i})^2} \tag{9}$$

We consider two optimization methods. In the **Sim-MonoObj** case, we use the CMA-ES [1] mono-objective optimisation algorithm, with the task of maximising the $S_{(e_1, e_2)}$ function. In the **Sim-MultiObj** case, we use the NSGA-II [13] multi-objective algorithm with three objectives to maximise. The first objective is a performance objective corresponding to the $S_{(e_1, e_2)}$ function. We also consider two other objectives used to guide the evolutionary process: one that promotes genotypic diversity [20] (defined by the mean euclidean distance of the genome of an individual to the genomes of the other individuals of the current population), the other encouraging behavioural diversity (defined by the euclidean distance between the O_e, T_e and D_e scores of an individual). In both methods, we use populations of 60 individuals (approximately twice the number of dimensions of the problem) and 300 generations. The **Sim-MonoObj** stabilises around the 50-th generation. The **Sim-MultiObj** stabilises around the 250-th generation. The linear speed v_i of the agents is not optimized, and is randomly drawn from the instantaneous speed distribution measured in the control experiment. It should be noted that evolutionary algorithms do not over-fit (as it is an optimization process), even if we use the same data (trajectories) for both training and testing.

3.2   Robot Implementation

The robot is driven by the model described in Sect. 2.2, after calibration. Robotic trials have a duration of 15 min, and are repeated 10 times. They involve one robot and four zebrafish. Every 333 ms, we integrate the tracked positions of the four fish into the model, and compute the target position of a fifth agent. We then control the robot to follow this target position by using the biomimetic movement patterns described in [4,9].

3.3   Model Performance Analysis and Experimental Validation

We assess the similarity between the results from the calibrated cases (**Sim-MonoObj**, **Sim-MultiObj** and **Biohybrid**) and those of the **Control** case by using the similarity measure defined in Sect. 3.1. The similarity scores are shown in Table 1.

Using information about zones occupation and probabilities of transition from one zone to another, we define a finite state machine corresponding to the behavioural attractors dynamics of the entire agent population. The resulting finite state machines obtained from the **Control** and **Biohybrid** cases are shown in Fig. 4. The probability of presence of an agent in each part of the arena is presented in Fig. 5A. Examples of agents trajectories are found in Fig. 5B.

The best-performing individuals of the **Sim-MonoObj** and **Sim-MultiObj** cases display distributions of inter-individual distances that are relatively close to those of the **Control** case, which suggests that these models can convincingly exhibit fish tendency to aggregate. However, of the two cases performed in simulation, only **Sim-MultiObj** is capable of displaying zones dynamics (occupation

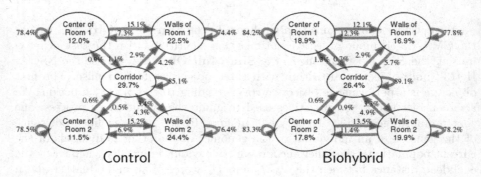

Fig. 4. Ethogram as finite state machine corresponding to the behavioural attractors for all agents. Each zones drive the agents into the corresponding behavioural attractor. Thus, agents modulate their behaviour in each zone as if they enter into a specific behavioural state. Here we show the resulting transition probabilities obtained after optimisation and implementation as robotic controllers (biohybrid) based on the experimental observations (control). The number in each state corresponds to the proportion of time agent spend in this state. The numbers on the arrows correspond to the transition probabilities between zones with a time-step of 1/3 s.

Table 1. Similarity scores between the best-performing individuals of the three calibrated cases and the **Control** case used as reference, as defined in Sect. 3.1. We consider three standard features to characterise the collective behaviour exhibited in each case. **Occupation** corresponds to the probability of presence of the agent in each zone. **Transitions** corresponds to the probabilities of an agent to transition from one zone to another. **Inter-individual distances** corresponds to the distribution of inter-individual distances between all agents in a specific zone. The fitness function is computed as the geometric mean of these scores.

	Sim-MonoObj	Sim-MultiObj	Biohybrid
Occupation	0.57	0.97	0.89
Transitions	0.76	0.81	0.88
Interindiv. dists	0.90	0.87	0.89
Fitness	0.73	0.88	0.89

of the zones, and transition probabilities from one zone to the others) similar to the **Control** case. This suggests that multi-objective optimisation is required to handle the conflicting dynamics present in fish collective behaviour.

The robot of the **Biohybrid** case is driven by a controller using our model with the parameters of the best-performing individual obtained in the **Sim-MultiObj**. The results of the **Biohybrid** case correspond to those of the **Sim-MultiObj** case. The ethogram of the **Biohybrid** case (cf Fig. 4) shows an increased preference for the centre of the rooms compared to the **Control** case. This could be explained by our current lower level robotic implementation of wall-following behaviour that could still be sub-optimal.

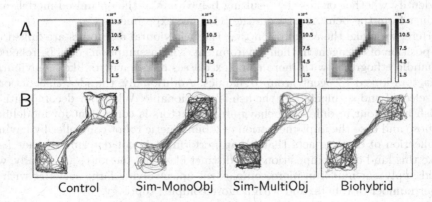

Fig. 5. Panel A Probabilities of presence in each part of the arena, for all cases. **Panel B** Examples of trajectories over a duration of 2 min (1800 frames). In the Biohybrid case, the robot is in black.

4   Discussion and Conclusion

Collective behaviour models often focus on collective motion in homogeneous unbounded environment. Here we present a multi-level model that is space-dependent with individuals that behave in a context-dependent way. We make the hypothesis that the type of behaviour displayed by the agents depends on their position in the environment. This allows us to segment our environment into several characteristic zones, each corresponding to a particular behavioural attractor, matching different types of agent behaviour.

We present a methodology to calibrate this model to correspond to the collective dynamics exhibited by fish in the experiments. This calibration process is challenging, as it involves a trade-off between social tendencies (group formation), and response to the environment (wall-following, exploration). Moreover, our model encompasses the notion of behavioural attractors, allowing agents to exhibit several different behaviours depending on the context. Our methodology is able to cope with this trade-off by using multi-objective optimisation.

However, this calibration methodology could still be improved: the similarity measure we use to compare two cases only takes into account three aspects of collective behaviours corresponding to behavioural attractors, and aggregation dynamics. Other behavioural aspects could also be relevant at the level of collective dynamics and can be considered: *e.g.:* agent groups aspects, residence time in a zone, at the level of the individuals *e.g.:* agent trajectory aspects, curvature of trajectories, etc. Moreover, in relation to the environment *e.g.:* the distance of an agent to the nearest wall could also be taken into account. Alternatively, it would be possible to perform the calibration without defining a similarity measure explicitly, using a method similar to [17], by co-evolving simultaneously the parameters of the models and classifiers. These classifiers would be trained to identify whether or not the resulting behaviours of the optimised models are distinct from the behaviours from the reference experiments.

Here, we make the assumption that the behavioural attractors are linked to the position of the agent in their environment. This assumption could be relaxed, to handle ethograms with more complex classes of behaviours like behavioural attractors linked to agent group dynamics. Additionally, the idea that actions are selected and segmented by the fish is questionable. While our decomposition of fish behaviour in different behavioural attractors is convenient for modelling purpose and ease the implementation of a biomimetic robot controller by having a collection of discrete acts that it can perform, it is not determined that fish make this kind of decomposition into distinct elements (actions) [6]. Finally, we could apply our model in more complex set-up, involving large societies with a larger number of robots, and with a more complex topology.

Acknowledgement. This work was funded by EU-ICT project 'ASSISIbf', no. 601074.

References

1. Auger, A., Hansen, N.: A restart CMA evolution strategy with increasing population size. In: The 2005 IEEE Congress on Evolutionary Computation, vol. 2, pp. 1769–1776. IEEE (2005)
2. Bonnet, F., Binder, S., de Oliveria, M., Halloy, J., Mondada, F.: A miniature mobile robot developed to be socially integrated with species of small fish. In: 2014 IEEE International Conference on Robotics and Biomimetics (ROBIO), pp. 747–752. IEEE (2014)
3. Bonnet, F., Cazenille, L., Gribovskiy, A., Halloy, J., Mondada, F.: Multi-robots control and tracking framework for bio-hybrid systems with closed-loop interaction. In: 2017 IEEE International Conference on Robotics and Automation (ICRA). IEEE (Forthcoming)
4. Bonnet, F., Cazenille, L., Seguret, A., Gribovskiy, A., Collignon, B., Halloy, J., Mondada, F.: Design of a modular robotic system that mimics small fish locomotion and body movements for ethological studies. Int. J. Adv. Rob. Syst. **14**(3), 1729881417706628 (2017)
5. Bonnet, F., Rétornaz, P., Halloy, J., Gribovskiy, A., Mondada, F.: Development of a mobile robot to study the collective behavior of zebrafish. In: 2012 4th IEEE RAS & EMBS International Conference on Biomedical Robotics and Biomechatronics (BioRob), pp. 437–442. IEEE (2012)
6. Botvinick, M.: Multilevel structure in behaviour and in the brain: a model of fuster's hierarchy. Philos. Trans. R. Soc. Lond. B Biol. Sci. **362**(1485), 1615–1626 (2007)
7. Cazenille, L., Bredeche, N., Halloy, J.: Multi-objective optimization of multi-level models for controlling animal collective behavior with robots. In: Wilson, S.P., Verschure, P.F.M.J., Mura, A., Prescott, T.J. (eds.) Living Machines 2015. LNCS, vol. 9222, pp. 379–390. Springer, Cham (2015). doi:10.1007/978-3-319-22979-9_38
8. Cazenille, L., Bredeche, N., Halloy, J.: Automated optimisation of multi-level models of collective behaviour in a mixed society of animals and robots. arXiv preprint arXiv:1602.05830 (2016)
9. Cazenille, L., Collignon, B., Bonnet, F., Gribovskiy, A., Mondada, F., Bredeche, N., Halloy, J.: How mimetic should a robotic fish be to socially integrate into zebrafish groups? Bioinspiration & Biomimetics (Forthcoming)
10. Collignon, B., Séguret, A., Halloy, J.: A stochastic vision-based model inspired by zebrafish collective behaviour in heterogeneous environments. R. Soc. Open Sci. **3**(1), 150473 (2016)
11. Collignon, B., Séguret, A., Chemtob, Y., Cazenille, L., Halloy, J.: Collective departures in zebrafish: profiling the initiators. arXiv preprint arXiv:1701.03611 (2017)
12. Correll, N., Schwager, M., Rus, D.: Social control of herd animals by integration of artificially controlled congeners. In: Asada, M., Hallam, J.C.T., Meyer, J.-A., Tani, J. (eds.) SAB 2008. LNCS, vol. 5040, pp. 437–446. Springer, Heidelberg (2008). doi:10.1007/978-3-540-69134-1_43
13. Deb, K., Pratap, A., Agarwal, S., Meyarivan, T.: A fast and elitist multiobjective genetic algorithm: NSGA-II. IEEE Trans. Evol. Comput. **6**(2), 182–197 (2002)
14. Deza, M., Deza, E.: Dictionary of Distances. Elsevier, Amsterdam (2006)
15. Halloy, J., Sempo, G., Caprari, G., Rivault, C., Asadpour, M., Tâche, F., Said, I., Durier, V., Canonge, S., Amé, J.: Social integration of robots into groups of cockroaches to control self-organized choices. Science **318**(5853), 1155–1158 (2007)

16. Knight, J.: Animal behaviour: when robots go wild. Nature **434**(7036), 954–955 (2005)
17. Li, W., Gauci, M., Gross, R.: Turing learning: a metric-free approach to inferring behavior and its application to swarms. arXiv preprint arXiv:1603.04904 (2016)
18. Lopez, U., Gautrais, J., Couzin, I.D., Theraulaz, G.: From behavioural analyses to models of collective motion in fish schools. Interface Focus **2**(6), 693–707 (2012)
19. Mondada, F., Halloy, J., Martinoli, A., Correll, N., Gribovskiy, A., Sempo, G., Siegwart, R., Deneubourg, J.: A general methodology for the control of mixed natural-artificial societies, Chap. 15. In: Kernbach, S. (ed.) Handbook of Collective Robotics: Fundamentals and Challenges, pp. 547–585. Pan Stanford (2013)
20. Mouret, J., Doncieux, S.: Encouraging behavioral diversity in evolutionary robotics: an empirical study. Evol. Comput. **20**(1), 91–133 (2012)
21. Patricelli, G.L.: Robotics in the study of animal behavior. In: Breed, M.D., Moore, J. (eds.) Encyclopedia of Animal Behavior, pp. 91–99. Greenwood Press Westport, CT (2010)
22. Pérez-Escudero, A., Vicente-Page, J., Hinz, R.C., Arganda, S., de Polavieja, G.G.: idTracker: tracking individuals in a group by automatic identification of unmarked animals. Nat. Methods **11**(7), 743–748 (2014)
23. Schmickl, T., Bogdan, S., Correia, L., Kernbach, S., Mondada, F., Bodi, M., Gribovskiy, A., Hahshold, S., Miklic, D., Szopek, M., Thenius, R., Halloy, J.: ASSISI: mixing animals with robots in a hybrid society. In: Lepora, N.F., Mura, A., Krapp, H.G., Verschure, P.F.M.J., Prescott, T.J. (eds.) Living Machines 2013. LNCS, vol. 8064, pp. 441–443. Springer, Heidelberg (2013). doi:10.1007/978-3-642-39802-5_60
24. Séguret, A., Collignon, B., Cazenille, L., Chemtob, Y., Halloy, J.: Loose social organisation of abstrain zebrafish groups in a two-patch environment. arXiv preprint arXiv:1701.02572 (2017)
25. Sumpter, D.J., Mann, R.P., Perna, A.: The modelling cycle for collective animal behaviour. Interface Focus **2**(6), 764–773 (2012)
26. Vaughan, R., Sumpter, N., Henderson, J., Frost, A., Cameron, S.: Experiments in automatic flock control. Rob. Auton. Syst. **31**(1), 109–117 (2000)
27. Zabala, F., Polidoro, P., Robie, A., Branson, K., Perona, P., Dickinson, M.: A simple strategy for detecting moving objects during locomotion revealed by animal-robot interactions. Curr. Biol. **22**(14), 1344–1350 (2012)

Spiking Cooperative Stereo-Matching at 2 ms Latency with Neuromorphic Hardware

Georgi Dikov[1,2], Mohsen Firouzi[1,3], Florian Röhrbein[2,3], Jörg Conradt[1,3], and Christoph Richter[1,3(✉)]

[1] Neuroscientific System Theory,
Department of Electical and Computer Engineering,
Technical University of Munich, 80333 Munich, Germany
`c.richter@tum.de`
[2] Robotics and Embedded Systems, Department of Informatics,
Technical University of Munich, 85748 Garching, Germany
[3] Bernstein Center for Computational Neuroscience Munich, Munich, Germany

Abstract. We demonstrate a spiking neural network that extracts spatial depth information from a stereoscopic visual input stream. The system makes use of a scalable neuromorphic computing platform, SpiNNaker, and neuromorphic vision sensors, so called silicon retinas, to solve the stereo matching (correspondence) problem in real-time. It dynamically fuses two retinal event streams into a depth-resolved event stream with a fixed latency of 2 ms, even at input rates as high as several 100,000 events per second. The network design is simple and portable so it can run on many types of neuromorphic computing platforms including FPGAs and dedicated silicon.

Keywords: Correspondence problem · Dynamic vision sensor (DVS) · Event-based vision · Event-based computation · Neuromorphic computing · PyNN · Spiking neural networks · Stereopsis

1 Introduction

Many algorithms, especially in computer vision, cite a biological inspiration. The level of detail and abstraction with regard to their actual biological counterpart—often the retina and visual cortex—varies widely. The prospects of the approach, though, are well understood: Natural vision systems are, apart from very few exceptions, still unmatched in terms of performance and efficiency. Taking inspiration from biology conventionally means to first learn how nature solves a particular task, and then emulate the working principle with machines. While in an ideal world we could just copy the biological system exactly, in practice there is a gaping mismatch between biological and technical vision systems, e.g. between the asynchronous operation of neurons in the retina and visual cortex, and the synchronous operation of frame-based cameras and sequential computers. Consequently the need to accommodate a particular algorithm for the latter devices,

© Springer International Publishing AG 2017
M. Mangan et al. (Eds.): Living Machines 2017, LNAI 10384, pp. 119–137, 2017.
DOI: 10.1007/978-3-319-63537-8_11

which represent the canonical computer vision tool set, forces us to abstract from biology substantially.

Recent advances in neuromorphic technology allow us to take the idea of biologically inspired vision and computation to a new and unprecedented level of fidelity. They enable us to set up and emulate large networks of spiking neurons *in-silico* efficiently [45] and feed those networks with input from *silicon retinas* [5], which are event-based vision sensors also known as dynamic vision sensors [22] (DVS). Neuromorphic computing brings with it a new programming paradigm: Unlike traditional programming where a set of sequential instructions constitute an algorithm, a spiking neural network (SNN) is defined by its network topology along with the properties of its neurons and synapses [45].

A clear merit of the spiking-neural approach towards solving computer vision problems lies in the way in which information is processed and represented. Since the computational units are neurons which interchange the same type of signals (action potentials, *spikes*), SNNs could easily be nested, stacked or combined in any other form, so that an output of one network would serve as an input to another. This property brings with itself an abstraction capability and therefore has huge potential in information processing. SNNs can also be emulated efficiently on specialized hardware ranging from high-level programmable field-programmable gate arrays [2] (FPGA), over customized many-core systems based on standard microprocessors [12], to dedicated digital [27], analog [38], or mixed signal [1,32] very-large-scale-integrated circuits. On such hardware architectures spiking networks process data in a massively parallel and globally asynchronous, event-based manner. In many cases this implies deterministic processing times and low latency, which is highly beneficial for many engineering problems, like real-time control tasks.

In this work we demonstrate a spiking neural network able to extract visual depth information from two retinal input streams at a fixed latency of 2 ms. Our algorithm can achieve this low latency, because unlike frame-based or early event-based stereo matching algorithms [19,36,39] it works fully asynchronous without any event-buffering or intermediate frame construction. It is based on biologically plausible principles, originally devised by David Marr and Tomaso Poggio [26]. While those principles have previously been applied in hardware [25] and software [11,30,47] systems for stereopsis, the present system is the first to do so using only spiking neurons. It uses simple leaky integrate-and-fire neurons with static inhibitory and excitatory synapses. Compared to biologically plausible algorithms based on a binocular energy model [29,41,42] our system is significantly simpler and shallower which is crucial for the low latency it can achieve. The network is written in PyNN [4,37], its design is highly portable and scalable. It runs in real-time on SpiNNaker, a neuromorphic computer based on a mesh network of low-power ARM processing cores [12,13]. Apart from biological neural networks, this is the first scalable implementation of a spiking neural network able to perform asynchronous low-latency stereo-matching at realistic input event rates.

2   Material and Methods

2.1   Frame- and Event-Based Stereo Matching

Stereo matching or the "correspondence problem" is the problem of finding a pair of corresponding pixels or patches in a set of stereo images. Having the matching objects or pixels, we can calculate the depth which is inversely proportional to the disparity—the displacement along the epipolar lines—between the matches.

The geometrical characteristics of the stereo cameras can be formulated to map a single pixel of one image into a set of possible corresponding pixels in the other image [17]. In classic computer vision there are two general approaches for matching pixel pairs: area-based and feature-based matching [3]. The use of dynamic vision sensors introduces a separate clue that can be used for matching: time [19,36]. Just as with its biological counterpart, the asynchronous operation of the silicon retina opens the possibility to match events based on their individual time of appearance. However, due to inherent sensor variations and spatial aliasing, event coincidence alone is not reliable enough for event matching and has to be combined with additional constraints in order to achieve a reasonable matching quality [19,30,36].

2.2   Spiking Cooperative Network for Stereo Matching

Cooperative computing basically refers to algorithms with distributed local computing elements which interact with each other according to a set of predetermined constraints. To provide a unique solution for a given problem the algorithm's dynamics should reach a stable point given the inputs.

David Marr and Tomasio Poggio were first to propose a cooperative network to address the stereo matching problem [26]. This network can be thought of as a three-dimensional array of cells, where each cell encodes a belief in matching a heterolateral pair of pixels from the left and the right images. Since pixel pairs have to satisfy the epipolar-geometric constraints, the size of the network grows asymptotically as the cube of the sensors one-dimensional resolution. In order to limit the number of false matches Marr and Poggio suggested that assumptions based on the physical properties of common objects and geometrical configurations of the stereo cameras should be made [26]. First, that rigid bodies are cohesive, i.e. their surfaces are smooth and therefore should be represented by a smooth disparity map. This principle is called *within-disparity continuity* and enforces a weak potentiation between neighboring cells that represent the same disparity. Second, that each pixel in one of the images should be paired with at most one corresponding pixel in the opposite image. This *cross-disparity uniqueness* can be interpreted as a strong negative interaction between certain cells, i.e. they should inhibit each other. In this case, the epipolar geometry can determine the precise patterns of inhibition. A third constraint is *compatibility* between events that are to be fused. In the context of event-based vision the polarity of events or the magnitude of the contrast change could be used to further constrain the number of possible matches.

The temporal correlation of DVS events combined with the physical constraints discussed above have previously been used in non-spiking networks for event-based stereo-fusion, e.g. [30] or [11]. Here, we extend this approach into a spiking neural network, which we deploy as a real-time simulation on the SpiNNaker platform. As we will show SNNs are a very natural substrate to implement cooperative networks of this kind. We represent the physical and geometrical constraints by excitatory and inhibitory synaptic connections. The emerging neural activity constitutes the network's belief in correct matching. The three-dimensional structure of the network can be thought of as a stack of two-dimensional arrays (layers) of neurons. Figure 1(a) shows one such layer along with the input connections from the left and right retina pixels. Similarly to the cooperative network of [11], the topological location of the neurons determines which pixels from the sensors will be fed in and hence the disparity they are sensitive to. The cell parameters and synaptic weights are tuned in such a way that if spikes from the left and right retina coincide in a narrow time window (typically 1–5 ms) the cell will receive an over-threshold excitation stimulus and will consequently spike. The aforementioned physical constraints and the corresponding connectivity patterns are similarly implemented via excitatory and inhibitory synaptic projections as shown in Fig. 1(b). This process is visualized in 2 steps shown in Fig. 1(c) and (d), respectively. The inhibitory weights corresponding to Marr's uniqueness constraint are tuned such that the inhibitory input removes the excitation previously added by the retinal input. The neighborhood excitation corresponding to Marr's continuity constraint should be weak compared to the retinal input. The main effect of the former is to pre-activate the respective cell in order to spike before its peers in response to the latter. We choose the respective excitatory weight to be 1/10 of the retinal input weights.

The geometrical constraints of the stereo setup allow only for horizontal disparity detection (along the epipolar lines in a parallel aligned camera setup), which is independent of the vertical object position in the image. The network structure reflects this property with a translational symmetry of its layers along the y (vertical) axis; all layers are completely identical both in topological and parametric sense.

2.3   Preventing Homolateral Matching

If we naively assigned a single leaky-integrate-and-fire neuron to every point in the cooperative network (Fig. 1), such a neuron could in principle be triggered to spike by homolateral excitation alone. When a pixel in the left retina excites the neuron at a high frequency then even if the right retina does not provide a corresponding stimulus, e.g. during a partial or complete occlusion of the object from the right retina's point of view, the homolateral excitation from the left retina can trigger disparity sensitive neurons to spike and hence signal an arbitrary disparity. For symmetry reasons the described behavior will also be present if the right retina sends input stimuli and the left remains silent.

In order to prevent this behavior we developed a neural mechanism that ensures that only a pair of heterolateral pixel events could eventually be

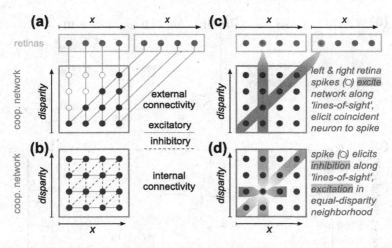

Fig. 1. Network topology of the cooperative stereo-matching network. **(a)**, **(b)** Connectivity in terms of excitatory and inhibitory connections: neurons/cells (dots) are framed into populations and interconnected by synaptic projections (lines). **(c)**, **(d)** Inhibition and excitation pattern for an exemplary pair of retinal input spikes **(c)** leading to a match, i.e. spike, in the cooperative network **(d)** which in turn triggers internal inhibition and excitation according to Marr's physical constraints.

considered as corresponding. Figure 2 sketches a neural micro-ensemble implementing this mechanism and its basic working principle. The micro-ensemble consists of one disparity-sensitive collector neuron C, and two blocker neurons B_l, B_r which are inserted between the retina cells R_l, R_r and C. The blocker neurons B are inhibitory interneurons. In the ipsilateral pathways $R_l \rightarrow B_l$, $R_r \rightarrow B_r$ they relay retina spikes from R_l, R_r in order to inhibit C, whereas retina spikes bypassing B_l, B_r excite C. The respective input weights w of C are matched such that the inhibitory (B $\rightarrow$ C) and excitatory (R $\rightarrow$ C) inputs cancel out: $|w_{B\rightarrow C}| = |w_{R\rightarrow C}|$. The respective delays d are matched such that the spike propagation along the R $\rightarrow$ B $\rightarrow$ C and R $\rightarrow$ C pathways take the same amount of time: $d_{R\rightarrow B} + \tau_B + d_{B\rightarrow C} = d_{R\rightarrow C}$, where τ_B represents the neuronal spike propagation time inside B. Consequently, a single ipsilateral retina spike, e.g. from R_l, has zero net effect on C's membrane potential. A concurrent spike from the contralateral retina R_r, however, suppresses the blocker neuron B_l via an inhibitory contralateral projection $R_r \rightarrow B_l$ and therefore disinhibits C.

So collectively B_l and B_r serve as ipsilateral inhibitors and suppress any one-sided propagation of repeated input stimuli to the collector neuron. Due to contralateral disinhibition any pair of heterolateral and time-wise correlated events will still result in an overall positive, above-threshold excitation of C and lead to a spike. Both cases are depicted as representative spike raster plots and measured membrane potentials in Fig. 2(b), (c), respectively. The latency of the network, which is governed by the micro-ensemble parameters, can also be read

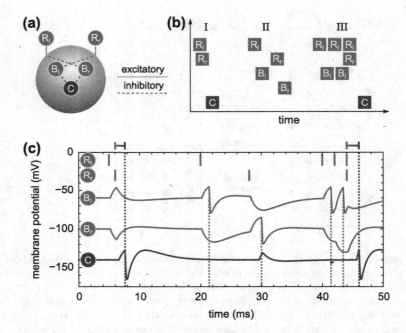

Fig. 2. Topology and working principle of neural micro-ensembles to prevent homolateral self-matching. **(a)** Inhibitory interneurons, "Blockers" B_l, B_r are inserted between retina pixels R_l, R_r and the disparity sensitive collector neuron C. **(b)** 3 typical spike sequences. (I) Left and right retina spikes arrive in quick succession and cause C to spike. (II) If R_r spikes too late after R_l, C does not spike. (III) Repeated retinal stimulation from R_l activates blockers B_l which annihilate R_l's excitatory input to C. Only once a contralateral spike from R_r adds additional excitatory C input and inhibits B_l, C spikes. **(c)** Measured membrane potentials of B_l, B_r, C (traces offset vertically for clarity) during cases I, II, III in that temporal order. Vertical bars indicate retina events, dashed lines mark neuronal spikes. The rulers on top of the graph mark the 2 ms delay between retina input and network output.

from Fig. 2(c) as the delay between a retina event and its consequent C neuron spike: it is 2 ms.

Note that all of the previously described synaptic connections within the cooperative network (Fig. 1) are between C neurons, i.e. a C neuron spike excites neighboring C neurons in the same disparity layer and inhibits other C neurons along the 2 retinal input projections.

2.4   Input Conditioning and Network Output

The neural micro-ensemble can successfully pair events which are separated by several milliseconds in time. The time window can be adjusted by means of neuron and simulation parameters, predominantly by the synaptic delays within the blocker micro-ensembles and the membrane time constant of C neurons.

By design, however, the neural network comprises neurons with biologically plausible time constants which are typically measured in milliseconds. This implies that any reaction time or operation of the network happens on a millisecond time scale as well. Event-based vision sensors, in contrast, can exhibit sub-millisecond reaction times and inter-event-intervals [22,31]. We therefore preprocess the DVS events before propagating them to the network. This preprocessing limits the interval between consecutive events from every individual pixel to a minimum value of 1 ms. The preprocessing is either done on a host computer or on the eDVS microcontroller [28].

The output of the network, similarly to that of event-based vision sensors, is an asynchronous stream of events in the form of spikes generated by the disparity-sensitive C neurons. Since each plausible pixel combination from the left and right images corresponds to a unique C neuron, the topological location of each spiking C neuron encodes the detected pixel pair and hence the disparity. The latter may be represented as a positive or negative integer, depending on the choice of reference pixel. In the experiments presented throughout this paper only positive disparity values will be used. In our epipolar geometry set-up both cameras are oriented in parallel, so the smallest disparity value, $d = 0$, corresponds to very distant objects.

2.5   Event Polarity and Compatibility Constraint

We are not making use of the polarity of events, i.e. we treat all input events from the DVS equally. While a naive realization of the compatibility constraint mentioned in Subsect. 2.4 would simply use 2 separate networks for the 2 event polarities, on second sight the merits of this approach are questionable. Imagine a gray pen moving from a black into a white background region! The polarity of events corresponding to the moving object would obviously change at the background transition, and it would change at different times in the left and right retina, because of their differing fields of view. While our merged-polarity approach can detect the correct polarity along the entire trajectory, the split-polarity approach would not be able to fuse the events during the transition from black to white background. Moreover, since the 2 separate networks do not cooperate, there would be no "hand-over" of the spatial neural activity corresponding to the moving pen from one network to the other.

2.6   Network Size and System Overview

The network size N, i.e. the number of neurons in our network, depends on the number of retinal input pixels in x and y directions $x_{\max}, y_{\max}$, the disparity range which is to be detected $[d_{\min}, d_{\max}]$, and the network layout. Because of the symmetry along the y-axis we only need to count the number of cells within a single layer n and then multiply by y_{max}: $N = n y_{\max}$. As can be seen in Fig. 1(a) the layers are trapezoids with a height of $h = d_{\max} - d_{\min} + 1$, a base line of $a = x_{\max} - d_{\min}$ and a top line of $b = x_{\max} - d_{\max}$ cells. Note that x, y

are counted starting at 1 and d is counted from 0. The total number of cells in each layer therefore is

$$n = \frac{a+b}{2}h = (x_{\max} - \frac{d_{\min} + d_{\max}}{2})(d_{\max} - d_{\min} + 1)$$

With a sensor size of $x_{\max} = y_{\max} = 128$, we would therefore need $N = 1056768 \approx 2^{20}$ cells to cover the full $[0, 127]$ disparity range. If we use the micro-ensembles described in Fig. 2, each of the N cells comprises 3 neurons. The resulting 3.15 million neurons would need at least 12,288 SpiNNaker cores, i.e. 723 SpiNNaker chips or 15 SpiNN-5 boards with the current SpiNNaker-PyNN implementation "sPyNNaker 3.0.0". To fit the network onto our locally available system of 6 SpiNN-5 boards, we typically reduce the network size for our experiments to at most $x_{\max} = y_{\max} = 106$, $d_{\max} = 32$. The prefiltered retina events are either injected via preprogrammed spike source arrays [37] or streamed live into the network via a host PC [33] or custom interface hardware [6]. The output spikes are either streamed out live or recorded into SpiNNaker's SDRAM and read out after the simulation has finished.

3 Results

3.1 Synthetic Test

In the first experiment, artificially generated data (devoid of any jitter or noise) is used to evaluate and verify the network performance, e.g. in comparing the expected and detected disparities. The stimulus in the experiment is produced by three points, whose trajectories in space describe the letters N, S and T. Each of the letters is lying in a different plane parallel to the sensor-plane (perpendicular to the z-axis), thus providing 3 traces at 3 different, constant disparities. In order to test the ability of the algorithm to detect different disparities at the same time, the 3 moving targets draw the respective letters in parallel. The accumulated traces as seen by the left and right retinas are shown in Fig. 3(a). The network output is visualized in Fig. 3(b) as a false color map, with x and y denoting retinal coordinates and the color encoding the detected disparity. A supplementary video (https://figshare.com/s/0d9fb146149b832ed8ec) shows an animated version of the figure that visualizes the course of input and respective output events.

Figure 3(c) displays the time course of the emitted spikes, i.e. the detected disparities as a raster plot. The true input disparities between the left and right retinas can be read from Fig. 3(a). They correspond closely to the detected disparities which are 12, 8, and 3 for the letters N, S, and T, respectively. For the letter N there are 71 events registered by each of the virtual cameras which result in 61 matches, 51 of which lie on the ground truth disparity $d = 12$ and 10 on $d = 13$. This inconsistency in perceived depth is present only when the diagonal segment of the letter N is drawn, which is due to the discretization of the synthetic images according to the camera resolutions, distances and other parameters. The wriggly trace of the letter S results in 16 successful matches at

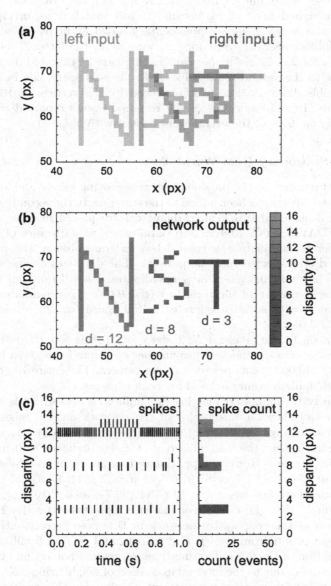

Fig. 3. Synthetic input data and corresponding network output. **(a)** Network input as events accumulated from the left (green pixels) and right (red pixels) DVS. The 3 letters are drawn concurrently by 3 pointers moving along the letter outlines. **(b)** Accumulated network output as the color coded median disparity per pixel. **(c)** Network output as a raster plot and histogram indicating, respectively, the time course and count of neural spikes with respect to their associated disparity. (Color figure online)

$d = 8$ (3 at $d = 9$) out of about 36 events per camera. One event has been falsely assigned to $d = 16$, because it was matched to an input event belonging to the letter T (see supplementary video at https://figshare.com/s/0d9fb146149b832ed8ec). Except for this mismatch, the letter T is perfectly matched; for a total of 22 events per camera there are 21 detected disparities all corresponding to the ground truth value of $d = 3$. This again, can be explained by the negligible discretization error produced by the vertical and horizontal letter segments. In summary 86% of the recorded events reported the ground truth disparity d_0, 99% of the events were correct within $d_0 \pm 1$.

3.2　Quasi-Stationary Real-World Test

The simulated pointers in the previous test were moving slowly and at constant speed. No noise or jitter has been added to the stimulus. In the second experiment two identical 3-blade fans (0.35 m in diameter) are placed in front of a DVS camera pair (DAVIS 240C, 10.0 cm horizontal offset) at a distance of 1.8 m and 2.8 m. In order to eliminate any time delays in transmission, the input data is pre-recorded and injected into the network. The data was down-sampled by discarding events with uneven x or y coordinates. The distant fan is already running at the beginning of the recording ($t = 0.0$ s), the proximate fan is set into motion at $t = 1.5$ s and accelerates to full speed (ca. 20 revolutions per second) by $t \approx 3.0$ s.

In comparison to the artificial NST data set, in this test the fast spinning fins of the fans generate a much larger number of events for a given time span: approximately 140,000 events per second, per camera. The entire 10 s long scene consists of 2.87 million events in total for both cameras.

Consequently, the number of matches (2.5 million) is also orders of magnitude larger. Figure 4(c) shows a raster plot of the network output spikes, i.e. the detected disparities over the 10.0 s of simulation time. In close correspondence to the geometric set-up the vast majority of spikes indicate a disparity of 7 for the distant fan, and 10 for the proximate fan. 96.7% of the events around the distant fan are at $d = 7$ or 8 (97.0% at $d = 7 \pm 1$). 95.9% of the events around the proximate fan are at $d = 10$ or 11 (96.7% at $d = 10 \pm 1$). It should be noted, though, that the number of matches represented by the 2.5 million recorded spikes of the cooperative network in this case exceeds either of the individual event counts from the left (1.4 million) and right (1.5 million) retina. So a large fraction of the 2.5 million matches represents not retina event pairs, but indirect matches due to the internal excitation of neighboring cells according to the within-disparity-continuity constraint. On its own this excitation is too weak to force any cell to spike (given the rate limit of the retina input), but it can pre-charge a cell such that a single cell-local retina event is sufficient to trigger an output spike.

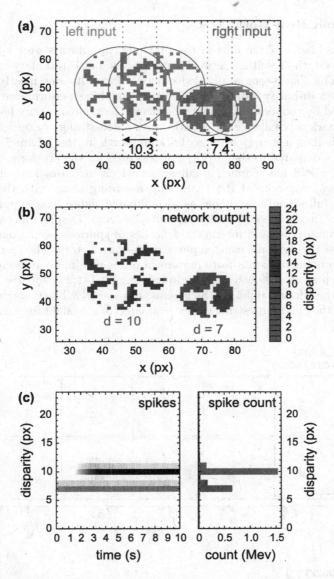

Fig. 4. Recorded input data (2 3-blade spinning fans; right, distant fan running continuously; left, proximate fan started at $t = 1.5$ s) and corresponding network output. **(a)** Network input as events accumulated from the left (green pixels) and right (red pixels) DVS over a 4.0 ms time interval starting at $t = 2000.0$ ms, when the left, proximate fan is accelerating. **(b)** Accumulated network output over a similar time interval 2.0 ms later, as the color coded median disparity per pixel. **(c)** Network output as raster plot and histogram indicating, respectively, the time course and count (in millions of events) of neural spikes with respect to their associated disparity. The acceleration of the proximate fan is clearly observable as a fading in of spikes at $d = 10$, starting after $t \approx 1.5$ s. (Color figure online)

3.3   Dynamic Real-World Test

Although the blades of the fans in the previous experiments were spinning at
very high speed, the resulting network activity was quasi stationary in terms of
disparities. The spatio-temporally dense network activity was mostly confined
to few distinct disparity values and the good matching performance therefore
mainly caused by cooperation among neurons at the same disparity level. In our
third benchmark we challenge the network with a non-stationary scene: a pendulum (tennis ball on a string) swinging back and forth in front of an eDVS stereo
camera pair. Compared to the DAVIS cameras used for the previous recordings,
the compact eDVS has a much smaller lens which increases lens distortions.
Figure 5 shows snapshots of the 10.0 s long recording along with the network
output. The full scene is available as an animated figure in a supplementary
video (https://figshare.com/s/0d9fb146149b832ed8ec). Despite the lens distortion, background noise and intermittent flashes of spurious events our network
is able to fuse the dynamic event input streams satisfactory. The raster plot in
Fig. 5(c) reflects the back-and-forth movement of the pendulum at around 0.6 Hz.
The network produced 99,686 matches in response to 141,252 and 261,307 input
events from the left and right eDVS cameras, respectively. The matching performance along the pendulum string can be evaluated in a straight forward manner,

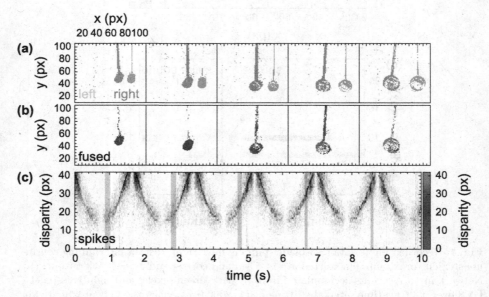

Fig. 5. Recorded input data (tennis-ball pendulum) and corresponding network output. **(a)** Network input from the left (green pixels) and right (red pixels) DVS during
different time windows as indicated by blue bands in (c), **(b)** Network output as disparity events (color code: see color bar in (c)), accumulated in the same respective
time window. **(c)** Network output as raster plot of neural spikes with respect to their
associated disparity. (Color figure online)

because of the particular geometric configuration of the scene. The pendulum swings almost exactly in a plane oriented vertically and along one line-of-sight of the left camera ($x_{\text{left}} = 91 \pm 1$), so the x-coordinate of any input event of the right camera that is related to the string should represent a disparity of $d = x_{\text{right}} - x_{\text{left}} = x_{\text{right}} - 91$. 88.3% of the events at $y > 57$ are within ± 2 pixels of this disparity value. If we assume the pendulum to be a flat object parallel to the image plane, then an extrapolation of the line representing the pendulum string should yield its expected disparity, which in this case depends linearly on the y coordinate. 62.2% of all events are within ± 2 pixels of that metric (52.2% within $d \pm 1$). The relatively bad matching performance is caused by a systematic error: In this dynamic scene the network tends to fuse the ball images mirror-symmetrically (right ball edge in right image to left ball edge in left image). This behaviour can be counteracted by object-level matching or a simple ordering-constraint, but we are only going to discuss those strategies in a separate paper.

We note that the same neuron, weight, and connectivity parameters were used for all three benchmark settings. The only difference were the network dimensions, which we accommodated to the fit the given scene[1].

3.4 Comparison to Non-spiking Implementation

In order to give a comparison to non-spiking implementations, we processed the very same data sets on a conventional personal computer (PC) using the non-spiking [11] variation of the algorithm written in Matlab. It emulates a cooperative network of cells with the same topology as its spiking counterpart. The cells cooperate with each other according to the same basic rules, but the potentiation and inhibition as well as the time decay of their "activation" are programmed explicitly. Furthermore there are no mechanisms to prevent homo-lateral matching. Comparisons to other event-based stereo matching algorithms, e.g. [19,30,36,39], are hindered by the fact that those references provide neither the input data nor their source code.

4 Discussion

4.1 Matching Performance

The spiking and non-spiking implementations give similar results in terms of detected disparity values for the synthetic as well as the real-world data sets. While our spiking implementation could handle all three scenes with the same weight and connectivity parameters, the non-spiking network needed major para-meter adaptations to process the different input scenes correctly.

---

[1] All code (including network and neuron parameters) and data sets neces-sary to reproduce the experiments are available at https://github.com/gdikov/SpikingStereoMatching.

Major differences lie in the scaling of computation time (and thus latency) with the number of events as well as in the number of detected stereo pairs. The latter is due to different detection strategies: The non-spiking Matlab implementation outputs the disparity corresponding to the cell with the highest activation score, so for almost every input event a corresponding disparity is reported. The spiking SpiNNaker implementation, in contrast, disregards any sub-threshold-potential activation of the cooperative network and counts only C neuron spikes as detected disparities. This typically yields one output spike for every pair of left and right input spikes. Only in the case of a stationary and repetitive input pattern like the one with the 2 fans the number of output spikes can approach the sum of left and right input spikes. Table 1 summarizes the results and the performances of both implementations. For each of the experiments the total run time, the maximum latency and matching capabilities are compared.

Table 1. Comparison of 2 different implementations of cooperative stereo-matching on 3 different tests

Dataset	NST	Pendulum	Fans
Input events (L/R)	1310/1290	141k/261k	1.38M/1.49M
Implementation	*Non-spiking (MATLAB)*		
Total number of detected pairs	2210	85.7k	2.86M
Total run time (s)	0.375	85.9	704.2
Implementation	*Spiking (SpiNNaker)*		
Total number of detected pairs	1020	99.6k	2.54M
Total run time (s)	10.0	10.0	10.0

4.2 System Scalability

The PC/Matlab implementation utilized 2 x86 cores (Intel Core i5, 3.3 GHz). The SpiNNaker/PyNN implementation utilized up to 5000 ARM cores (ARM968, 200 MHz). The PC implementation processes events sequentially with memory access as its main bottleneck and therefore the runtime increases roughly linearly with the number of input events. The SpiNNaker implementation runs highly parallel with a fixed latency between event input and network output of 2 ms regulated by the synapse parameters.

The parallel design of the SpiNNaker implementation has natural advantages over the sequential design of the PC implementation. First, it runs in real time and second, it is scalable: For any given input resolution, a suitable SpiNNaker machine can be set up to accommodate the required network size while retaining the real-time operation mode. Previous non-spiking cooperative network implementations are typically memory-bound and can handle on the order of 1000 events per second [11,30]. Our spiking implementation benefits from SpiNNaker's distributed local memory architecture [12]—its throughput is limited mainly by SpiNNaker's input–output bandwidth of about 10,000 events per second using a standard Ethernet interface [33] and well over 100,000 events per second using our custom hardware interface [6].

4.3   Power Consumption and Latency

It should be noted, though, that already a 3-board SpiNNaker machine consumes about 90 W. So in terms of power consumption the SpiNNaker implementation of our stereo-fusion network is difficult to use in mobile or autonomous settings, let alone on small flying drones. In such settings advanced commercial solutions like the Movidius Myriad-2 vision processing unit (http://www.movidius.com/) or the Stereolabs ZED stereo camera (http://stereolabs.com) are ostensibly superior. Both systems work with frame based image sensors at rates up to 60 Hz, which translates into latencies of around 20 ms or more. Our network provides a significantly lower latency of 2 ms. It is asynchronous from its sensor front-end to its neurocomputing back-end, meaning it can emit a depth-resolved event 2 ms after the associated physical event. This asynchronous, low-latency stereo-vision processing can enable high-speed control applications that have hitherto been impossible with visual input alone.

The size and power draw of the system can be improved by several means, most importantly by running our portable algorithm on more efficient platforms. The relatively high power consumption of our present implementation is mainly due to the fact that (1) SpiNNaker's processing cores are fabricated in a dated 130 nm process. And (2) SpiNNaker is simulating 3 leaky-integrate-and-fire neurons with exponential synapse dynamics for every disparity sensitive cell. Each cell is updated at every time step (0.2 ms), so this constitutes a sizable overhead. A specialized neuron model (e.g., with simpler dynamics or directly implementing the gating done by the blocker micro-ensembles) would reduce the resource demands in terms of core utilization and power consumption markedly. Furthermore, the fact that the synaptic weights are highly homogeneous could be exploited to eliminate expensive accesses to the shared SDRAM and simulate up to 2048 (instead of 255) neurons per core [40]. Such approaches would, however, break the portability of the PyNN code. At this point in history it is rational to favor portability over any platform specific optimization, since future neuromorphic platforms which are being developed in multiple research labs around the globe, will improve the efficiency of spiking-neural-network emulation by many orders of magnitude. The IBM TrueNorth chip, for example, can emulate 1 million spiking neurons at a power budget of <100 mW [27] using dedicated digital logic. Future chips, which might as well be based on analog silicon circuits or novel electronic materials [18,24] will inevitably improve this much further. Portable and scalable network designs like the one presented here are essential in order to benchmark and compare different neuromorphic systems [7,23] and therefore provide an important contribution towards their development process.

5   Conclusion and Outlook

Neuromorphic architectures like TrueNorth and SpiNNaker have previously been shown to be highly efficient image processing platforms [20,27,40,44]. Here we have shown event-based stereo-vision as another promising application. We have demonstrated a portable spiking neural network written in PyNN [4] that solves the correspondence or stereo-matching problem given event-based input from a pair of dynamic vision sensors [22]. Previous solutions were either limited by computational resources [11,30] or required custom hardware designs [8,9]. Our method does not need any event buffering and therefore achieves a very low latency of 2 ms, even at input rates as high as several 100,000 events per second. The network design is scalable to any input and depth resolution and runs in real-time on SpiNNaker. At the same time the network is portable, so the same network could be made to run on much more power-efficient systems like FPGAs [2], dedicated silicon [27] as well as future neuromorphic systems. Therefore, our work suggests a viable method to perform real-time stereo-matching also in mobile, wearable settings, e.g., in assistive devices for visually impaired people [15]. It also serves as a useful building block for robotics projects, which are increasingly utilizing neural algorithms and controllers [35,43]. Especially in the former example low latency is an obvious requirement and enables very natural interaction schemes [10].

In order to improve our neural model, it can be augmented with additional recurrence and feedback mechanisms that have been found in the visual cortex [14,16,34]. Our system could be further improved by utilizing modern high-resolution event-based sensors [21,31,46]. In addition to event-based vision those sensors feature traditional greyscale or color imaging via active pixel sensors on-chip, so they can be used to combine event-based and frame-based stereo-matching. Such a hybrid stereo-matching scheme could combine the speed and low latency of event-based stereo-fusion with the robustness and accuracy of established frame-based methods.

Acknowledgements. We thank S. Temple and the SpiNNaker Manchester team for their invaluable hardware, software and support. We also acknowledge I. Krawczuk and L. Everding for fruitful discussions, technical assistance with benchmarks and power measurements as well as for help in obtaining good stereo-DVS datasets. The research leading to these results has received funding from the European Union Seventh Framework Programme (FP7/2007–2013) under grant agreement no. 720270 (Human Brain Project) and the Bundesministerium für Bildung und Forschung via grant no. 01GQ0440 (Bernstein Center for Computational Neuroscience Munich).

References

1. Benjamin, B.V., Gao, P., McQuinn, E., Choudhary, S., Chandrasekaran, A.R., Bussat, J.M., Alvarez-Icaza, R., Arthur, J.V., Merolla, P.A., Boahen, K.: Neurogrid: a mixed-analog-digital multichip system for large-scale neural simulations. Proc. IEEE **102**(5), 699–716 (2014)

2. Cheung, K., Schultz, S.R., Luk, W.: NeuroFlow: a general purpose spiking neural network simulation platform using customizable processors. Front. Neurosci. **9**(516), 1–15 (2016)
3. Davies, E.: 3D vision and motion. In: Machine Vision. Signal Processing and its Applications, 3 edn., p. 443. Morgan Kaufmann, Burlington (2005)
4. Davison, A., Brüderle, D., Eppler, J., Kremkow, J., Muller, E., Pecevski, D., Perrinet, L., Yger, P.: PyNN: a common interface for neuronal network simulators. Front. Neuroinform. **2**, 11 (2009). http://journal.frontiersin.org/article/10.3389/neuro.11.011.2008
5. Delbruck, T., Linares-Barranco, B., Culurciello, E., Posch, C.: Activity-driven, event-based vision sensors. In: Proceedings of 2010 IEEE International Symposium on Circuits and Systems, pp. 2426–2429, May 2010
6. Denk, C., Llobet-Blandino, F., Galluppi, F., Plana, L.A., Furber, S., Conradt, J.: Real-time interface board for closed-loop robotic tasks on the SpiNNaker neural computing system. In: International Conference on Artificial Neural Networks (ICANN), Sofia, Bulgaria, pp. 467–474, September 2013. http://mediatum.ub.tum.de/doc/1191903/90247.pdf
7. Diamond, A., Nowotny, T., Schmuker, M.: Comparing neuromorphic solutions in action: Implementing a bio-inspired solution to a benchmark classification task on three parallel-computing platforms. Front. Neurosci. **9**, 491 (2016). http://journal.frontiersin.org/article/10.3389/fnins.2015.00491
8. Domínguez-Morales, M., Jimenez-Fernandez, A., Paz, R., López-Torres, M.R., Cerezuela-Escudero, E., Linares-Barranco, A., Jimenez-Moreno, G., Morgado, A.: An approach to distance estimation with stereo vision using address-event-representation. In: Lu, B.-L., Zhang, L., Kwok, J. (eds.) ICONIP 2011. LNCS, vol. 7062, pp. 190–198. Springer, Heidelberg (2011). doi:10.1007/978-3-642-24955-6_23
9. Eibensteiner, F., Kogler, J., Scharinger, J.: A high-performance hardware architecture for a frameless stereo vision algorithm implemented on a FPGA platform. In: Proceedings of the IEEE Conference on Computer Vision and Pattern Recognition Workshops, pp. 623–630 (2014)
10. Everding, L., Walger, L., Ghaderi, V.S., Conradt, J.: A mobility device for the blind with improved vertical resolution using dynamic vision sensors. In: IEEE HealthCom 2016, Munich, Germany, September 2016
11. Firouzi, M., Conradt, J.: Asynchronous event-based cooperative stereo matching using neuromorphic silicon retinas. Neural Process. Lett. **43**(2), 311–326 (2016)
12. Furber, S.B., Galluppi, F., Temple, S., Plana, L.A.: The SpiNNaker project. Proc. IEEE **102**(5), 652–665 (2014)
13. Furber, S.B., Lester, D.R., Plana, L.A., Garside, J.D., Painkras, E., Temple, S., Brown, A.D.: Overview of the spinnaker system architecture. IEEE Trans. Comput. **62**(12), 2454–2467 (2013)
14. Georgieva, S., Peeters, R., Kolster, H., Todd, J.T., Orban, G.A.: The processing of three-dimensional shape from disparity in the human brain. J. Neurosci. **29**(3), 727–742 (2009)
15. Ghaderi, V.S., Mulas, M., Santos Pereira, V., Everding, L., Weikersdorfer, D., Conradt, J.: A wearable mobility device for the blind using retina-inspired dynamic vision sensors. In: 2015 37th Annual International Conference of the IEEE Engineering in Medicine and Biology Society (EMBC), pp. 3371–3374, August 2015
16. Grossberg, S., Howe, P.D.: A laminar cortical model of stereopsis and three-dimensional surface perception. Vis. Res. **43**(7), 801–829 (2003)
17. Hartley, R., Zisserman, A.: Multiple View Geometry in Computer Vision. Cambridge University Press, Cambridge (2003)

18. Jany, R., Richter, C., Woltmann, C., Pfanzelt, G., Förg, B., Rommel, M., Reindl, T., Waizmann, U., Weis, J., Mundy, J.A., et al.: Monolithically integrated circuits from functional oxides. Adv. Mater. Interfaces **1**(1) (2014)

19. Kogler, J., Humenberger, M., Sulzbachner, C.: Event-based stereo matching approaches for frameless address event stereo data. In: Bebis, G., et al. (eds.) ISVC 2011. LNCS, vol. 6938, pp. 674–685. Springer, Heidelberg (2011). doi:10.1007/978-3-642-24028-7_62

20. Layher, G., Brosch, T., Neumann, H.: Real-time biologically inspired action recognition from key poses using a neuromorphic architecture. Front. Neurorobot. **11** (2017). http://journal.frontiersin.org/article/10.3389/fnbot.2017.00013/full

21. Li, C., Brandli, C., Berner, R., Liu, H., Yang, M., Liu, S.C., Delbruck, T.: Design of an RGBW color VGA rolling and global shutter dynamic and active-pixel vision sensor. In: 2015 IEEE International Symposium on Circuits and Systems (ISCAS), pp. 718–721. IEEE (2015)

22. Lichtsteiner, P., Posch, C., Delbruck, T.: A 128 × 128 120 db 15 µs latency asynchronous temporal contrast vision sensor. IEEE J. Solid State Circ. **43**(2), 566–576 (2008)

23. Liu, Q., Pineda-Garca, G., Stromatias, E., Serrano-Gotarredona, T., Furber, S.B.: Benchmarking spike-based visual recognition: a dataset and evaluation. Front. Neurosci. **10**, 496 (2016). http://journal.frontiersin.org/article/10.3389/fnins.2016.00496

24. Lorenz, M., Rao, M.S.R., Venkatesan, T., Fortunato, E., Barquinha, P., Branquinho, R., Salgueiro, D., Martins, R., Carlos, E., Liu, A., Shan, F.K., Grundmann, M., Boschker, H., Mukherjee, J., Priyadarshini, M., DasGupta, N., Rogers, D.J., Teherani, F.H., Sandana, E.V., Bove, P., Rietwyk, K., Zaban, A., Veziridis, A., Weidenkaff, A., Muralidhar, M., Murakami, M., Abel, S., Fompeyrine, J., Zuniga-Perez, J., Ramesh, R., Spaldin, N.A., Ostanin, S., Borisov, V., Mertig, I., Lazenka, V., Srinivasan, G., Prellier, W., Uchida, M., Kawasaki, M., Pentcheva, R., Gegenwart, P., Granozio, F.M., Fontcuberta, J., Pryds, N.: The 2016 oxide electronic materials and oxide interfaces roadmap. J. Phys. D Appl. Phys. **49**(43), 433001 (2016). http://stacks.iop.org/0022-3727/49/i=43/a=433001

25. Mahowald, M.A., Delbrück, T.: Cooperative Stereo Matching Using Static and Dynamic Image Features, pp. 213–238. Springer, Boston (1989). doi:10.1007/978-1-4613-1639-8_9

26. Marr, D., Poggio, T.: Cooperative computation of stereo disparity. Science **194**(4262), 283–287 (1976)

27. Merolla, P.A., Arthur, J.V., Alvarez-Icaza, R., Cassidy, A.S., Sawada, J., Akopyan, F., Jackson, B.L., Imam, N., Guo, C., Nakamura, Y., et al.: A million spiking-neuron integrated circuit with a scalable communication network and interface. Science **345**(6197), 668–673 (2014)

28. Müller, G.R., Conradt, J.: A miniature low-power sensor system for real time 2D visual tracking of led markers. In: 2011 IEEE International Conference on Robotics and Biomimetics (ROBIO), pp. 2429–2434, December 2011

29. Ohzawa, I., DeAngelis, G.C., Freeman, R.D., et al.: Stereoscopic depth discrimination in the visual cortex: neurons ideally suited as disparity detectors. Science **249**(4972), 1037–1041 (1990)

30. Piatkowska, E., Belbachir, A.N., Gelautz, M.: Cooperative and asynchronous stereo vision for dynamic vision sensors. Meas. Sci. Technol. **25**(5), 1–8 (2014)

31. Posch, C., Matolin, D., Wohlgenannt, R.: A QVGA 143 dB dynamic range frame-free PWM image sensor with lossless pixel-level video compression and time-domain CDS. IEEE J. Solid State Circ. **46**(1), 259–275 (2011)

32. Qiao, N., Mostafa, H., Corradi, F., Osswald, M., Stefanini, F., Sumislawska, D., Indiveri, G.: A reconfigurable on-line learning spiking neuromorphic processor comprising 256 neurons and 128k synapses. Front. Neurosci. 9 (2015). http://www.ncbi.nlm.nih.gov/pmc/articles/PMC4413675/
33. Rast, A.D., et al.: Transport-independent protocols for universal AER communications. In: Arik, S., Huang, T., Lai, W.K., Liu, Q. (eds.) ICONIP 2015. LNCS, vol. 9492, pp. 675–684. Springer, Cham (2015). doi:10.1007/978-3-319-26561-2_79
34. Read, J.: Early computational processing in binocular vision and depth perception. Prog. Biophys. Mol. Biol. 87(1), 77–108 (2005)
35. Richter, C., Jentzsch, S., Hostettler, R., Garrido, J.A., Ros, E., Knoll, A.C., Röhrbein, F., van der Smagt, P., Conradt, J.: Musculoskeletal robots: scalability in neural control. IEEE Robot. Autom. Mag. 23(4), 128–137 (2016). doi:10.1109/MRA.2016.2535081
36. Rogister, P., Benosman, R., Ieng, S.H., Lichtsteiner, P., Delbruck, T.: Asynchronous event-based binocular stereo matching. IEEE Trans. Neural Netw. Learn. Syst. 23(2), 347–353 (2012)
37. Rowley, A.G.D., Stokes, A.B., Knight, J., Lester, D.R., Hopkins, M., Davies, S., Rast, A., Bogdan, P., Davidson, S.: PyNN on SpiNNaker software 2015.004, July 2015. http://dx.doi.org/10.5281/zenodo.19230
38. Schemmel, J., Brüderle, D., Grübl, A., Hock, M., Meier, K., Millner, S.: A wafer-scale neuromorphic hardware system for large-scale neural modeling. In: Proceedings of 2010 IEEE International Symposium on Circuits and systems (ISCAS), pp. 1947–1950. IEEE (2010)
39. Schraml, S., Schön, P., Milosevic, N.: Smartcam for real-time stereo vision-address-event based embedded system. In: VISApp (2), pp. 466–471 (2007)
40. Serrano-Gotarredona, T., Linares-Barranco, B., Galluppi, F., Plana, L., Furber, S.: ConvNets experiments on SpiNNaker. In: 2015 IEEE International Symposium on Circuits and Systems (ISCAS), pp. 2405–2408, May 2015
41. Shi, B.E., Tsang, E.K.: A neuromorphic multi-chip model of a disparity selective complex cell. In: Thrun, S., Saul, L.K., Schölkopf, P.B. (eds.) Advances in Neural Information Processing Systems, vol. 16, pp. 1051–1058. MIT Press, Cambridge (2004)
42. Shimonomura, K., Kushima, T., Yagi, T.: Binocular robot vision emulating disparity computation in the primary visual cortex. Neural Netw. 21(23), 331–340 (2008). Advances in Neural Networks Research: International Joint Conference on Neural Networks, IJCNN 2007, July 2007. http://www.sciencedirect.com/science/article/pii/S089360800700247X
43. Stewart, T.C., Kleinhans, A., Mundy, A., Conradt, J.: Serendipitous offline learning in a neuromorphic robot. Front. Neurorobot. 10, 1–11 (2016)
44. Sugiarto, I., Liu, G., Davidson, S., Plana, L.A., Furber, S.B.: High performance computing on SpiNNaker neuromorphic platform: a case study for energy efficient image processing. In: 2016 IEEE 35th International Performance Computing and Communications Conference (IPCCC), pp. 1–8, December 2016
45. Walter, F., Röhrbein, F., Knoll, A.: Neuromorphic implementations of neurobiological learning algorithms for spiking neural networks. Neural Netw. 72(C), 152–167 (2015)
46. Yang, M., Liu, S.C., Delbruck, T.: A dynamic vision sensor with 1% temporal contrast sensitivity and in-pixel asynchronous delta modulator for event encoding. IEEE J. Solid State Circ. 50(9), 2149–2160 (2015)
47. Zitnick, C.L., Kanade, T.: A cooperative algorithm for stereo matching and occlusion detection. IEEE Trans. Pattern Analy. Mach. Intell. 22(7), 675–684 (2000)

Development of Novel Foam-Based Soft Robotic Ring Actuators for a Biomimetic Peristaltic Pumping System

Falk Esser[1,2(✉)], Tibor Steger[1], David Bach[1,2], Tom Masselter[1], and Thomas Speck[1,2]

[1] Plant Biomechanics Group, Faculty of Biology,
Botanic Garden University Freiburg, Freiburg im Breisgau, Germany
Falk.esser@biologie.uni-freiburg.de
[2] FMF – Freiburg Materials Research Center, Freiburg im Breisgau, Germany

Abstract. Peristaltic pumping in nature allows for the transport of various media in a simple and secure way. Different types of peristaltic pumps exist in the application area of soft robotics. Most systems are based on pneumatic network (pneu-net) fluidic elastomer actuators or artificial muscle actuators. In this study the development of a pump actuated by foam-based, flexible, compliant and lightweight ring actuators is presented. Utilizing a custom built pump test bench the soft robotic ring actuators are characterized in terms of contraction rate and volumetric displacement. Furthermore we introduce a flexible and elastic soft robotic peristaltic pumping system as an alternative to conventional technical pumps.

Keywords: Foam-based ring actuators · Soft robotics · Peristaltic pumping system · Biomimetics

1 Introduction

Fluids are transported directionally by pumping systems in nature and technology. Examples for natural systems include vertebrate hearts, bird lungs, arthropod open circulatory systems and peristalsis of specific hollow organs [1–4]. For technical pumping systems, rotary and positive displacement pumps e.g. centrifugal pumps, rotary vane pumps and gear pumps are state-of-the-art [5].

In the present study, we focus on the implementation of peristalsis into a novel biomimetic pumping system. In preliminary work, we identified and characterized biological pumping systems and their biomimetic potential [4, 6]. The peristaltic principle was found to be the most suitable for a biomimetic implementation in terms of technical feasibility and innovation potential.

The implementation of peristalsis motion patterns and transport into soft robotic applications has been well covered in recent literature [7–10]. A pneumatic network (pneu-net) fluidic elastomer actuator (FEA) based on biology-inspired swallowing was developed for investigating the rheology of texture-modified foods [6]. Pneu-net actuators are elastomer-based actuators, with a network of channels and chambers

M. Mangan et al. (Eds.): Living Machines 2017, LNAI 10384, pp. 138–147, 2017.
DOI: 10.1007/978-3-319-63537-8_12

inside that are deformed by pressurized inflation. In 2015, Dirven *et al.* [6] presented a first soft robotic FEA ring actuator, which – pneumatically driven – could perform esophageal like peristalsis wave propagation along a conduit [8–11]. It consisted of silicone ring actuators with a four chamber system in a stiff frame. Another implementation of a peristaltic pump is used for powder conveyance in printers [12–14]. This pump is based on bowel peristalsis and actuated pneumatically by artificial rubber muscles. The actuator unit consists of a straight-fiber-type artificial muscle hold by stiff flanges.

An alternative to FEA actuators are foam-based soft actuators (FBSAs) which use open-celled elastic foam that is sealed airtight with an elastomer coating and outfitted with a strain-limiting layer on one side [15]. When pressure is induced, the FBSAs are able to bend and expand into the opposite direction of the strain limiting layer. The actuation does not need a stiff frame, and in contrast to the pneu-net FEAs the FBSAs are composed entirely of soft parts. A considerable advantage of using open-celled foam is that no complex air routing system inside the FBSAs is needed [15], as used in FEAs [8, 12], which simplifies the production of FBSAs to a molding-demolding process, in contrast to the complex FEA production incorporating sacrificial molds or complex 3D printed molds to produce an air routing system. Additionally, with the interiorly soft and flexible system of the FBSAs smaller systems may be designed and more complex geometries may be achieved.

The main goal of this study is, by converting the FBSAs into ring actuators, the biomimetic implementation of peristalsis into a flexible, silent, robust, energy efficient, space-saving and low cost technical application for the usage in combustion engines and cooling systems.

The biomimetic implementation was based on the human esophagus, in which the peristalsis motion pattern is produced by a continuous contraction of the circular musculature [1–4, 16], resembling a square wave (Fig. 1).

The peristalsis occlusion rates, in the range of 60 to 100% [3, 16–18], and the esophageal diameter of 20 mm and length of 200–260 mm [7] can be used as a target for the development of a peristalsis-based soft robotic actuator. The occlusion rate describes to which percentage the inner diameter of esophagus is reduced while the circular musculature contracts in the peristaltic motion.

2   Materials and Methods

2.1   Development of Foam-Based Soft Robotic Ring Actuators (FSRAs)

For biomimetic implementation of the biological principle foam-based soft actuators (FBSAs) as described by [15] were selected. Utilizing this material platform, we designed foam-based soft robotic ring actuator units (FSRAs). The novel FSRA units can be used to mimic the circular esophageal and intestinal muscles in their function, and form a stably functioning soft robotic foam-based peristalsis pump (FBPP) (Fig. 2). With a specific actuation pattern of eight actuator units, peristaltic transport could be achieved, which corresponds well to the biological role model.

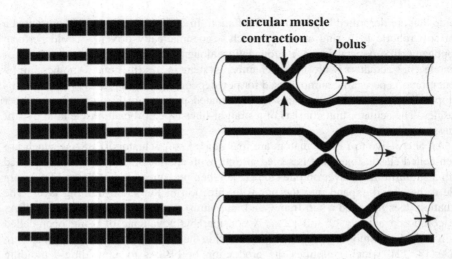

Fig. 1. Left: Simplified peristalsis motion pattern. Right: Simplified circular muscle contraction movement propelling bolus in the esophagus.

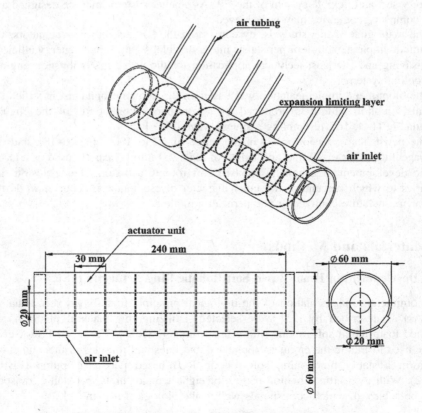

Fig. 2. Technical drawing of the soft robotic tube pump.

The inner diameter of the actuator foam units was chosen to approximate the inner conduit width of ca. 20 mm of the biological role model (Table 1). The cylindrical foam units were fabricated using mold casting. After casting the rings were coated to achieve a water-proof and oil resistant surface.

Table 1. Dimensions of FSRAs and the foam-based peristalsis pump (FBPP)

Dimensions	Single coated actuator unit	Coated tube pump (built of 8 actuator units)
Diameter [mm]	62	62
Inner diameter [mm]	18	18
Length [mm]	32	256
Conduit volume [mm³]	8143	65144

Different foam materials were investigated. The most suitable basis foam material was found to be open-celled, flexible and elastic self-expanding polyurethane (PU) foams as FlexFoam-iT!$^{®}$ 6 and FlexFoam-iT!$^{®}$ X (FF6 and FFX; KauPo Plankenhorn, e.K.). These were selected due to their low density, (FF6 = 96 kg/m³ and FFX = 160 kg/m³) and their good chemical, water and oil resistance. Additionally, FF6 and FFX were beneficial as they exhibit a short room temperature curing time ($\approx$2 h) and are highly compatible with the selected brushed coating (UreCoat$^{®}$, UC; KauPo Plankenhorn, e.K.). This coating also exhibits sufficient water and oil resistance [19].

For a batch production, 39 cm long casting molds of aluminum or polyte-trafluoroethylene (PTFE) were used to fabricate foam tubes. After curing the foam tubes were cut into three centimeter long pieces and sealed with UC in order to manufacture the FSRAs. For pressurization, the actuators were outfitted with air tubing and a strain-limiting layer, which consisted of duct tape (tesa extra Power$^{®}$ Universal; Tesa, SE) (Fig. 3(a)). This setup allows for inward expansion only, so that the inner diameter is decreased when air pressure is applied. For the tubing a five millimeter deep and six millimeter wide hole was cut in the outer surface of the foam rings and the air tubing was fixated air tight with the coating. For the strain-limiting layer of the single actuators, two layers of duct tape were applied.

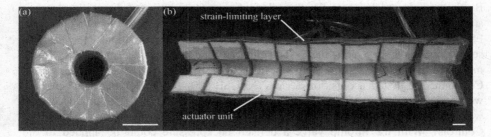

Fig. 3. (a) PU based FSRA coated with UC. (b) Cut open foam-based peristalsis pump (FBPP). Scale bars: 20 mm.

The foam-based peristalsis pump (FBPP) was formed through bonding of the eight pre-coated actuator units using a two millimeter thick layer of UC in between the individual actuators to form an air tight chamber (Fig. 3(b)). The outer surface of the tube was coated with a two millimeter thick layer of UC to achieve an air tight sealing. After coating the strain-limiting layer, i.e. duct tape, was applied.

The FBPP was operated with a specifically designed and custom-built pump test bench (Fig. 4). A square wave like opening and closing pattern of valves allowed for a formation of a peristaltic wave and pumping in the FBPP.

Fig. 4. Biomimetic pump test bench setup. Scale bar: 50 mm.

2.2   Pump Test Bench

The pump test bench for the biomimetic pumping system was built in cooperation with the AG Technik of the Institute for Biology II/III (Albert-Ludwigs-University Freiburg) (Fig. 4). It was used to investigate the contraction rates of the FSRAs and the pump performance of the FBPP.

The actuators and the pump were connected to the test bench via air tubing (Fig. 2). By the use of a micro-controller the solenoid valves are activated. For a simple peristalsis motion pattern, a continuous square wave motion was used to pressurize the FSRAs. The pressure inside the actuators (TruStability® Board Mount Pressure Sensors; Honeywell, Int. Inc.), the system pressure and the pressure in the fluid were measured by pressure sensors (JUMO Midas Type 401001; M.K. Juchheim GmbH & Co) connected to data acquisition devices (NI USB 6002 DAQs; National Instruments), which are in turn connected to a laptop with LabVIEW software. The produced flow was measured by a turbine flow sensor (FT-210 – Turbo Flow®; Gems™ Sensors & Controls). With the throttle valve a defined backpressure could be applied to the pump. The potentiometers controlled the actuation frequency (ranging from 0.2 to 10 Hz) by which the actuators were pressurized.

2.3 Determination of FSRA Contraction

The contraction rates of single FSRAs at different frequencies and pressures were tested by using the pneumatic system of the pump test bench. A video camera was mounted on a tripod perpendicular to the FSRAs, so that the contraction of the inner conduit could be filmed. The contraction i.e. the ratio between cross-sectional area of the conduit in contracted and uncontracted state, was then measured via Fiji (open-source software, version ImageJ.48q, https://fiji.sc/) [20]. The lower limit for the contraction was set to 20% as an insufficient flow rate is attained with a lower contraction, because the registered contractions were only marginal. Considering the inner volume of the FSRAs conduit, 60% contractions at 0.5 Hz would correspond to a volumetric displacement of 5.6 cm^3 and a theoretical flow rate of approx. 10 l/h. In contrast with a 20% contraction, 1.9 cm^3 and a theoretical flow rate of 3.4 l/h could be achieved, which is considered insufficient for pumps in combustion engines and cooling systems.

3 Results

Of the different tested foam materials for the FSRAs, the PU foam-based FFX FSRAs showed the best results. To identify the ideal foam material and combination of pressure and actuation frequency contraction tests for the FSRAs were performed with the pump test bench (Fig. 5). The burst pressure of the UC coating was determined as 2 bar. Exemplary results are shown in Fig. 5. At a pressure of 1 bar and frequencies of 0.5 to 0.55 Hz contractions in the range of 60% could be achieved. At frequencies over 1.5 Hz and pressures lower than 1 bar only contractions of fewer than 25% of the inner conduit could be achieved.

UC coated PU foam-based FSRAs achieved a contraction rate of 55% at 0.8 bar at 0.4 Hz (Fig. 6(a)). Also investigated was a variant with a polyethylene tube inside the conduit (Fig. 6(b)). However, no circular contraction was achieved with a tube inside.

3.1 Biomimetic Pump

The pump test bench was used for investigating the square wave like peristaltic actuation. In this simplified peristaltic motion pattern, the solenoid valves open consecutively with a short time delay (Fig. 7). Thereby all FSRAs inflate one after another.

A first test of the complete system could be performed with a FBPP consisting of five FSRAs (Fig. 8). The peak flow rate of 60 ml/min for the transported fluid (water) is achieved at a frequency of 0.8 Hz. An ideal pressure range between 0.5 bar and 1.4 bar and an optimal frequency range of 0.5 Hz to 1 Hz could be determined empirically (see Fig. 5).

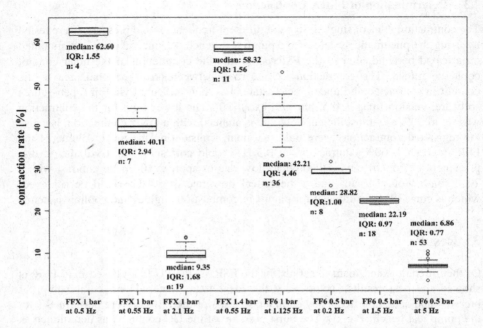

Fig. 5. Contractions of FSRAs at different frequencies and pressures. FFX foams achieve higher contractions than FF6 foams.

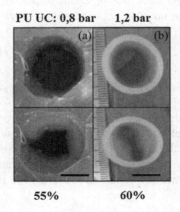

Fig. 6. Contractions of PU foam-based FSRAs with a strain-limiting layer on the outer ring surface. The measured contraction is given in percent. (a) PU foam-based FSRA coated with UC in uncontracted (upper image) and contracted state (lower image) at 0.4 Hz. (b) PU foam-based FSRA with inner PE tube in uncontracted (upper image) and contracted state (lower image). Scale bars: 10 mm.

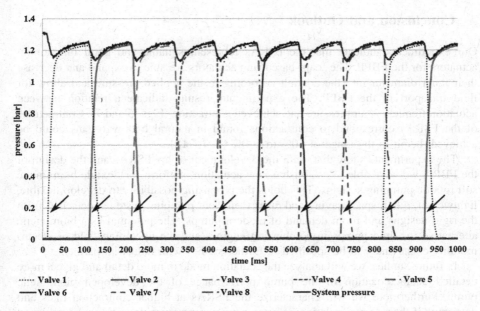

Fig. 7. Pressure built-up and peristaltic actuation sequence of the biomimetic pump test bench (without pump). System pressure was set to 1.3 bar, actuation frequency of a single valve was set to 1 Hz, and the overlapping activation pattern of the solenoid valves is visible (arrow).

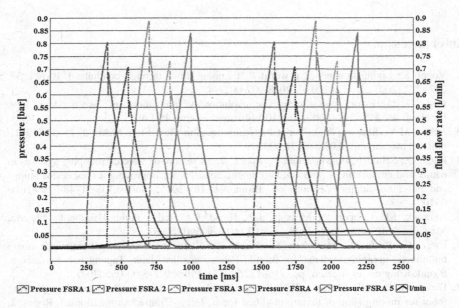

Fig. 8. Flow test of FBPP consisting of five FSRAs. Exemplarily, the five FSRA pressure curves and a measured flow rate at a frequency of 0.8 Hz are shown

4 Conclusion and Outlook

Our experiments indicate that the FSRAs represent suitable and good functioning actuators for the FBPP. The foam-based ring actuators are able to expand and decrease their inner diameter into the conduit according to the applied pressure and allow for fluid transport in the FBPP. The experimental results indicate a relation between actuation frequency, air pressure and achieved contraction (Figs. 5 and 7). Contractions of the FSRAs correspond to contractions found in natural hollow organs could be achieved, being in the range of 60% to 100% [3, 16–18].

The experiments show that, with the development of the FSRAs and the design of the FBPP, we were able to establish a new actuation platform for flexible biomimetic soft robotic pumping systems. This holds the promising possibility to develop flexible, lightweight, quiet, space-saving and energy efficient pumping systems, detached from the rigid designs and room demand of modern pumps. The potential as a biomimetic alternative for fluid pumps in combustion engines and the upcoming field of electro mobility exists and will be specified in further studies with industrial partners.

In future studies, we will analyze the actuation mode in more detail and give a more detailed characterization of the pump performance of the developed biomimetic pump. Furthermore we will characterize the FSRAs at higher contraction rates and determine if there is a correlation between actuation pressure, frequency and achieved contraction rates. Ultimately we would like to further develop the design into an advanced system, in which an electrically driven actuation is incorporated in the foam material, for a pump application in combustion engines and the growing field of electro mobility.

References

1. Vogel, S.: Living in a physical world X. Pumping fluids through conduits. J. Biosci. **32**, 207–222 (2007). doi:10.1007/s12038-007-0021-4
2. Pass, G.: Accessory pulsatile organs: evolutionary innovations in insects. Ann. Rev. Entomol. **45**, 495–518 (2000). doi:10.1146/annurev.ento.45.1.495
3. Jaffrin, M.Y., Shapiro, A.H.: Peristaltic pumping. Ann. Rev. Fluid Mech. **3**, 13–37 (1971). doi:10.1146/annurev.fl.03.010171.000305
4. Bach, D., Schmich, F., Masselter, T., Speck, T.: A review of selected pumping systems in nature and engineering - potential biomimetic concepts for improving displacement pumps and pulsation damping. Bioinspirat. Biomimet. **10** (2015). doi:10.1088/1748-3190/10/5/051001
5. Karassik, I.J., Cooper, P., Messina, J.P., Heald, C.C. (eds.) Pump Handbook, 4th edn. Mc Graw Hill, New York (2008). doi:10.1002/aic.690220632
6. Esser, F., Bach, D., Masselter, T., Speck, T.: Nature as concept generator for novel biomimetic pumping systems. In: Bionik: Patente aus der Natur, Tagungsbeiträge zum 8. Bionik-Kongress in Bremen, pp. 116–122 (2017). ISBN 978-3-00-055030-0
7. Dirven, S., Xu, W., Cheng, L.K., Allen, J., Bronlund, J.: Biologically-inspired swallowing robot for investigation of texture modified foods. Int. J. Biomechatron. Biomed. Robot. **2**, 163–171 (2013). doi:10.1504/IJBBR.2013.058719

8. Dirven, S., Xu, W., Cheng, L.K.: Sinusoidal peristaltic waves in soft actuator for mimicry of esophageal swallowing. IEEE/ASME Trans. Mechatron. **20**, 1331–1337 (2015). doi:10. 1109/TMECH.2014.2337291

9. Chen, F., Dirven, S., Xu, W., Li, X.: Large-deformation model of a soft-bodied esophageal actuator driven by air pressure. IEEE/ASME Trans. Mechatron. **22**, 81–90 (2017). doi:10. 1109/TMECH.2016.2612262

10. Zhu, M., Xu, W., Cheng, L.K.: Measuring and imaging of a soft-bodied swallowing robot conduit deformation and internal structural change using videofluoroscopy. In: 2016 23rd International Conference on Mechatronics and Machine Vision in Practice (M2VIP), pp 1–6 (2016). doi:10.1109/M2VIP.2016.7827312

11. Zhu, M., Xu, W., Cheng, L.K.: Esophageal peristaltic control of a soft-bodied swallowing robot by the central pattern generator. IEEE/ASME Trans. Mechatron. **22**, 91–98 (2017). doi:10.1109/TMECH.2016.2609465

12. Suzuki, K., Nakamura, T.: Development of a peristaltic pump based on bowel peristalsis using for artificial rubber muscle. In: 2010 IEEE/RSJ International Conference on Intelligent Robots and Systems (IROS), pp. 3085–3090 (2010). doi:10.1109/IROS.2010.5653006

13. Kimura, Y., Saito, K., Nakamura, T.: Development of an exsufflation system for peristaltic pump based on bowel peristalsis. In: 2013 IEEE/ASME International Conference on Advanced Intelligent Mechatronics (AIM), pp. 1235–1240 (2013). doi:10.1109/AIM.2013. 6584263

14. Yoshihama, S., Takano, S., Yamada, Y., Nakamura, T., Kato, K.: Powder conveyance experiments with peristaltic conveyor using a pneumatic artificial muscle. In: 2016 IEEE International Conference on Advanced Intelligent Mechatronics (AIM), pp. 1539–1544 (2016). doi:10.1109/AIM.2016.7576989

15. Mac Murray, B.C., An, X., Robinson, S.S., van Meerbeek, I.M., O'Brien, K.W., Zhao, H., Shepherd, R.F.: Poroelastic foams for simple fabrication of complex soft robots. Adv. Mater. **27**, 6334–6340 (2015). doi:10.1002/adma.201503464

16. Chen, F.J., Dirven, S., Xu, W.L., Bronlund, J., Li, X.N., Pullan, A.: Review of the swallowing system and process for a biologically mimicking swallowing robot. Mechatronics **22**, 556–567 (2012). doi:10.1016/j.mechatronics.2012.02.005

17. Walsh, J.H., Leigh, M.S., Paduch, A., Maddison, K.J., Philippe, D.L., Armstrong, J.J., Sampson, D.D., Hillman, D.R., Eastwood, P.R.: Evaluation of pharyngeal shape and size using anatomical optical coherence tomography in individuals with and without obstructive sleep apnoea. J. Sleep Res. **17**, 230–238 (2008). doi:10.1111/j.1365-2869.2008.00647.x

18. Brasseur, J.G.: A fluid mechanical perspective on esophageal bolus transport. Dysphagia **2**, 32–39 (1987). doi:10.1007/BF02406976

19. Smooth-on safety data sheets. In: Smooth-On, Inc. https://www.smooth-on.com/

20. Schindelin, J., Arganda-Carreras, I., Frise, E., Kaynig, V., Longair, M., Pietzsch, T., Preibisch, S., Rueden, C., Saalfeld, S., Schmid, B., Tinevez, J.-Y., White, D.J., Hartenstein, V., Eliceiri, K., Tomancak, P., Cardona, A.: Fiji: an open-source platform for biological-image analysis. Nat. Meth. **9**, 676–682 (2012). doi:10.1038/nmeth.2019

Introducing Biomimomics: Combining Biomimetics and Comparative Genomics for Constraining Organismal and Technological Complexity

Claudio L. Flores Martinez[✉]

Biozentrum Grindel, University of Hamburg, Martin-Luther-King Platz 3,
20146 Hamburg, Germany
claudio.flores.martinez@embl.de

"The burgeoning fields of genomic and metagenomic sequencing
and bioinformatics are based on the notion that informational bits
are literally vital."
From Matter to Life [1]

Abstract. Integrated genomics and transcriptomics data, together with the analysis of total protein and metabolite content of a given cell, is providing the basis for complex, multi-scale and dynamic models of cellular metabolism in health and disease. Accordingly, the functional triad of genomics, transcriptomics and metabolomics is regarded as a foundational methodology in systems biology. Opening up a never-before seen vista into the organization and dynamical evolution of cellular life at multiple scales of complexity, Omics-approaches are poised to facilitate discoveries in biomimetic design processes. In the following, the proposed merger of biomimetics with Omics-techniques will be called "Biomimomics". Focusing on comparative genomics, this paper will outline how ongoing work in the field is revising our understanding of early nervous system and synapse evolution in animals and, at the same time, promises to give insights into truly, i.e. evolutionarily-based, biomimetic neuromorphic computing architectures. We will show how a new kind of modular workflow based on a "Biomimomic Traceability Matrix" (BTM) can structure and facilitate both biomimetic design solutions and the discovery of universal principles underlying complexifying biological and technological systems.

Keywords: Comparative genomics · Nervous system evolution · Chemical computation

1 Introduction

1.1 From Functional to Comparative Genomics

Starting around the year 2000 the high-throughput sequencing revolution, exemplified by the new discipline of functional genomics, fundamentally changed the way scientists

© Springer International Publishing AG 2017
M. Mangan et al. (Eds.): Living Machines 2017, LNAI 10384, pp. 148–160, 2017.
DOI: 10.1007/978-3-319-63537-8_13

engage problems in genome evolution, human health and computational biology. Today, next-generation sequencing tools allow the relatively cheap, fast and precise characterization of the gene and regulatory motif complement of various model and non-model organisms. The generated data, assembled from automated bioinformatic pipelines [2], can then be used to compare genomes of species across vast phylogenetic distances, from viruses and bacteria to unicellular eukaryotes and multicellular organisms such as jellyfish, insects and mammals. In fact, comparative genomics was initially used to define the minimal gene set of the Last Universal Common Ancestor (LUCA) of all life [3] and allowed to constrain and, first on theoretical grounds, reverse engineer [4] an ancient minimally complex cell by producing a list of essential transcriptional and translational cellular parts.

Comparative genomics has yielded unique insights into the evolutionary mechanisms underlying large-scale evolutionary events, from the origin of life to the emergence of multicellularity and complex nervous systems, and therefore presents a powerful tool in understanding major transitions in biological complexity [5]. Aside from data related to genome content, next-generation sequencing techniques are now readily employed to produce genome-wide expression profiles of various RNA species. Genomics is thus complemented by the emerging field of transcriptomics.

In the following we will propose a new method in reverse engineering bio-complexity based on the merger of biomimetic design and the wealth of information contained in the sequenced genomes of organisms from the three domains of life at all levels of complex organization.

1.2   Early Animal and Nervous System Evolution

Three of the five extant superphyla of Metazoa (animals) possess nervous systems composed of neuronal cell types that contain subcellular specializations, synapses, needed for synaptic transmission: Ctenophora (comb jellies), Cnidaria (jellyfish) and Bilateria (Fig. 1.). The latter sister taxa are traditionally grouped into the "Eumetazoa" while the remaining two animal phyla, Porifera (sponges) and Placozoa (made up by a single species, *Trichoplax adhaerens*), are lacking recognizable neural structures. Currently it is a matter of intense debate whether neurons and complex nervous systems have a singular homologous evolutionary origin or emerged independently via convergent evolution within the animal kingdom.

Here we are exploring an emerging view which proposes that, in contrast, synapses have evolved three times independently within the polyphyletic Neuralia (encompassing Ctenophora, Cnidaria and Bilateria) during a synchronous macroevolutionary phase of nervous system complexification that occurred around 600 Ma ago (Fig. 1) [6].

At an archetypal vertebrate synapse the propagation of the electric signal is achieved by the fast release of canonical low-molecular-weight neurotransmitters (LMW-NTs) from the presynaptic terminal across the synaptic cleft. In the following discussion we define this kind of synaptic signaling involving LMW-NTs as fast chemical neurotransmission (FCN).

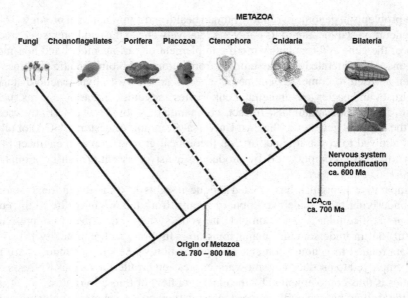

Fig. 1. Simplified phylogenetic tree of eukaryotes. Porifera, Placozoa, Ctenophora, Cnidaria and Bilateria together form the Metazoa (dashed lines denote unclear phylogenetic position). Extant nervous systems can be found in ctenophores, cnidarians and bilaterians. The origin of Metazoa can be approximately dated to 780–800 Ma and the cnidarian/bilaterian LCA (LCA$_{C/B}$) to ca. 700 Ma. The origin of Bilateria is approximated at 688 Ma (not shown). Red dots indicate convergent complexification events in early nervous systems occurring in the respective lineages. Only in the bilaterian branch occurred a "systemic overwrite" of metabolic and transcriptional networks related to fast chemical neurotransmission enabling higher-order neuronal complexity and emergent adaptive behavior (Color figure online).

A multitude of LMW-NTs belonging to different chemical classes are recognized in FCN: acetylcholine (ACh, the main neurotransmitter at the vertebrate neuromuscular junction); the biogenic amines serotonin (5-HT), dopamine (DA), octopamine (OA, widespread among invertebrates), norepinephrine (NE), epinephrine (Epi) and histamine (HA); and the main excitatory and inhibitory amino acid (EAA/IAA) neurotransmitters glutamate (Glu, EAA), its derivative *gamma*-Aminobutyric acid (GABA, IAA) and glycine (Gly, IAA) respectively.

2   Methods

We developed a novel organizational scheme, the Biomimomic Traceability Matrix (Fig. 2), which is modeled after the systems engineering tool from space mission design known as Science Traceability Matrix (STM) [7].

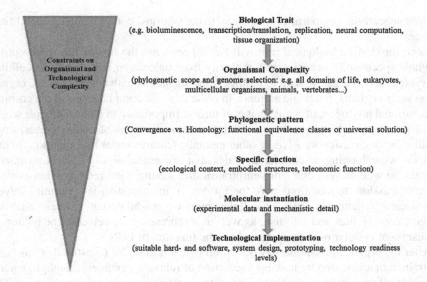

Fig. 2. Logic flow of a BTM. Each step of the workflow further constrains possible technological implementations for a given biological trait that is to be emulated by a biomimetic design solution.

STMs have become a standard tool in rapid mission architecture prototyping. They mainly serve to: "… provide the overview of what a mission will accomplish relative to high-level objectives suggested through Academy of science surveys, NASA road-maps, or program objectives. It provides a logical flow from these high level objectives through mission objectives, science objectives, measurement objectives, measurement requirements, instrument requirements and spacecraft and system requirements to data products and eventual publications. It is the one document that shows the relationship between all these key elements and the one document that provides the breadth needed to perform and document high level trades effecting science outcome and overall design" [7].

In a previous work we have generated a STM for an advanced robotic lander mission, the "Enceladus Explorer (EnEx)", bound for in situ exploration on Saturn's moon Enceladus. The main purpose of this specific STM was the biologically-oriented instrument selection for an autonomous melting probe that is part of the envisioned landing module [8, 9].

The logic flow of a BTM is arranged to create specific design solutions for technological implementations from generic biological traits (Fig. 2). These can be wide-spread across the tree of life, e.g. bioluminescence that is found in bacteria, unicellular eukaryotes and animals. In other cases they are only occurring at a certain level of organizational complexity, for instance with neural computation, which is exclusively present in the realm of animal complexity.

The first step in a design process based on BTMs is the definition of the biological trait that a technological system is supposed to be emulating. In our case let us choose the trait "neural computation". As we are interested in neuromorphic computing platforms we define neural computation as being based on specific cell types, i.e. neurons and their subcellular compartments essential to the neurons' computational capabilities, synapses.

Next, the desired biological trait will be mapped onto the tree of life representing the whole spectrum of organismal complexity from unicellular bacteria, to unicellular eukaryotes such as algae, and up to multicellular organisms with complex tissue organization such as plants, fungi and animals. In order to avoid complications in the ensuing computational phylogenetic analysis it is of utmost importance to make a precise selection of genomes that contain the needed information in terms of coding genes, transcription factors, regulatory RNA or other genomic features under investigation. In our case, we were looking at cellular modules that are essential to FCN: biosynthetic enzymes, as well as vesicular (for neurotransmitter loading) and reuptake transporters (for transmission termination). We performed a metazoan-wide genomic survey (encompassing genomes from unicellular protists, pre-bilaterian animals such as sponges, comb jellies and jellyfish, as well as a representative selection of bilaterian organisms) of cellular modules required for the function of FCN.

After the organismal complexity of the biological trait under investigation has been constrained in a first step by making a selection of relevant genomes, the phylogenetic analysis itself is commencing. This is one of the most work-intensive steps in the BTM process as bioinformatics experts together with evolutionary biologists have to determine the phylogenetic pattern of the biological trait under investigation. Broadly speaking, in this step of the analysis researchers are interested whether a certain biological trait arose via convergent evolution, i.e. independently and multiple times, or whether it has a singular homologous origin. The results of such an analysis can then demonstrate whether nature has found multiple solutions for a certain biological problem or whether one universal solution has been inherited to all members of a species, genus, family, order, class, phylum, kingdom or domain of organizational complexity. Generally, from a biomimetics perspective, convergent origins are immediately useful in further conceptualizing an alternate technological solution of a given problem as nature has already found multiple ways of solving it.

If a solution has arisen through convergent evolution, different organisms still perform specific functions with their respective biological trait. For example both plants and animals have turned multicellular. Nonetheless, each respective kingdom has a fundamentally different lifestyle, morphology and position in the planetary ecosystem. Thus, the specific function of a desired biological trait is examined based on its ecological context, embodied structures and teleonomic function (end-directedness).

Genomic analyses can lay the foundation for the experimental elucidation of the molecular underpinnings of a given biological trait. Often, considerable advances have been already made in this direction prior to the genomic analysis. For example, the synapses of bilaterian organisms have been studied for decades. However, since we now know that synapses within the animal kingdom most likely originated three times independently via convergent evolution, genomic analyses can help in narrowing down the differences in synapse organization. The main difference is constituted by the use of LMW-NTs in bilaterians vs. the use of neuropeptides in comb jelly and cnidarians.

Finally, with the last step of the BTM we arrive at putative technological implementations for the desired biological trait. For neuromorphic computing architectures we are mainly interested in the fast synapse of bilaterians since it is needed for higher cognitive abilities such as memory, associative thinking and abstract reasoning. This is why future research should focus on neuromorphic computational models that encompass the main difference between fast and "slow" synapses: the exploitation of highly-concentrated LMW-NTs diffusing into the synaptic cleft.

The technological implementation step should not be seen as a final and definitive solution for the biomimetic emulation of the desired biological trait but rather as a recommendation for further research & development work in the area of the derived target field. This starts new iterative rounds of technology-specific R&D efforts with soft- and hardware implementations, prototyping, comparisons to previously developed benchmarks, and technology readiness level assessments.

3   Results

3.1   Cellular Modules Needed for FCN Are Only Found in Bilateral Animals

Recent sequencing of the first comb jelly genomes together with functional evidence showed that the comb jellies' nervous system is strikingly different from those of Eumetazoa [10]. Especially surprising is the finding that Ctenophora are apparently using a distinct set of neurotransmitters when compared to Cnidaria and Bilateria, which are commonly thought to share fast transmitters.

Surprisingly, and contrary to previous studies, our findings (the computed phylogenetic trees are not shown here and will be published separately, we will focus on deriving general biomimetic insights from comparing animal genomes across a wide spectrum of organismal complexity) reveal Bilateria-specific orthologous subgroups absent in pre-bilaterian organisms that arose through gene duplication events in multiple protein families underlying fast synaptic signaling. This means that pre-bilaterian organisms only encode general purpose proteins in their genomes while bilateral organisms possess specialized enzymes and transporters for synthesizing LMW-NTs, package those into vesicles at the presynaptic terminal and take them up again after the transmission is completed.

Based on our analysis we conclude that fast synaptic signaling using classical neuro-transmitters originated along the stem line leading toward the Urbilaterian and, consequently, is a synapomorphy (a lineage-specific evolutionary innovation) of bilateral animals (Fig. 1). We propose that major neurotransmitter systems employed across Metazoa are not homologous, an unexpected finding that points toward the convergent evolution of complex neuronal cell types in animals.

We then applied our findings onto a BTM scheme (Table 1) in order to deduce biomimetic insights from the performed genomic analysis. For our phylogenomic bioinformatics pipeline we chose the adequate selection of genomes, i.e. from nearly all available sequenced animal genomes (including animals that do not have neurons such as sponges and *Trichoplax*). To compare the genomes of animals across each other reference genomes that are not stemming from the animal kingdom have to be included into the analysis as well ("outgroups"). Therefore, we added genomes of the animals' closest unicellular relatives, the choanoflagellates, into our pipeline.

Table 1. BTM for neural computation (blue), synaptic complexity/soft robotics (green) and eusociality/swarm intelligence (purple). Relevant target fields are marked in red.

Biological Trait	Organismal Complexity	Phylogenetic Pattern	Specific Function	Molecular Instantiation	Technological Implementation
Neural Computation (based on neurons and synapses)	Animals (with unicellular protists as outgroup)	Convergence (independent origins of neurons in three of five animal phyla)	Fast synapse (bilateral organisms)	LMW-NTs	Neurotransmitter fields
			Slow endocrine like (comb jellies and cnidarians)	Neuro-Peptides	Purely action-Potential based computation?
Synapse complexity (soft robotics)	Bilateral organisms (+ octopus genome)	Convergence (independent protein family expansion)	Cell-cell adhesion, early nervous system development (vertebrates)	Protocadherin Family	Neural circuit Formation?
			Unknown (cephalopods)	Protocadherin Family	Distributed brain, soft body, autonomous limbs, synaptic connectivity?
Eusociality / swarm behavior	Insects (Hymenoptera)	Convergence (at least 11 times, multiple gene families and molecular functions)	Small colony (bumble bee)	Signal transduction	Chemical cues
			Giant colony (super organism, ants)	Gland development	
			Beehive	Carbohydrate metabolism	

The pattern we obtained from our analysis is summarized above (Fig. 1). In short, neurons in each respective animal phylum that has a nervous system are using a distinct set of neurotransmitters (LMW-NTs, Bilateria) and neuropeptides (Ctenophora and Cnidaria). Fast synapses that are associated with centralized nervous systems and advanced cognitive abilities are only found in bilateral organisms.

3.2 Soft Robotics and Swarm Behavior

In addition, we included some relevant findings of recent genomic studies (Table 1) [11, 12] into our BTM approach. In the case of synaptic complexity (soft robotics) further research on the independent expansions in the proto-cadherin protein family (in invertebrates and vertebrates) are of particular interest. Regarding swarm intelligence future analyses should increase the scope of organismal complexity to constrain the physiology underlying swarm behavior.

4   Discussion

4.1   Comparison of Comb Jelly, Cnidarian and Bilaterian Synapses

A convergence of multiple lines of evidence places the emergence of embodied cognition, associative learning and complex sensory adaptations into the Cambrian period [13, 14]. This transition has most likely occurred exclusively in bilateral organisms (although multiple times independently via convergent evolution in arthropods, mollusks and vertebrates) and involved the function of FCN modules for increased synaptic complexity [15].

Based on ultrastructural, morphological and system-wide functional differences we hypothesize that the neurons and synapses of Ctenophora, Cnidaria and Bilateria are distinct evolutionary routes, deriving from the deep homology of a non-neuronal paracrine signalling system inherited from the last common ancestor of Metazoa (LCA_{MZ}), leading toward functionally analogous solutions performing specialized neurosecretory roles within the context of multicellular animal complexity. Most significant in regard to structural specificities across pre-bilaterians and Bilateria is the absence of well-developed presynaptic terminals and non-invaginating spine synapses in the former [16].

Cnidarian and comb jelly pre-synapses rather resemble endocrine cells from bilaterians than their more morphologically complex bilaterian FCN counterparts. Secretory cell-like synapses in ctenophores and cnidarians possibly represent convergent, lineage-specific adaptations for more rapid, neuronal transmission (functionally quite similar but not closely related to cells capable of FCN in Bilateria) that derived independently from a more general non-neuronal paracrine signaling system already present in LCA_{MZ}.

The presence of action potential-based synaptic transmission in the simple nerve nets of comb jellies and cnidarian jellyfish, which use neuropeptides produced by protein biosynthesis instead of classical low-molecular-weight neurotransmitters, led us to hypothesize that fast transmitters played a decisive role in the complexification of central nervous systems in bilateral organisms by expanding the informational capacities of synapses to include a novel type of bioenergetically favorable chemical computation via dense and highly concentrated molecular neurotransmitter clouds located inside the synaptic cleft.

4.2   Neurotransmitter Fields and Synapse Complexification

The idea of neurotransmitter-mediated computation is corroborated by the theoretical work of Douglas S. Greer on "neurotransmitter fields" [17, 18], which posits that neurotransmitter fields differ from neural fields in the underlying principle that the state variables are not the neuron action potentials but the chemical concentration of neurotransmitters in the extracellular space.

The neurotransmitter field model is not only supported by, firstly, our own comprehensive phylogenetic analysis and, second, our novel BTM workflow but, in theory, makes it possible for a small number of neurons, even a single neuron, to "establish an association between an arbitrary pattern in the input field and an arbitrary output pattern"

[19], thereby allowing a single layer of neurons to perform the computation of a two-layer neural network.

Neurotransmitter fields and biomimetic computation, as first put forward by Greer and expanded on by Greer and Tuceryan (see [18] for a comprehensive technical report), work according to the following three main principles:

(1) Neurotransmitters (LMW-NTs, e.g. ACh or 5-HT) are released into the synaptic cleft from the presynaptic terminal and subsequently mix in the confined space. The final concentration of transmitters within the synapse equals the sum of the amounts contributed by each presynaptic terminal. Dendritic spine receptors (DR) located at the postsynapse as well as presynaptic terminal autoreceptors (AR) detect the presence of transmitters within the synaptic cleft and elicit downstream signalling cascades.

(2) The postsynaptic response, β, can be defined as a function of the transmitter concentration, g (instead of mean firing rates based on electric action potentials). If the presynaptic neuron releases a fixed amount of transmitter into the synaptic cleft, defined as Δg, the effect on the postsynaptic neuron will be different when the concentration is near the central inflection point of a sigmoidal transfer function, as compared to the postsynaptic response near its tails.

(3) In the composite mathematical model discrete neural networks are a special case of more abstract and general neurotransmitter fields underlying biological computation. Dendritic and axonal trees are connecting the discrete neurons with the continuous neurotransmitter fields. The composite model assumes that the intracellular state of each neuron is separate (discrete) from the environment (creating individual computational units), whereas the surrounding extracellular state is continuous and in contact with all nearby cells. Accordingly, the most appropriate mathematical model of neural computation contains both discrete and continuous elements. In this scenario discrete nodes represent individual cells and continuous images map onto the extracellular molar concentrations of LMW-NTs. The neurotransmitter fields may encode physical quantities like force, mass, energy, or location, defined on continua such as space, time, or frequency. By producing a topological alignment of multiple input images multisensory integration and the construction of complex operations become possible.

Since the ability to associate images is the basis for learning relationships in vision, hearing, tactile sensation, and kinetic motion, neurotransmitter field algorithms are promising to generate innovative and more biology-based – biomimetic rather than just bio-inspired – technological implementations in neuromorphic computing than purely spike-dependent models [20].

4.3 Major Synthetic Transition in Artificial Synapse Design

In order to initiate an evolutionary "major synthetic transition" in biomimetic computation [21] we have to comprehend in more detail how nervous systems initially evolved and developed (Fig. 3). By using comparative genomics we can begin to reverse engineer the first fast synapse that originated more than 600 Ma ago [15] and subsequently complexified during the Cambrian. The lessons learned can then be applied to the design process of our own technological implementations, which require a deeper physiological understanding of the neural substrate of learning and chemical computation within the brain. The evolution of both biological and technological synapses (in neuromorphic systems) appears to be constrained by a single energy–time minimization principle that may govern the design of many complex systems that process energy, materials and information [22].

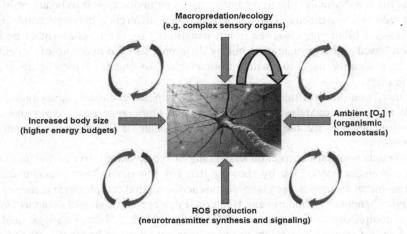

Fig. 3. Adaptive spatio-temporal information organizer based on fast chemical synapses. Four major selective pressures could have shaped the emerging nervous system of early bilaterians (blue arrows): ecological changes including the onset of macropredation, rising ambient oxygen concentrations effecting overall organismic oxygen homeostasis and endogenous reactive oxygen species (ROS) production due to neurotransmitter synthesis and signaling, as well as increased body sizes. Green arrows indicate reciprocal modulatory relationships between respective selective pressures. For example, increased body sizes influence predatory trophic modes. These, in turn, feedback on the evolution of body sizes. Emergent neural structures based on fast chemical synapses act as a superordinate adaptive spatio-temporal information organizer that integrates external and internal signals. The activity of this organizer initiates the central feedback layer of early nervous system evolution (grey arrow): self-referential informational selection on nascent microcircuits and integrated large-scale distributed networks conferring higher cognitive abilities and metabolic control (Color figure online).

The origins of functional FCN coincide with both the emergence of bilateral organization and nascent nervous systems and, therefore, chemical computation can be considered a driving force in advanced information processing and tissue complexification [23]. Using a biomimomic approach we were able to extract pivotal genomic information concerning fundamental principles that are underlying neuronal computation at a higher level of complexity in bilateral organisms. Consequently, future neuromorphic architectures should incorporate models of chemical computation at complex synapses.

5 Conclusions

The preliminary conclusion from our biomimomic perspective on synapse evolution is the realization that LMW-NTs constitute a key element in the complexification of early nervous systems. Simple nerve nets, employing bioenergetically expensive neuropeptides, in gelatinous comb jellies and jellyfish are capable of generating action potential-based synaptic transmission. However, their computational capacity remains limited. Synaptic plasticity and complex behavior appear to be an exclusive trait of bilateral organisms. The major functional difference between comb jelly, cnidarian and bilaterian synapses is not constituted by the mechanism of action-potential based signal propagation but by differences in the make-up of molecular transmitter identity used to switch from electric to chemical messaging at the synaptic cleft.

Further, neurotransmitters are not only acting as chemical messengers, but embody a computational device in their own right. Our comprehensive phylogenetic analysis underscores the largely overlooked relevance of Greer's neurotransmitter field concept.

Our results revise the current understanding of synaptic and nervous system evolution in two major ways. First by showing that key functional low-molecular-weight neurotransmitter systems are not homologous across animal complexity, i.e. the neuralized phyla Ctenophora, Cnidaria and Bilateria. Future research should avoid to look at synaptic complexity in Ctenophora and Cnidaria through a "bilaterian lens" and we expect that fast transmission might be actually present in these lineages – albeit being instantiated by different functionally equivalent molecular tool-kits. The extremely sudden emergence of fast synapses in Bilateria with the parallel deployment of transmitters and gap junctions in non-neural contexts opens up new ways of thinking on early nervous system and tissue evolution [24].

Second, a functional closure of metabolic and transcriptional networks, forming a superordinate autocatalytic set, led to the emergence of a higher-order functional unit capable of synaptic plasticity and homeostasis in fast chemical neurotransmission systems of early bilaterians. Thus, our findings on the emergence of synaptic complexity are not only relevant to evolutionary biology and comparative genomics but also to neuroscience as a whole, medicine of neurodegenerative diseases and neuromorphic computing.

Refining the biomimomic approach based on BTMs promises to generate a systems engineering perspective on biomimetics. BTMs logically connect foundational research questions and reverse engineering procedures dealing with bio-complexity. Biomimomics unlocks the wealth of hidden information stored away in the constantly increasing number of published genomes. In the same way that TRIZ, the problem solving methodology for engineering, has turned biological [25], we have fused biomimetics with genomics to use nature's patents for designing and constructing the living machines of the future.

References

1. Walker, S.I., Davies, P.C.W., Ellis, G.F.R. (eds.): From Matter to Life: Information and Causality, p. 1. Cambridge University Press, Cambridge (2017)
2. Huerta-Cepas, J., et al.: eggNOG 4.5: a hierarchical orthology framework with improved functional annotations for eukaryotic, prokaryotic and viral sequences. Nucleic Acids Res. **44**(D1), D286–D293 (2016)
3. Koonin, E.V.: Comparative genomics, minimal gene-sets and the last universal common ancestor. Nat. Rev. Microbiol. **1**(2), 127–136 (2003)
4. Csete, M.E., Doyle, J.C.: Reverse engineering of biological complexity. Science **295**(5560), 1664–1669 (2002)
5. Koonin, E.V.: The logic of chance: the nature and origin of biological evolution, xii edn, p. 516. Pearson Education, Upper Saddle River (2012)
6. Liebeskind, B.J., et al.: Complex homology and the evolution of nervous systems. Trends Ecol. Evol. **31**(2), 127–135 (2016)
7. Weiss, J.R., Smythe, W.D., Wenwen, L.: Science traceability. In: 2005 IEEE Aerospace Conference (2005)
8. Flores Martinez, C.L.: Convergent evolution and the search for biosignatures within the solar system and beyond. Acta Astronaut. **116**, 394–402 (2015)
9. Konstantinidis, K., et al.: A lander mission to probe subglacial water on Saturn's moon enceladus for life. Acta Astronaut. **106**, 63–89 (2015)
10. Moroz, L.L., et al.: The ctenophore genome and the evolutionary origins of neural systems. Nature **510**(7503), 109–114 (2014)
11. Albertin, C.B., et al.: The octopus genome and the evolution of cephalopod neural and morphological novelties. Nature **524**(7564), 220–224 (2015)
12. Kapheim, K.M., et al.: Social evolution. Genomic signatures of evolutionary transitions from solitary to group living. Science **348**(6239), 1139–1143 (2015)
13. Verschure, P.F.: Synthetic consciousness: the distributed adaptive control perspective. Philos. Trans. R. Soc. Lond. B Biol. Sci. **371**(1701), 20150448 (2016)
14. Feinberg, T.E., Mallatt, J.: The evolutionary and genetic origins of consciousness in the Cambrian period over 500 million years ago. Frontiers Psychol. **4**, 667 (2013)
15. Bronfman, Z.Z., Ginsburg, S., Jablonka, E.: The transition to minimal consciousness through the evolution of associative learning. Frontiers Psychol. **7**, 1954 (2016)
16. Petralia, R.S., et al.: The diversity of spine synapses in animals. NeuroMol. Med. **18**, 497 (2016)
17. Greer, D.S.: Neurotransmitter fields. In: Sá, J.M., Alexandre, L.A., Duch, W., Mandic, D. (eds.) ICANN 2007. LNCS, vol. 4669, pp. 19–28. Springer, Heidelberg (2007). doi: 10.1007/978-3-540-74695-9_3
18. Greer, D.S., Tuceryan, M.: Neurotransmitter Field Theory: A Composite Continuous and Discrete Model 2010. Department of Computer and Information Science, Purdue University. http://cs.iupui.edu/~tuceryan/tech-reports/TR-CIS-0315-10.pdf
19. Greer, D.S.: Images as symbols: an associative neurotransmitter-field model of the brodmann areas. In: Gavrilova, M.L., Tan, C.J.K., Wang, Y., Chan, K.C.C. (eds.) Transactions on Computational Science V. LNCS, vol. 5540, pp. 38–68. Springer, Heidelberg (2009). doi: 10.1007/978-3-642-02097-1_3
20. Aur, D.: From neuroelectrodynamics to thinking machines. Cogn. Comput. **4**(1), 4–12 (2012)
21. Sole, R.: The major synthetic evolutionary transitions. Philos. Trans. R. Soc. Lond. B Biol. Sci. **371**(1701), 20160175 (2016)

22. Moses, M., et al.: Energy and time determine scaling in biological and computer designs. Philos. Trans. R. Soc. Lond. B Biol. Sci. **371**(1701), 20150446 (2016)
23. Newman, S.A.: Form and function remixed: developmental physiology in the evolution of vertebrate body plans. J. Physiol. **592**(11), 2403–2412 (2014)
24. Baluška, F., Levin, M.: On having no head: cognition throughout biological systems. Frontiers Psychol. **7**, 902 (2016)
25. Vincent, J.F., et al.: Biomimetics: its practice and theory. J. R. Soc. Interface **3**(9), 471–482 (2006)

Effects of Locomotive Drift in Scale-Invariant Robotic Search Strategies

Carlos Garcia-Saura$^{(\boxtimes)}$, Eduardo Serrano, Francisco B. Rodriguez, and Pablo Varona

Grupo de Neurocomputación Biológica, Escuela Politécnica Superior, Universidad Autónoma de Madrid, Madrid, Spain
carlos.garciasaura@uam.es

Abstract. Robots play a fundamental role in the exploration of environments that are harmful to humans or animals: robotic probes can reach deep into the earth's crust, explore our oceans, traverse high radiation areas, navigate in outer space, etc. The harsh conditions and large amounts of uncertainty of these environments can complicate the use of global positioning systems, and in some cases robots have to depend exclusively in local information as external position landmarks are not available. *Lévy walks* are increasingly studied as effective solutions in these exploratory contexts. The superdiffusive (dispersive) properties of these forms of random walks are often exploited by many animal species, in particular when tackling search problems that have uncertainty. Based on experimentation with low-cost mobile robots, this work has characterized how long-term motion drift (which is inherent to search contexts that lack global positioning systems) can have an effect in the overall characteristics of Lévy trajectories. The results show that Lévy-based searches can be robust and maintain superdiffusive properties for some ranges of motion drift parameters that are closely related to the scale of the search problem. Locomotive drift seems to act effectively as a long-term truncation parameter that could be corrected or even incorporated during the design of a search task.

Keywords: Lévy flight · Bio-inspired random walks · Exploration · Mobile robots · Scale invariance · Motion drift · Odor sensor

1 Scale-Free Search Problems

Tasks such as exploration and monitoring, surveillance, search and rescue, etc., are increasingly being undertaken by robots [1–4]. In many applications, basic algorithms to drive each search strategy are sufficient when there is availability of information regarding the environment, the robot(s), or even the target(s). However, uninformed searches with higher levels of uncertainty (e.g. regarding the operational space, the location of the targets and the robots, battery duration, sensory ranges, etc.) demand other solutions.

© Springer International Publishing AG 2017
M. Mangan et al. (Eds.): Living Machines 2017, LNAI 10384, pp. 161–169, 2017.
DOI: 10.1007/978-3-319-63537-8_14

Lévy walks are random motion patterns consisting of a combination of clusters of short steps with longer spaces in between. This organization is repeated across all scales, resulting in a fractal structure with no characteristic scale and dispersive properties that arise from an inherent probability distribution that is heavy-tailed. These properties make Lévy walks more adequate for search purposes than other walk strategies such as Brownian motion because they reduce visits to previously explored regions [5]. Lévy walks are also known as Lévy flights particularly in the cases where each step is discrete and instantaneous, instead of a continuous motion along each point of the trajectory. These scale-free random walks are studied in a wide variety of applications in different domains ranging from biology, physics and economics [6]. In particular, they are used to describe effective search strategies under uncertainty for several animal species (for a review see [7,8]).

Recently, several robotic works have employed bio-inspired Lévy walk strategies to address the problem of finding sparsely distributed targets in high uncertainty environments with single robots or robot swarms [9–16]. One example of such problems is the localization of odor sources in the environment using mobile robots equipped with gas sensors; uncertainty can arise from the latency of the sources, the distribution of odor plumes, etc. In these studies, the Lévy walk strategy is typically built without taking into account aspects of the motor plant and/or the walk implementation, which can result in a search pattern with burdensome deviations from the corresponding theoretical Lévy distribution (see Fig. 1).

As these walk strategies are increasingly studied for search purposes under high uncertainty conditions, it is important to emphasize that these deviations might lead to less effective exploration. A similar problem can be encountered when Lévy walks are combined with other search strategies to build the motion pattern, for instance as a backup behavior when there is not enough information available to perform directed search.

In this paper we characterize the problem of locomotive drift in scale-free searches, and provide a comparison between a pure Lévy trajectory and the actual motion performed by a robot with different levels of drift. Among the methodology used in the analysis, we have employed a technique recently purposed in [17] to assess the comparison of random walks with curvature across multiple scales. Our study highlights a tight relation between locomotive drift and scale, and points to the necessity of taking into account drift (and other factors that affect motion) when designing bio-inspired foraging strategies into real world problems.

2   Robotic Platform and Drift Model

The robot used in this work is a GNBot[1], which is part of a low-cost platform originally designed to study the task of odor localization with bio-inspired cooperative motion strategies [18]. One example of such problems is illustrated in

---

[1] https://github.com/carlosgs/GNBot/wiki.

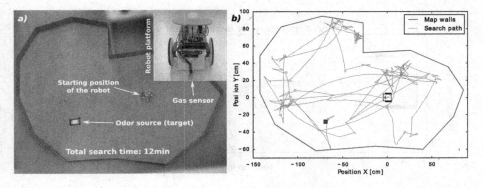

Fig. 1. *Example implementation of a bio-inspired search strategy in a context with uncertainty.* The experimental setup is illustrated in Fig. 1a and uses a GNBot equipped with a low-cost odor sensor (TGS2600). The goal is to locate a target made of a piece of cotton impregnated with a volatile compound (ethanol). The ground truth trajectory shown in Fig. 1b was recorded with a ceiling camera by placing a color marker on top of the robot. In this experiment, the search strategy is based on a Lévy walk and defined as a set of uniform in-place rotations combined with linear trajectories whose lengths vary according to a Lévy distribution. Note that the segments that were supposed to be linear are actually performed as arcs due to drift in yaw. See also a video in https://www.youtube.com/watch?v=HuAor0oZO_s

Fig. 1 as an odor search task. The GNBot enables to evaluate a range of search behaviors in mobile robots under laboratory conditions, and study how sensory information can be incorporated to improve the quality of searches when there is uncertainty (i.e. limited range or set of sensory resources, the lack of a map of the environment, etc.).

Each robot can perform arbitrary locomotion in the 2D plane by using two actuators attached to the wheels. These are digitally-controlled continuous-rotation servomotors mechanically coupled to each wheel and commanded by the main microcontroller. Hence directionality of motion is achieved by means of differential drive: the heading of each robot is determined by the ratio between the rotational velocities of its two wheels.

Setting the same velocity for each wheel should yield a forward motion. In practice, any inaccuracies such as variable wheel diameter, fluctuations in motor velocities, or even physical effects such as tire slipping, etc., will directly translate into a motion that diverges from a straight line (see Fig. 2). These effects could be reduced with the incorporation of additional sensors such as wheel encoders or gyroscopes [19]. Although drift could be corrected with external landmark tracking, global reference systems are often expensive or even unavailable in some situations (i.e. deep sea exploration, high radiation environments, etc.).

Figure 3 shows the characterization of a basic form of locomotion drift. This naive model is useful to illustrate the inaccuracies that arise from assuming accurate linear trajectories while implementing scale-free locomotion strategies. The studied ranges of drift (2 to 32°/m) are representative for the values that

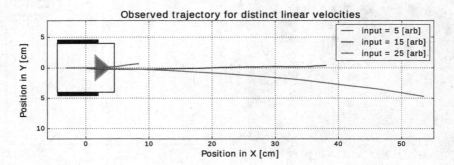

Fig. 2. *Motion drift in an open-loop robot controller.* The robot was commanded to describe a straight path by setting both motors to the same velocity parameter. Each trajectory was recorded with a ceiling camera for 3 s. The input speed setting has arbitrary units since both actuators are uncalibrated servo-motors and do not have a direct mapping with real world velocities. It can be observed that there is translational drift that varies for each speed parameter, so straight paths are performed as arcs of different radiuses.

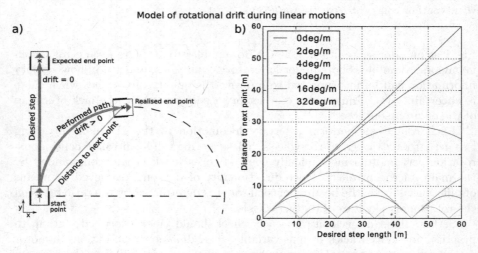

Fig. 3. *Distortion of straight paths due to locomotive drift.* The model used for simulating Lévy trajectories with motion drift approximates linear translations -steps- as arcs (panel *a*). The drift parameter is specified as the amount of rotation that the robot incorporates after moving forward a given distance -measured as the longitude of the performed arc- and its units are degrees per meter (deg/m). One of the assumptions of the model is that drift is constant for a given robot velocity. Panel *b* shows the relation between the length of the desired steps and the actual displacements that are observed when measuring only the stopping positions of the robot. At large scales the simulated robot will move in perfect circles due to the simplicity of the model.

have been observed during experimentation: from an uncalibrated open-loop motion controller shown in Fig. 1 ($\delta \lesssim 32°/m$) to the use of a gyroscope for increased precision ($\delta \approx 2°/m$) [19].

3   Results

The drift observed in the GNBot platform (Figs. 1 and 2) serves to illustrate a common problem in mobile robotics: with open-loop position control, perfect straight motions are not possible (not at every scale). This is highly relevant for some bio-inspired algorithms that depend on fractal properties, as drift at any scale can have great impact in the overall trajectories. Hence we see the necessity of assuming that some sort of motion drift will always be present, in order to incorporate it into search algorithms that are robust to uncertainty.

3.1   Distortion of a Lévy Walk in a Robot Platform

During a Lévy walk a robot moves to each new position with *steps* of length l that vary along every scale, and between each step there is an in-place rotation obtained from a uniform distribution. In particular, the length of each step is drawn from a heavy-tailed probability density function $f(l) = \frac{\alpha-1}{l_{min}} \left(\frac{l}{l_{min}} \right)^{-\alpha}$ where α is the *exponent* or *scaling* parameter and l_{min} is the lower bound parameter [8]. The scaling parameter must lie in the range $2 < \alpha < 3$ in order to achieve superdiffusive properties. In our case we choose $\alpha = 2.5$ and $l_{min} = 10\,cm$ as the scaling and bound parameters, respectively.

Motion drift is then incorporated using the model from Fig. 3. Linear steps are translated into arcs of radiuses defined by drift. A series of simulations were run with these parameters, generating motion sequences for different values of drift. These sequences were analyzed by measuring the distances between each of the resting points of the robot (Fig. 3) and calculating their probability density. This has been represented in Fig. 4. Drift becomes apparent in the Lévy statistics as a restriction over the maximum length of each step.

It could be argued that the truncations observed in Fig. 4 might be introduced artificially, as the length of a trajectory is calculated using only the end points of each step. For instance a robot could perform a number of consecutive short steps that align with a particular direction, that would still count as individual segments rather than a single large one.

3.2   Divergence at Multiple Scales

A different analysis is possible with the *rescaling method* recently proposed by Tromer et al. [17]. This method relies in the fractal properties of superdiffusive patterns, which preserve the same general shape at multiple scales. *Zooming* in or out should always yield similarly shaped patterns. This is quantified with a

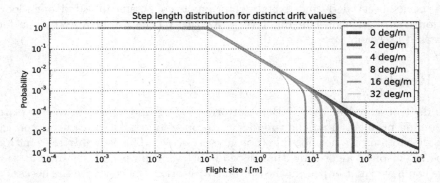

Fig. 4. The graph shows the cumulative distribution function for Lévy walks of size l (distances between each resting point during a search path, the points where the robot stops to rotate in-place). Translational drift δ has the effect of limiting the maximum length that a robot can travel in a particular step $l_{max} \sim 1/\delta$. Traces show a sharp transition at $l = 10^{-1}$ m that corresponds to the minimum step value used in the Lévy generator ($l_{min} = 10\ cm$). Steps below this value occur only for $\delta \neq 0$ and are caused by long-range walks where the robot describes one or more circles and ends up near the starting point.

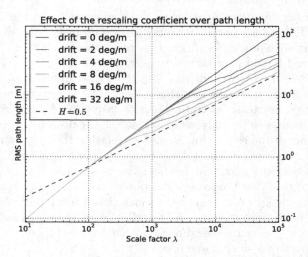

Fig. 5. *Divergence from superdiffusion at large scales*. Given a list of robot positions along a trajectory, the rescaling coefficient λ is defined in order to generate a coarse-grain resample of the trajectory. For each value of λ, one sample out of λ is taken into account for the RMS path length calculation (Root Mean Square of distances between coordinates). This allows to directly compare the diffusivity of different motion strategies across all scales [17]. In this case each positive value of drift yields trajectories that, at some scale, diverge from an "ideal" Lévy walk ($\delta = 0°/m$). $H = 0.5$ is the slope of diffusivity expected from pure Brownian motion.

coarse-grain resample of the trajectories (see Fig. 5) that overcomes the masking effect that is introduced when multiple short steps are arranged to form longer walks.

The rescaling method is a powerful tool to analyze and compare the scale invariance of many types of bio-inspired searches. Figure 5 illustrates that Lévy walks can maintain superdiffusivity for some ranges of drift and scale, but eventually slow down their dispersion rate into classic Brownian behavior. This can be reflected in the effectiveness of the search but may not become apparent with a visual inspection of the trajectories (see Fig. 6).

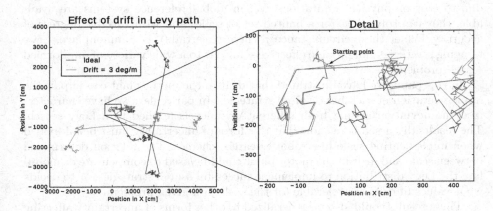

Fig. 6. *Search path with motion drift diverges from a pure* Lévy *walk.* Two simulations were ran for drift values $[0, 3]°/m$ while leaving the rest of parameters unchanged (these were specified in Sect. 3.1). By using the same seed for the random number generator, it is possible to visually compare both paths: a desired, ideal Lévy walk (blue), and the distorted trajectory that is performed due to inaccuracies in the locomotive plant (green). Both searches diverge at large scales (left panel) and arcs are appreciated for the path with drift, while this effect is not as evident at smaller scales (right panel). However, differences in the statistical properties of each trajectory are not obvious and require the analysis shown in Fig. 5. The drift value was selected to illustrate the problem with real-world data from experiments with the GNBot robot [19]. Drift model was explained in Fig. 3. This simulation is equivalent to a single run of a GNBot that starts with a full battery and moves continuously for approximately 2 h at a speed of 10 cm/s. (Color figure online)

The results shown in Figs. 4, 5 and 6 indicate that even subtle amounts of drift can perturb the characteristics of Lévy walks by limiting their diffusivity at some scales. This needs to be considered when implementing the sort of bio-inspired search strategies that rely in scale invariance, as is often the case for search problems with high degrees of uncertainty.

Our analysis also enables to determine the ranges of locomotive drift that can maintain superdiffusive properties during a search task, in the cases where the scale of the search problem is known a priori.

4   Conclusion and Discussion

Robotic search under uncertainty is a problem that arises in different domains: ocean, lunar, meteor and planetary exploration, search and rescue operations in specific contexts (indoor, underwater, aerial, underground, dense forest locations), etc. In such domains, the sources of uncertainty might be related to the inability to fully or partially characterize the location of the robot(s), the environmental features and/or the search targets, which could move, appear and disappear with unknown latencies, etc. For instance, absolute positioning systems require reliable landmarks or external beacons that may not be an option due to cost or physical limitations. When global reference systems are available, they can solve drift for a limited set of scales and with variable degrees of accuracy. Under these circumstances, where uncertainty is commonplace, Lévy foraging strategies can provide effective solutions to a wide variety of uninformed search problems.

In this paper we have illustrated that drift in the motor plant can impair the implementation of such bio-inspired strategies. In particular, we have characterized the deviation arising from different drift levels applied to a Lévy search. The results discussed here show that this form of uncertainty must be addressed when implementing scale-free robotic search schemes. The relation determined between scale and drift in the motor plant could be used to compensate or modulate the Lévy distribution to implement successful search strategies, or to decide what values of drift are tolerable for the scale of a given problem.

These results could also be generalized to other forms of uncertainty affecting locomotion. For instance the same analysis would be valid to assess the effect of environmental limitations such as walls or other obstacles, and/or interaction between multiple robots within the same search space, etc.

With our implementation of Lévy walks, decision making takes place at discrete intervals and in two dimensions. In the animal kingdom these movement patterns arise from continuous interactions between complex neurophysical systems and the real world. However, in the absence of adequate external landmarks, animal walks could also suffer from motion biases similar to the drift effects that have been illustrated (Figs. 4, 5 and 6) so the work reported here is also relevant to understand the conditions for effectiveness in animal search strategies.

Acknowledgements. We acknowledge support from MINECO/FEDER DPI2015-65833-P, TIN2014-54580-R (http://www.mineco.gob.es/).

References

1. Marjovi, A., Marques, L.: Multi-robot olfactory search in structured environments. Robot. Auton. Syst. **59**(11), 867–881 (2011)
2. Hu, J., Xu, J., Xie, L.: Cooperative search and exploration in robotic networks. Unmanned Syst. **01**(01), 121–142 (2013)
3. Sugiyama, H., Tsujioka, T., Murata, M.: Real-time exploration of a multi-robot rescue system in disaster areas. Adv. Robot. **27**(17), 1313–1323 (2013)

4. Hollinger, G.A., Yerramalli, S., Singh, S., Mitra, U., Sukhatme, G.S.: Distributed data fusion for multirobot search. IEEE Trans. Robot. **31**(1), 55–66 (2015)
5. Shlesinger, M.F., Klafter, J.: Lévy walks versus Lévy flights. In: Stanley, H.E., Ostrowsky, N. (eds.) On Growth and Form, pp. 279–283. Springer, Dordrecht (1986)
6. Zaburdaev, V., Denisov, S., Klafter, J.: Lévy walks. Rev. Mod. Phys. **87**(2), 483–530 (2015)
7. Humphries, N.E.M., Queiroz, N., Dyer, J.R.M., Pade, N.G., Musyl, M.K., Schaefer, K.M., Fuller, D.W., Brunnschweiler, J.M., Doyle, T.K., Houghton, J.D.R., Hays, G.C., Jones, C.S., Noble, L.R., Wearmouth, V.J., Southall, E.J., Sims, D.W.: Environmental context explains Lévy and brownian movement patterns of marine predators. Nature **465**(7301), 1066–1069 (2010)
8. Reynolds, A.: Liberating Lévy walk research from the shackles of optimal foraging. Phys. Life Rev. **14**, 59–83 (2015)
9. Nurzaman, S.G., Matsumoto, Y., Nakamura, Y., Koizumi, S., Ishiguro, H.: 'Yuragi'-based adaptive mobile robot search with and without gradient sensing: from bacterial chemotaxis to a levy walk. Adv. Robot. **25**(16), 2019–2037 (2011)
10. Mohanty, P.K., Parhi, D.R.: Cuckoo search algorithm for the mobile robot navigation. In: Panigrahi, B.K., Suganthan, P.N., Das, S., Dash, S.S. (eds.) SEMCCO 2013. LNCS, vol. 8297, pp. 527–536. Springer, Cham (2013). doi:10.1007/978-3-319-03753-0_47
11. Stevens, T., Chung, T.H.: Autonomous search and counter-targeting using Levy search models. In: 2013 IEEE International Conference on Robotics and Automation, pp. 3953–3960 (2013)
12. Sutantyo, D., Levi, P., Moslinger, C., Read, M.: Collective-adaptive Lévy flight for underwater multi-robot exploration. In: 2013 IEEE International Conference on Mechatronics and Automation (IEEE ICMA 2013) (2013)
13. Fioriti, V., Fratichini, F., Chiesa, S., Moriconi, C.: Levy foraging in a dynamic environment extending the Levy search. Int. J. Adv. Robot. Syst. **12**(7), 98 (2015)
14. Fricke, G.M., Hecker, J.P., Cannon, J.L., Moses, M.E.: Immune-inspired search strategies for robot swarms. Robotica **34**(08), 1791–1810 (2016)
15. Katada, Y., Nishiguchi, A., Moriwaki, K., Watakabe, R.: Swarm robotic network using Lévy flight in target detection problem. Artif. Life Robot. **21**(3), 295–301 (2016)
16. Mohanty, P.K., Parhi, D.R.: Optimal path planning for a mobile robot using cuckoo search algorithm. J. Exper. Theor. Artif. Intell. **28**(1–2), 35–52 (2016)
17. Tromer, R.M., Barbosa, M.B., Bartumeus, F., Catalan, J., da Luz, M.G.E., Raposo, E.P., Viswanathan, G.M.: Inferring Lévy walks from curved trajectories: A rescaling method. Phys. Rev. E **92**(2), 22147 (2015)
18. García-Saura, C., Borja Rodríguez, F., Varona, P.: Design principles for cooperative robots with uncertainty-aware and resource-wise adaptive behavior. In: Duff, A., Lepora, N.F., Mura, A., Prescott, T.J., Verschure, P.F.M.J. (eds.) Living Machines 2014. LNCS, vol. 8608, pp. 108–117. Springer, Cham (2014). doi:10.1007/978-3-319-09435-9_10
19. Garcia-Saura, C.: Self-calibration of a differential wheeled robot using only a gyroscope and a distance sensor. CoRR, abs/1509.02154 (2015)

Simulation of Human Balance Control Using an Inverted Pendulum Model

Wade W. Hilts[1]([⊠]), Nicholas S. Szczecinski[2], Roger D. Quinn[2], and Alexander J. Hunt[1]

[1] Department of Mechanical and Materials Engineering, Portland State University, Portland, OR, USA
whilts@pdx.edu
[2] Department of Mechanical and Aerospace Engineering, Case Western Reserve University, Cleveland, OH, USA

Abstract. Human balance control is a complex feedback system that must be adaptable and robust in an infinitely varying external environment. It is probable that there are many concurrent control loops occurring in the central nervous system that achieve stability for a variety of postural perturbations. Though many engineering models of human balance control have been tested, no models of how these controllers might operate within the nervous system have yet been developed. We have created a synthetic nervous system that provides Proportional-Derivative (PD) control to a single jointed inverted pendulum model of human balance. In this model, angular position is measured at the ankle and corrective torque is applied about the joint to maintain a vertical orientation. The neural network computes the derivative of the angular position error, which allows the system to maintain an unstable equilibrium position and provide corrections at perturbations. This controller demonstrates the most basic components of human balance control, and will be used as the basis for more complex controllers and neuromechanical models in future work.

1 Introduction

Understanding how the human body processes sensory information and reacts to disturbances causing instability is crucial to developing solutions to proprioceptive disorders and other diseases that affect the human balance control system. The human balance system can be broken down into 6 fundamental components: Biomechanical constraints, cognitive processing, control of dynamics, orientation in space, sensory strategies, and movement strategies [1].

The primary factors in human balance control is derived from joint angles, force sensors, vestibular feedback, and visual feedback [2]. Ankle joint feedback is particularly important as it responds to changes in body orientation very quickly. It has been observed that proprioceptive joint angle, vestibular orientation and visual cues contribute to the corrective maneuvers to maintain balance [1, 3, 4]. The weighting of these factors depends on what is available to the nervous system at the time of disturbance, and can change drastically under varying conditions. For example, blindfolded

© Springer International Publishing AG 2017
M. Mangan et al. (Eds.): Living Machines 2017, LNAI 10384, pp. 170–180, 2017.
DOI: 10.1007/978-3-319-63537-8_15

subjects with vestibular loss were completely unable to recognize freefall conditions when the platform they stood on tilted while maintaining a constant ankle joint ankle [3].

Human balance during quiet stance is frequently and effectively modeled as a multi-link inverted pendulum in the sagittal plane, with the single link inverted pendulum being the most basic [5–8]. In this model, the feet are fixed to the floor and a torque is provided about the ankles to keep the center of mass (CoM) from falling over. With this model, it was found that torque about the ankles is well described by a time delayed controller based on center of mass feedback from vertical alignment above the ankles [8]. This controller consists of proportional torque in response to how far the actual position is from the desired position and how fast the position is moving (PD control). This activation is time delayed because of neural processing time. Ankle torque is further augmented by a long term positive force feedback [3], improving model fit to human subject data.

These models, however, are pure engineering models that do not postulate how the nervous system performs the control and are unable to fully explain experimental data. Controllers developed from a synthetic nervous system allows potential benefits for better matching experimental data as well as providing an explanation for how the nervous system is capable of achieving the effective control signals. Individual control functions to be implemented in a single artificial neural network. Synthetic nervous systems can be designed to mimic coordinated muscle movements and have been successfully applied to controlling a variety of modeled animals in simulation and robotics [9–13]. Often setting parameter values in these networks is a complex task of hand tuning, however, effective tools for applying these networks have the ability of engineers to effectively have recently been developed [14].

In this work, we demonstrate a synthetic nervous system for control of inverted pendulum as a model of human balance. The synthetic nervous system operates by taking input from the joint angle and produces corrective torque to stabilize the system and maintain balance. This control architecture may provide insight into how the human nervous system processes ankle angle input.

2   Methods

2.1   Mechanical Model

The simulation of the single jointed inverted pendulum in MATLAB was simulated as a non-linear, unconstrained system. In the case of an ill-fitting controller, the pendulum quickly exhibits unstable behavior.

The inverted pendulum was designed to match similar characteristics of the human body [3]. The physical system was described by the following differential equation:

$$J\frac{d^2\theta}{dt^2} = mgh \cdot \sin(\theta) + T_c, \tag{1}$$

where J represents the moment of inertia of the body, m is the mass, g is acceleration due to gravity, h is the center of mass height, T_c is the corrective torque and θ is the ankle joint angle. Angular velocity dependent damping forces were not included in the initial model.

2.2   Synthetic Nervous System

The synthetic nervous system is constructed to produce corrective torque base on proportional feedback of the error in desired position and velocity of the joint as was modeled in the work by Peterka [8]. To control the inverted pendulum system, a synthetic nervous system was constructed in a custom MATLAB simulation environment [14]. The proposed control system takes a single input, the inverted pendulum's angular position, and outputs a corrective torque that is applied at the angle joint. In order to emulate a PD controller, the neurons and synapses being simulated must be assigned specific characteristics and connections. Information is encoded in the neurons' resting potentials, and is transmitted via synaptic connections. The membrane voltage of a neuron can be approximated by the following "leaky integrator" model:

$$C_m \frac{dV}{dt} = I_{leak} + I_{syn} + I_{app} \tag{2}$$

Where C_m is the membrane capacitance, V is membrane voltage, and various current sources and sinks, I_x, are summed. For the neural arithmetic used in this paper, the sodium and potassium currents are not considered ($I_{NaP} = 0$). The leakage current drives the membrane to its resting potential at a rate proportional to the potential difference between its resting potential and current membrane voltage as represented in Eq. (3) below:

$$I_{leak} = G_m \left(E_r - V \right) \tag{3}$$

Where G_m is the membrane conductance. Neurons can transmit information via synapses, this input current, I_{syn}, is defined in Eq. (4) below:

$$I_{syn} = \sum_{i=1}^{n} G_{s,i} \left(E_{s,i} - V \right) \tag{4}$$

Where the synapse conductance can be described by a piecewise function:

$$G_{s,i} = \begin{cases} 0 \ if \ V_{pre} < E_{lo} \\ g_{s,i} \cdot \dfrac{V_{pre} - E_{lo}}{E_{hi} - E_{lo}} \ if \ E_{lo} < V_{pre} < E_{hi} \\ g_{s,i} \ if \ V_{pre} > E_{hi} \end{cases} \tag{5}$$

Equation (5) parametrizes the range over which postsynaptic neurons receive increasing current from presynaptic neurons, after which the synapse is saturated at its maximum conductance, $g_{s,i}$. E_{lo} and E_{hi} are the lower and upper thresholds of this conductance activation range.

The applied current in (2), I_{app}, is an external stimulus current. For the purpose of this simulation, the external stimulus current is injected into a neuron to represent outside information, such as the angular position of the inverted pendulum model.

To more easily describe neural arithmetic operations mathematically, we employ several simplifying definitions:

$$U = V - E_{lo} \tag{6}$$

$$R = E_{hi} - E_{lo} \tag{7}$$

$$\Delta E_{s,i} = E_{s,i} - E_{r,post} \tag{8}$$

Where U represents the membrane potential above the resting potential, R is the range over which I_{syn} is increasing, E_{lo} is the resting potential, and $\Delta E_{s,i}$ is the potential difference between the synaptic equilibrium potential and the postsynaptic neuron's resting potential.

2.2.1 Neural Addition

Neurons can perform elementary addition, where two presynaptic neurons resting potentials can be summed and reflected in a postsynaptic neuron. Synapses terminating in a triangle are excitatory, whereas the shaded circular terminals are inhibitory synapses (Fig. 1).

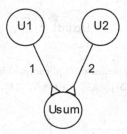

Fig. 1. Schematic illustration of a neural addition subnetwork, $U_{sum} = U_1 + U_2$. Optimal parameters are: $g_{1,2} = 0.115\,\mu s, \Delta E_1 = \Delta E_2 = 194\,mV, R = 20\,mV$

2.2.2 Neural Subtraction

Subtraction is a very important calculation for closed loop control systems with feedback. Neurons can be structured to compute subtractive arithmetic, which is used in control systems to calculate the error between desired state and measured state (Fig. 2).

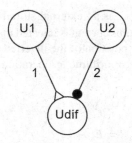

Fig. 2. Schematic illustration of a neural subtractive subnetwork, $U_{dif} = U_1 - U_2$. Optimal parameters are: $g_s = 1.0\,\mu s$, $\Delta E_1 = -\Delta E_2 = 40\,mV$, $R = 20\,mV$

2.2.3 Neural Multiplication

Multiplication is required to compute the PD controller's response to the error in angular position and its derivative. A subnetwork of four neurons is required to compute the product of two inputs (Fig. 3).

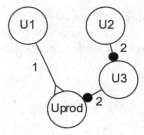

Fig. 3. Schematic illustration of a neural multiplication subnetwork, $U_{prod} = U_1 \cdot U_2$. Optimal Parameters are: $g_1 = 0.115\,\mu s$, $g_2 = 20\,\mu s$ $\Delta E_1 = 194\,mV$, $\Delta E_2 = -1\,mV$, $R = 20\,mV$

2.2.4 Neural Differentiation

A neural subnetwork can compute the derivative of an input neuron by routing the input neuron's membrane voltage to two separate neurons with different membrane capacitances. This difference in membrane capacitance effectively creates a response time delay between the neurons. Subtracting these neurons results in an approximation similar to a backward difference function (Fig. 4).

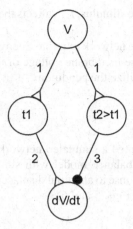

Fig. 4. Schematic illustration of a neural derivation subnetwork, where optimal parameters are: $g_1 = 0.115\,\mu s, g_2 = g_3 = 1\,\mu s, \Delta E_1 = 194\,mV, \Delta E_2 = 20\,mV, \Delta E_3 = -20\,mV, R = 20\,mV$

2.3   Assembling Subnetworks to Create a Neural PD Controller

The elementary subnetworks outlined can be strategically assembled into a PD controller (Fig. 10). Since neural arithmetic is abstracted by the physical constraints in the neuron simulation, there are a few complexities that must be added to the network. The neural subtraction algorithm functions most efficiently (with minimum error), when the subtractive elements are computing in one direction. That is, it is best to compute a negative difference with a separate circuit than a positive difference. For this reason, the neural network is broken into two separate regions. The input, pendulum angle, is converted to a current and used as an input to separate error calculations one after being negated (Neurons 1 and 15). When the error is positive, Neuron 1 gets a positive current while Neuron 15 gets a negative current, and when the error is negative, Neuron 1 gets a negative current while Neuron 15 gets a positive current. We defined $\theta = 0$ to be in the upright position and the target position for the control system. Neurons 1 and 15 feed into separate PD controllers that output either a clockwise or counter clockwise corrective torque, depending on the pendulums position and velocity. Since the derivative circuit also employs a subtractive element, the negative derivative terms are fed into the opposing error circuit. Physically, this corresponds to a drop in corrective torque when the pendulum is swinging quickly towards the equilibrium position. This element of a PD controller is what makes it able to stabilize an inverted pendulum system while a simple P controller cannot [4]. A simple P controller will not make any attempt to slow down the pendulum until it has already passed its unstable equilibrium point, often at a considerable velocity.

There are several points within the network where external communication must occur. The pendulum angular position, θ, is mapped to the input signal neurons, thus an angle value must be transformed into a varying applied tonic stimulus current. Additionally, the Kp and Kd terms of the controller must be defined by applying a constant stimulus to the neuron. This tonic stimulus, I_{app}, affects the membrane voltage as demonstrated in Eq. (2).

The two output neurons of the network must have converters, analogous to a classical controller's power amplifier. The membrane voltage of the output neurons is scaled to provide sufficient torque to stabilize the pendulum.

3   Results

We have successfully implemented a simulated network of neurons that can stabilize the inverted pendulum human balance model with similar performance quality as a classical PD controller. The response to an initial displacement from equilibrium of each controller type is shown below (Figs. 5 and 6).

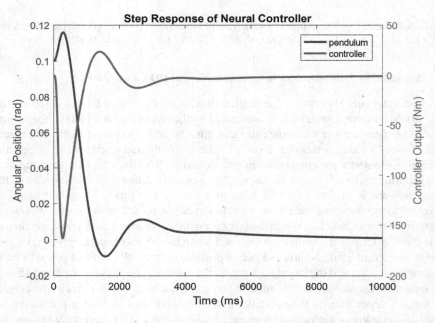

Fig. 5. The neural PD controller's response to an initial displacement of 0.1 radians

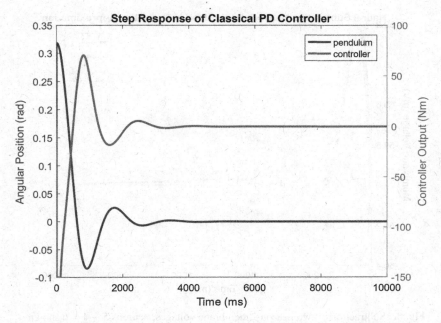

Fig. 6. A classical PD controller's response to an initial displacement of 0.1 radians

The individual neuron membrane voltages within the subnetworks are performing their tasks properly. Examples of each subnetwork in action are shown in the figures below (Figs. 7, 8 and 9).

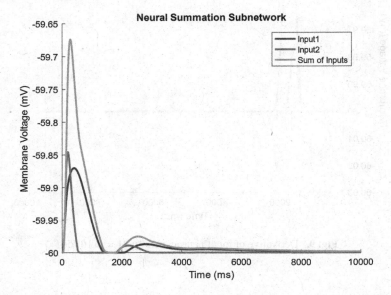

Fig. 7. Summation of two neurons' membrane voltages, neurons $10 + 13 = 14$ shown

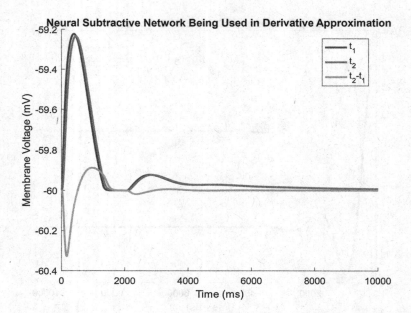

Fig. 8. Subtraction of two neurons' membrane voltages, neurons $5 - 4 = 6$ shown

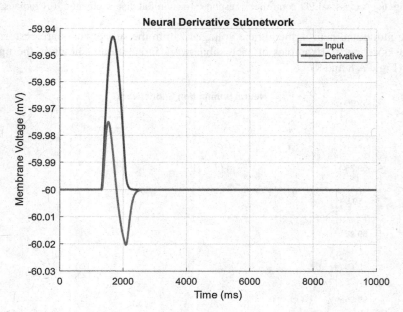

Fig. 9. Derivative of input neuron, neuron $21 = d(16)/dt$

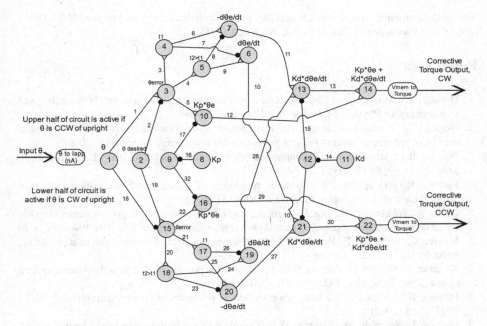

Fig. 10. Simulated network of neurons and synapses performing PD control

4   Discussion

Human balance has been observed to utilize ankle angle feedback as part of its multi-layered control structure [1, 3, 15]. While the exact control model is still unknown, piecing together individual components of balance using neurons may provide insight into full-scale balance behavior. The results of this experiment suggest that a network of neurons can perform single-input, single-output PD control to stabilize a human-like model. This PD control architecture can be expanded to include other sensory inputs that the human brain has access to, such as visual and vestibular senses [8].

This network is also ready to provide input to a more physically realistic model of human balance by controlling agonist and antagonist muscle pairs rather than ideal motor control at the joints. Neurons 14 and 23 are readily adapted into motor neurons which can individually control the muscles with little to no modification of the network. The neuron and synapse information can be input into the software Animatlab, which has 3D visualization and simulation capabilities [16].

By continuing to construct a controller model for human balance in a piecemeal fashion, we can perform increasingly complex experiments by varying neuron and synapse parameters, adjusting the weights of individual portions of the control and applying memory neurons to store information of successful and failed iterations. This could lead to a closed loop learning design, and provide insights to how the nervous system is capable of effectively adapting its behavior to such a large variety of perturbations.

Acknowledgements. The authors would like to acknowledge the support by the NASA Office of the Chief Technologist, Grant Number NNX12AN24H.

References

1. Horak, F.B., Macpherson, J.M.: Postural orientation and equilibrium. In: Terjung, R. (ed.) Comprehensive Physiology. Wiley Inc., Hoboken (1996)
2. Horak, F.B.: Postural orientation and equilibrium: what do we need to know about neural control of balance to prevent falls? Age Ageing **35**(Supplement 2), ii7–ii11 (2006)
3. Peterka, R.J.: Simplifying the complexities of maintaining balance. IEEE Eng. Med. Biol. Mag. **22**(2), 63–68 (2003)
4. Peterka, R.J., Loughlin, P.J.: Dynamic regulation of sensorimotor integration in human postural control. J. Neurophysiol. **91**(1), 410–423 (2004)
5. Chiba, R., Takakusaki, K., Ota, J., Yozu, A., Haga, N.: Human upright posture control models based on multisensory inputs; in fast and slow dynamics. Neurosci. Res. **104**, 96–104 (2016)
6. Maurer, C., Mergner, T., Peterka, R.J.: Multisensory control of human upright stance. Exp. Brain Res. **171**(2), 231 (2006)
7. Mergner, T., Maurer, C., Peterka, R.J.: A multisensory posture control model of human upright stance. Prog. Brain Res. **142**, 189–201 (2003). Elsevier
8. Peterka, R.J.: Sensorimotor integration in human postural control. J. Neurophysiol. **88**(3), 1097–1118 (2002)
9. Hunt, A.J., Schmidt, M., Fischer, M.S., Quinn, R.D.: A biologically based neural system coordinates the joints and legs of a tetrapod. Bioinspir. Biomim. **10**(5), 055004 (2015)
10. Szczecinski, N.S., Chrzanowski, D.M., Cofer, D.W., Moore, D.R., Terrasi, A.S., Martin, J.P., Ritzmann, R.E., Quinn, R.D.: MantisBot: a platform for investigating mantis behavior via real-time neural control. In: Wilson, S.P., Verschure, P.F.M.J., Mura, A., Prescott, T.J. (eds.) LIVINGMACHINES 2015. LNCS, vol. 9222, pp. 175–186. Springer, Cham (2015). doi: 10.1007/978-3-319-22979-9_18
11. Hunt, A.J., Szczecinski, N.S., Andrada, E., Fischer, M.S., Quinn, R.D.: Using animal data and neural dynamics to reverse engineer a neuromechanical rat model. Biomim. Biohybrid Syst. Living Mach. **2015**, 211–222 (2015)
12. Szczecinski, N.S., Brown, A.E., Bender, J.A., Quinn, R.D., Ritzmann, R.E.: A neuromechanical simulation of insect walking and transition to turning of the cockroach Blaberus discoidalis. Biol. Cybern. **108**(1), 1–21 (2013)
13. Li, W., Szczecinski, N.S., Hunt, A.J., Quinn, R.D.: A neural network with central pattern generators entrained by sensory feedback controls walking of a bipedal model. In: Lepora, N.F.F., Mura, A., Mangan, M., Verschure, P.F.M.J.F.M.J., Desmulliez, M., Prescott, T.J.J. (eds.) Living Machines 2016. LNCS, vol. 9793, pp. 144–154. Springer, Cham (2016). doi: 10.1007/978-3-319-42417-0_14
14. Szczecinski, N.S., Hunt, A.J., Quinn, R.: Design process and tools for dynamic neuromechanical models and robot controllers. Biol. Cybern. **111**(1), 105–127 (2017)
15. Bosco, G., Poppele, R.E.: Representation of multiple kinematic parameters of the cat hindlimb in spinocerebellar activity. J. Neurophysiol. **78**(3), 1421–1432 (1997)
16. Cofer, D.W., Cymbalyuk, G., Reid, J., Zhu, Y., Heitler, W.J., Edwards, D.H.: AnimatLab: a 3D graphics environment for neuromechanical simulations. J. Neurosci. Methods **187**(2), 280–288 (2010)

Reducing Versatile Bat Wing Conformations to a 1-DoF Machine

Jonathan Hoff[1,2]([✉]), Alireza Ramezani[1], Soon-Jo Chung[3], and Seth Hutchinson[1,2]

[1] Coordinated Science Laboratory, Urbana, USA
jehoff2@illinois.edu
[2] Department of Electrical and Computer Engineering,
University of Illinois at Urbana-Champaign (UIUC), Urbana, IL 61801, USA
[3] Graduate Aerospace Laboratories (GAL), California Institute of Technology,
Pasadena, CA 91125, USA

Abstract. Recent works have shown success in mimicking the flapping flight of bats on the robotic platform Bat Bot (B2). This robot has only five actuators but retains the ability to flap and fold-unfold its wings in flight. However, this bat-like robot has been unable to perform folding-unfolding of its wings within the period of a wingbeat cycle, about 100 ms. The DC motors operating the spindle mechanisms cannot attain this folding speed. Biological bats rely on this periodic folding of their wings during the upstroke of the wingbeat cycle. It reduces the moment of inertia of the wings and limits the negative lift generated during the upstroke. Thus, we consider it important to achieve wing folding during the upstroke. A mechanism was designed to couple the flapping cycle to the folding cycle of the robot. We then use biological data to further optimize the mechanism such that the kinematic synergies of the robot best match those of a biological bat. This ensures that folding is performed at the correct point in the wingbeat cycle.

Keywords: Aerial robotics · Bats · Biologically-inspired robots · Kinematics

1 Introduction

Bats have a complex wing mechanism that has over 40 Degrees of Freedom (DoF) [16]. This structure allows them to actively fold and unfold their wings in the period of a wingbeat. It is thought that bats improve flight efficiency by folding their wings during the upstroke of the wingbeat cycle, which in turn reduces the moment of inertia of the wings [15]. Energy is thus saved by reducing the moment of inertia of the wings during the upstroke. Negative lift may be decreased as well. Bahlman *et al.* [1] studied the cost of flight by designing an articulated wing mechanism and testing both fixed wing positions and folding during the upstroke. Folding the wing did indeed lower the cost of flight: it used less power, and negative lift during the upstroke was reduced. While net thrust

© Springer International Publishing AG 2017
M. Mangan et al. (Eds.): Living Machines 2017, LNAI 10384, pp. 181–192, 2017.
DOI: 10.1007/978-3-319-63537-8_16

is reduced, negative lift is decreased (net lift is increased) and there is a decrease in power consumption because of the reduced inertial and aerodynamic costs.

Robots with flapping wings have shown success in this area of folding and unfolding the wings during flight. A one-way folding mechanism allowed for passive folding during the upstroke of a micro aerial vehicle (MAV) [9]. Similarly, Wissa et al. [19] reduced power consumption and improved lift of an ornithopter by inserting a compliant spine on the wings such that they were passively morphed during the upstroke. Another study designed wings that unfold in an outward sweeping motion from the centrifugal accelerations of wing flapping [18].

The folding of wings during the upstroke is an important aspect of bat flight. This capability was integrated into the design of the robotic Bat Bot (B2) in attempt to mimic the behavior of bat flight while maintaining a reduced complexity [11–14]. B2's morphing wings allow it to change its wing inertia, reduce wing area, and modulate the tension of the membrane.

In recent works, we presented an approach to match the kinematic synergies of a biological bat to those of B2 [7]. This entailed optimizing the geometric structure of the armwing as well as the actuator trajectories such that the synergies derived from principal component analysis (PCA) match those of the biological bat. Despite the drastic difference in DoFs, B2 was able to replicate the motion of flapping and folding-unfolding found in biological bats.

However, experimental testing showed that B2 could not match the rapid folding-unfolding of the armwing within the time interval of a wingbeat. The DC motors were incapable of driving the spindle mechanisms at the same frequency as the flapping motion. Thus, the wings could not be folded during the upstroke of the wingbeat period.

We address the hardware problem in this paper by coupling the flapping and folding motions of the wings with a four-bar crank-rocker mechanism. The new closed loop kinematic chain has one DoF operated by the main brushless DC motor (BLDC). This drives the flapping motion of the wings and now also operates the folding-unfolding motion of the wing. The methods in [7] are used to find the optimal-dimensions of the designed mechanism and the actuator trajectory of the BLDC motor. The dimensions provide the best matching of the two most dominant principal components of B2 to those of a biological bat. We describe in detail the design process of the coupling mechanism in Sect. 2, optimization of the mechanism in Sect. 3, and the simulation results in Sect. 4. Concluding remarks are made in Sect. 5.

2   Mechanism Design

Systems of linkages are often designed to couple the motion of different mechanisms. This creates a synergy because there is coordinated movement. Synergies were first defined as cooperative activations of different muscle groups to simplify control for the central nervous system [2]. The human hand has over 20 DoFs [8], but many grasping tasks have DoFs that are coupled in movement. Research in grasping has found postural synergies in the human hand, i.e. movements in which joints are coupled in their behavior [17]. The two most dominant synergies have been implemented on a robotic hand using only two actuators and a system of pulleys [3]. A more recent study simplified the control of 19 DoFs with a single actuator using 'soft synergies' [4].

Studies in flapping flight have also attempted to reduce the number of actuators on aerial vehicles to reduce weight and add simplicity to the design. In a way, these are synergies, or couplings between the different movements. Coupling mechanisms have frequently been used for design of MAVs that mimic insect flight. A four-bar linkage was designed for an insect-like MAV to generate the figure eight pattern of the wing tip observed in insect flight using the planar figure eight pattern of the mechanism [20]. Conn *et al.* [5] designed a parallel crank-rocker mechanism with a single actuator that couples flapping and pitching. Pitching was adjusted by changing the phase of the parallel mechanism. The flight patterns of the wings of dragon flies were captured on a MAV using only one rotary actuator and a modified slider-crank mechanism such that wing rotations and flapping were coupled [6].

In a similar way, we considered the task of generating the folding-unfolding motion of the wing from the crank mechanism driving flapping. Currently, this motion of each wing is driven by a separate DC motor that moves a spindle mechanism. This in turn adjusts the position of B2's forelimb. The new mechanism should seamlessly integrate with the current crank design such that no modifications need to be made for the flapping configuration. In addition, the spindles and DC motors driving them should be replaced by sliders such that the folding can be driven by the crank mechanism. The mechanism should thus convert motion from the crank to linear motion of the new slider.

We used a four-bar linkage to convert the circular motion of the crank to motion of a rocker. The rocker arm in turn drives the slider (i.e. the spindle). This mechanical system has one DoF. The crank angle q_C determines the complete motion of the flapping of B2 as well as the folding movement, replacing the two DC motors that drove the spindles. It should be noted that the DC motors can be introduced in cooperation with this coupling mechanism such that minor adjustments can be made in flight to the wing positions, though this will not be addressed in the paper.

The parameters defining the coupling mechanism are shown in Fig. 1. The shoulder joint j_0 is offset distances of s_y and s_z in the $-y$ and z directions from the crank center j_3. The length of the crank radius is r_c. These are fixed parameters and will not be optimized. Link l_1 connects the end of the crank arm to the shoulder with ball-and-socket joints. Joint j_1 is a distance of r_s from j_0.

The four-bar crank-rocker mechanism consists of the drive link r_c, the coupler link l_2, the rocker l_3, and the base dimension $\sqrt{w_x^2 + w_z^2}$. Link l_2 couples the flapping motion produced by the crank arm to the rocker arm. Revolute joint j_4 is offset distances of w_x and w_z in the $-x$ and $-z$ directions from j_3.

The rocker arm is of length $l_3 + l_4$. The end of the rocker (j_6) is fixed to the slider at joint j_7 by link l_5. Ball-and-socket joints make up the two connections. Revolute joint j_8 is a distance l_6 from j_7 and is constrained along the slider in

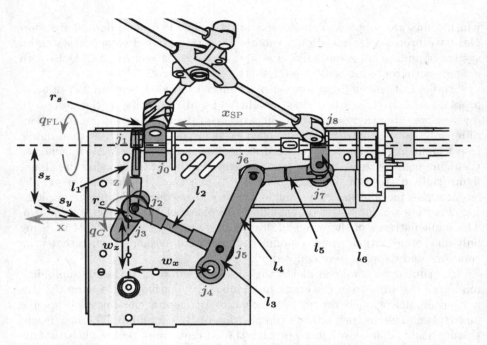

Fig. 1. Flapping and folding coupling mechanism. Red variables denote changing positions of angles and distance. Black variables give the parameters of the structure, and blue variables denote joint locations. Joints shaded brown experience both rotation and translation, and joints in white (j_0, j_3, and j_4) rotate only. The coordinate system is colored gray, with the **y** axis pointing out of the page. This system has just one DoF: the crank angle q_C is directly controlled by the BLDC motor, and this in turn characterizes the flapping angle q_{FL} and the spindle position x_{SP}. (Color figure online)

the $\pm x$ directions. Link l_6 can also rotate about the slider and is coupled to the flapping motion by the forelimb structure. All of the above listed lengths can be be combined to give the vector $\mathcal{C}$.

The mechanism has one DoF, and thus a set of kinematic constraints must be imposed to enforce this. The constraint equations for this mechanism can be solved analytically because it is a closed-loop kinematic chain and has one DoF. First, the constraint equations governing the linkage driving the flapping of the wing are solved to find the flapping angle q_{FL}. This is the angle between the **xy** plane and the wing.

The link l_1 is projected onto the **xy**, **xz**, and **yz** planes, and the equations for these projections are combined as

$$2l_1^2 = l_{1,xy}^2 + l_{1,xz}^2 + l_{1,yz}^2. \tag{1}$$

Algebraic manipulations produce the equation of form

$$A\cos q_{FL} + B\sin q_{FL} + C. \tag{2}$$

The terms A, B, and C are functions of q_C and $\mathcal{C}$. This is equivalent to Freudenstein's Equation, and thus q_{FL} can be solved for as a quadratic by making trigonometric substitutions for $\sin q_{FL}$ and $\cos q_{FL}$.

Next, the planar four-bar linkage shown in Fig. 1 can be solved. The drive link is r_c, and q_C is the driving angle. The closed loop kinematics are written as

$$\overrightarrow{j_3 j_2} + \overrightarrow{j_2 j_5} + \overrightarrow{j_5 j_4} + \overrightarrow{j_4 j_3} = 0. \tag{3}$$

This means that the sum of the vectors around the linkage equals zero, i.e. the loop is constrained to be closed. This relation is also described by

$$Rot_y\left(\phi_3\right)\begin{bmatrix} r_c \\ 0 \\ 0 \end{bmatrix} + Rot_y\left(\phi_2\right)\begin{bmatrix} l_2 \\ 0 \\ 0 \end{bmatrix} + Rot_y\left(\phi_5\right)\begin{bmatrix} l_3 \\ 0 \\ 0 \end{bmatrix} + Rot_y\left(\phi_4\right)\begin{bmatrix} \|w\| \\ 0 \\ 0 \end{bmatrix} = 0, \tag{4}$$

where $\|w\| = \sqrt{w_x^2 + w_z^2}$ and ϕ_2, ϕ_3, ϕ_4, and ϕ_5 are the respective angular positions of joints j_2, j_3, j_4, and j_5. The angle ϕ_4 is fixed at $\phi_4 = 2\pi - \text{atan}\,\frac{w_z}{w_x}$ and Rot_y is the rotation matrix about the $\mathbf{y}$ axis. Also, ϕ_3 is equivalent to the crank angle q_C. These equations can be analytically solved by using algebraic manipulations to achieve the form of Eq. (2). The spindle position (or slider position) x_{SP} can then be determined from the resulting value of ϕ_5. This is the distance between j_0 and j_8.

2.1 Forelimb Design [13]

These equations give a mapping from q_C and $\mathcal{C}$ to q_{FL} and x_{SP}. From here, the forward kinematics in [7] give the positions of points on the forelimb of B2.

B2's forelimb is a three-link mechanism with a humeral link, a radial link, and a carpal plate. This mechanism is connected by revolute joints, and it is uniquely defined by three biologically meaningful angles: the shoulder retraction-protraction q_{RP}, the elbow flexion-extension q_{FE}, and the carpal plate abduction-adduction q_{AA}. This mechanism is constrained to one DoF by adding several extra links and fixing the radial link to the spindle.

The joint angles are directly controlled by the spindle position x_{SP}, i.e. the distance between the shoulder joint and the radial link joint. Linear movements of the radial link joint along the spindle toward the shoulder result in forward rotation of the shoulder joint. This in turn forces the elbow to extend. The carpal plate is pushed away from the body in response to this extension. The digits of B2 are thin carbon fiber rods that are secured to the carpal plate, and these are pushed outward as a result of the movement of the carpal plate. These rods are flexible, and they introduce the passive DoFs of abduction-adduction and flexion-extension of the digits. The digits lack joints, and thus their motion is dependent on the movement of the carpal plate. The wing as a whole has one actuated DoF with several passive DoFs. This actuated DoF controls the three biologically meaningful angles.

3   Optimization [7]

The dimensions of the coupling mechanism and forelimb as well as the trajectory of q_C should be selected such that B2's kinematic motion best matches that of the biological bat. We use optimization to select the dimensions and the actuator trajectory. A brief summary of the methods are provided here. For a more thorough explanation, readers should see [7] as a reference. The optimization in this paper differs slightly from that presented in [7]: the structure of B2's forelimbs has been improved, marker selection was adjusted, the objective function was modified, and several extra constraints were added.

B2 was originally designed to mimic *Rousettus aegyptiacus*, though data for this bat was not available. In this study, we used data for *Tadarida brasiliensis*. *Rousettus aegyptiacus* is much larger than *Tadarida brasiliensis*, thus we linearly scaled the data for *Tadarida brasiliensis* such that its wingspan matches that of *Rousettus aegyptiacus* [10]. Additionally, the coordinate system is centered on the anterior sternum marker, the **x** axis passes through the two sternum markers, the **y** axis points to the left wing, orthogonal to gravity, and the **z** axis points up. This is the body-referenced coordinate system used by [16].

The shoulder marker and nine markers of the digits on the biological bat were selected for comparison. Complementary markers on B2 were chosen by projecting these biological bat markers onto the digits of B2 at each point in time over a wingbeat period to get the closest points on B2's digits to those on the biological bat's digits. These two sets of markers were then used in the optimization formulation.

The crank angle is parameterized based on the angular frequency ω and the phase ϕ as

$$q_C\left(t_i\right) = \omega t_i + \phi. \tag{5}$$

These are combined into the vector $\mathcal{A}_C = \begin{bmatrix} \omega & \phi \end{bmatrix}^\top$. Besides l_1, the crank assembly is not optimized. Crank radius r_c, shoulder radius r_s, and shoulder offsets s_y and s_z are left unchanged because the crank is already tuned to provide maximum torque for flapping while not overexerting the motor. Even small changes in lengths could lead to hardware issues. The rest of the parameters are optimized and are combined into the vector $\bar{\mathcal{C}}$.

The optimization routine for selected optimized variable $\mathcal{X}$ is formulated as a constrained nonlinear optimization problem to minimize the objective function

$$\mathcal{J}(\mathcal{X}) = \left\| \hat{\mathcal{M}}_r\left(\mathcal{X}\right) - \mathcal{M}_r \right\|_F^2. \tag{6}$$

Matrix $\mathcal{M}_r$ is derived from the data matrix $\mathcal{M}$. The columns of $\mathcal{M}$ are the xyz coordinates of each of the ten markers, and the rows are the time sample over a wingbeat cycle. We performed dimensionality reduction on $\mathcal{M}$ using PCA to use only the first two kinematic synergies (principal components) of the biological bat data, giving the matrix $\mathcal{M}_r$. The matrix $\hat{\mathcal{M}}_r$ is similarly derived for B2.

The objective function implements a sum of squared differences between the points on B2 and the biological bat reconstructed from PCA using the Frobenius norm.

The main optimization routine is separated into three subroutines in which the crank angle coefficients $\mathcal{A}_C$, the mechanism parameters $\bar{\mathcal{C}}$, and the forelimb parameters $\bar{\mathcal{P}}$ are individually optimized. The vectors $\bar{\mathcal{C}}$ and $\bar{\mathcal{P}}$ contain only the parameters being optimized, whereas $\mathcal{C}$ and $\mathcal{P}$ contain all of the parameters describing the mechanism and the forelimb.

A set of inequality and equality constraints help shape the optimization problem. These constraints differ depending on the choice of $\mathcal{X}$. For $\mathcal{X} = \mathcal{A}_C$, the equality constraint $q_C(t_1) - q_C(t_n) \in \{0, \pm 2\pi, \pm 4\pi, \cdots\}$ enforces periodicity of the motor cycle.

Several extra constraints are necessary for $\mathcal{X} = \bar{\mathcal{C}}$. First, the Grashof conditions for a four-bar mechanism are introduced. In order to drive the folding-unfolding motion, the crank must be able to spin freely, but the driving arm should be a rocker. Thus, the conditions for a crank-rocker mechanism were introduced as constraints to the optimization routine. These are given by the equations $-r_4 - r_2 + r_1 + r_3 < 0$, $-r_3 - r_4 + r_1 + r_2 < 0$, and $-r_3 - r_2 + r_1 + r_4 < 0$, where $r_1 = r_c$, $r_2 = l_2$, $r_3 = l_3$, and $r_4 = \sqrt{w_x^2 + w_z^2}$. In addition, the top of the rocker (j_6) is restricted from passing above the shoulder line (j_0-j_8), else it will interfere with the spindle mechanism and the membrane. The spindle position x_{SP} is also restricted to the range $x_{min} \leq x_{SP} \leq x_{max}$, which is adjusted based on mechanical limitations. The mechanism parameters are restricted with upper and lower bounds as $l_k \leq \bar{\mathcal{C}}_k \leq u_k$, $k = 1, \ldots, 8$.

Constraints are necessary for $\mathcal{X} = \bar{\mathcal{P}}$. The angles between the digits of B2 are forced not to overlap by the constraint $\gamma_1 \leq \gamma_2 \leq \gamma_3$. Additionally, the wing area of B2 is prevented from dropping below that of *Rousettus aegyptiacus* [10]. This is necessary such that the resulting structure of B2 can provide enough lift. The optimized variables also have upper and lower bounds as $l_k \leq \bar{\mathcal{P}}_k \leq u_k$, $k = 1, \ldots, 12$.

4   Simulation Results

We run the optimization routine separately for q_C, $\bar{\mathcal{C}}$, and $\bar{\mathcal{P}}$. The main optimization routine is iterated four times. The flapping angle trajectory q_{FL} can be compared to the biologically meaningful angles q_{RP}, q_{FE}, and q_{AA} describing the folding mechanism of the wing in order to observe the differences in phase and amplitude. Figure 2 shows the evolution of these angles over a wingbeat period. The offset of each angle has been removed to provide a comparison of the phases and amplitudes of the trajectories. The angles q_{RP} and q_{AA} are in phase with each other as expected. They decrease when the wing is extending and increase when it is folding. The angle q_{FE} is 180° out of phase, and thus moves opposite to these two angles. The three angles are out of phase with the flapping angle q_{FL}. The wing initially extends at the very end of the upstroke to prepare for the downstroke, reaches maximum extension near the end of the downstroke, and begins to fold at the tail end of the downstroke and all through the upstroke.

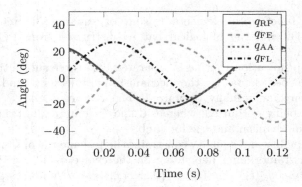

Fig. 2. Trajectories of the biologically meaningful angles q_{RP}, q_{FE}, q_{AA}, and q_{FL} over a wingbeat cycle. The trajectories have been centered about the origin to remove offsets in order to give a better comparison. These angles are all coupled by the motion of the crank angle q_C.

The angles are also compared to those of the biological bat data in Fig. 3. First, the angles of the biological bat are centered about the origin by subtracting the mean from each. Second, for each angle, the trajectory is divided by its largest amplitude such that the trajectory remains between -1 and 1. This is likewise done for B2's angles. The angles of B2 have much different offsets that those of the biological bat, and B2's angles have significantly larger amplitudes (except q_{FL}, which has a lower amplitude). This is as expected because B2's forelimbs have a different topological structure than that of a biological bat due to the extra links added to constrain it. Thus, centering and normalizing the angles provides a way to characterize the changing behavior of the sets of angles on the same scale. The figure shows that the changing behavior of the flapping angle q_{FL} is almost identical to that of the biological bat, and the folding-unfolding motion is also comparable to this motion of the biological bat because of the matching of the angles q_{RP} and q_{FE}. The wrist angles are less similar as there are high frequency oscillations of q_{AA} that are not present in B2. Though it cannot be seen in this figure, the flapping angle has a lower amplitude than the biological bat because the parameters that adjust the magnitude of the flapping angle are not optimized. As a result, the amplitude cannot increase to match the flapping amplitude of the biological bat. In practice, we have found that a larger amplitude results in hardware issues.

Figure 4 gives the phase plots of the four angles for B2 and the biological bat. Angular velocities are generated from the position data by taking differences between adjacent angles and dividing by the sampling period. Similar to the above procedure, the angular positions and velocities are centered about the origin through mean subtraction and normalized between -1 and 1. For the same reasons as above, this allows for better comparison of the behavior of position versus velocity of B2 and the biological bat even though it sacrifices magnitudes of these results. It can be seen from the phase plots that the states of B2 and

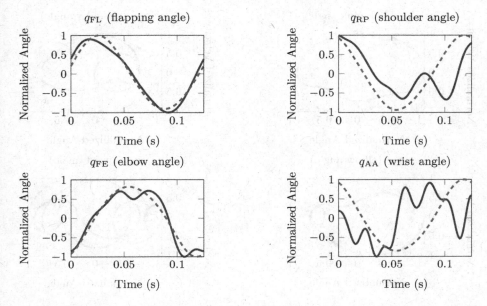

Fig. 3. Normalized biologically meaningful angles of B2 are compared to those of the biological bat over one wingbeat cycle. These angles are centered about the origin, and then normalized between -1 and 1, allowing for better comparison between the biological bat's angles (blue lines) and B2's angles (dotted red lines). The amplitudes of B2's angles (except q_{FL}) are significantly larger than the biological bat's when not normalized. All of these angles are coupled in B2 and move in response to the crank position q_C. (Color figure online)

the biological bat behave in a very similar manner, especially for the case of q_{FL}. The wrist angle q_{AA} of the biological bat has several oscillations within the wingbeat cycle. This is also present in q_{RP} and q_{FE} but less pronounced.

Information can also be gained by analyzing the results from PCA. The principal components themselves give the directions of motion of each point on the wing. When considering only one component, the point is constrained to movement on a line. When considering two components, the point resides in a space spanned by the two components. Thus the point can move in two dimensions by taking linear combinations of the principal components. Different directions of motion are determined by taking different weights of the two components and adding them together. These weights are equivalent to the projection of the original data onto the principal components.

This is realized in Fig. 5. The data markers from the biological bat are projected onto the principal components to give the temporal weights over time, i.e. how the magnitudes of the principal components vary over the course of a wingbeat cycle. The plot shows the evolution of the weights of the two most dominant principal component over a wingbeat. We can further understand the relation between flapping and folding here because the first component is equivalent

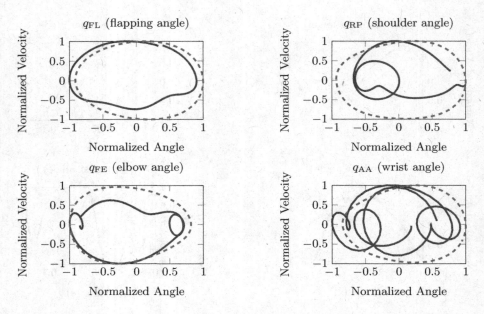

Fig. 4. Normalized phase plots of angular velocities versus angular positions of the four biologically meaningful angles in B2 and the biological bat. These angular positions and velocities are centered about the origin, and then normalized between −1 and 1, allowing for better comparison between the phase plots of the biological bat (blue lines) and B2 (dotted red lines). The amplitudes of B2's angles (except q_{FL}) are significantly larger than the biological bat's when not normalized. (Color figure online)

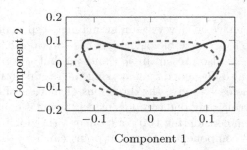

Fig. 5. Marker position data for both the biological bat and B2 are projected onto the two most dominant principal components. Component 1, the flapping direction, is plotted against component 2, the folding-unfolding direction. B2 is shown in red, and the biological bat is blue. (Color figure online)

to the direction of flapping, and the second gives the direction of the folding-unfolding of the wing. Studying these weights of these two directions shows how far the wing folds at some certain point in the flapping cycle. Even though flapping and folding are coupled, the resulting motion is still quite similar to the behavior of the biological bat.

5   Conclusion

Coupling the flapping and folding-unfolding motions of a robotic bat with a mechanism can yield behavior that is consistent with that found in biological bats. The dimensions of this mechanism have been selected via optimization such that the basic kinematic synergies of wing flapping and folding-unfolding of the robot closely match those of a biological bat. The folding-unfolding of the wing can be parameterized based on the flapping angle because the two motions are coupled, and thus the flapping angle alone can give the position of the wing.

We have reduced the DoFs of the forelimbs of B2 from three to one by introducing this coupling mechanism. This further simplifies the design of the robot, but it removes control inputs that are helpful in stabilizing B2 during flight and are necessary for banking maneuvers. The controller makes adjustments to the spindle positions during flight to reduce or increase the surface area of the wings in order to improve roll stability. The hindlimbs can also stabilize for roll, but they have more significant effects on the pitch angle of B2.

However, future modifications could allow for the mechanism to couple flapping and folding as well as for the DC motors to make spindle adjustments during flight. This would ensure folding-unfolding within a wingbeat as well as independent control of each spindle to adjust the wings.

Acknowledgments. We would like to thank the team of graduate and undergraduate students from aerospace, electrical, computer, and mechanical engineering departments at the University of Illinois at Urbana-Champaign for their contribution to construct the initial prototype of B2.

The biological motion capture data set was provided by Dr. Kenneth Breuer and Dr. Sharon Swartz from Brown University. We would like to thank them in their assistance with this, as well as José Iriarte-Díaz for compiling the data.

This work was supported by NSF Grant 1427111.

References

1. Bahlman, J.W., Swartz, S.M., Breuer, K.S.: Design and characterization of a multi-articulated robotic bat wing. Bioinsp. Biomim. **8**(1), 016009 (2013)
2. Bernstein, N.A.: The Coordination and Regulation of Movements. Pergamon Press, Oxford (1967)
3. Brown, C.Y., Asada, H.H.: Inter-finger coordination and postural synergies in robot hands via mechanical implementation of principal components analysis. In: IEEE/RSJ International Conference on Intelligent Robots and Systems (IROS), pp. 2877–2882 (2007)
4. Catalano, M.G., Grioli, G., Farnioli, E., Serio, A., Piazza, C., Bicchi, A.: Adaptive synergies for the design and control of the Pisa/IIT SoftHand. Int. J. Robot. Res. **33**(5), 768–782 (2014)
5. Conn, A., Burgess, S., Ling, C.: Design of a parallel crank-rocker flapping mechanism for insect-inspired micro air vehicles. Proc. Inst. Mech. Eng Part C J. Mech. Eng. Sci. **221**(10), 1211–1222 (2007)

6. Fenelon, M.A., Furukawa, T.: Design of an active flapping wing mechanism and a micro aerial vehicle using a rotary actuator. Mech. Mach. Theor. **45**(2), 137–146 (2010)
7. Hoff, J., Ramezani, A., Chung, S.-J., Hutchinson, S..: Synergistic design of a bio-inspired micro aerial vehicle with articulated wings. In: Robotics: Science and Systems (RSS) (2016)
8. Lin, J., Wu, Y., Huang, T.S.: Modeling the constraints of human hand motion. In: IEEE Workshop on Human Motion, pp. 121–126 (2000)
9. Mueller, D., Gerdes, J.W., Gupta, S.K.: Incorporation of passive wing folding in flapping wing miniature air vehicles. In: ASME Mechanism and Robotics Conference, pp. 797–805 (2009)
10. Norberg, U.M., Rayner, J.M.: Ecological morphology and flight in bats (Mammalia; Chiroptera): wing adaptations, flight performance, foraging strategy and echolocation. Philos. Trans. Royal Soc. B: Biolog. Sci. **316**(1179), 335–427 (1987)
11. Ramezani, A., Chung, S.-J., Hutchinson, S.: A biomimetic robotic platform to study flight specializations of bats. Sci. Robot. 2(3) (2017)
12. Ramezani, A., Shi, X., Chung, S.-J., Hutchinson, S.: Lagrangian modeling and flight control of articulated-winged bat robot. In: IEEE/RSJ International Conference on Intelligent Robots and Systems (IROS), pp. 2867–2874 (2015)
13. Ramezani, A., Shi, X., Chung, S.-J., Hutchinson, S.: Bat Bot (B2), a biologically inspired flying machine. In: IEEE International Conference on Robotics and Automation (ICRA), pp. 3219–3226 (2016)
14. Ramezani, A., Shi, X., Chung, S.-J., Hutchinson, S.: Modeling and nonlinear flight controller synthesis of a bat-inspired micro aerial vehicle. In: AIAA Guidance, Navigation, and Control Conference, p. 1376 (2016)
15. Riskin, D.K., Bergou, A., Breuer, K.S., Swartz, S.M.: Upstroke wing flexion and the inertial cost of bat flight. Proc. Royal Soc. Lond. B: Biolog. Sci **279**(1740), 2945–2950 (2012)
16. Riskin, D.K., Willis, D.J., Iriarte-Díaz, J., Hedrick, T.L., Kostandov, M., Chen, J., Laidlaw, D.H., Breuer, K.S., Swartz, S.M.: Quantifying the complexity of bat wing kinematics. J. Theor. Biol. **254**(3), 604–615 (2008)
17. Santello, M., Flanders, M., Soechting, J.F.: Postural hand synergies for tool use. J. Neurosci. **18**(23), 10105–10115 (1998)
18. Stowers, A.K., Lentink, D.: Folding in and out: passive morphing in flapping wings. Bioinsp. Biomim. **10**(2), 025001 (2015)
19. Wissa, A., Tummala, Y., Hubbard, J., Frecker, M.: Passively morphing ornithopter wings constructed using a novel compliant spine: design and testing. Smart Mater. Struct. **21**(9), 094028 (2012)
20. Zbikowski, R., Galinski, C., Pedersen, C.B.: Four-bar linkage mechanism for insect-like flapping wings in hover: Concept and an outline of its realization. Trans. ASME: J. Mech. Des. **127**(4), 817–824 (2005)

Mathematical Modeling to Improve Control of Mesh Body for Peristaltic Locomotion

Yifan Huang[1]([✉]), Akhil Kandhari[1], Hillel J. Chiel[2], Roger D. Quinn[1], and Kathryn A. Daltorio[1]

[1] Department of Mechanical and Aerospace Engineering,
Case Western Reserve University, Cleveland, USA
{yxh649,kam37}@case.edu
[2] Department of Biology, Neurosciences and Biomedical Engineering,
Case Western Reserve University, Cleveland, USA

Abstract. In this work, we built a kinematic simulation model for our worm robot, which does peristaltic locomotion. We studied the construction of our robot's mesh-rhombus structure and the structural behavior in response to the actuator controls and simulated them in MATLAB. With some kinematic assumptions, we can model changes in body shape. Friction, gravity, internal forces are not directly modeled, however a single correction factor can be used to align the simulation and hardware progress. New control methods are found based on this model, which reduced the motion slip on the robot. In future work, this simulation can help us control and design future mesh-based robots.

Keywords: Soft robotics · Earthworm-inspired kinematics · Simulation · Cable actuation

1 Introduction

A mesh structure is common to many biologically inspired robots that locomote with earthworm-like peristalsis. Mesh body patterns are found in designs with diverse actuation (individual servomotors that actuate cables [1], reduced actuation schemes with a single drive motor [2], shape memory alloys [3], electroactive polymers and compressed air [4]). In each case, the mesh contributes a mechanical coupling between local decreases in body diameter to local increases in body length (and vice versa), similar to the effect of hydrostatics in earthworms [5].

In developing such robots, it has become clear that the actuation of different segments requires coordination in order to achieve soft-body locomotion. Control strategies have been determined with numerical optimization algorithms [6]. However, we have hypothesized that the key to maximizing speed is to reduce required slip between contact points, by elongating one part of the robot at exactly the same rate as another part of the robot retracts [2, 7]. In a simplified simulation, we have shown this to be effective, even if the robot encounters an obstacle such as narrowing in a pipe [7]. We have typically implemented this on the robot as a traveling wave in which the segments first elongate with a constant actuator speed, and then retract with equal and opposite actuator speed.

© Springer International Publishing AG 2017
M. Mangan et al. (Eds.): Living Machines 2017, LNAI 10384, pp. 193–203, 2017.
DOI: 10.1007/978-3-319-63537-8_17

However, especially in soft structures, actuator position does not linearly correspond to changes in length.

Therefore, we are developing a 3D simulation of the mesh body structure. This will provide a theoretical basis for calibrating straight-line locomotion and turning behaviors to minimize required slip. Furthermore, the kinematic simulation presented here will be a first step to full kinetic simulations that respond to deformation due to gravity, uneven terrain, and constrained space navigation. Accurate body simulations will help us design new robots at different scales and the controllers they need for applications in pipe inspection, medicine, and search and rescue.

Others have modeled worm-like locomotion. Non-mesh-based robots have been modeled in straight line walking [6], turning [8], spring-mass compression-wave motion [13] and motion in compliant environments [10]. A hollow mesh was abstracted to single-rhombus models in [7]. However, this model does not permit close analysis of 3D positions and deformations (or permit turning). Models for soft actuators and structures, such as this model of fiber-wrapped pneumatic actuators [9], often require nonlinear elasticity approaches.

Here, we show that kinematic assumptions for the mesh are sufficient for modeling the local changes to body shape in oiled plate experiments. However when locomoting on wood, a higher friction substrate, simple friction assumptions are insufficient without a correction factor. Nonetheless, this simulation permits adjustments in actuation that mitigate length changes between contact points. This will be valuable in designing controllers that reduce slip and in considering design parameters of other robots that utilize mesh structure to mimic earthworm peristalsis.

2   Method

The mesh structure of our robot is constructed from rhombuses, which are formed from flexible tubing connected at vertex pieces (named as "nodes" in model). The whole structure can be divided into six segments, and each segment has two cables in both left and right sides linking two actuators and six vertex pieces (Fig. 1).

Locomotion is achieved by pulling the cables through actuators and thus changing the aspect ratio of the rhombuses. The vertex pieces in our robot permit relative rotation but not translation (e.g. 'pin joints') [1, 11, 12]. Specifically, the pin joint permits 1DOF planar rotation between opposing lengths of tubing. This allows changes in rhombus aspect ratio in one part of the body to cause changes in nearby rings, much like the joint in scissors or a pantograph. The flexibility of the tubing and the play in the 3D printed joints makes this transmission imperfect and permits adjacent segments to have differing diameters. In the simulation, we are modeling the rhombus side lengths as a constant. However, we are permitting rotation between all four segments that meet at a vertex location. This simplifies the model: the tubes bend only at the node rather than all along their length and the vertexes are assumed to perfectly permit all rotation.

The state of the robot can thus be completely represented by a four-dimensional matrix $[N_s \times 3 \times N_r \times 3]$ of all the node's positions, where in our robot's case, $N_s = N_r = 6$.

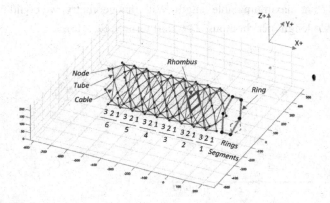

Fig. 1. The whole worm is divided into six segments and each segment contains three rings. Each ring consists of six nodes, which have three coordinates X, Y and Z. The definition of a ring was a hexagon (in the case of $N_r = 6$), which is illustrated at the most right of the figure (see also Fig. 3.a). Ring-2 is the central ring in each segment and includes the actuators and cables. Every four adjacent nodes connected by four tubes, form a rhombus as highlighted in the figure (see also Fig. 2.a). Every six rhombuses form a ring.

Relationship between cable length and ring dimensions

First, we will consider how the cable affects the central ring in each segment. The cable is a string which is always kept in tension by springs attached to the body, which also provide the restoring force [12]. In the robot, the maximum ring diameter is determined by the minimum included angle θ_L between two adjacent tubes which the vertex pieces permit. This corresponds to a maximum cable length (defined as the side-length of the hexagon ring in Fig. 1). Turning the actuator spools the cable, decreasing the cable length and making the ring contract. Then, reversing the actuator increases the cable length to expand the ring back to maximum diameter to form peristaltic motion.

Figure 2(a) details the rhombus structure. At the maximum cable length, the ring is a regular hexagon. Each rhombus is not planar but has a fold in the middle to bend around the ring. Such folds make the folding angle β the exterior angle of a regular dodecagon rather than a hexagon: $\beta = 2\pi/(2N_r) = \pi/N_r$ (Figs. 2(a) and 3(a)). The rhombus angle θ

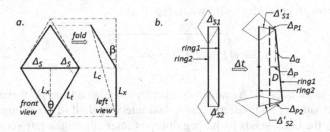

Fig. 2. (a) is the rhombus structure as highlighted in Fig. 1. (b) is the position change of two dorsal-view rings over Δt. Two rhombuses of (a) are placed at both sides for better illustration.

(Fig. 2(a)) is θ_L at maximum cable length. With this geometry, we could calculate the maximum cable length per rhombus, L_c^{max} (left term in Eq. (1)).

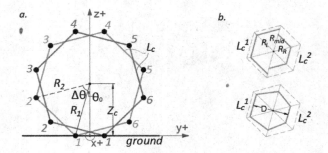

Fig. 3. (a) is an example of two hexagon rings. Two colors represents two rings' position statuses. Black dots denoted the nodes. (b) shows the linearization of nodes' positions while having a bias between two sides. (Color figure online)

As the actuator turns, the total cable length decreases as a function of the actuator position angle, φ, which can be expressed as $\varphi = \dfrac{\pi}{30} \displaystyle\int_0^t \Omega dt$, where Ω is the actuator speed in *rpm*. In the robot, each half of the ring is controlled by one cable, thus $N_r/2$ (in our case 3) segments are directly affected by each actuator. Given the spool radius and tube length, R_s and L_t respectively, and assuming that each rhombus in the half ring has uniform rhombus angle, the cable length per rhombus at position angle φ (or time t) is:

$$L_c = \underbrace{2L_t \cos\frac{\theta_L}{2}\cos\frac{\pi}{2N_r}}_{maximum} - \underbrace{\frac{2R_s}{N_r}\varphi}_{decrement} \tag{1}$$

Ring spacing as a function of ring size
One of the most critical aspects of the mesh to simulate is the way the changes in diameter affect local changes in body length. The detail of our simulation permits us to determine how the diameter of each ring can be used to calculate the spacing or interval, Δ_S, between adjacent rings. For the rings that included actuators, the adjacent interval Δ_S is determined through geometry as shown in Fig. 2(a).

$$\Delta_S = \sqrt{L_t^2 - \left(L_c/2\cos\frac{\pi}{2N_r}\right)^2} \tag{2}$$

For a non-cable interval, Δ_S cannot be directly calculated from L_c, therefore is linearized from its anterior and posterior cable-laid intervals (half of the summation).

In turning, the body bends by having different intervals on the left versus the right (Fig. 2(b)). Given initial left and right intervals Δ_{S1} and Δ_{S2} at time t, actuator speeds on the left and right result in new intervals Δ'_{S1} and Δ'_{S2} at time $t + \Delta t$. Since Δt is very small the angle change over time is also very small. Thus, all Δ_S, Δ'_S and ΔP can be approximated to the same direction. The progress distance ΔP and turning angle $\Delta \alpha$ of ring-centers after each time step Δt can then be expressed as:

$$\Delta_{P1,P2} = \Delta'_{S1,S2} - \Delta_{S1,S2}$$
$$\Delta P = \left(\Delta_{P1} + \Delta_{P2}\right)/2$$
$$\Delta \alpha = arctan\frac{\Delta_{P1} - \Delta_{P2}}{D}$$

$$(3)$$

Individual Node positions

Finally based on ring size and intervals, we can calculate the positions of all the nodes within the mesh. The nodes of adjacent rings are offset by half of $\Delta \theta$ as shown by the red and green hexagons in Fig. 3(a). This is represented in the model by two ring statuses (one in which node 1 is on the left and one in which node 1 is on the right of the center) that are assigned to alternating rings.

Neglecting any sagging of the structure due to gravity, we assume that the center of each ring remains at constant height Z_C above the ground. At least two segments are always controlled to this dimension, thus the other segments are assumed to be able to be lifted off the ground. On the physical robots, we find this assumption holds better for segments in the center of the robot, and slightly less well for the segments on the ends, which might drag if cantilevered out too far.

Inside one ring, the position angle and center distance of each node is denoted as θ_n and R_n. We can then calculate those values through geometry:

$$Z_c = \frac{L_t}{2tan\theta_0}cos\frac{\theta_L}{2}, \theta_n = \lambda\theta_0 + (n-1)\Delta\theta, R_{L,R} = \frac{L_c^{1,2}}{2sin\frac{\Delta\theta}{2}}$$

$$(4)$$

Where $\theta_0 = \pi/2N_r$, $\Delta\theta = 2\pi/N_r$. $\lambda = \pm 1$ represents the rings' two position statuses. R_n's are further linearized between R_L and R_R (in the case of $N_r = 6$ in Fig. 3(b), $R_{mid} = \left(R_L + R_R\right)/2$). The ring diameter D, which is required in the ring-interval calculation, is also defined in Fig. 3(b).

Once we have Zc, θ_n and R_n, all nodes' local coordinates can be calculated. In order to position those nodes in the global coordinate system, we need the ring's center global-coordinates (X_C, Y_C, Z_C) whose (X_C, Y_C) are tracked after every time step as explained in ring-spacing section, and the ring's direction angle α. Given the perpendicular direction to the ring-plane was the ring's direction, we set axis $X+$ the initial direction, i.e. $\alpha = 0$ along axis $X+$. Then the nth ($n = 1 \sim N_r$) node of a ring whose center and direction angle were (X_C, Y_C, Z_C) and α respectively can be expressed as:

$$\begin{bmatrix} X_n \\ Y_n \\ Z_n \end{bmatrix} = \begin{bmatrix} cos\alpha & -sin\alpha & 0 \\ sin\alpha & cos\alpha & 0 \\ 0 & 0 & 1 \end{bmatrix}\begin{bmatrix} 0 \\ -R_n sin\theta_n \\ -R_n cos\theta_n \end{bmatrix} + \begin{bmatrix} X_c \\ Y_c \\ Z_c \end{bmatrix}$$

$$(5)$$

Soft Body Locomotion

The above equations define the shape of the body, and here we will show how the simulation estimates how those shape changes result in position changes (soft body locomotion). These assumptions will provide a simplification from our previous dynamic simulation [7].

First, we will determine the number of segments at the maximum diameter at a particular time step. If there is only one, or one contiguous group of segments, that segment or contiguous group of segments is an anchor and is fixed in space. Segments with smaller diameters are permitted to freely expand and contract until there are more than one segment at the maximum diameter. In such cases, we assumed that no slip would occur where there was the largest number of contiguous segments. Thus, since here we are testing the robot with a single wave that travels back to front down the body of the robot, we assume that the front segments slip until the wave passes the center of the body, when the rear segments slip.

We note that this is a clear oversimplification. On a real robot, sometimes both segments will slip, sometimes neither will slip (and the body will deform in response to the friction forces). Our goal is to use the simulation to determine how to control the actuators in order to balance elongation and retraction between maximum diameter rings. In this case, there will be no slip at any of the maximum diameter rings, so it will not matter which was chosen as the anchor point.

The below equation tells how shape changes are related to the position changes and how correction factors during motion are implemented into the simulation.

$$
\left\{
\begin{array}{l}
\alpha^{+\prime} = \alpha^+ + \kappa\Delta\alpha \\
\begin{bmatrix} X_C^{+\prime} \\ Y_C^{+\prime} \\ Z_C^{+\prime} \end{bmatrix} =
\begin{bmatrix} cos(\kappa\Delta\alpha) & -sin(\kappa\Delta\alpha) & 0 \\ sin(\kappa\Delta\alpha) & cos(\kappa\Delta\alpha) & 0 \\ 0 & 0 & 1 \end{bmatrix}
\left(\begin{bmatrix} X_C^+ \\ Y_C^+ \\ Z_C^+ \end{bmatrix} - \begin{bmatrix} X_C \\ Y_C \\ Z_C \end{bmatrix} \right)
+ \begin{bmatrix} X_C \\ Y_C \\ Z_C \end{bmatrix} + \mu\Delta P \begin{bmatrix} cos\alpha^- \\ sin\alpha^- \\ 0 \end{bmatrix}
\end{array}
\right. \tag{6}
$$

Subscript 'c' indicates the center of ring. Superscript '+' or '−' indicates the ring after or before current ring. The prime superscript indicates that it is after Δt. κ and μ are correction factors for turning and progressing. $\Delta\alpha$ and ΔP are obtained from Eq. (3). The results of this equation (ring direction and ring center coordinates) will be further used in Eq. (5).

3   Results

Structural consistency between robot and simulation

In order to test the precision of simulation, we did some experiments on a modified version of our robot CMMWorm [1, 11], CMMWorm-S [15]. In the first experiment, the robot was placed on an oiled glass to minimize the friction, and we only control one segment while leaving its neighboring segment at specific diameters. Then the distance between those two segments was measured to evaluate progress, and the bending angle was measured to evaluate turning. This experiment shows how well this simulation accords with the real robot when no friction is considered. See Fig. 4.

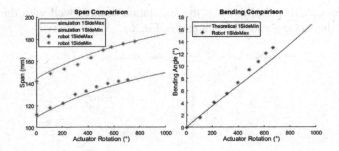

Fig. 4. Structural comparison between robot and simulation for progress and turning (Color figure online)

The blue line and asterisks in Fig. 4-left represent the experiment in which the neighboring segment remains expanded. Based on the limiting included angle between tubes, the actuator is theoretically able to turn 997°, but the tube bending stiffnesses limit contraction above 720°. The orange line and asterisks in Fig. 4-left represent the experiment in which the neighboring segment remains contracted. In Fig. 4-right, we keep the neighboring segment expanded, and only contract one of the other side's actuators. The resulting bias will cause the segment to turn, and the bending angle are measured. Those measured data are all close to the simulated line.

Simulation error during motion and the correction factor

Another experiment was done on a wooden surface to see how much the difference was between the ideal simulation model and the robot in real conditions. In this experiment, the same control methods were used in both real robot and simulation. Actuator speeds were also the same (70 *rpm*). Some specific frames during the motion are extracted and illustrated in Fig. 5.

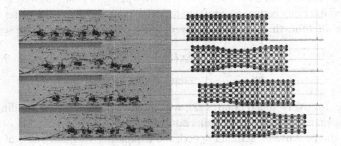

Fig. 5. Robot and simulation frames during motion.

The experiment's results in the left part of Fig. 6 show that the friction plays an important role in motion. Segments progress less during each wave in simulation than in real condition. However, it also demonstrates some consistency of when segments tends to progress, slip back or remain stationary. Therefore, we have a correction factor in the code, which is a factor in the calculation of progress distance ΔP, to minimize the influences from friction. In this case, the simulation accords well with the real robot when setting the factor $\mu = 0.16$ (right part of Fig. 6).

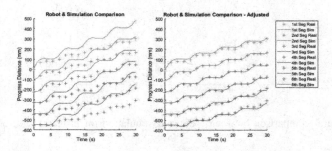

Fig. 6. Progress distance comparison between robot and simulation in both adjusted and unadjusted situations.

Control to reduce slip

In our previous work, (for example for our endoscope model [4], CMMWorm [1] and Fabric Worm [14]) the actuators were controlled at maximum speed without calibration based on the structure. These can be thought of as square waves. This "square" control means expanding one segment at the same actuator speed as its neighboring contracting segment for a constant duration. However, slip will occur in this method because the length is not a linear function of the actuation angle, i.e. the slopes of the span (Fig. 4-left) are nonlinear. However, the simulation can be used to take these differences into account to improve the control waveform shape.

While controlling two neighboring segments (one contracting and the other expanding), keeping the total span between contact points constant should reduce slip. That means one segment's control method will be determined by the other segment. From Eqs. (1) and (2), the interval Δ_S is a known function of actuator position angle φ. By setting the sum of the intervals between segments to a constant, and then taking the derivative, we can derive the following constraint between the speeds of the two active segments:

$$\frac{\partial \Delta_S(\varphi_1)}{\partial \varphi_1}\Omega_1 + \frac{\partial \Delta_S'(\varphi_2)}{\partial \varphi_2}\Omega_2 = 0 \tag{7}$$

Therefore, for every speed control Ω_1, we can always have the responding Ω_2 from this equation, such that the span remains constant.

To make the best use of actuator speed, the expanding segment starts at maximum speed, $\Omega_1 = 70\ rpm$ (Fig. 7). The corresponding Ω_2 starts out less but eventually reaches the maximum speed, intersecting the Ω_1 line. After this time, if Ω_1 was still driven at the maximum speed, Ω_2 would need to exceed the maximum speed to keep the constant span. Therefore, after the intersection point, we decrease Ω_1 such that Ω_2 operates at maximum speed. Due to symmetry, speed lines before and after the intersection point are mirror images.

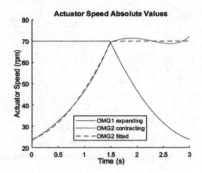

Fig. 7. Improved control of actuator speeds.

Finally, in order to implement on the real time robot controller, a second order regression (parabola) is fit to the curve using Matlab polyfit. The resulting speed controls are:

$$\Omega_1 = \begin{cases} 70 & (t \le 1.495s) \\ 14.53t^2 - 95.58t + 179.80 & (1.495s < t \le 2.99s) \end{cases}$$

$$\Omega_2 = \begin{cases} 14.53t^2 + 8.69t + 23.92 & (t \le 1.495s) \\ 70 & (1.495s < t \le 2.99s) \end{cases}$$

$$(8)$$

It should be noted that the parabolic fit appears almost indistinguishable from the original line (compare Ω_2 and Ω_2-fitted in Fig. 7 before the intersection). The higher-order effects required to cancel out these differences in the faster segment are show as the differences between Ω_2 and Ω_2-fitted after the intersection in Fig. 7. We do not attempt to implement these higher order effects because we expect them to be absorbed by the compliance of the body [1].

In this control method, there will always be an actuator operating at maximum speed, the span will theoretically remain constant, and both segments will reach its maximum and minimum diameters respectively.

We implemented this improved control into our simulation and compared it to our previous square wave control. Figure 8 shows that the improved control method allows a more stable span. With the square wave control, the range of the span is 10 mm, whereas with the improved wave control, the span only changes by 1.5 mm. Thus if the robot was rigid and the ground was low friction, this error could result in 85% reduction in slip.

A potential drawback of this control waveform is that the maximum span is less and the total time required is longer. However, the energy efficiency of this control may be higher than for the square control.

Figure 9 shows the result of span measurements on the robot on oiled glass (N = 2). On the robot, the actuators may not be able to follow the desired curve perfectly, and friction and structural stiffness can have unaccounted affects, so the improved span is not as consistent as it is in the simulation. However, compared to our previous square wave control, the range of the span has reduced from 23 mm (140–163 mm) to 16 mm (138–154 mm) (30% reduction), which proves the validity of this new control.

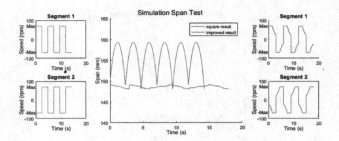

Fig. 8. Simulation span test. Left two graphs are previous square wave controls, right two graphs are improved wave controls.

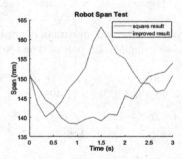

Fig. 9. Robot span test.

4 Conclusion

The consistency of structural behavior between robot and simulation proved our model's reasonability. The correction factor can also offset the error during motion caused by friction and body stiffness. These two facts ensure a precise motion simulation, from which we can extract various kinds of data that are difficult to measure from robot, e.g. trajectory, bias and slip. A control method to better reduce slip is also calculated based on the model. Additional factors can also be included in the simulation for increased accuracy as needed.

References

1. Horchler, A.D., Kandhari, A., Daltorio, K.A., Moses, K.C., Ryan, J.C., Stultz, K.A., Kanu, E.N., Andersen, K.B., Kershaw, J., Bachmann, R.J., Chiel, H.J., Quinn, R.D.: Peristaltic locomotion of a modular mesh-based worm robot: precision. Compliance Friction Soft Robot. **2**(4), 135–145 (2015)
2. Boxerbaum, A.S., Horchler, A.D., Shaw, K.M., Chiel, H.J., Quinn, R.D.: 2012 worms, waves and robots. In: Proceedings of the IEEE International Conference on Robotics and Automation, pp. 3537–3538 (2012)

3. Seok, S., Onal, C.D., Cho, K.-J., Wood, R.J., Rus, D., Kim, S.: Meshworm: a peristaltic soft robot with antagonistic nickel titanium coil actuators. IEEE/ASME Trans. Mechatron. **18**(5), 1485–1497 (2013)
4. Mangan, E.V., Kingsley, D.A., Quinn, R.D., Chiel, H.J.: 2002 Development of a peristaltic endoscope. In: Proceedings of the IEEE International Conference on Robotics and Automation, pp. 347–352 (2002)
5. Gray, J., Lissmann, H.W.: Studies in animal locomotion VII. locomotory reflexes in the earthworm. J. Exp. Biol. **15**, 506–517 (1938)
6. Fang, H., Li, S., Wang, K.W., Xu, J.: On the periodic gait stability of a multi-actuated spring-mass hopper model via partial feedback linearization. Nonlinear Dynamics (2015)
7. Daltorio, K.A., Boxerbaum, A.S., Horchler, A.D., Shaw, K.M., Chiel, H.J., Quinn, R.D.: Efficient worm-like locomotion: slip and control of soft-bodied peristaltic robots. Bioinspir. Biomim. **8**(3), 035003 (2013)
8. Omori, H., Nakamura, T., Iwanaga, T., Hayakawa, T.: Development of mobile robots based on peristaltic crawling of an earthworm. In: Abdellatif, H. (ed.) Robotics Current and Future Challenges, InTech (2010)
9. Connolly, F., Walsh, C.J., Bertoldi, K.: Automatic design of fiber-reinforced soft actuators for trajectory matching. PNAS **114**, 51–56 (2017)
10. Zarrouk, D., Sharf, I., Shoham, M.: Analysis of earthworm-like robotic locomotion on compliant surfaces. In: International Conference on Robotics and Automation (ICRA) (2010)
11. Horchler, A.D., et al.: Worm-like robotic locomotion with a compliant modular mesh. In: Wilson, S., Verschure, P., Mura, A., Prescott, T. (eds.) Living Machines 2015. LNCS, vol. 9222, pp. 26–37. Springer, Cham (2015). doi:10.1007/978-3-319-22979-9_3
12. Kandhari, A., Horchler, A.D., Zucker, G.S., Daltorio, K.A., Chiel, H.J., Quinn, R.D.: Sensing contact constraints in a worm-like robot by detecting load anomalies. In: Lepora, N., Mura, A., Mangan, M., Verschure, P., Desmulliez, M., Prescott, T. (eds.) Living Machines 2016. LNCS, vol. 9793, pp. 97–106. Springer, Cham (2016). doi:10.1007/978-3-319-42417-0_10
13. Ross, D., Lagogiannis, K., Webb, B.: A model of larval biomechanics reveals exploitable passive properties for efficient locomotion. In: Wilson, S., Verschure, P., Mura, A., Prescott, T. (eds.) Living Machines 2015. LNCS, vol. 9222, pp. 1–12. Springer, Cham (2015). doi: 10.1007/978-3-319-22979-9_1
14. Mehringer, A.G., Kandhari, A., Chiel, H.J., Quinn, R.D., Daltorio, K.A.: An integrated compliant fabric skin softens, lightens, and simplifies a mesh robot. Living Machines (accepted) (2017)
15. Kandhari, A., Huang, Y., Daltorio, K.A., Chiel, H.J., Quinn, R.D.: Longitudinal stiffness affects forward locomotion while circumferential stiffness affects turning in a soft bodied earthworm like robot (2017, in preparation)

Neuronal Distance Estimation by a Fly-Robot Interface

Jiaqi V. Huang[✉] and Holger G. Krapp

Department of Bioengineering, Imperial College London, London SW7 2AZ, UK
j.huang09@imperial.ac.uk

Abstract. The ability of an autonomous robot to avoid collisions depends on distance estimates. In this paper, we focus on open-loop responses of the identified directional-selective H1-cell in the fly brain, recorded in animals that are mounted on a 2-wheeled robot. During oscillatory forward movement along a wall clad with a pattern of vertical stripes on one side of the robot, the H1-cell periodically increases and decreases its spike rate, where the response amplitude depends on two parameters: (i) the turning radius of the robot, and (ii) the distance between the wall and the mean forward trajectory of the robot. For small turning radii, we found a monotonic relationship between the H1-cell's spike rate and wall distance. Our results suggest that, given a known turning radius, the responses of the H1-cell could be used in a negative feedback-loop to control the average forward trajectory of an autonomous robot that avoids collisions with potential obstacles in its environment.

Keywords: Motion vision · Brain machine interface · Blowfly · H1-cell · Collision avoidance · Distance estimation

1 Introduction

Autonomous robotic platforms require on-board sensing and processing capabilities to avoid any collisions with potential obstacles. Although radar, sonar and vision-based sensors provide range information, the application of those technologies is often limited to large scale systems with sufficient payload and power resources.

Distance estimation of obstacles is not only an engineering challenge, but also concerns most animal species. Nature has evolved several ways of estimating distance including echolocation in bats [1], sonar in aquatic mammals [2], or mechanisms based on vision in both vertebrates and invertebrates [3]. In contrast to man-made sensors, however, most biological mechanisms are more adaptive to the current environmental conditions, have greater sensitivity, smaller scale and significantly lower power consumption [4].

Collision avoidance is one of the key conditions to serve our ultimate goal, which is the development of a closed-loop experimental platform to investigate multisensory integration in flies. We mounted a miniaturized recording platform on an autonomous two-wheeled robot and studied the response properties of a directional-selective inter-neuron in the blowfly motion vision pathway, the H1-cell [5–8]. The neuronal activity, or spike rate, of this identified cell is meant to be used as visual feedback to steer the

© Springer International Publishing AG 2017
M. Mangan et al. (Eds.): Living Machines 2017, LNAI 10384, pp. 204–215, 2017.
DOI: 10.1007/978-3-319-63537-8_18

robot, effectively creating a hybrid Fly-Robot Interface (FRI). Distance estimation will be essential for the implementation of a collision avoidance system which enables autonomous robot movements without bumping into potential obstacles. The question is: does the neuronal activity of the H1-cell contain sufficient information to estimate distance?

The H1-cell of blowflies has been studied for decades and its sensitivity to wide-field horizontal motion is well established [9–17]. This spiking cell is responding to horizontal optic flow, where back-to-front and front-to-back motion increase and decrease H1-cell activity, respectively [11]. In recent studies, we have implemented a wall-following algorithm on a mobile robot and found that the spike rate of the H1-cell was dependent on the turning radius of the robot's trajectory. Turning radius proved to be an important parameter as it is related to the fraction of the H1-cell receptive field exposed to horizontal motion in its preferred back-to-front direction [8] and therefore modulates the spike rate of the cell.

In this paper, we use the FRI and focus on studies of the H1-cell activity under open-loop conditions during a combination of rotational and translational body movements. We characterized the relationship between H1-cell spike rate and lateral wall distance, which is crucial for the implementation of a closed-loop collision avoidance system.

2   Methods

2.1   Experiment Setup

The experiments were carried out under light-controlled conditions within an arena of $640 \times 740 \times 1040$ mm enclosed by black fabric attached to an aluminum scaffold (Fig. 1). A striped pattern (wavelength = 30 mm, as seen from the centre) was fitted to one of the long arena walls covering a height of 456 mm. To keep the robot on a straight

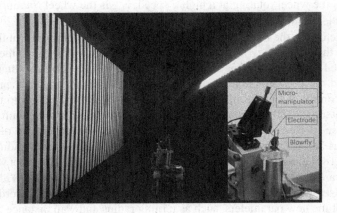

Fig. 1. Setup of the experimental arena. The stripe pattern (left) and the acrylic board plus guiding line and the robot on top are shown. Only a LED light source is mounted on the right of the arena. (For scale: the diameter of the robot is 10 cm.) Inset: the mobile recording platform on top of the robot. (For details, see [5])

trajectory on average, a black tape was fixed to a piece of transparent acrylic board (8 × 224 × 1000 mm) on the arena floor, parallel to the wall, serving as a guiding line for the robot's on-board infrared line detecting sensors. Two stripes of LED were mounted opposite to the striped pattern at a height of 430 mm providing a pattern contrast of 84% (black: 1.5 cd/m^2, white: 17.5 cd/m^2).

The FRI system was actuated by a two-wheeled robot (Pololu© m3pi), carrying the extracellular recording platform. Extracellular H1-cell activity was recorded from a blowfly mounted in the centre of the platform, 150 mm above the floor. The assembly was similar to the one used in a previous wall-following experiment [8].

To access the dependence of the spike rate on distance and turning radius we performed open-loop experiments where the robot was oscillating along the guiding lines parallel to the vertical stripe-patterned wall. The chosen radii of the turns were: 5, 10, 15, 20, 25 cm, respectively. The distances from the guiding line to the patterned wall were: 10, 15, 20, 25, 30 cm respectively. Altogether 25 different combinations of wall distance (Dw) and turning radius (Rt) were tested while the H1-cell activity was recorded. During most of the experiments, H1-cell responses were measured for 10 s of oscillatory robot movement, repeated either 5 (N = 4) or 3 (N = 1) times per condition. The sequence of stimulus conditions tested was reversed in some experiments to test for stationarity of the average neuronal responses and potential adaptation effects over the duration of the experiments.

The different turning radii were pre-programmed on the on-board robot processor (mbed NXP LPC1768). They were calculated based on the speed of the left and right wheels, along with the radius of the robot platform. The speed of one wheel was always kept maximum (35% PWM duty cycle), while the other wheel speed was calculated by:

$$\frac{v_{low}}{v_{high}} = \frac{R_t - R_r}{R_t + R_r}, \tag{1}$$

where v_{high} is the wheel rotating at a higher speed; v_{low} is the wheel rotating at a lower speed; R_t is the turning radius and R_r is the radius of the robot (5 cm).

The trajectory of the robot during the experiment was programmed in three stages to guarantee the consistency of the procedure (cf. flow chart Fig. 2): (i) the robot was initially placed on the guiding line close to, and facing the starting point. After the power was switched on, the robot started running a self-calibration routine, then found the guiding line on the floor and moved straight along the guiding line up to the starting point where it turned 180° clockwise. (ii) The robot was moving forward with alternating anti-clockwise (preferred direction) and clockwise turns (null direction) of one given turning radius and wall distance. After reaching the end point on the guiding line, the robot turned 180° anti-clockwise (to prevent cable tangling). (iii) The robot was moving straight forward all the way back to the starting point, and then stopped (Video: https://youtu.be/hW-LQhkfeTs). The three stages were repeated for neural recording before the change of the new parameters, such as turning radius and wall distance.

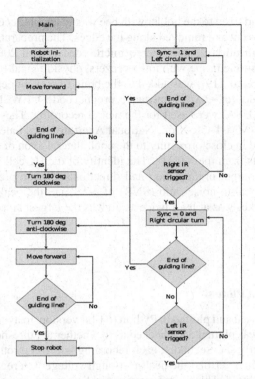

Fig. 2. Flow chart of robot software controlling the action of the FRI.

On the recording platform, a tungsten electrode was used to monitor extracellular H1-cell spiking activity which was fed into a customised miniaturized preamplifier with a nominal gain of 10k. The amplified signals were sent to a PC for off-line data analysis, together with a synchronization signal, produced by the on-board infrared sensors for temporally aligning the neural responses with the turns of the robot.

The data was recorded by a data acquisition card (NI USB-6215, National Instruments Corporation, Austin, TX, USA) at a rate of 20k Hz, and stored on the hard disk of a PC.

2.2 Blowfly Preparation

The preparation of blowfly has been the same as described in previous accounts [5–8]. In brief: For the experiments, we used female blowflies, *Calliphora vicina*, between 4 and 11 days old. Wings were immobilised by applying bee wax to the hinges. Legs and proboscis were removed to reduce unwanted muscular movements which could degrade the quality of the recordings. Bee wax was used to seal cuts and open wounds to prevent desiccation. The head of the fly was fixed to a custom-made fly holder using bee wax, after adjusting its orientation according to the "pseudopupil methods" [18]. The thorax was bent ventrally and fixed to the holder with bee wax to enable access to the rear head, where the cuticle was cut and removed along the edges. The preparation was performed under optical magnification using a stereo microscope (Stemi 2000, Zeiss©). After removal of fat and muscle tissue with fine tweezers, physiological Ringer solution (for recipe see Karmeier et al. [19]) was added to the brain to prevent desiccation.

Tungsten electrodes (of ~3 MΩ impedance, product code: UEWSHGSE3N1M, FHC Inc., Bowdoin, ME, USA) were used for the neural recording. They were mounted on a micromanipulator (MM-1-CR-XYZ, National Aperture, Inc., Salem, NH, USA) and their tips were placed in close proximity to the contralateral axon of the H1-cell, which receives visual inputs from the left eye. The identity of the H1-cell was verified based on its unique input-output organization and directional motion preferences [11]. All recordings had a signal-to-noise-ratio (SNR) of >2:1, i.e. the peak amplitude of the recorded H1-cell spikes was at least twice as high as the largest amplitude of the background noise.

3   Results

3.1   Recorded Data Plotted

The data were processed and plotted in Python (64-bit version, to avoid memory errors). Firstly, a recording was plotted and checked as to whether the threshold value for spike detection was correctly set. Secondly, each recording was run through a Python script for spike sorting, which verified the peak-to-trough voltage difference and duration of a detected action potential. Thirdly, the synchronization signals were delayed by 90 ms, a value estimated from data analysis. The signal needs to be shifted because of an inertia-based time delay between the sync pulse and the actual change of the direction of rotation of the robot. Finally, the mean spike rate was calculated in each time window initialized by the synchronization signal (Fig. 3D, green numbers on top). And the spike rates shown in the middle of a recording (Fig. 3D, 185 Hz) were picked as result to compile curves in Fig. 4. The spike rate in the middle indicating that the robot was in the centre of the arena and there was an equal fraction of the striped pattern in front and behind the robot. This was for the development of closed-loop control algorithm in the future, when the robot is in a long corridor along a continuous vertical stripe pattern.

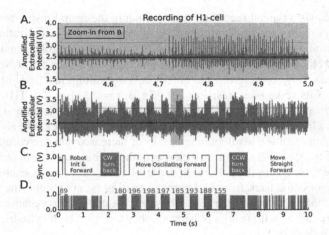

Fig. 3. Recorded data of blowfly H1-cell during a robot movement. (B, C, D) The robot started the clockwise turn at the starting point around the 2nd second, and moved oscillatory forward while the sync signal was toggling, and an anti-clockwise turn at the end point was around the 7th second. (A) A zoom-in potential trace from t = 4.5–5 s of subplot B. The black line is the reference voltage of a single supplier amplifier which is 2.5 V. The cyan line is the threshold level for spike sorting. (B) The amplified extracellular potential trace of the H1-cell. Grey area is zoomed into subplot A. (C) the synchronization signal from a robot digital pin, which toggles between 0 V and 3.3 V. (D) The individual spike occurrence (red) and 90 ms compensated sync signal (green). The green numbers on the top of each green pulses are the local spike rates. Here, the spike rate in a green window right before the 5th second (185 Hz) was used to compile Fig. 4. (Color figure online)

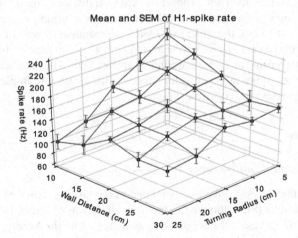

Fig. 4. The H1-cell spike rate as a function of turning radius and wall distance. Each blue point is averaged from the mean spike rates of the insects which recorded in the anti-clockwise turn (preferred direction) in the middle of the trajectory with a particular set of parameters (wall distance and turning radius). The red bars are the correlated standard error of means. (N = 5, n = 3, 5, 5, 5, 5) (Color figure online)

Figure 3 shows a typical recording trace including the neural responses during the different phases of an experiment as described above. The robot initialization started at the first synchronization pulse (green traces in Fig. 3). While the robot rotated counter-clockwise and clockwise on the spot trying to find the guiding line on the floor, the H1-cell was stimulated in its preferred and null direction resulting in a maximum increase and decrease of its spike rates, respectively (Fig. 3, t = 0.5–1 s). The H1-cell activity recorded for data analysis started from the second high synchronization signal, where the consecutive high and low states indicate phases of counterclockwise and clockwise turns of the robot resulting in alternating phases of increased and decreased spike rate, respectively (Fig. 3, t = 2–7 s). During the onset of the sync signal, the robot was making anti-clockwise turn, where the pattern on the left-hand side wall was exciting the H1-cell. When robot came back from end point to starting point straight, the black cloth generates no contrast to inhibit the H1-cell, where the cell fired spontaneously. Those responses were in line with results previously obtained with both cylindrical [12] and planar stimulus devices [8].

3.2   Dependence of H1-Cell Spike Rate on Turning Radius and Wall Distance

Figure 4 shows the average H1-cell spike rate as well as the standard error of the mean obtained during counter-clockwise phases in the middle of the oscillatory robot trajectory recorded in 5 flies, plotted against turning radius and wall distance.

The spike rate of H1-cell shows an inversely proportional dependence on turning radius for a wall distance of 10 cm. As the wall distance increases, both the peak and trough of the spike rates in the curve are reduced, and the curve becomes flat.

The spike rate is also inversely proportional to wall distances for a turning radius of 5 cm. But it becomes proportional to wall distances for a turning radius of 25 cm. In between, when Rt is 15 cm, the spike rate seems independent of wall distances.

One segment of the curves where Rt = 5 and Dw = 10–25 appears to be linear, where the spike rate is decreasing at a rate of around 20 Hz per 5 cm. The inversely proportional relationship between spike rate and wall distance could be used to implement a simple feedback control system for collision avoidance.

4   Discussion

4.1   Spike Rate and Wall Distance

The finding that the H1-cell spike rate is related to the turning radius of the robot is in line with earlier results [8]. Although the previous study was different, in that it focused only on the turning radius of the robot as a parameter while the average wall distance was kept constant.

It is interesting to see an inversely proportional relationship between spike rate and wall distance at small turning radii but a proportional relationship between the neuronal response and wall distance at large turning radii, between wall distances from 10 to 20 cm (Fig. 5). To our knowledge, these relationships have neither been predicted nor measured before. There are several parameters involved which could have an impact on the H1-cell spike rate in this experiment, including height, contrast and temporal frequency of the pattern.

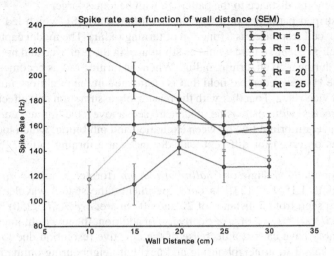

Fig. 5. H1-cell spike rate as a function of wall distance for different turning radii (see inset). For a turning radius of 15 cm, the spike rate becomes independent of wall distance. This may be due to an equilibrium between stimulation of the receptive field in the preferred (excitatory) and null (inhibitory) direction. (Data replotted from Fig. 4)

In contrast to earlier studies on dendritic gain control in blowfly directional-selective interneuron [20], we found the spike rate to be a linear function of pattern height within a range of $\pm15°$ to $\pm55°$ vertical extension [Huang and Krapp unpublished data]. Pattern height, however, might not be a crucial parameter in our current experiments because along elevation the pattern covered a $0°$ to $+45°$ section of the H1-cell receptive field [11], above the eye equator.

Pattern contrast is one of the most important parameters in the motion vision pathway [21]. But because of the limited spatial resolution of the photodetector array in the compound eye, the contrast edges become increasingly blurred with distance [22]. The optical limitations of the compound eye could potentially explain why, independent of the turning radius, the spike rate of the H1-cell is no longer modulated for wall distances >25–30 cm (Fig. 5).

The responses in the fly motion vision pathway also depend on the temporal frequency of the pattern which is defined by the ratio of the angular velocity and the spatial wavelength composition of the stimulus pattern. In previous studies, the H1-cell spike rate was reported to peak at around 1–4 Hz temporal frequency, where a vertically striped pattern was generated either by LEDs [12] or space-continuous sheets of paper attached to the walls of a cylinder [5]. For an animal that is placed in the centre of a cylinder the spatial frequency distribution of a periodically striped pattern would be homogeneous. However, when an equivalent pattern is displayed on a flat surface, for instance along the walls of an arena as used in our experiments, the perceived spatial frequency distribution becomes inhomogeneous. At the smallest distance between the eyes and the pattern the spatial frequency is lowest but perceptually increases for viewing directions where the distance to the patterned wall becomes larger.

Vertical striped patterns, were used in an earlier study [8], which led to a model describing H1-cell responses as a function of turning radius. The model captured which parts of the cell's receptive field would be stimulated in the preferred and null-direction, respectively, during a given turning radius. When the turning radius becomes larger, the fraction of the H1-cell receptive field that is stimulated in the inhibitory null-direction increases and vice versa. Together with the reasonable assumption that effective pattern contrast decreases with increasing distance (see above) the tuning curves would converge to an equilibrium line between excitation and inhibition, which looks exactly like responses observed for different wall distances at a turning radius, Rt of 15 cm (Fig. 5).

The interommatidial angle of *Challiphora vicina* (former: *Challiphora erythroce-phala*) is around $1.1°–1.3°$ [23]. In our experiments, the spatial wavelengths of the striped pattern seen from a distance of 25 and 30 cm are: $arctan(7.5/250) * 2 = 3.43°$ and $arctan(7.5/300) * 2 = 2.86°$, respectively. In addition, the low light intensity in the rig will probably have caused a decrease of the effective resolution due to the spatial integration of signal from neighbouring and next but neighbouring ommatidia [21]. It is likely, therefore, that the strongest contribution to the H1-cell responses was based on those parts of the striped pattern where the ratio between the effective interommatidial angle and the spatial wavelength were still in a range not prone to spatial aliasing and response reversal of the EMDs [21]. The flat part of the curves in Fig. 5 for wall distances between 25–30 cm may reflect a regime where spatial aliasing is about to kick in.

After all, it's not too surprising to see the relationship between spike rate and wall distance is partially proportional (Fig. 5, Rt = 5–10; Dw = 10–25) and partially inversely proportional (Fig. 5, Rt = 20–25; Dw = 10–25). And the fly could, in principle, exploit the fact that certain turning radii would enable the estimation of relative distance based on the spike rate of the H1-cells. A similar active process in the visual estimation of distance had been proposed for blowflies flying in a confined space performing an alternating pattern of body saccades followed by inertia-based translational drift during which HS-cells estimate relative distance [24, 25]. Here we believe, by using H1-cell and turning radius calculated from its body properties, e.g.: the power of the left and right wings, and the tip to tip width of wings (see formula 1), a blowfly could actively estimate the distance from distant obstacles which produces contrast.

4.2 A Potential Model for Collision Avoidance Control

If the turning radius is fixed to 5 cm, the inversely proportional dependence of spike rate on wall distance is almost linear over a range from 10 to 25 cm. This relationship could be developed into a collision avoidance control algorithm for our FRI and potentially other autonomous robotic system.

A straightforward method would be to implement an inverse model and set up a threshold spike rate at, for instance, 200 Hz, where the robot should be 15 cm away from one of the walls in our experimental arena. The robot could be programmed to move on an oscillatory forward trajectory, switching between clockwise and anti-clockwise quarter-turns, while measuring the spike rate. If the spike rate is higher than 200 Hz, it means the robot is less than 15 cm from the wall, and the period of the next directional turn should be expanded to steer the robot away from the wall. The extra period of that turn should not be involved in concurrent spike rate measurement. This may either be achieved by simply disabling the integration of the spike rate during the extra turning phase, or by subtracting the expected H1-cell spike rate in response to the extra turn. The latter solution would be in line with recent findings in the motion vision pathway of the fruit fly demonstrating the use of efference copies during spontaneous turns to avoid the fly being trapped by its own optomotor reflexes [26]. With this method, the robot should be able to move in the middle of the track in our experimental arena, keeping a minimum distance of 15 cm to the walls on either side.

5   Summary

In this research, we have described the relationship between H1-cell spike rate and wall distance for different turning radii during an oscillating forward movement. Our results will be instrumental when implementing a vision-based collision avoidance system for a closed-loop autonomous FRI to study fly multisensory integration.

Acknowledgments.   The authors would like to thank Naomi Ho for calibrating the robot firmware, Dr. Caroline Golden for improving the proofreading the manuscript. This work was partially supported by US AFOSR/EOARD grant FA8655-09-1-3083 to HGK.

References

1. Fenton, M.B.: Natural history and biosonar signals. In: Popper, A.N., Fay, R.R. (eds.) Hearing by Bats, pp. 37–86. Springer, New York (1995). doi:10.1007/978-1-4612-2556-0_2
2. Au, W.W.L.: The Sonar of Dolphins. Springer, New York (1993). doi:10.1007/978-1-4612-4356-4
3. Hartenstein, V., Reh, T.A.: Homologies between vertebrate and invertebrate eyes. In: Moses, D.K. (ed.) Drosophila Eye Development, pp. 219–255. Springer, Heidelberg (2002). doi:10.1007/978-3-540-45398-7_14
4. Sarpeshkar, R.: Ultra Low Power Bioelectronics: Fundamentals, Biomedical Applications, and Bio-Inspired Systems. Cambridge University Press, Cambridge (2010)
5. Huang, J.V., Krapp, H.G.: Miniaturized electrophysiology platform for fly-robot interface to study multisensory integration. In: Lepora, N.F., Mura, A., Krapp, H.G., Verschure, P.F.M.J., Prescott, T.J. (eds.) Living Machines 2013. LNCS, vol. 8064, pp. 119–130. Springer, Heidelberg (2013). doi:10.1007/978-3-642-39802-5_11
6. Huang, J.V., Krapp, H.G.: A predictive model for closed-loop collision avoidance in a fly-robotic interface. In: Duff, A., Lepora, N.F., Mura, A., Prescott, T.J., Verschure, P.F.M.J. (eds.) Living Machines 2014. LNCS, vol. 8608, pp. 130–141. Springer, Cham (2014). doi:10.1007/978-3-319-09435-9_12

7. Huang, J.V., Krapp, H.G.: Closed-loop control in an autonomous bio-hybrid robot system based on binocular neuronal input. In: Wilson, S.P., Verschure, P.F.M.J., Mura, A., Prescott, T.J. (eds.) Living Machines 2015. LNCS, vol. 9222, pp. 164–174. Springer, Cham (2015). doi:10.1007/978-3-319-22979-9_17

8. Huang, J.V., Wang, Y., Krapp, H.G.: Wall following in a semi-closed-loop fly-robotic interface. In: Lepora, N.F.F., Mura, A., Mangan, M., Verschure, P.F.M.J., Desmulliez, M., Prescott, T.J. (eds.) Living Machines 2016. LNCS, vol. 9793, pp. 85–96. Springer, Cham (2016). doi:10.1007/978-3-319-42417-0_9

9. Strausfeld, D.N.J.: The atlas: sections through the brain. In: Strausfeld, D.N.J. (ed.) Atlas of an Insect Brain, pp. 57–115. Springer, Heidelberg (1976). doi:10.1007/978-3-642-66179-2_7

10. Hausen, K.: Functional characterization and anatomical identification of motion sensitive neurons in the lobula plate of the blowfly Calliphora erythrocephala. Z. Naturforsch. **31**, 629–633 (1976)

11. Krapp, H.G., Hengstenberg, R., Egelhaaf, M.: Binocular contributions to optic flow processing in the fly visual system. J. Neurophysiol. **85**, 724–734 (2001)

12. Jung, S.N., Borst, A., Haag, J.: Flight activity alters velocity tuning of fly motion-sensitive neurons. J. Neurosci. **31**, 9231–9237 (2011)

13. Longden, K.D., Krapp, H.G.: Octopaminergic modulation of temporal frequency coding in an identified optic flow-processing interneuron. Front. Syst. Neurosci. **4**, 153 (2010)

14. Maddess, T., Laughlin, S.B.: Adaptation of the motion-sensitive neuron H1 is generated locally and governed by contrast frequency. Proc. R. Soc. Lond. B Biol. Sci. **225**, 251–275 (1985)

15. Lewen, G.D., Bialek, W., de Ruyter van Steveninck, R.R.: Neural coding of naturalistic motion stimuli. Netw. Bristol Engl. **12**, 317–329 (2001)

16. Huston, S.J., Krapp, H.G.: Nonlinear integration of visual and haltere inputs in fly neck motor neurons. J. Neurosci. **29**, 13097–13105 (2009)

17. Parsons, M.M., Krapp, H.G., Laughlin, S.B.: A motion-sensitive neurone responds to signals from the two visual systems of the blowfly, the compound eyes and ocelli. J. Exp. Biol. **209**, 4464–4474 (2006)

18. Franceschini, N.: Pupil and pseudopupil in the compound eye of drosophila. In: Wehner, R. (ed.) Information Processing in the Visual Systems of Anthropods, pp. 75–82. Springer, Heidelberg (1972). doi:10.1007/978-3-642-65477-0_10

19. Karmeier, K., Tabor, R., Egelhaaf, M., Krapp, H.G.: Early visual experience and the receptive-field organization of optic flow processing interneurons in the fly motion pathway. Vis. Neurosci. **18**, 1–8 (2001)

20. Borst, A., Egelhaaf, M., Haag, J.: Mechanisms of dendritic integration underlying gain control in fly motion-sensitive interneurons. J. Comput. Neurosci. **2**, 5–18 (1995)

21. Buchner, E.: Behavioural analysis of spatial vision in insects. In: Ali, M.A. (ed.) Photoreception and Vision in Invertebrates, pp. 561–621. Springer, New York (1984). doi: 10.1007/978-1-4613-2743-1_16

22. Chahl, J.S.: Range and egomotion estimation from compound photodetector arrays with parallel optical axis using optical flow techniques. Appl. Opt. **53**, 368–375 (2014)

23. Land, M.F.: Visual acuity in insects. Annu. Rev. Entomol. **42**, 147–177 (1997)

24. Lindemann, J.P., Kern, R., van Hateren, J.H., Ritter, H., Egelhaaf, M.: On the computations analyzing natural optic flow: quantitative model analysis of the blowfly motion vision pathway. J. Neurosci. **25**, 6435–6448 (2005)

25. Bertrand, O.J.N., Lindemann, J.P., Egelhaaf, M.: A bio-inspired collision avoidance model based on spatial information derived from motion detectors leads to common routes. PLoS Comput. Biol. **11**, e1004339 (2015)
26. Kim, A.J., Fitzgerald, J.K., Maimon, G.: Cellular evidence for efference copy in Drosophila visuomotor processing. Nat. Neurosci. **18**, 1247–1255 (2015)

Bio-inspired Robot Design Considering Load-Bearing and Kinematic Ontogeny of Chelonioidea Sea Turtles

Andrew Jansen[1], Kevin Sebastian Luck[2], Joseph Campbell[2], Heni Ben Amor[2], and Daniel M. Aukes[3($\boxtimes$)]

[1] School of Life Sciences, Tempe, AZ, USA
majanse1@asu.edu
[2] The School of Computing, Informatics, and Decision Systems Engineering, Tempe, AZ, USA
{ksluck,jacampb1,hbenamor}@asu.edu
[3] The Polytechnic School, Arizona State University, Mesa, AZ, USA
danaukes@asu.edu

Abstract. This work explores the physical implications of variation in fin shape and orientation that correspond to ontogenetic changes observed in sea turtles. Through the development of a bio-inspired robotic platform – CTurtle – we show that (1) these ontogenetic changes apparently occupy stable extrema for either load-bearing or high-velocity movement, and (2) mimicry of these variations in a robotic system confer greater load-bearing capacity and energy efficiency, at the expense of velocity (or vice-versa). A possible means of adapting to load conditions is also proposed. We endeavor to provide these results as part of a theoretical framework integrating biological inquiry and inspiration within an iterative design cycle based on laminate robotics.

Keywords: Bio-inspired robots · Turtles · Locomotion · Mobile robots · Kinematics · Rapid-prototyping · Laminates · Granular media · Fabrication · Design

1 Introduction

The use of robotics to answer questions in biology is a well-established paradigm which offers benefits to both fields. For biologists, the ability to study repeatable physical systems is an attractive option, even if such systems replicate only a small part of the biological analog. Robotic platforms can be modified quickly to test a wide range of morphologies and behaviors, and sensors can be mounted both in-situ and in the surrounding environment to determine the effect of morphological and behavioral changes on the body and to the world. Such platforms have made it possible to understand more about the locomotion of caterpillars, geckos, and sea-turtles, to name a small selection [3,15,20]. A thorough review is provided by Ijspeert [9].

© Springer International Publishing AG 2017
M. Mangan et al. (Eds.): Living Machines 2017, LNAI 10384, pp. 216–229, 2017.
DOI: 10.1007/978-3-319-63537-8_19

For roboticists, such collaborations offer insights into robotic design strategies that takes into account knowledge of how species' adaptations make them suited for certain activities or environments. Many have found that such insights successfully transfer to robotic designs inspired by, for example, cockroaches [1,4,5,8,10], geckos [19], bees [13], and sea turtles [12]. Such insights lead both to improved robotic designs and to a better understanding of biological systems.

These studies are often made possible through technological and manufacturing innovations which facilitate the rapid design and fabrication of robotic systems. Many of the platforms cited above make use of rapid prototyping techniques such as 3D printing [20], multi-material laminate fabrication processes [4,8,13], or iterative processes such as Shape Deposition Manufacturing (SDM) [5,19]. Such methods enable the manufacturing of monolithic systems where subcomponents exhibit vastly different material properties and performance due to the targeted placement of rigid and soft materials.

We propose a workflow for bio-inspired robotics in Fig. 2, where design inspiration is drawn from biological systems, then the resulting prototype is used as a physical, manipulable analogue to further investigate the properties of that

Fig. 1. Three generations of a sea turtle-inspired robot.

Fig. 2. Our workflow and design cycle for biomimicry. We begin by examining the natural history of the study organisms (Chelonioidea), from which we draw inspiration for an initial design iteration. This design is simplified and customized using an iterative approach to optimize functionality. We then iterate over numerous parameters of the design, leading to further questions about emergent behaviors and physical principles in both the robotic and the biological systems. Further investigation of the biological system then leads to additional biological observation and mimicry in design variables not considered in the previous design cycle. Turtle image courtesy of the Florida Fish and Wildlife Conservation Commission (photo by Meghan Koperski) and drawing by Dodd Jr [6]

system. Using this theoretical framework, we present our work exploring the ontogenetic differences in sea turtle morphology and locomotion using a crawling robot as a bio-analogue. We outline in detail (1) the implications of the natural history of sea turtles for our design, and (2) the implications of the observed experimental performance of the robot for sea turtle biology. This process is made possible through the design and fabrication of a robotic analog, CTurtle, which permits us to explore the connection between morphology, load-carrying capacity, and performance on a simplified system which can be thoroughly and repetitively tested (Fig. 1).

2   Background: Movement of Turtles

The selection of sea turtles is motivated by three aspects of their locomotion that we identified as promising for this particular application. First, sea turtles are capable of effective movement through unstable media, without loss of traction or sinking; the broad surface area, upturned plastron, and "crutching" motion of the body and fins avert sinking or digging into the substrate, preventing the animal from getting stuck under normal conditions [17,21]. Secondly, the low center of mass of the turtle and intermittent contact with the ground make the body inherently stable and difficult to overturn; as such, only two limbs are needed to generate forward thrust, making this form of locomotion simple to mimic, easy to manipulate, and amenable to our laminate manufacturing approach. Finally, the unique kinematic behavior of sea turtles can either enable rapid movement of the hatchlings or permit the large and exceedingly heavy adults [up to 915 kg in *Dermochelys coriacea* (Vandelli, 1761)] to move more slowly through granular media, but under considerable load [7,22].

While there are many differences between species, terrestrial locomotion of sea turtles does exhibit some common variation as a function of development, especially when comparing the hatchling and adult phases of life [6,14,16,17,21]. During the hatchling phase, the young, comparatively lightweight turtles favor speed over load-bearing capacity to escape predation [6,14,17,21]. In particular, terrestrial locomotion in hatchlings is characterized by use of the palmar and plantar surfaces of the limbs to compact loose media and generate forward motion, with the arms relatively straight [6,14,17,21]. Compared to adults, hatchlings have more flexible fins to induce substrate compaction while minimizing limb slippage [14,15]. Furthermore, the body shape of hatchling sea turtles is comparatively narrow, and proportionately lighter than adults, with relatively long limbs; their morphology and gait permits some species to elevate the body fully above ground during motion [6,14,16].

By contrast, the older, proportionately sturdier, adult turtles move more slowly, especially in heavier species [6,17,21]. Adult Chelonioid terrestrial locomotion is generally characterized by the use of the humerus and radial edge of fin to elevate and advance the body, as if moving on crutches [6,17,21]. This occurs with alternating ventral extension and flexion of the humerus, while the radius remains at an approximate 80–90° angle to the humerus, thus crawling on the elbows [6,17,21]. Paired movement of both forelimbs in this "crutching" motion elevates center of mass to reduce friction [17,21]. Compared to hatchlings, the fins of adult sea turtles are much stiffer, more muscular, and comparatively shorter for swimming [6,17,21]. This fin morphology, combined with the relatively broad body and paired forelimb motion, greatly reduces the speed of terrestrial locomotion in adults, but enables them to carry a much heavier load [6,7,17,21,22]. Juveniles, although rarely observed on land, have been documented using either of these gaits (with no known intermediate state); adults learn unique swimming and crawling gaits to compensate for developmental differentiation in the fins, which become optimized for hydrodynamic propulsion [6,17,18,21].

We believe that the observed decrease in relative speed and increased load-bearing capacity of terrestrial locomotion of adult sea turtles are the result of using a shorter lever arm (palm is distal to humerus) to elevate and advance the body. In this case, the shorter moment arm provides more direct support of center of mass due to the position of humerus; conversely, the longer moment arm of the hatchling fins provides greater apical velocity. From these observations, we predict that a sand crawling robot can be modified to maximize its load-bearing capacity or velocity by changing the design of its fins to mimic these particular aspects of sea turtle ontogeny. Hatchlings are known for their high energetic needs for rapid escape to the ocean during "hatchling frenzy" [6]. However, it is unknown whether the energy efficiency of these competing forms of locomotion are influenced by their kinematic properties or are determined solely by age, size, and muscle development. The issue of optimizing motion under load for travel distance and energy consumption is critical in battery operated robots.

Based on the reasoning and biological observations presented above, we evaluate the following hypotheses: (1) In keeping with the results of Mazouchova et al. [15], we hypothesized that limb flexibility aids in substrate compaction and enhances forward motion. (2) We hypothesize that rotation of the limb about the humeral angle with age, as seen in adult locomotion, reduces the moment arm of the forelimb, increasing load bearing capacity at the expense of velocity. (3) Given the shorter moment arm of the forelimbs, we hypothesize that adult style locomotion will be more energy efficient than that of hatchlings.

3   Design and Fabrication of the Robotic Sea Turtle

To test our hypotheses, we mimic the change in fin orientation and usage seen during sea turtle development using a turtle-inspired robot which uses three detachable fin designs, where an ellipsoid representing the fin is rotated about the lateral angle of the attachment point, as seen in Fig. 3b, with each design roughly corresponding to a specific developmental state. (1) The first design was an unrotated (longitudinal) ellipsoidal fin that mimics the relatively straight hatchling fin configuration. (2) The second fin design is identical in shape to the first, but is rotated of laterally by 90° (transverse). This rotation created an L-shaped base that is similar to the humeral angle of adult sea turtle fins during crawling. **(3)** The third design was a fin rotated laterally by 45° (diagonal). This configuration does not correspond to any known position adopted by a sea turtle, but rather, simulates an ad-hoc intermediate configuration between the two previous states. The length and width of the fins was standardized such that the ratio of fin length to body width was equal to that of an adult *Caretta caretta* (Linnaeus, 1758), for experimental consistency and to avoid having to redesign the body.

For each fin design, we developed a predictive kinematic model to examine what differences in motion and fin shape might have affected their experimental trajectories. In the experiments, we applied successively heavier loads to the crawler to assess fin performance with a hand-coded controller (see Sect. 3.1). In particular, we measured the total distance traveled with two repetitions of fin motion while (1) varying the angle of the fin in relation to the body, (2) changing the composition and stiffness of the fins, and (3) imposing a load either on the front or back end of the robot. We additionally measured power consumption for the adult and hatchling inspired fins, to address our third hypothesis. Based on the results of this work, we consider the implications of our data as a physical analogue for understanding the ontogeny of terrestrial locomotion in Chelonioidea, and briefly explore possible design changes to allow our robot to dynamically respond to imposed loads via alteration of fin angle.

3.1   Kinematic Modeling

Explicit kinematic models were generated based on the paired four-bar mechanisms (see Fig. 3a) used to elevate the arm of CTurtle and actuate the fin, which are reasonably accurate compared to the actual motion recored in Fig. 4a and b. These models (Fig. 5a and b) show a shortened fulcrum based on fin rotation about the lateral attachment point, especially for the transverse fin. Figure 5b clearly shows in the lateral (see also Fig. 4a and b) and anterior views that the humeral angle of the transverse fin penetrates the substrate (ground at $z = 0$)

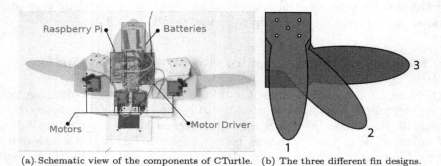

(a) Schematic view of the components of CTurtle.   (b) The three different fin designs.

Fig. 3. The final robotic design CTurtle inspired by sea turtles. (b) highlights the three different fin designs tested on the robot: (1) longitudinal fin, (2) intermediate fin and (3) transverse fin.

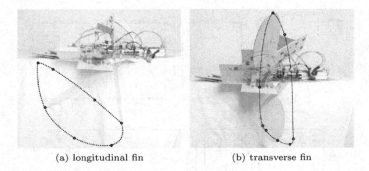

(a) longitudinal fin   (b) transverse fin

Fig. 4. Lateral views of apical fin motion for (a) the hatchling inspired longitudinal fin, and (b) the adult inspired transverse fin.

at approximately 2/3 of the depth attained by longitudinal fin, and with only 1/2 of the maximum forward stroke length.

The open-loop controllers used to generate the motor commands were hand-designed sinusoidal functions offset by 180° to allow sweeping of the fin during the downstroke of the arm, followed by resetting of the fin during the arm upstroke. The sweeping angles of the fin and total identical number of commands given to each motor for each fin design were identical; the time per command and total time per cycle were also identical for each motor and between fin designs (i.e. radial velocity). Thus, rather than try to calculate velocity, we refer to distance traveled per cycle, because each cycle uses the same number of angle commands and uses the same amount of time. In addition, because all fin designs use exactly the same span of angles during the compression of substrate, we are able to effectively eliminate the effect of varying gear ratio as a possible confounding effect on force transmitted to substrate. The importance of this consistency can be seen in Fig. 5c, which shows that transmission efficiency (with respect to gear ratio) increases with motor angle, providing more effective compression of substrate as the stroke progresses.

The only difference was the magnitude of the downstroke, which was adjusted for each fin design to maximize ground penetration and distance traveled, while minimizing backplowing of the substrate on upstroke. Consequently, in the longitudinal fin, compression of the substrate occurred near the fin apex, similar to hatchling sea turtles [14,21]. In the transverse fin, substrate compression happened more evenly along the radial edge of the fin and especially near the extended basal portion (analogous to the humeral angle), similar to adult sea turtles [6,17,21].

These kinematics imply that, compared to the longitudinal fin, the transverse fin has shorter (Fig. 5c), shallower strokes, with comparatively low apical velocity (due to shorter effective radius of the fin). However, due to the shorter moment arm that creates this pattern, transverse fin was predicted to provide greater force per stroke, and reduced lifting of the body, given the posterior arrangement

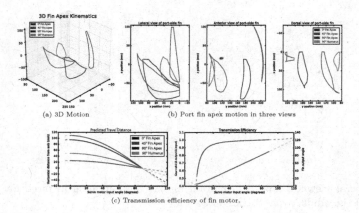

(a) 3D Motion   (b) Port fin apex motion in three views

(c) Transmission efficiency of fin motor.

Fig. 5. (a) Three-dimensional motion of port fin apex, viewed laterally from robot mid-line. (b) Motion of port fin apex (and humeral angle of transverse fin) throughout entire stroke; viewed from lateral, anterior, and dorsal views, respectively. (c) Output angle of port fin with respect to arm-mounted servo (fin motor), as well as gear ratio of 4-bar mechanism to which these angles belong; upper panel indicates predicted x-axis coordinate of the fin apex. Dashed lines indicate full range of motion of servo motor, while thicker line represents joint angles used in experiments. In all figures, robot arm is centered at origin, with the predicted fin downstroke generated by 0–90° motor angles.

of components (Fig. 3a). Given the reduced lifting, we predicted greater energy efficiency for transverse fin, due to reduced energy wasted on upward motion.

4   Experiments

Two different experiments were conducted to investigate the hypotheses made in Sect. 2. The first experiment is designed to evaluate hypothesis 1 by measuring the locomotion performance of CTurtle with fins of different stiffness. The second experiment addresses hypotheses 2 and 3 by measuring locomotion performance and energy efficiency of fins with varying rotational angle inspired by their biological counterparts.

4.1   Experimental Setup

All experiments were conducted with the CTurtle robotic platform shown in Fig. 3a. In order to guarantee reproducibility of results, the robot was powered with an external 5 V, 2 A power source as opposed to the onboard batteries in the schematic. Experiments were performed in a simulated sand environment consisting of poppy seeds, in order to avoid the detrimental effect that actual sand has on equipment. The similar granularity between poppy seeds and sand makes it a suitable replacement, despite the difference in density – 0.54 $\frac{g}{ml}$ to

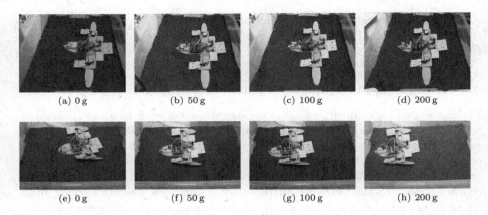

Fig. 6. Final position after the execution of two complete gait cycles with different additional weights carried by the robotic device. In the upper row (a–d) the transverse fin design is used and in the lower row (e–h) the longitudinal fin design is used.

$1.46 \frac{g}{ml}$, respectively. Unlike previous work [12], the motor commands for a gait cycle were derived from a hand-coded combination of sinusoidal functions with a shift of $180°$ between the functions for vertical and horizontal movements. Joint angles relative to the middle position are given by

$$a_{1,4} = \cos(180° + \frac{t}{T} \cdot 360° \cdot 2) \cdot 60,$$

$$a_{2,3} = \sin(180° + \frac{t}{T} \cdot 360° \cdot 2) \cdot m + o \tag{1}$$

with $a_{1,4}$ being the joint angles for horizontal movements and $a_{2,3}$ for vertical movements. For each of the three different fin designs the vertical movements were adapted to achieve an optimal movement for each fin design with no additional load. For both longitudinal and intermediate fins, the magnitude m was 20 while the transverse required a magnitude of 60 to lift the fin high enough. The offset o for the longitudinal, intermediate and transverse were 10, -15 and -40, respectively.

4.2   Measuring the Effect of Fiberglass Reinforcement

For this experiment, we created two sets of fins that vary in rigidity. Flexible, pliant fins were created with a 3-layer laminate consisting of two 6-ply paper layers held together by a 1-ply adhesive layer. Rigid fins were created by reinforcing the 3-layer laminate design with two additional layers of a fiberglass coating (as well as two additional adhesive layers) resulting in a 7-layer laminate. Each set consists of a longitudinal fin and a transverse fin (Fig. 3b) in order to determine whether rigidity performance is affected by the rotational angle of the fin.

The fins were evaluated by attaching them to CTurtle and measuring how far it traveled in the simulated sand environment after executing two complete gait cycles. Each evaluation was performed five times to capture the mean and standard deviation. This process was repeated for load weights varying from 0 g to 100 g, with the weight placed near the front of the robot.

4.3   Evaluating Performance of Different Fin Angles

For the second experiment, we again measured the travel distance of CTurtle in a simulated sand environment, but for all fin designs (Fig. 3b) and for a larger range of load weights. Furthermore, this experiment explored the effect of center of mass on locomotion performance. Evaluations were performed with weights varying from 0 g to 300 g placed at the rear of the robot – in effect moving the center of mass to the rear of CTurtle – and again with weights varying from 0 g to 200 g placed at the front of the robot – forcing the center of mass further forward. All evaluations were performed with rigid, fiberglass-reinforced fins for two complete gait cycles and five repetitions.

An additional set of evaluations was performed to test the energy efficiency of the transverse fin and longitudinal fin designs. The current consumption of all four motors powering CTurtle's limbs were measured with a DC current sensor operating at 60 Hz while two complete gait cycles were executed. The mean and standard deviation for three repetitions were captured.

5   Results

The results of the first experiment, shown in Fig. 7 indicate that the rigid, fiberglass-reinforced fins yield longer travel distance than the flexible, paper fins across all tested loads. The results for the second experiment are shown in Figs. 6 and 8 and show that the transverse fin configuration enables CTurtle to travel

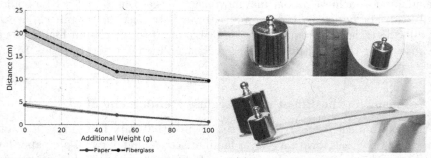

(a) Performance with additional front-weight   (b) Front and side view of bending fins.

Fig. 7. (a) Evaluation of the performance of rigid fins (fiberglass-reinforced laminate) and flexible fins (paper laminate). The graph shows the mean and standard deviation of five executions for each data point. (b, c) Fiberglass-reinforced (orange) and paper (blue) fins with the same bending radius for 100 g (orange) and 20 g (blue) of weight. (Color figure online)

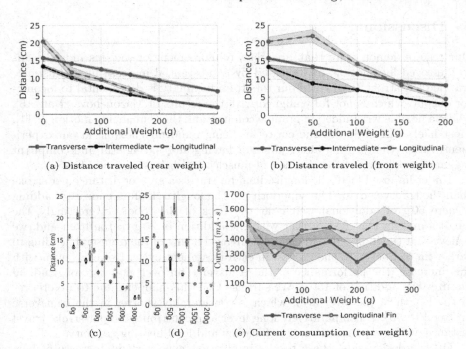

(a) Distance traveled (rear weight)

(b) Distance traveled (front weight)

(c)

(d)

(e) Current consumption (rear weight)

Fig. 8. Performance of CTurtle for the three different fin designs with an additional payload weight of 0–300 g. (a) and (b) shows the distance traveled with varying weights placed at the rear and front of CTurtle's frame, respectively. Each data point represents the mean of five evaluations with the shaded area indicating the standard deviation. (c) and (d) indicates the median, 25%, and 75% quantile for the same evaluations as (a) and (b). (e) shows the mean current consumption of three evaluations for varying weights placed at the rear of CTurtle's frame.

longer distances than the other fin designs with a heavier load. The longitudinal fin design yields the greatest distances for low loads, however, the performance degrades more rapidly than that of the longitudinal fin design with an increasing load, falling behind when the load is 50 g (Fig. 8a) or 150 g (Fig. 8b) depending on the center of mass. The intermediate fin design performs the worst, yielding the shortest traveled distance for all but the highest loads.

Another interesting result is the measured current consumption for the different fin designs, as shown in Fig. 8e. The transverse fin yielded lower current consumption as the additional rear-weight increased, reaching a low of 1191 mA·s for 300 g of additional weight. The longitudinal fin, on the other hand, displayed an increasing current consumption trend as the rear-weight increased with 1462 mA·s at the same weight; this is approximately a 22% increase over the transverse fin at the maximum tested load.

6   Discussion

Our results demonstrate that fiberglass reinforcement of the fins indeed leads to improved locomotion, as indicated by a substantial increase in the distance traveled per stroke (Fig. 7). In our robot, fin flexibility is detrimental to locomotor output, which is not fully aligned with the results of Mazouchova et al. [15], where a flexible wrist aided in granule compaction, thus enhancing motion [2,11]. Most likely this is due to the entire fin being uniformly flexible in our experiment, rather than flexing along joints or having alternating stiff and compliant regions due to varying distribution of muscles and bones in actual fins.

Under imposed load, the longitudinal fin travels a greater distance per stroke than the transverse fin. The superiority of the longitudinal fin ends after adding a mere 50 g of additional weight at the back of the robot. Interestingly, the longitudinal fin travelled further with the addition of 50 g to the front end; we believe that the imposed load caused decreased lifting of the front end, leading to earlier fin contact with the ground in each movement cycle. However, even with this increase, the performance of the transverse fin overtakes the longitudinal fin after the addition of 150 g. We surmise that this happens due to differences in the length of the lever arm, which is considerably shorter in the transverse fin (see Fig. 5b). A comparatively long lever arm contributes to per-stroke travel distance at the expense of energy efficiency and load-bearing capacity.

On the design side, these results imply that even a slight increase of battery capacity will require the use of the transverse fin for effective locomotion unless weight redistribution is considered. In general, the results suggest that improved locomotion and energy efficiency is possible with anterior redistribution of battery weight. This, in turn, could be the input into the next iteration of the proposed design pipeline.

In addition to possibly informing future decisions within the next design cycle, these differences in locomotive performance can improve our understanding of sea turtle biology. Hatchling sea turtles have longer limbs in proportion to their body, and, more importantly, compact substrate at a more distal location (palmar surface) on the fore-limbs than adults [6,14,16,17,21]. This gives them a longer moment arm at the shoulder joint (scapula to palmar surface). Based on our results, an elongate moment arm and relatively low weight should enable hatchlings to cover ground at a higher velocity per stroke than adults, *before* considering gait. A comparison of the energy consumption rates of the different fins suggests that this may occur at the expense of increased energy expenditure compared to adults; this observation is supported by high lactic acid production (for rapid energy production) characteristic of hatchling metabolism [6,18].

Adult sea turtles compact substrate and support their body weight at the anatomical equivalent of the elbow, and have proportionately shorter fins than hatchlings (scapula to humeral angle) [6,16,17,21]. With the transverse fin, our robot was able to move substantially heavier loads across the substrate than with the longitudinal fin, apparently at a considerable cost-savings in energy consumed per movement cycle (approx. 20% at 300 g). Per-cycle energy expenditure decreases with increased load with the transverse fin, but increases with load using the longitudinal fin. Notably, this appears to be an intrinsic advantage due to fin morphology rather than purely a result of gait asymmetry (paired motion of forelimbs) as all sets of fins utilized an asymmetric gait in our experiments. We therefore infer that by propping up the body with both forelimbs and using the humerus to bear weight, adult sea turtles are able to carry a heavier mass (relative to body size) than their immature counterparts.

Surprisingly, a fin rotated at an intermediate angle of 45° was inferior to both of the other fins in distance traveled and load-bearing capacity. The 45° rotation would approximately correspond to an enforced 45° abduction of the humerus; this configuration *is not known* to be used in sea turtle locomotion. However, the inferior locomotive capabilities of this arm position suggests that this might not exist at all because its use would result in poor performance. Although we did not collect energy consumption data for the intermediate fin, we posit that anatomically this position would be disadvantageous and would require the arm to actively support the body using constant tension of the pectoralis major rather than passive support on the rigid humerus. This finding corroborates hypotheses reported in the literature that adult gaits are learned after reaching developmental thresholds of fin differentiation and weight gain [6,17,21].

In summary, these biological inferences lead us to believe that we could generate a new design that can reconfigure itself to alternatively maximize velocity or minimize total energy expenditure for a given load. This reconfiguration could be as simple as rotating the fin about a set point upon loading, or by shortening the length of the fin according to the magnitude of the load.

7   Conclusion

This paper explores the relationship between (1) an observed ontological shift in Chelonioid locomotion that we use to inform a robotic design cycle, and (2) experimental results from that design iteration, which lead further inferences into sea turtle biology. We have found strong similarities between the locomotive behavior of our bio-inspired robot and what occurs in nature; namely, a longer moment arm generally exhibits higher per-cycle travel distance and energy consumption, but relatively poor load bearing capacity compared to a shorter effective moment arm. We hypothesize, in turn, that we can design a fin that dynamically reconfigures to optimize its velocity or energy use for an imposed load. This design/exploration process can be fed back into additional biological questions. In future work we could examine the interplay between fin surface area and locomotion under varying loads, both in sea turtles and their robotic analogues, beginning the biological investigation and robotic design cycle anew.

References

1. Altendorfer, R., Moore, N., Komsuoglu, H., Buehler, M., Brown Jr., H., McMordie, D., Saranli, U., Full, R., Koditschek, D.E.: RHex: a biologically inspired hexapod runner. Auton. Robots **11**(3), 207–213 (2001)
2. Askari, H., Kamrin, K.: Intrusion rheology in grains and other flowable materials. Nat. Mater. **15**(12), 1274–1279 (2016)
3. Autumn, K., Dittmore, A., Santos, D., Spenko, M., Cutkosky, M.: Frictional adhesion: A new angle on gecko attachment. J. Exp. Biol. **209**(Pt. 18), 3569–3579 (2006)
4. Birkmeyer, P., Peterson, K., Fearing, R.S.: A dynamic 16g hexapedal robot. In: 2009 IEEE/RSJ International Conference on Intelligent Robots and Systems, pp. 2683–2689 (2009)
5. Cham, J.G., Bailey, S.A., Clark, J.E., Full, R.J., Cutkosky, M.R.: Fast and robust: hexapedal robots via shape deposition manufacturing. Int. J. Rob. Res. **21**(10–11), 869–882 (2002)
6. Dodd Jr., C.K.: Synopsis of the biological data on the loggerhead sea turtle caretta caretta (linnaeus 1758) (1988)
7. Eckert, K.L., Luginbuhl, C.: Death of a giant. Mar. Turt. Newsl. **43**, 2–3 (1988)
8. Hoover, A., Steltz, E., Fearing, R.: An autonomous 2.4g crawling hexapod robot. In: 2008 IEEE/RSJ International Conference on Intelligent Robots and Systems, pp. 26–33. IEEE (2008)
9. Ijspeert, A.J.: Biorobotics: using robots to emulate and investigate agile locomotion. Science **346**(6206), 196–203 (2014)
10. Kingsley, D., Quinn, R., Ritzmann, R.: A cockroach inspired robot with artificial muscles. In: 2006 IEEE/RSJ International Conference on Intelligent Robots and Systems, pp 1837–1842. IEEE (2006)
11. Li, C., Zhang, T., Goldman, D.I.: A terradynamics of legged locomotion on granular media. Science **339**, 1408–1412 (2013)
12. Luck, K., Campbell, J., Jansen, M., Aukes, D.M., Ben Amor, H.: From the lab to the desert: fast prototyping and learning of robot locomotion. In: Robotics: Science and Systems Conference (RSS 2017)
13. Ma, K.Y., Chirarattananon, P., Fuller, S.B., Wood, R.J.: Controlled flight of a biologically inspired. Insect-Scale Robot. Sci. **340**(6132), 603–607 (2013)
14. Mazouchova, N., Gravish, N., Savu, A., Goldman, D.I.: Utilization of granular solidification during terrestrial locomotion of hatchling sea turtles. Biol. Lett. **6**, 398–401 (2010)
15. Mazouchova, N., Umbanhowar, P.B., Goldman, D.I.: Flipper-driven terrestrial locomotion of a sea turtle-inspired robot. Bioinspiration Biomimetics **8**(2), 026, 007 (2013)
16. Pritchard, P., Mortimer, J.: Taxonomy, external morphology, and species identification. Res. Manag. Tech. Conserv. Sea Turtles **4**, 21 (1999)
17. Renous, S., Bels, V.: Comparison between aquatic and terrestrial locomotions of the leatherback sea turle (dermochelys coriacea). J. Zool. **230**(3), 357–378 (1993)
18. Smith, K.U., Daniel, R.S.: Observations of behavioral development in the loggerhead turtle (caretta caretta). Science **104**, 154–156 (1946)
19. Spenko, M., Trujillo, S., Heyneman, B., Santos, D., Cutkosky, M., Kim, S., Spenko, M., Trujillo, S., Heyneman, B., Santos, D., Cutkoskly, M.R., Cutkosky, M.: Smooth vertical surface climbing with directional adhesion. IEEE Trans. Rob. **24**(1), 65–74 (2008)

20. Umedachi, T., Vikas, V., Trimmer, B.A.: Softworms: the design and control of non-pneumatic, 3D-printed, deformable robots. Bioinspiration Biomimetics **11**(2), 025001 (2016)
21. Wyneken, J.: Sea turtle locomotion: mechanics, behavior, and energetics. In: Lutz, P.L. (ed.) The Biology of Sea Turtles, pp. 168–198. CRC Press (1997)
22. Zug, G.R., Parham, J.F.: Age and growth in leatherback turtles, dermochelys coriacea (testudines: Dermochelyidae): a skeletochronological analysis. Chelonian Conserv. Biol. **2**, 244–249 (1996)

Feather-Inspired Sensor for Stabilizing Unmanned Aerial Vehicles in Turbulent Conditions

Christos Kouppas[1]([✉]), Martin Pearson[2], Paul Dean[3], and Sean Anderson[3]

[1] Loughborough University, Loughborough, UK
C.Kouppas@lboro.ac.uk
[2] University of the West England, Bristol, UK
Martin.Pearson@uwe.ac.uk
[3] University of Sheffield, Sheffield, UK
{P.Dean,S.Anderson}@sheffield.ac.uk

Abstract. Stabilizing unmanned aerial vehicles (UAVs) in turbulent conditions is a challenging problem. Typical methods of stabilization do not use feedforward information about the airflow disturbances but only UAV attitude feedback signals, e.g. from an inertial measurement unit. The novel proposal of this work is the development of a feather-inspired sensor and feedforward controller that transforms from sensed turbulent airflow to feedforward control action for improving the stability of the UAV. The feedforward controller was based on fuzzy logic, combined in a feedforward-feedback loop with a standard PID control system. An experimental rig based on a one degree of freedom helicopter plant (elevation only) was developed to evaluate the potential of the sensor and control algorithm. Evaluation results showed reduction of disturbance using the fuzzy feedforward-feedback scheme, under turbulent airflow, versus a classical feedback PID-controlled system.

Keywords: UAV · Feather-inspired sensor · Turbulence sensor · Turbulence reduction · Feedback/feedforward fuzzy controller

1 Introduction

Unmanned Aerial Vehicles (UAVs) are widely used in many disciplines, from entertainment to the defence sector. In particular, small, or micro, UAVs (that weigh less than 1 kg), have great potential to impact on civilian tasks such as crop surveying, infrastructure inspection, and search and rescue missions [3]. However, small UAVs suffer from instability in turbulent conditions [10]. Currently, feedback sensors such as inertial-measurement units (IMUs), or vision, are commonly used in attitude stabilization [5,7,18]. The main disadvantage of these sensors is that they react to change in the UAV attitude after the disturbance takes effect. Recently, there has been growing interest in biologically-inspired 'phase advanced sensors', that can sense turbulence and compensate for

---

C. Kouppas—PhD Student in Computer Science at Loughborough University.

© Springer International Publishing AG 2017
M. Mangan et al. (Eds.): Living Machines 2017, LNAI 10384, pp. 230–241, 2017.
DOI: 10.1007/978-3-319-63537-8_20

the disturbance before it takes effect, using feedforward control [11]. The aim of this paper is the development of such a feedforward control scheme, using a feather-inspired sensor.

Bioinspiration potentially has something to offer UAV design for stabilization in turbulent conditions: various flying animals have developed tactile receptors for sensing flow to aid flight control, including mechanoreceptors surrounding the follicles of bird feathers [2], mechanoreceptors activated by microscopically small, stiff hairs on bat wings [15] and hair activated mechanoreceptors in flying insects [17]. The observation of these sensory systems in flying animals has motivated the development of a number of bioinspired flow sensors for UAVs [1,9,11,14]. Currently, such designs are in their early stages, and have received too limited investigation to ascertain the optimality of any particular approach. There is still much scope, therefore, for the development of new ideas in this area.

The novel contribution of this paper is the design, implementation and testing of a feather-inspired sensor and control scheme. The design of the feather-inspired sensor in this paper is based on a tactile sensor previously used for fluid flow sensing in underwater robotics [13] and tactile sensing in mobile robots [16]. The feather-inspired sensor is intended to measure turbulence disturbances acting on the UAV and therefore must be used in conjunction with a feedforward controller.

Feedforward control with a measured disturbance can be performed using a type of inverse plant compensation, e.g. with adaptive neural networks [8]. Fuzzy logic can also be used in feedforward control schemes [4]. In this paper, a feedforward-feedback control scheme is developed, where a fuzzy controller is used in the feedforward path, in combination with a PID feedback controller.

To test and evaluate the feather-inspired sensor and control scheme, a one degree-of-freedom (1 DOF) rig was developed, based on the elevation system of a small UAV copter. The advantage of this system compared to, e.g. a quadcopter, is that control design and evaluation is simpler and more rapid, whilst real-world context is retained. A benchmark comparison was made between three control schemes: 1. feedback control using the PID controller (referred to as "Reference"), 2. feedback control using fuzzy-PID (referred to as "Fuzzy") and 3. feedforward-feedback control using fuzzy-PID and the feather-inspired sensor (referred to as "FuzzyS"). This design and evaluation is based on an investigation conducted in [6]. The results show an improvement using the feather-inspired sensor, demonstrating the important potential in this type of bioinspired control scheme.

2 Methods

2.1 Experimental Rig

In order to evaluate turbulence stabilization for UAVs an experimental rig was developed based on a 1 DOF helicopter system, controlling elevation only. The main components were: a two-blade fan, a square metal beam, a counter-weight and an elevation sensor (a potentiometer on the beam to measure angle), as can be seen in Fig. 1. The system had one input (for the fan voltage) and one output (for the beam angle sensor). The box was connected to a multifunction data acquisition (DAQ) device from National Instruments, model USB-6008. The DAQ was connected to a personal computer (PC) and was controlled with the LabVIEW software application.

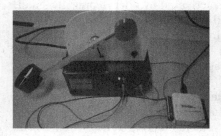

Fig. 1. Experimental setup of the elevation system.

2.2   Design of a Feather-Inspired Turbulence Sensor

The novel feather sensor that was developed for this investigation was firstly developed as a tactile sensor in a whisking rat-like robot [16]. It was externally made from an artificial shaft and a follicle. The main parts were enclosed in the artificial follicle case and consisted of the polyurethane bearing, an insert, a magnet and a tri-axis Hall-effect sensor integrated circuit [16] as can be seen in Fig. 2. The insert connected the magnet with the shaft, which was a long plastic conical rod with length of 5 cm. The insert was locked in the follicle case with a polyurethane bearing in order to be waterproof and offer flexibility to turn $10°$ around the x and y axes. Movement around those axes was read from the Hall-effect sensor and produced a voltage that was measured using the DAQ described above.

For the purpose of the current project a synthetic feather was attached at the shaft thus, the new area of the shaft increased from $0.5\,\mathrm{cm}^2$ to $10\,\mathrm{cm}^2$. The feather area is proportionally related to the forcing input from the turbulence that activates the sensor. The turbulent airflow was produced by a simple wind tunnel: four fans which were connected edge to edge to form a wall (Fig. 3).

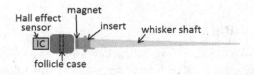

Fig. 2. Schematic diagram of the artificial whisker.

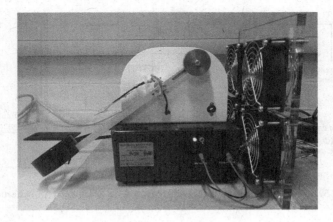

Fig. 3. The experimental setup with the fan wall in the right and the feather sensor parallel to the elevation beam.

Fig. 4. Single sensor (left), on top of the beam and two sensors (right), one on top and the other beneath the elevation beam.

Two physical arrangements of the feather sensor schemes were tested: 1. using one feather sensor only, mounted above the beam, and 2. using two feather sensors, mounted above and below the beam (Fig. 4). When using two feather sensors, the signals from the sensors were added to fuse them into a single measure of relative airflow above and below the beam.

2.3   PID Feedback Control Design

In order to design a PID controller using a mathematical model, a system identification was performed on the experimental rig. The system was initialized in open loop with an elevation angle close to zero, then input-output data were recorded with a bandwidth limited excitation signal. Model selection and parameter estimation was performed using direct continuous-time transfer function methods (implemented via the Matlab system identification toolbox). The selected model of the identified system was implemented in Simulink to optimize the design of the PID feedback controller for the real system.

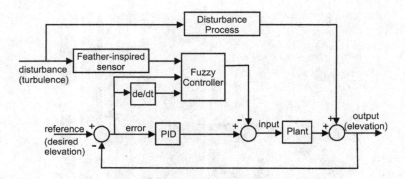

Fig. 5. Diagram of the fuzzy Feedforward-Feedback control scheme.

The PID feedback controller was optimized by minimising the mean squared error of the desired output value against the output of the system. The Simulink environment was used to simulate the one DOF elevation system (Fig. 1). The PID controller was a central feedback controller which was aiming to stabilize the elevation, by minimizing the error of the system. The simulated controller was imported into LabVIEW (National Instruments), to be evaluated. An experiment was performed to compare the real PID control system to the simulation. The real system behaved similarly to the identified model, as simulated in Simulink, therefore the PID controller was used directly, without re-tuning online. (Due to space restrictions in this paper we avoid reporting quantitative results of the system identification and PID tuning, which follow standard methods.)

2.4   Feedforward-Feedback Fuzzy Control Scheme

The fuzzy controller was combined in a feedforward-feedback control scheme to stabilize the system and cancel out the turbulence disturbances. The controller was designed using the Mamdani-type of fuzzy system [12]. The fuzzy controller had three inputs (error, derivative of error and feather sensor signal) and one output (control input voltage to the plant). The inputs and outputs were divided in five fuzzy membership functions (Big Negative, Negative, Zero, Positive, Big Positive) and the rule base that was designed can be seen in Table 1. The rules were heuristically written to follow a logic flow that input in one direction causes a response in the opposite direction. Figure 5 shows the block diagram of the system with the fuzzy feedforward-feedback controller.

Table 1. Fuzzy rules

Rule	Input 1 (error)	Input 2 (error_dot)	Input 3 (disturbance)	Output (voltage)
1	Big Negative	ANY	ANY	Zero
2	Big Positive	ANY	ANY	Zero
3	Positive	Positive	Zero	Big Negative
4	Positive	Zero	Zero	Negative
5	Positive	Negative	Zero	Zero
6	Positive	Negative	Positive	Negative
7	Positive	Zero	Positive	Big Negative
8	Positive	Positive	Positive	Big Negative
9	Positive	Positive	Negative	Negative
10	Positive	Zero	Negative	Zero
11	Positive	Negative	Negative	Zero
12	Zero	Positive	Zero	Negative
13	Zero	Zero	Zero	Zero
14	Zero	Negative	Zero	Positive
15	Zero	Negative	Positive	Zero
16	Zero	Zero	Positive	Negative
17	Zero	Positive	Positive	Big Negative
18	Zero	Positive	Negative	Zero
19	Zero	Zero	Negative	Positive
20	Zero	Negative	Negative	Big Positive
21	Negative	Negative	Negative	Big Positive
22	Negative	Zero	Negative	Big Positive
23	Negative	Positive	Negative	Positive
24	Negative	Positive	Positive	Zero
25	Negative	Zero	Positive	Zero
26	Negative	Negative	Positive	Positive
27	Negative	Negative	Zero	Big Positive
28	Negative	Zero	Zero	Positive
29	Negative	Positive	Zero	Zero

3 Results

3.1 Experimental Results Comparing Control Schemes in Non-turbulent and Turbulent Conditions

In the first set of experiments, three different control schemes were designed: 1. a PID feedback controller (labelled Reference), 2. a PID-fuzzy feedback controller (labelled Fuzzy) and 3. a fuzzy-PID feedforward-feedback controller using the feather inspired sensor (labelled FuzzyS). The purpose of testing the fuzzy-PID controller in both feedback and feedforward-feedback mode was to check whether any control improvement was due to the feedforward action, driven by the feather-inspired sensor. Experiments were repeated five times for each controller, with both one and two feather sensors. Performance was measured over two different sections: Sect. 1 was without turbulence and Sect. 2 was with turbulence.

The first experiments established the reference MSE for both sections with the feather sensor fully deactivated thus, the system was controlled only with feedback. Each control scheme was initialized in simulation (Fuzzy rule-base was heuristically written) and fine-tuned on the real system (gains were adjusted with a step experiment response). Table 2 summarizes the MSE for each control scheme and as noted in the last row, the fuzzy controller with the activated feather-inspired sensor showed significant reduction of the noise from turbulent airflow (Sect. 2 of the data). Reduction was up to 55% for one sensor and 13% for two sensors. However, the MSE of the Sect. 1 (under negligible airflow) was increased for the single sensor up to 180%, due to low airflow around the sensor which was handled as disturbance from the controller, but was decreased for the double sensor up to 70%. The rig was modified slightly while the second sensor was added therefore, MSE between the Single and the Double orientation cannot be compared. The best controller, overall, was the fuzzy controller with the feather sensor enabled (FuzzyS).

Table 2. MSE with the feather-like sensor as a turbulence controller.

Setup	Single			Double		
	Section 1	Section 2	Sum	Section 1	Section 2	Sum
Reference	0.0597	0.5352	0.5949	0.0358	0.5388	0.5746
FuzzyS	0.1080	0.2435	0.3515	0.0102	0.4697	0.4799
Fuzzy	0.0636	0.2914	0.3550	0.0148	0.5035	0.5183
Minimum	Reference	FuzzyS	FuzzyS	FuzzyS	FuzzyS	FuzzyS

3.2 Experimental Results Comparing Control Schemes in Turbulent Conditions at Varying Angles of Elevation

To test the controllers in a more extensive experiment, a benchmark was created with three sections, with the desired elevation set to $0°$ (Sect. 1), $10°$ (Sect. 2) and $-10°$ (Sect. 3) for 60 s per section. Benchmarking was performed on the system with three controllers: PID feedback (Reference), fuzzy-PID feedback (Fuzzy) and fuzzy-PID feedforward-feedback with the feather-inspired sensor (FuzzyS). The control gains were initialized in simulations and then fine-tuned in LabVIEW.

Due to the stochastic nature of the disturbance signal, each benchmark was repeated ten times. The experiments were also repeated with both one and two

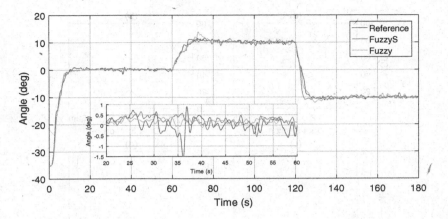

Fig. 6. Representative smooth response of benchmark setup with one feather sensor. (Color figure online)

feather sensor arrangements (see Methods). The MSE of each section was averaged to produce a single representative value for each control scheme. The MSE of each section was calculated with two different methods: **1.** averaging the squared error of each sample using the raw data (labelled 'Raw' in tables of results) and **2.** averaging using the filtered output response to remove high frequency noise of the sensor electrical reading (labelled 'Smoothed' in tables of results).

Figure 6 shows the smooth response with a single sensor. The system was able to maintain a stable steady state, even with turbulence. The control schemes without the use of the feather sensor (blue and yellow graphs) show an increased number of rapid transients. On the other hand, the controller with the enabled feather sensor (red graph) shows a smoother response with more damped oscillations.

The performance results of the three control schemes across each of the three elevation angles (Sects. 1, 2, and 3) using one feather sensor only, are shown in terms of MSE and the standard deviation (SD) in Fig. 7(a)–(b). The FuzzyS controller without the feather sensor (Fuzzy) shows an improvement only in the first section of the raw signal. On the contrary, the FuzzyS controller with the sensor (FuzzyS) have similar or better results than Reference response.

In order to analyse only the performance of the turbulence cancellation, the step response of each section (first 10 s) was removed from the data and MSE values re-calculated. The results are presented in Fig. 7(c)–(d): the FuzzyS controller with the sensor enabled shows overall improvement up to 22%, on summed MSE of all the sections. The fuzzy controller with the disabled feather sensor shows improvement but is not as consistent as the one with the enabled sensor.

After testing one feather sensor, the second feather sensor was added below the elevation beam in order to investigate the use of two sensors. Figure 8 shows the response of the system during the benchmark with two feathers. As previously, the Reference response (blue graph) has more sharp edges than the

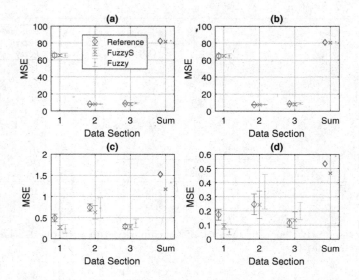

Fig. 7. Control performance using one feather sensor across ten experiments, in terms of MSE and SD (error bars denote one SD). The three data sections correspond to the three angles of elevation tested, i.e. $-10°$, $0°$, and $10°$. (a) Results obtained using all raw data. (b) Results obtained using all smoothed data. (c) Results obtained using raw data from steady-state portion only. (d) Results obtained using smoothed data from steady-state portion only.

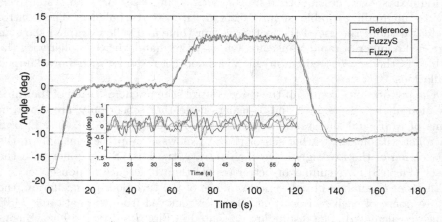

Fig. 8. Representative smooth response of benchmark setup with two feather sensors. (Color figure online)

response of the setups with the FuzzyS controller. However, with the two sensors, the system was slower in the step response (in Fig. 8 the blue line demonstrates a slightly faster response, at time 70 and 130 s). Specifically, in the third section, MSE of both fuzzy control schemes is much greater than the Reference because the PID response is stabilizing the system faster, this can also be seen in Fig. 9(a)–(b).

In order to analyse turbulence disturbance rejection without the complication of tracking elevation angle changes, Fig. 9(c)–(d) shows raw and smoothed MSE and SD of each section without the step response part, i.e. analyzing the steady-state part of the response for turbulence disturbances only. The results show that the FuzzyS controller improves the performance. However, the results of the controller with enabled feather sensor were not as impressive as the first results that were reported in Sect. 3.1. The Fuzzy controller with two enabled sensors was better by 7% compared to the Reference controller on raw data. On smoothed data the FuzzyS controller was worse by 37%. These results demonstrate that there might be some advantage to the FuzzyS controller but that further investigation is needed.

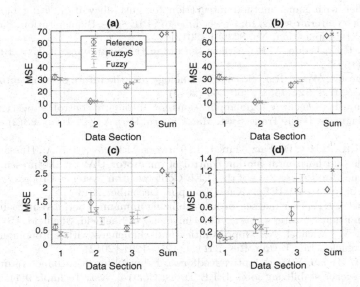

Fig. 9. Control performance using two feather sensors across ten experiments, in terms of MSE and SD (error bars denote one SD). The three data sections correspond to the three angles of elevation tested, i.e. $-10°$, $0°$, and $10°$. (a) Results obtained using all raw data. (b) Results obtained using all smoothed data. (c) Results obtained using raw data from steady-state portion only. (d) Results obtained using smoothed data from steady-state portion only.

4   Summary

In conclusion, a pilot study of a feather-inspired sensor for UAV stabilization in turbulence has been presented. The sensor was developed and used in conjunction with a feedforward-feedback fuzzy-PID control scheme. The key idea was to use the feather sensor to provide a disturbance measurement of turbulence, in advance of turbulence affecting the attitude of the UAV, i.e. a feedforward signal. The sensor and control scheme was evaluated in experiment on a one degree of freedom, lab based, helicopter elevation rig, and benchmarked against PID feedback control, and fuzzy-PID feedback. The sensor showed some promising results, using both one and two feathers. These initial results suggest that feather-inspired sensors may open up new opportunities for feedforward turbulence compensation in UAVs. Future research should focus on the use of sensors arranged in an array around the body of the system and testing on quadcopter or other UAV system in flight.

References

1. Blower, C.J., Lee, W., Wickenheiser, A.M.: The development of a closed-loop flight controller with panel method integration for gust alleviation using biomimetic feathers on aircraft wings. In: Proceedings of SPIE 8339, Bioinspiration, Biomimetics, and Bioreplication, No. 833901 (2012)
2. Brown, R.E., Fedde, M.R.: Airflow sensors in the avian wing. J. Exp. Biol. **179**, 13–30 (1993)
3. Floreano, D., Wood, R.J.: Science, technology and the future of small autonomous drones. Nature **521**(7553), 460–466 (2015)
4. Isermann, R.: On fuzzy logic applications for automatic control, supervision, and fault diagnosis. IEEE Trans. Syst. Man Cybern. Part A Syst. Hum. **28**(2), 221–235 (1998)
5. Joyo, M.K., Hazry, D., Faiz Ahmed, S., Tanveer, M.H., Warsi, F.A., Hussain, A.T.: Altitude and horizontal motion control of quadrotor UAV in the presence of air turbulence. In: Proceedings of the IEEE Conference on Systems. Process & Control (ICSPC), pp. 16–20. IEEE, Kuala Lumpur, December 2013
6. Kouppas, C.: Feather-like sensor for stabilizing unmanned aerial vehicles in turbulent conditions. Master's thesis, The University of Sheffield, Sheffield, UK (2016)
7. Kumar, V., Michael, N.: Opportunities and challenges with autonomous micro aerial vehicles. Int. J. Rob. Res. **31**(11), 1279–1291 (2012)
8. Lin, C.L., Hsiao, Y.H.: Adaptive feedforward control for disturbance torque rejection in seeker stabilizing loop. IEEE Trans. Control Syst. Technol. **9**(1), 108–121 (2001)
9. Mohamed, A., Abdulrahim, M., Watkins, S., Clothier, R.: Development and flight testing of a turbulence mitigation system for micro air vehicles. J. Field Rob. **33**, 639–660 (2016)
10. Mohamed, A., Clothier, R., Watkins, S., Sabatini, R., Abdulrahim, M.: Fixed-wing MAV attitude stability in atmospheric turbulence, part 1: suitability of conventional sensors. Prog. Aerosp. Sci. **70**, 69–82 (2014)
11. Mohamed, A., Watkins, S., Clothier, R., Abdulrahim, M., Massey, K., Sabatini, R.: Fixed-wing MAV attitude stability in atmospheric turbulence - part 2: investigating biologically-inspired sensors. Prog. Aerosp. Sci. **71**, 1–13 (2014)
12. Passino, K.M., Yurkovich, S.: Fuzzy Control. Addison Wesley Longman Inc., Menlo Park (1998)

13. Rooney, T., Pearson, M.J., Pipe, T.: Measuring the local viscosity and velocity of fluids using a biomimetic tactile whisker. In: Wilson, S.P., Verschure, P.F.M.J., Mura, A., Prescott, T.J. (eds.) Living Machines 2015. LNCS, vol. 9222, pp. 75–85. Springer, Cham (2015). doi:10.1007/978-3-319-22979-9_7
14. Shen, H., Xu, Y., Dickinson, B.: Fault tolerant attitude control for small unmanned aircraft systems equipped with an airflow sensor array. Bioinspiration Biomim. 9(4), 046015 (2014)
15. Sterbing-D'Angelo, S., Chadha, M., Chiu, C., Falk, B., Xian, W., Barcelo, J., Zook, J.M., Moss, C.F.: Bat wing sensors support flight control. Proc. Nat. Acad. Sci. 108(27), 11291–11296 (2011)
16. Sullivan, J.C., Mitchinson, B., Pearson, M.J., Evans, M., Lepora, N.F., Fox, C.W., Melhuish, C., Prescott, T.J.: Tactile discrimination using active whisker sensors. IEEE Sens. J. 12(2), 350–362 (2012)
17. Taylor, G.K., Krapp, H.G.: Sensory systems and flight stability: what do insects measure and why? Adv. Insect Physiol. 34, 231–316 (2007)
18. Xu, Y., Luo, D., Xian, N., Duan, H.: Pose estimation for UAV aerial refueling with serious turbulences based on extended Kalman filter. Optik 125(13), 3102–3106 (2014)

Towards Identifying Biological Research Articles in Computer-Aided Biomimetics

Ruben Kruiper[✉], Julian F.V. Vincent,
Jessica Chen-Burger, and Marc P.Y. Desmulliez

Heriot-Watt University, Edinburgh EH14 4AS, Scotland, UK
rk22@hw.ac.uk

Abstract. When solving engineering problems through biomimetic design, a lack of knowledge of biology often impedes the translation of biological ideas into engineering principles. Specific challenges are the identification, selection and abstraction of relevant biological information. The use of engineering terminology to search for relevant biological information is hypothesised to contribute to the adventitious character of biomimetics. Alternatively, a holistic approach is proposed where a division is made between the analysis of biological research papers and the decomposition of the engineering problem. The aim of a holisitic approach is to take into account the importance of context during analogical problem solving and provide a theoretical framework for the development of Computer-Aided Biomimetics (CAB) tools. Future work will focus on the development of tools that support engineers during the analysis of biological research papers and modelling of biological systems by providing relevant biological knowledge.

Keywords: Biomimetics · Computer-Aided Biomimetics (CAB) · Trade-offs · Problem-solving

1 Introduction

Biomimetics aims to solve engineering problems through abstraction, transfer and application of knowledge from biological systems, processes, materials etc. (Fayemi et al. 2014). Over the last decades research in biomimetics has rapidly expanded in engineering and related subjects, such as robotics and materials sciences (Lepora et al. 2013). However, while a plethora of biomimetics design methods and tools have been proposed, solving engineering problems through biomimetics remains adventitious and serendipitous (Jacobs et al. 2014; Vincent 2016; Wanieck et al. 2017). Figure 1 visualizes which steps differentiate a generic biomimetics process from a generic problem solving process (Fayemi et al. 2014). Besides abstraction, transfer and application, there are differences in the generation and the selection of alternative solutions. Importantly, the (1) identification of possible candidates of biological models, (2) the selection of the relevant models and (3) their abstraction may be expected to take place primarily in the biological domain. Exactly these steps - identifying, selecting and abstracting biological

© Springer International Publishing AG 2017
M. Mangan et al. (Eds.): Living Machines 2017, LNAI 10384, pp. 242–254, 2017.
DOI: 10.1007/978-3-319-63537-8_21

knowledge relevant to a given engineering problem - are challenging for someone who is not familiar with biology (Vattam and Goel 2013a; Fayemi et al. 2015a).

Fig. 1. Generic sequential classical problem solving process (Massey and Wallace 1996) and biomimetics problem solving process. Based on Fayemi et al. (2014).

The most common approach to support the steps of identification, selection and abstraction is through the application of what can be regarded as the 'function bridge' (Helfman Cohen and Reich 2016), where engineering functions are used to classify and describe biological systems. However, the notions of functions in engineering is not the same as that in biology (Perlman 2009; Artiga 2016). Although both artefactual and biological systems may be explained in terms of function, the latter are characterised by dynamic, cyclic, hierarchical processes that rely on information from various systemic levels. Fundamental differences between biology and engineering (Fish and Beneski 2014) therefore complicate the automated extraction of engineering information from biological texts (Mizoguchi et al. 2016).

During the manual abstraction step, a lack of biological knowledge often leads to wrong interpretations or oversimplification of biological functions (Stricker 2006; Helms et al. 2009). Understanding why a biological system is organised as it is, has been obscured through changing environmental conditions and evolutionary requirements. Therefore, Fayemi et al. (2015b) argue that function is not a suitable starting point for the identification and selection of potential biological analogies, as well as for abstraction. Computer-Aided Biomimetics (CAB) tools aim to support manual abstraction of engineering information from biological texts, by extracting and providing within-domain biological information.

To overcome the pitfalls inherent in the use of functions, our research proposes to use trade-offs as a starting point for identification, a central concept in biology that indicates a dialectical relation between traits (Garland 2014). At a sufficiently high abstraction level, technical and biological trade-offs may be mapped to one another (Vincent 2016). Trade-offs provide an initial mapping that may be expected to require further filtering to select candidate biological systems. Validating abstractions of biological systems requires contextual knowledge, such as environmental interactions and properties of e.g. biological structures and materials (Kaiser et al. 2014).

The following Sect. 2 introduces the challenges of CAB according to literature. Section 3 elaborates on the use of engineering terminology to overcome these challenges and Sect. 4 emphasises the importance of context and analysis of context. Section 5 introduces the proposed approach to CAB, which requires future work to focus on relevant computational techniques such as knowledge graphs and relation extraction.

2   CAB Requirements According to Literature

Several approaches based on the use of databases have been proposed that aim to support engineers through the provision of biological knowledge presented in a terminology that is easy to understand for engineers (Vattam et al. 2010; Sartori et al. 2010). The most well-known example is AskNature (Deldin and Schuhknecht 2014), which was found to increase novelty of generated solutions without decreasing the technical feasibility (Vandevenne et al. 2016a). However, considering the large amount of documented biological knowledge (Vandevenne et al. 2016b), manually created databases will never be exhaustive. Due to this inherent limitation on size, and because entering new cases into a database can be arbitrary and effortful, methods and algorithms aimed at supporting biomimetics should be scalable (Vandevenne et al. 2011).

Approaches that apply Natural Language (NL) analysis can alleviate the inherent limitations of manually created databases by taking advantage of existing repositories of biological knowledge. Shu and Cheong (2014) explored NL analysis for biomimetics using an introductory course book to biology, but the scalability of this approach towards larger repositories remains to be proven. Recent efforts use biological research papers (Kaiser et al. 2014) to automate annotation of biological texts in engineering terminology (Rugaber et al. 2016) and improve the identification of relevant analogies using latent semantics (Vandevenne et al. 2016b). The underlying assumption of this approach is that biological research papers comprehensively cover all documented biological knowledge and describe up-to-date specific expert knowledge (Kaiser et al. 2012; Vandevenne et al. 2011).

When searching for relevant biological research papers, the main challenges are finding, recognising and understanding relevant information sources (Vattam and Goel 2011; Vattam and Goel 2013b). These challenges correspond to the steps of identification, selection and abstraction of biological models as shown in Fig. 1, and take place in the biological domain. Familiarity with biology and biological terminology eases these challenges. On the other hand, a lack of biological knowledge impedes abstraction (Helfman Cohen and Reich 2016) and problem solving in general. Accordingly, a common finding in literature is that a holistic, iterative approach benefits biomimetics (Kruiper et al. 2016).

Regarding the transfer and implementation steps, a variety of models have been proposed to represent biological knowledge for biomimetics, e.g. SAPPhIRE models (Chakrabarti 2014), a Living System Theory approach (Fayemi et al. 2015b) and Structure-Behaviour-Function models (Rugaber et al. 2016). Although differences exist between these, each type of model may be expected to be useful in abstracting and transferring knowledge (Durand et al. 2015). The process of modelling itself helps rationalising thought and developing understanding (Schön 1983; Brereton 2004). Table 1 provides an overview of the recommendations for CAB tools according to literature.

Table 1. Overview of the challenges Computer-Aided Biomimetics tools should aim to overcome and the related themes of common mistakes and recurring findings. Based on Kruiper et al. (2016) and Vattam and Goel (2011, 2013a).

Themes	Challenges
Scalability	Ability to integrate large amounts of biological knowledge to support biomimetics processes
Formalisation Transfer impediments Validation Analogies	**Identification** of possibly relevant information sources, out of all existing information sources, based on a query. Reduce the amount of time spent browsing query results
Transfer impediments Validation Analogies	**Selection** of information sources that seem most relevant within the set of possible information sources. Improve the ability to recognize the content in the results
Transfer impediments Validation Abstraction	**Abstraction** in biomimetics is "*the process of refining the biological knowledge (design solutions) to some working principles, strategies or representative models that explain the biological solution and could be further transferred to the target application*" (Helfman Cohen and Reich 2016). This encompasses moving from 'understanding the biological terminology' towards 'using appropriate methods for describing and decoding biological principles'
Holistic approach	Ability to alternate between problem decomposition and analogical reasoning, simultaneously expanding the designers' knowledge required for validation

3   Engineering Terminology in CAB

Considering the desired scalability of CAB tools an approach that applies NL analysis of biological research papers seems reasonable. However, assuming that biology research papers are written using domain-specific terminology, this approach limits the use of engineering terminology in identifying relevant biological information. Reasons include the different words used in biological and engineering research papers, as well as differences in the semantic meaning of words and core concepts like function as explained earlier.

3.1   Differences in Biological and Engineering Terminology

Nagel et al. (2014) notes that domain knowledge is required to understand biological '*flows*' (as cited from Pahl and Beitz 2007) of materials, energy and information. Having to look up each term is tedious and disrupts the thought process, or rather the '*flow*' (Csikszentmihalyi 1990). Figure 2 displays a word cloud that provides some intuition on the type of terms that engineers may not be familiar with. Crucially, unknown terms will not be useful during identification or selection of biological research papers. Terminological differences between biological texts and engineering texts therefore render direct keyword-based search inadequate (Vattam and Goel 2011).

Fig. 2. Word cloud generated from eleven biological research papers. Two sets of eleven research papers belonging to the biological and engineering domain respectively were used from the Elsevier OA-STM-Corpus (2015). The words in the vocabulary represent all words in the biology papers, minus those words occurring in the engineering papers. Although the sets of documents are limited in size and do not cover all topics of both domains, the resulting terms provide an intuition of biological terminology that engineers may not be familiar with, such as haploinsufficiency, hepatocyte, allograft, placode, microglia, caspase etc.

Vattam and Goel (2011) suggest that, in absence of biological knowledge, selection is specifically limited to *semantic similarity*. However, similarity and relevance of higher-order relations are neglected, e.g. taxonomical and entity-relations. *Semantic similarity* here is the overlap in number of similar words used in a search query and found in the retrieved documents. In this sense, *semantic similarity* is a measure of frequent association without necessarily overlapping in the semantic meaning of the words.

To improve the identification and selection of relevant biological systems Vattam and Goel (2013b) suggest annotating biological research papers with engineering terminology, which may also ease the understanding of biological analogies. Rugaber et al. (2016) describe a CAB system that automatically annotates biological research papers with Structure-Behaviour-Function (SBF) models. Functions are represented in verb-object format and extracted by matching functional verbs from a controlled vocabulary that is based on the Functional Basis (Hirtz et al. 2002) and the Biomimicry Taxonomy of AskNature. Behaviours are extracted using syntactic patterns that are a subset of the patterns by Khoo et al. (1998, 2000). Structures are terms matched against a vocabulary of biological structures based on part of an ontology for biomimetics – a formal and explicit representation of knowledge – created by Vincent (2014, 2016).

The assumption by Rugaber et al. (2016) is that SBF models can robustly represent biological systems. However, automatically extracted annotations will only be as useful as the type and quality of information they model. Using vocabularies of engineering functions to annotate and retrieve biological research papers does not extend key-word based search. The added value then lies in the quality of extracted behaviours and structures – the systemic context of the functions.

3.2   Different Semantics and Core Concepts

While SBF models may be used to represent a biological system, differences in the semantic meaning of words introduce an important issue in automatically extracting engineering information from biological research papers. Vandevenne et al. (2016b) circumvent this problem by clustering terms frequently occurring together in documents, either in the biological or engineering domain. The resulting manually labelled concepts, 300 so-called Organism Aspects, are extracted from 8,011 biological research papers. In a previous work, the authors extract 300 Product Aspects from 155,000 patents. Cross-domain associations are provided, based on the similarity between the concept vectors, representing the occurrence of domain-specific terms.

The cross-domain associations enable the matching of *semantically similar* groups of biological terms to groups of engineering terms. Based on the terms occurring in biological research papers, the papers can then be annotated with *semantically similar* engineering concepts. Therefore, using pre-specified engineering terminology – the Product Aspects – relevant biology research papers can be identified. A keyword like the verb 'to float' may, for example, return a paper that mentions buoyancy several times. However, the occurrence of *semantically similar* words does not necessarily improve the selection, or indicate the existence, of analogies relevant to the engineering problem.

In analogical problem solving, relations between individual parts of a system are dominant aspects (Markman and Gentner 2000; Gentner and Kurtz 2006; Verhaegen et al. 2011). Representations that capture systemic relations, e.g. functional representations like SBF models, can thus support analogical reasoning. Functions in general may be used to capture analogies between individual parts of biological systems and engineering systems. The usefulness of such analogous functions in Design-by-Analogy is reflected in the commonly used 'function-bridge' in biomimetics. However, in the case where CAB systems focus on text-processing of biological research papers fundamental differences between the semantic meaning of concepts like biological and engineering functions have to be taken into account. The same applies to the semantic meaning of words associated to such concepts; well-known functional verbs for example may not convey the same meaning in engineering and biology (Nagel et al. 2014).

From a teleological point of view, biological functions and engineering functions are different (Artiga 2016). To denote the intention-neutrality of biological functions, a static view of a biological process may be considered a *role* within the context of a system (Chandrasekaran and Josephson 2000). Generally a function may be regarded as the capacity to perform a behaviour within a given context (Mizoguchi et al. 2016).

4   Context and Analysis

Kaiser et al. (2014) show that verbs associated with engineering functions are not always present in biological research papers. In case functional terms are found in biological research papers, they often co-occur with terms describing environmental characteristics that may influence the function. To exemplify the importance of context, consider the self-cleaning functionality known as the *lotus effect*. This function, or rather property, exhibited by a variety of plants and insects, is based on superhydrophobicity that can be introduced through a variety of microscopic structures (Meyers 2015; Barthlott and Neinhuis 1997). In plant leaves, these structures are mainly formed by epicuticular wax crystalloids, of which the shape is determined by specific compounds in the wax. In contrast to marsh and water plants the wettability of surfaces is noted to be of little importance to plants originating from Mediterranean-type habitats or subtropical regions. "*Here, trichomes or waxes are most probably involved in the regulation of the radiation budget and, therefore, indirectly in temperature control*" (Neinhuis and Barthlott 1997). Thus, although the self-cleaning functionality of surfaces may be attributed to superhydrophobicity of a surface, this functionality depends on a variety of traits at multiple hierarchical levels. Furthermore, depending on external properties, the same trait can vary statically over phylogenetic distance or dynamically over ecological similarity. The differences between the surface structures found on leaves of various plant species are strongly correlated to their wettability and thus the self-cleaning property.

On the other hand, over-reliance on e.g. the self-cleaning function of superhydrophobic surfaces may obscure other functionalities based on the same principle. Superhydrophobicity can also enable floating capacity, e.g. in water striders legs (Feng et al. 2007). Furthermore, Cicada orni combine hydrophobicity with the anti-reflective property known as the moth-eye effect in a multi-layered nanostructure (Dellieu et al. 2014). Similarly the swim bladder of a fish is well known as an organ that provides buoyancy while swimming underwater. Simultaneously, the swim bladder is a structure that includes a lumen. Some species of fish use a swim bladder to improve precision in sensing water pressure (Taylor et al. 2010) and some to support sound production and hearing (Millot et al. 2011). The same structure or process may thus be involved in various functional properties.

The contextual variables on which a functional property depends can greatly influence the functional capacity. While an engineer may be interested only in a single '*function*' of a biological system, neglecting context often renders a direct transfer to engineering impossible (Inkermann et al. 2011). Due to the tight coupling between properties, biological systems can hardly be seen as parts associated with functions (Fayemi et al. 2015b). Considering knowledge transfer as the goal of modelling biological systems in terms of technical systems, searching for biological research papers using engineering functions requires that some form of abstraction is already performed.

5   Proposed CAB Design Approach

The biological and engineering terminology and semantics are inherently different. As a result, the use of engineering terminology to search over or automate annotation of biological texts is unreliable. While engineering functions may suffice to describe a design requirement independent of context, biological functions are tightly interrelated to context and a direct transfer may not be possible. Accordingly, the transfer of knowledge between both domains is noted to actually happen between representations of the biological and technical system (Sartori et al. 2010). Therefore, as displayed in Fig. 3, the proposed approach to CAB focuses on supporting engineers in representing biological systems of interest.

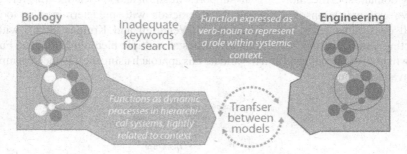

Fig. 3. Rather than searching for biological research papers using engineering terminology (top: function-bridge), the proposed approach aims to support the modelling of biological systems (bottom: taking context into account).

TRIZ, a set of tools for solving engineering problems creatively, has received a fair amount of attention in biomimetics (Vincent and Mann 2002; Vincent et al. 2006; Bogatyrev and Bogatyrev 2015; Fayemi et al. 2014; Vandevenne et al. 2015; Helfman Cohen and Reich 2016). TRIZ theory prescribes a high level of abstraction during problem decomposition, e.g. using contradictions to denote a trade-off or dialectic relation between parameters of components in a technical system (Cavallucci et al. 2009). In TRIZ these abstract contradictions can be used to direct an engineer towards abstract solution routes. Similarly, according to Vincent (2016) trade-offs can be used to classify biological solution routes based on the abstract parameters involved. Trade-offs can thus support the identification of relevant biological systems without the need to encode biological information in specific theoretical models. As a result, the proposed approach offers freedom over theoretical models used to represent biological knowledge, but is limited to either highly abstract or biological terminology.

In modelling the relations between properties, interdependence exists between (1) the available biological knowledge, (2) the abstraction of engineering knowledge from a source of biological information and (3) the transfer of knowledge to a given engineering application. Hence, as noted in Table 1, an iterative approach is beneficial to biomimetics. Using theoretical models to capture contextual variables in various representations helps developing design understandings (Brereton 2004). In support of a holistic design process, we propose that CAB tools focus on indicating the relations, semantic concepts and named entities required to understand biological strategies in their respective context.

In a holistic biomimetic design process, communicating 'raw' design ideas throughout the design process, as well as intuitive exploration, supports rationalising thought (Wendrich 2012). Although the search is for analogies, surprising properties and differences may be expected to provide considerable heuristic power in biomimetic problem solving (Bensaude-Vincent 2011; Salgueiredo 2013). Figure 4 displays how cross-domain knowledge transfer is facilitated by a continuous loop of communication, reflection-in-action, reflection-on-action and reflection-in-practice (Schön 1983). Hence, in the proposed biomimetic design approach the designer is constantly representing domain-specific knowledge to support validation and knowledge transfer. Such iterative externalisation of ideas and the interaction with predetermined or loosely defined constraints leads to novel insights (Wendrich and Kruiper 2016), without neglecting multi-functionality and interrelations at multiple hierarchical levels. Future work will now implement and aim to validate this approach using the advanced computer tools available to us.

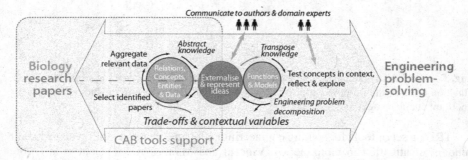

Fig. 4. Overview of the proposed approach to biomimetics supported by Computer-Aided Biomimetics (CAB) tools. CAB tools here aim to support a holistic, iterative approach to the search for biological data.

6   Conclusion

Several prevalent challenges in biomimetic problem solving processes are related to a lack of biological knowledge. In overcoming these challenges, computational tools may access the knowledge captured in biological research papers. Using biological research papers as information source limits the use of engineering terminology for search and automated annotation. However, the terminology used in biological research papers captures specific contextual knowledge. Such contextual variables can greatly influence functional capacities of a biological system. Modelling these contextual variables and inter-relations in biological systems leads to understanding the necessary knowledge for abstraction. A holistic approach is proposed that supports iteratively exploring biological information against predetermined engineering constraints.

In the proposed approach Computer-Aided Biomimetics (CAB) tools focus on extracting contextual variables and relation from biological research papers. Specifically relations between individual parts of a system are of interest to analogical problem solving. Various theoretical models may be useful in representing biological systems to support validation and knowledge transfer. Future work will focus on the extraction, retrieval and representation of knowledge in CAB tools to support our approach. The aim is to provide common sense biological background knowledge and identify trade-offs between abstract parameters.

Acknowledgements. The authors are grateful to Dr. Ing. Robert E. Wendrich and Dr. Rupert Soar for their advice. This PhD research is funded by the EPSRC Centre for Doctoral Training in Embedded Intelligence and the School of Mathematical and Computer Sciences at Heriot-Watt University, Edinburgh, Scotland, UK.

References

Artiga, M.: New perspectives on artifactual and biological functions. Appl. Ontol. **11**, 1–9 (2016)

Barthlott, W., Neinhuis, C.: Purity of the sacred lotus, or escape from contamination in biological surfaces. Planta **202**(1), 1–8 (1997). doi:10.1007/s004250050096

Bensaude-Vincent, B.: A cultural perspective on biomimetics. In: Nature, pp. 1–12. InTech (2011). doi:10.5772/10546

Bogatyrev, N., Bogatyrev, O.A.: TRIZ-based algorithm for Biomimetic design. Procedia Eng. **131**, 377–387 (2015)

Brereton, M.: Distributed cognition in engineering design: negotiating between abstract and material representations. In: Goldschmidt, G., Porter, W.L. (eds.) Design representation, pp. 83–103. Springer, London (2004). doi:10.1007/978-1-85233-863-3_4

Cavallucci, D., Rousselot, F., Zanni, C.: Linking contradictions and laws of engineering system evolution within the triz framework. Creat. Innov. Manage. **18**(2), 71–80 (2009). doi:10.1111/j.1467-8691.2009.00515.x

Chakrabarti, A.: Supporting analogical transfer in biologically inspired design. In: Goel, A., McAdams, D., Stone, R. (eds.) Biologically Inspired Design, pp. 201–220. Springer, London (2014). doi:10.1007/978-1-4471-5248-4_8

Chandrasekaran, B., Josephson, J.R.: Function in device representation. Eng. Comput. **16**, 162–177 (2000)

Csikszentmihalyi, M.: FLOW: The Psychology of Optimal Experience. HarperCollins, New York (1990)

Deldin, J.-M., Schuhknecht, M.: The AskNature database: enabling solutions in biomimetic design. In: Goel, A., McAdams, D., Stone, R. (eds.) Biologically Inspired Design, pp. 17–27. Springer, London (2014). doi:10.1007/978-1-4471-5248-4_3

Dellieu, L., Sarrazin, M., Simonis, P., Deparis, O., Vigneron, J.P.: A two-in-one superhydrophobic and anti-reflective nanodevice in the grey cicada Cicada orni (Hemiptera). J. Appl. Phys. **116**(2), 24701 (2014). doi:10.1063/1.4889849. American Institute of Physics

Durand, F., Helms, M., Tsenn, J., McTigue, E., McAdams, D.A., Linsey, J.S.: Teaching students to innovate: evaluating methods for bioinspired design and their impact on design self efficacy. In: 27th International Conference on Design Theory and Methodology, vol. 7 (2015). doi:10.1115/DETC2015-47716

Elsevier OA-STM-Corpus: An open access corpus of scientific, technical, and medical content, 11 January 2015. https://github.com/elsevierlabs/OA-STM-Corpus

Fayemi, P. E., Maranzana, N., Aoussat, A., Bersano, G.: Bio-inspired design characterisation and its links with problem solving tools. In: Proceedings of DESIGN 2014, pp. 173–182 (2014)

Fayemi, P.-E., Maranzana, N., Aoussat, A., Bersano, G.: Assessment of the biomimetic toolset—design spiral methodology analysis. In: Chakrabarti, A. (ed.) ICoRD'15 – Research into Design Across Boundaries Volume 2. Smart Innovation, Systems and Technologies, vol. 35. Springer, New Delhi (2015a). doi:10.1007/978-81-322-2229-3_3

Fayemi, P.-E., Maranzana, N., Aoussat, A., Chekchak, T., Bersano, G.: Modeling biological systems to facilitate their selection during a bio-inspired design process. In: DS 80-2 Proceedings of the 20th International Conference on Engineering Design (ICED 2015), Milan, Italy, vol. 2, pp. 225–234 (2015b)

Feng, X.Q., Gao, X., Wu, Z., Jiang, L., Zheng, Q.S.: Superior water repellency of water strider legs with hierarchical structures: experiments and analysis. Langmuir 23(9), 4892–4896 (2007). doi:10.1021/la063039b

Fish, F.E., Beneski, J.T.: Evolution and bio-inspired design: natural limitations. In: Goel, A.K., McAdams, D.A., Stone, R.B. (eds.) Stone Biologically Inspired Design, Chap. 12, pp. 287–312. Springer, London (2014). doi:10.1007/978-1-4471-5248-4_12

Garland Jr., T.: Trade-offs. Curr. Biol. 24(2), R60–R61 (2014)

Gentner, D., Kurtz, K.J.: Relations, objects, and the composition of analogies. Cogn. Sci. 30(4), 609–642 (2006). doi:10.1207/s15516709cog0000_60

Helfman Cohen, Y., Reich, Y.: Biomimetic Design Method for Innovation and Sustainability. Springer, Cham (2016). doi:10.1007/978-3-319-33997-9

Helms, M., Vattam, S.S., Goel, A.K.: Biologically inspired design: process and products. Des. Stud. 30(5), 606–622 (2009). doi:10.1016/j.destud.2009.04.003

Hirtz, J., Stone, R.B., Mcadams, D.A., Szykman, S., Wood, K.L.: A functional basis for engineering design: reconciling and evolving previous efforts. Res. Eng. Des. 13, 65–82 (2002)

Inkermann, D., Stechert, C., Löffler, S., Vietor, T.: A 24. IASTED Technology Conferences/705: ARP/706: RA/707: NANA/728: CompBIO, March 2016. http://doi.org/10.2316/P.2010.706-031

Jacbos, S.R., Nichol, E.C., Helms, M.E.: 'Where are we now and where are we going?' The BioM innovation database. J. Mech. Des. 136(11), 111101 (2014)

Kaiser, M.K., Hashemi Farzaneh, H., Lindemann, U.: An approach to support searching for biomimetic solutions based on system characteristics and its environmental interactions. In: Proceedings of International Design Conference, DESIGN, vol. DS 70, pp. 969–978 (2012)

Kaiser, M.K., Hashemi Farzaneh, H., Lindemann, U.: BioSCRABBLE - the role of different types of search terms when searching for biological inspiration in biological research articles. In: International DESIGN Conference, Dubrovnik, Croatia, pp. 241–250 (2014)

Khoo, C.S.G., Chan, S., Niu, Y.: Extracting causal knowledge from a medical database using graphical patterns. In: Proceedings of 38th Annual Meeting ofthe Association for Computational Linguistics, pp. 336–343 (2000)

Khoo, C.S.G., Kornfit, J., Oddy, R.N., Myaeng, S.H.: Automatic extraction of cause-effect information from newspaper text without knowledge-based inferencing. Literary Linguist. Comput. 13, 177–186 (1998)

Kruiper, R., Chen-Burger, J., Desmulliez, M.P.Y.: Computer-aided biomimetics. In: Lepora, Nathan F.F., Mura, A., Mangan, M., Verschure, P.F.M.J., Desmulliez, M., Prescott, T.J. (eds.) Living Machines 2016. LNCS, vol. 9793, pp. 131–143. Springer, Cham (2016). doi: 10.1007/978-3-319-42417-0_13

Lepora, N., Verschure, P., Prescott, T.: The state of the art in biomimetics. Bioinspir. Biomim. 8, 1–11 (2013). doi:10.3905/jpm.1974.408489

Markman, A.B., Gentner, D.: Structure mapping in the comparison process. Am. J. Psychol. 113(4), 501–538 (2000). doi:10.2307/1423470

Massey, A., Wallace, W.: Understanding and facilitating group problem structuring and formulation: mental representations, interaction, and representation aids. Decis. Support Syst. 17(4), 253–274 (1996)

Meyers, A.: Surface allows self-cleaning, 10 December 2015. AskNature: https://asknature.org/strategy/surface-allows-self-cleaning/#.WMlpAm-LSUk

Millot, S., Vandewalle, P., Parmentier, E.: Sound production in red-bellied piranhas (Pygocentrus nattereri, Kner): an acoustical, behavioural and morphofunctional study. JEB **214**, 3613–3618 (2011). doi:10.1242/jeb.061218

Mizoguchi, R., Kitamura, Y., Borgo, S.: A unifying definition for artifact and biological functions. Appl. Ontol. **11**(2), 129–154 (2016). doi:10.3233/AO-160165

Nagel, J.K.S., Stone, R.B., McAdams, D.A.: A thesaurus for bioinspired engineering design. In: Goel, A., McAdams, D., Stone, R. (eds.) Biologically Inspired Design, Chap. 4, pp. 63–94. Springer, London (2014). doi:10.1007/978-1-4471-5248-4_4

Neinhuis, C., Barthlott, W.: Characterization and distribution of water-repellent, self-cleaning plant surfaces. Ann. Bot. **79**(6), 667–677 (1997). doi:10.1006/anbo.1997.0400

Pahl, G., Beitz, W., Feldhusen, J., Grote, K.H.: Engineering Design: A Systematic Approach. Springer, London (2007)

Perlman, M.: Changing the mission of theories of teleology: DOs and DON'Ts for thinking about function. In: Functions in Biological and Artificial Worlds, pp. 17–36 (2009)

Rugaber, S., Bhati, S., Goswami, V., Spiliopoulou, E., Azad, S., Koushik, S., Kulkarni, R., Kumble, M., Sarathy, S., Goel, A.K.: Knowledge extraction and annotation for cross-domain textual case-based reasoning in biologically inspired design. In: Goel, A., Díaz-Agudo, M., Roth-Berghofer, T. (eds.) Case-Based Reasoning Research and Development (ICCBR). LNCS, vol. 9969. pp. 342–355. Springer, Cham (2016). doi:10.1007/978-3-319-47096-2_23

Salgueiredo, C.F.: Modeling biological inspiration for innovative design. In: 20th International Product Development Management Conference, Paris, France, June 2013

Sartori, J., Pal, U., Chakrabarti, A.: A methodology for supporting "transfer" in biomimetic design. Artif. Intell. Eng. Des. Anal. Manuf. **24**(4), 483–506 (2010). doi:10.1017/S0890060410000351

Schön, D.A.: The Reflective Practitioner, How Professionals Think in Action. Basic Books, Inc., New York (1983). ISBN: 0-465-06878-2

Shu, L.H., Cheong, H.: A natural language approach to biomimetic design. In: Goel, A., McAdams, D., Stone, R. (eds.) Biologically Inspired Design, pp. 29–62. Springer, London (2014). doi:10.1007/978-1-4471-5248-4_3

Stricker, H.M.: Bionik in der Produktentwicklung unter der Berücksichtigung menschlichen Verhaltens. Ph.D. thesis Lehrstuhl Für Produktentwicklung, TU Munich (2006)

Taylor, G.K., Holbrook, R.I., de Perera, T.B.: Fractional rate of change of swim-bladder volume is reliably related to absolute depth during vertical displacements in teleost fish. J. Roy. Soc. Interface **7**, 1379–1382 (2010). doi:10.1098/rsif.2009.0522. The Royal Society

Vandevenne, D., Verhaegen, P.-A., Dewulf, S. Duflou, J.R.: A scalable approach for the integration of large knowledge repositories in the biologically-inspired design process. In: International Conference on Engineering Design, ICED 2011 (2011)

Vandevenne, D., Verhaegen, P.-A., Dewulf, S., Duflou, J.R.: Product and organism aspects for scalable systematic biologically-inspired design. Procedia Eng. **131**, 784–791 (2015)

Vandevenne, D., Pieters, T., Duflou, J.R.: Enhancing novelty with knowledge-based support for biologically-inspired design. Des. Stud. (2016a). doi:10.1016/j.destud.2016.05.003

Vandevenne, D., Verhaegen, P.-A., Dewulf, S., Duflou, J.R.: SEABIRD: scalable search for systematic biologically inspired design. Artif. Intell. Eng. Des. Anal. Manuf. **30**, 78–95 (2016b)

Vattam, S., Wiltgen, B., Helms, M., Goel, A.K., Yen, J.: DANE: fostering creativity in and through biologically inspired design. In: Taura, T., Nagai, Y. (eds.) Design Creativity 2010, vol. 8, pp. 115–122. Springer, London (2011). doi:10.1007/978-0-85729-224-7_16

Vattam, S.S., Goel, A.K.: Foraging for inspiration: understanding and supporting the online information seeking practices of biologically inspired designers. In: Proceeding of the ASME IDETC/CIE 2011, 28–31 August 2011

Vattam, S.S., Goel, A.K.: An information foraging model of interactive analogical retrieval. In: Proceedings of 35th Annual Meeting of Cognitive Science Society (2013a)

Vattam, S.S., Goel, A.K.: Biological solutions for engineering problems: a study in cross-domain textual case-based reasoning. In: Proceedings of 21st International Conference on Case Based Reasoning 2013, pp. 343–357 (2013b). doi:10.1007/978-3-642-39056-2_25

Verhaegen, P.A., D'Hondt, J., Vandevenne, D., Dewulf, S., Duflou, J.R.: Identifying candidates for design-by-analogy. Comput. Ind. **62**, 446–459 (2011)

Vincent, J.F.V., Mann, D.L.: Systematic technology transfer from biology to engineering. Philos. Trans. Ser. A Math. Phys. Eng. Sci. **360**(1791), 159–173 (2002). doi:10.1098/rsta.2001.0923

Vincent, J.F.V., Bogatyreva, O.A., Bogatyreva, N.R., Bowyer, A., Pahl, A.-K.: Biomimetics: its practice and theory. J. Roy. Soc. Interface **3**(9), 471–482 (2006)

Vincent, J.F.V.: An ontology of biomimetics. In: Goel, A., McAdams, D., Stone, R. (eds.) Biologically Inspired Design, pp. 269–286. Springer, London (2014). doi: 10.1007/978-1-4471-5248-4_11

Vincent, J.F.V.: The trade-off: a central concept for biomimetics. Bioinspired, Biomimetic and Nanobiomaterials, (2016). doi:10.1680/jbibn.16.00005

Wanieck, K., Fayemi, P.-E., Maranzana, N., Zollfrank, C., Jacobs, S.: Biomimetics and its tools. Bioinspir. Biom. Nanobiomater. (2017). doi:10.1680/jbibn.16.00010

Wendrich, R.E.: Multimodal interaction, collaboration, and synthesis in design and engineering processing. In: Proceedings of International Design Conference, DESIGN, vol. DS 70, pp. 579–588 (2012)

Wendrich, R.E., Kruiper, R.: Keep it real: on tools, emotion, cognition and intentionality in design. In: Proceedings of International Design Conference, DESIGN, Dubrovnik, Croatia, 16–19 May 2016

Deep Dynamic Programming: Optimal Control with Continuous Model Learning of a Nonlinear Muscle Actuated Arm

Andrew G. Lonsberry$^{(\boxtimes)}$, Alexander J. Lonsberry, and Roger D. Quinn

Department of Mechanical and Aerospace Engineering,
Case Western Reserve University,
10900 Euclid Ave., Cleveland, OH 44106, USA
{agl10,ajl17,roger.quinn}@case.edu
http://biorobots.case.edu/

Abstract. We outline a new technique for on-line continuous model learning control and demonstrate its utility by controlling a simulated 2-DOF arm actuated by 6 muscles as well as on an inverted pendulum. Work presented is part of an effort to develop controllers for human appendages rendered inoperable by paralysis. Computerized control provides an alternative to neural regeneration by means of electric muscle stimulation. It has been demonstrated that paralyzed individuals can regain self-powered mobility via use of external muscle controllers. A barrier to proliferation of the technology, is the difficulty in control over the living system which is highly nonlinear and unique to each individual. Here we demonstrate a novel, continuous model learning technique to simultaneously learn and control continuous, non-linear systems. The technique expands upon vanilla Q-learning and dynamic programming. Unlike typical Q-learning, where the action-value function updates are only for the most recent set of states visited and stored in memory, the method presented also generates updates to the action-value function for unvisited state-space and state-space visited far in the past. This is made feasible by giving the agent the ability to continually learn and update explicit local models of the environment and of itself, which we encapsulated in a set of deep neural networks.

Keywords: Optimal control · Deep learning · Continuous model learning

1 Introduction

Paralysis due to spinal cord injury (SCI) results in degeneration to almost every organ system that threatens overall health and well-being and compromises a productive life style [4,5]. Neuro-prostheses utilizing functional neuro-muscular stimulation (FNS) provides the ability to stand and walk [6–8] which can positively impact the overall health of persons with SCI [9,10]. However, walking with FNS alone results in excessive upper extremity effort to maintain balance

© Springer International Publishing AG 2017
M. Mangan et al. (Eds.): Living Machines 2017, LNAI 10384, pp. 255–266, 2017.
DOI: 10.1007/978-3-319-63537-8_22

and rapid muscle fatigue [11]. Wearable walking robots or powered exoskeletons can facilitate standing and stepping motions [12]. However difficulties remain in application of these powered devices particularly in the form of dynamic balance and fall prevention. The underlying problem is a lack of adequate controls which is impeded by accurately generating a system model for every unique individual.

Model-free reinforcement learning approaches can operate without or only with limited information about a given system model, and do not make an explicit effort to understand the underlying system dynamics. These reinforcement learning approaches are applicable to the control problem presented here. A popular approach of this type is Q-learning [13], which is an off-policy learning strategy that learns the action-value function based on the reward at the current state and the discounted return of the expected future reward assuming a greedy action policy. The algorithm has been proven to converge upon an optimal solution. Scaling this algorithm to operate on real life problems, deep neural networks are used to save the action-value function. These systems are referred to as Deep Q Networks (DQN) [15], and inherently utilize generalization capability of these networks to save the optimal action choice at each state. While these systems work well in discrete action space, they are less functional in continuous action space. A system that utilizes the strengths of Q-learning for continuous action space is Deep Deterministic Policy Gradients (DDPG) [14] which use actor-critic networks to update the policy based on Q-learning value iteration. Though both these techniques are powerful, they depend on fully observable state-action pairs that are within a limited time horizon from the reward signal. The result of this is that a majority of the state-space and action space explored will go unused during the update and information will be lost. Due to this, systems such as DQN and DDPG can struggle with problems that have very intermittent reward signal, such as the problem at hand. To alleviate these issues, we deploy a system that leverages the strengths of the DQN and DDPG while utilizing all the state-action space observed to create intelligent exploration and quick control.

We present a new method for continuous model learning control, called Deep Dynamic Programming (DDP), which expands upon the model-free, action-value structure and dynamic programming. DDP expands past model-free learning by creating continuously updating forward and inverse single step predictive models that are encapsulated by a set of neural networks. These networks are used in conjunction with backwards induction to generate updates to the action-value function for state-space regions outside of recent history, such as regions never explored before or those explored in the distant past. This inclusion of continuous environmental modeling and generalization of the action-value function provides for rapid generation of control policies as well as continual error reduction and control refinement. We demonstrate the approach on a simulated, redundantly actuated 2-DOF arm driven by six muscles, as well as an inverted pendulum. While not identical to the human-assisted system, this is a representative, model to show applicability to control over power-assisted human appendages (Fig. 1).

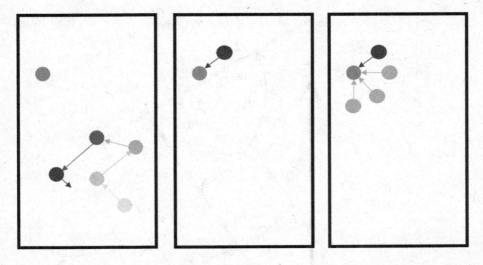

Fig. 1. Visual description of DDP: LEFT shows the agent (Blue) randomly exploring space with limited rewards (Green), the process occurs in both the presented method and other cited approaches. CENTER shows typical RL scheme where only the most recent states (1 state) are remembered and used for system updates. RIGHT demonstrates the method here described where the agent backwards solves from a reward state, for other partially observable states-action pairs (Orange), outside of recent history. (Color figure online)

2   Methods

2.1   Arm and Muscle Model

Arm Kinematics. A musculo-skeletal model with redundancies, based on the model in [1], is used in simulation and is constituted by two serial links with four mono-articular muscles and two bi-articular muscles to replicate the upper human arm. The notation that follows is widely taken from [1]. This simulate arm, while a simplification of am actual, is used to demonstrate the feasibility of the developed approach.

The model is developed under the assumption that motion is limited to a single plane and that gravity does not have affect. All frictional effects at the joints or between soft tissue is ignored. All muscles are assumed to be linearly connected, and the influence of curved connection paths is not included. The change in muscle mass position and inertia terms during activation is also omitted for simplicity. A depiction of the model is shown in Fig. 2, wherein the orientation of the skeletal portions is given by $\boldsymbol{\theta} = (\theta_1, \theta_2)^T$ where we constrain the angles reasonably to $\theta_{1,2} \in [\pi/6, -\pi/6]$. The position of the arm's end-point $\boldsymbol{x} = (x, y)^T$ in Cartesian coordinates is given as,

$$\boldsymbol{x} = \begin{pmatrix} L_1 cos(\theta_1) + L_2 cos(\theta_1 + \theta_2) \\ L_1 sin(\theta_1) + L_2 sin(\theta_1 + \theta_2) \end{pmatrix}. \tag{1}$$

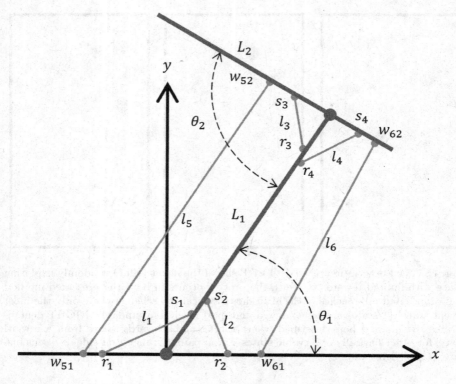

Fig. 2. Two-DOF arm comprised of two skeletal links (blue), 4 mono-articular muscles (orange), and two bi-articular muscles (green). (Color figure online)

The lengths of the muscles $l(\theta) = (l_1, l_2, l_3, l_4, l_5, l_6)^T$ is calculated as follows,

$$l(\theta) = \begin{pmatrix} (r_1^2 + s_1^2 + 2r_1s_1cos(\theta_1)^{1/2} \\ (r_2^2 + s_2^2 - 2r_2s_2cos(\theta_1)^{1/2} \\ (r_3^2 + s_3^2 + 2r_3s_3cos(\theta_2)^{1/2} \\ (r_4^2 + s_4^2 - 2r_4s_4cos(\theta_2)^{1/2} \\ (w_{51}^2 + w_{52}^2 + L_1^2 + 2w_{51}L_1cos(\theta_1) \\ + 2w_{52}L_1cos(\theta_2) + 2w_{51}w_{52}cos(\theta_1 + \theta_2))(1/2) \\ (w_{61}^2 + w_{62}^2 + L_1^2 - 2w_{61}L_1cos(\theta_1) \\ - 2w_{62}L_1cos(\theta_2) + 2w_{51}w_{52}cos(\theta_1 + \theta_2))(1/2) \end{pmatrix} \quad (2)$$

where for mono-articular muscles $i = 1, 2, 3, 4$, the lengths l_i are calculated using r_i and s_i which are the distances between the center of the joint and insertion point, and where for the bi-articular muscles $i = 5, 6$, the lengths are calculated using the distances between joint center and point of insertion w_{i1} and w_{i2}. The rate of change of muscle length with respect to time can be found by taking the time-derivate of (2),

$$\dot{l}(\boldsymbol{\theta}) = \boldsymbol{W}^T \dot{\boldsymbol{\theta}} \tag{3}$$

where $\boldsymbol{W}^T = \nabla_{\boldsymbol{\theta}} l(\boldsymbol{\theta})$ is the Jacobian matrix. The relationship between the contractile forces produced by the muscles $\boldsymbol{F}_m = (F_1, F_2, F_3, F_4, F_5, F_6)^T$ and the torque produced about the joints is,

$$\boldsymbol{\tau} = \boldsymbol{W} \boldsymbol{F}_m \tag{4}$$

Non-linear Muscle Model. Many muscle models have been created to predict both passive and active muscle forces in bio-mechanical simulation. Many of such models omit mass and inertia terms. Perhaps the most common are Hill-type [3] models which are one dimensional and generate force between origin and insertion points. Typical, Hill-type muscles are composed of a contractile element incorporating force-length and force-velocity dependencies, along with a serial and a parallel elastic element. Model inputs are typically muscle length, contraction velocity, and neural muscle stimulation.

Our musculo-skeletal model here is adapted from [1] wherein we strictly use the muscle model therein described. We description of the model here is breif as the focus is on the mechanism of learning the system model and the control thereof is the focal point. For more information see [1]. The muscle model implemented is of the Hill-type, however the serial-elastic element is ignored as it is assumed to have near infinite stiffness. Therefore the model is simplified to only the force-velocity relationship expressed as,

$$(f + a)(\dot{l} + b) = b(f_o + a) \tag{5}$$

where f is the tensile force produced by the muscle, $\dot{l}$ is said muscle's contraction velocity, f_o is the maximum tensile force applied, a is the heat constant, and b is the the rate constant of energy liberation. Research [2] has found that muscles will exert more tensile force in lengthening rather than shortening phase, where then the force can be represented as,

$$f(\bar{\alpha}, \dot{l}) = \begin{cases} \frac{.9l_o}{0.9l_o + |\dot{l}|}(\bar{\alpha} - \frac{0.25}{0.9l_o}\bar{\alpha}\dot{l}) & \text{if } \dot{l} \geq 0 \\ \frac{.9l_o}{0.9l_o + |\dot{l}|}(\bar{\alpha} - \frac{2.25}{0.9l_o}\bar{\alpha}\dot{l}) & \text{if } \dot{l} < 0 \end{cases} \tag{6}$$

where $\bar{\alpha} = f_o \times \alpha$ where $\alpha \in [0, 1]$ is muscle activation level. Muscles also have intrinsic dampening which can be added to (6),

$$f(\bar{\alpha}, \dot{l}) = p\{\bar{\alpha} - (\bar{\alpha}c + c_o)\dot{l}\}, \tag{7}$$

where,

$$p = \frac{.9l_o}{0.9l_o + |\dot{l}|}, \tag{8}$$

and,

$$c = \begin{cases} \frac{0.25}{0.9l_o} > 0 & \text{if } \dot{l} \geq 0 \\ \frac{2.25}{0.9l_o} > 0 & \text{if } \dot{l} < 0 \end{cases} \tag{9}$$

Now the forces produced by the muscles can be related to the system in matrix form,

$$F_m = P\bar{\alpha} - P\{A(\bar{\alpha})C + C_o\}i \qquad (10)$$

where F_m is a vector of muscle forces, P is a matrix with values of p across the diagonal and zeros everywhere else, $\bar{\alpha}$ is a vector of activations, $A(\bar{\alpha})$ is a matrix with the values of α across the diagonal and zeros everywhere else, and C and C_o are matrices with the values of c_i and c_{0i} respectively along the diagonal and zeros everywhere else.

Dynamic Model. Using well-known Lagrangian in classical dynamics, we produce the dynamic model of the system as follows,

$$\begin{bmatrix} H_{11} & H_{12} \\ H_{21} & H_{22} \end{bmatrix} \begin{bmatrix} \ddot{\theta}_1 \\ \ddot{\theta}_2 \end{bmatrix} + \begin{bmatrix} -h\dot{\theta}_2 & -h(\dot{\theta}_2 + \dot{\theta}_1) \\ h\dot{\theta}_1 & 0 \end{bmatrix} \begin{bmatrix} \dot{\theta}_1 \\ \dot{\theta}_2 \end{bmatrix} = \begin{bmatrix} \tau_1 \\ \tau_2 \end{bmatrix} = WF_m \qquad (11)$$

where,

$$\begin{aligned} H_11 &= m_1 l_{c1}^2 + I_1 + m_2(l_2^2 + l_{c2}^2 + 2l_1 l_{c2}cos(\theta_2)) + I_2 \\ H_{12} &= H_{21} = m_2 l_1 L_{c2}cos(\theta_2) + m_2 l_{c2}^2 + I_2 \\ h &= m_2 l_1 l_{c2}sin(\theta_2) \end{aligned} \qquad (12)$$

The parameter values used are in part adapted from [1] and are summarized in Table 1 and all values of $c_{oi} = 0.2$. The values of the muscle rest lengths are given by $l_o = l(\theta_0)$ where $\theta_o = [\pi/2, \pi/2]^T$.

Table 1. The parameter values

Model parameters	
L_1	0.31 (m)
L_{c1}	0.165 (m)
m_1	1.93 (kg)
I_1	0.0141 (kg m^2)
L_2	0.27 (m)
L_{c2}	0.135 (m)
m_2	1.32 (kg)
I_2	0.0120 (kg m^2)

2.2   Learning the Model and It's Controller

DDP is best described as a sum of its components: (1) an ANN that learns the action-value function, in our experiments the Q-learning algorithm was used (2)

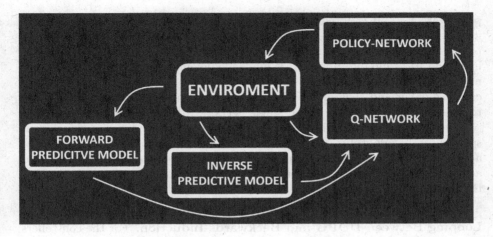

Fig. 3. Flow chart of information through the DDP layout

an ANN that learns predictive single step forward trajectories through state-space, (3) a set of ANN's that learn to predict single step backward (inverse) trajectories through state-space, and (4) an ANN that learns the control policy. Components (1), (2), and (3) can all be run in parallel to each other (Fig. 3).

The first component is an ANN that learns the action-value function, as similarly done in Q-learning (13). Network inputs are the current state and action, and network output is expected action-value. The network is trained by passing the square of the difference between the predicted Q value and the actual Q value.

$$Q(x_t, u_t) \leftarrow Q(x_t, u_t) + \alpha(r_{t+1} + \lambda * max_{u'}(Q(x_{t+1}, u_t)) - Q(x_t, u_t)) \quad (13)$$

The second component is an ANN to learn forward, 1-step prediction where inputs are the current state x_t and control input u_t, and the output is the ·expected next state x_{t+1}. The network is trained by passing the square error between the predict state next state and the actual state next state. This network is used to validate outputs from the inverse predictive networks and forward path planning.

The third component is a set of ANNs that learn to predictive 1 time-step backwards, where network inputs are the current state x_t and control input u_{t-1}, and the outputs are the expected former state x_{t-1}. In the specific case of the system here the control terms $u = \alpha = [\alpha_1, \alpha_2, \alpha_3, \alpha_4, \alpha_5, \alpha_6]^T$. It is important to note that finding networks to perform this mapping may not be feasible as there may be an infinite number of solutions. We help resolve this problem by dividing up state-space, and making separate networks to model the inverse dynamics in said section of state-space.

A separate ANN models each sub-space wherein each sub-space has specific properties. A sub-space here is defined such that a final state x_t, given the control u_{t-1} at the former time-step $t - 1$, uniquely transitions from one, and only one, prior state x_{t-1}. It is noted that this does not hold for the control terms, there may be an infinite number of controls that induce the transition from one state to the other. Thus for each sub-space, a separate ANN is trained where it takes in x_t and u_{t-1} and outputs x_{t-1}. While added memory for multiple networks is not wanted, because each network is only handling a sub-space, these networks can be smaller and thus require smaller computation on the forward pass and to invert as compared to a larger, single network to handle all of state-space.

The fourth component is the policy network, where the input is the current state x_t and the output is a real valued action u_t chosen from the continuous action space. This network is trained by passing the square error of the predicted action to the most optimal action chosen by the inverse networks, or by moving the parameters in the direction of the Q network gradient.

Looping Between DDPG and Backwards Induction. For the controllers developed here operating on a continuous action space, policy network learning takes place in two different phases. The first phase is DDPG, where the system navigates through state-space and updates the policy network by moving the parameters in the direction of the gradient of the Q network with respect to the loss. While this occurs the inverse prediction networks are learning the underlying dynamics of the agent and the environment, with the goal of finding local or global patterns that can be generalized across regions of state-space and implemented into the policy network iteratively.

The second phase of learning is backwards induction with the inverse prediction networks. When an or all the individual inverse prediction networks reach an error threshold within a local or global area, they become active for use in policy network learning. Depending on the current state that the system is in, the networks are used to create optimal trajectories backwards. For our purposes and examples here, there is only a single reward state, which is the desired pose of the arm or the swinging pendulum. From this pose the system initially performs backwards induction, updating the Q network and policy network until the predicted compound error exceeds limitations. The system then returns to the DDPG loop and continues to update the action-value function purely on observable state-space. The system repeats this cycle continually refining the controller on each pass. Exploration of the agent is done either by optimal selection from the policy network, or by a series of intelligent heuristics that monitor the error in the forward and inverse networks.

3   Results

Nonlinear Inverted Pendulum. Before application to the simulated arm with muscles, we first applied the approach on the an inverted pendulum modeled using MATLAB and the ode45 solver. For initial data collection of the pendulum moving around in its environment, the pendulum was randomly initialized between 0 and 2pi, random torque values were sampled between a max and a min and the torques acted on the system were chosen at random between a time max and a time min.

Two networks were trained in the bijective state space areas of the pendulum. These networks were small and learned the global transformation through state-space after only a few iterations. A larger network was trained to model the full forward dynamics. The large forward dynamics model was used occasionally during the backwards induction process to check the validity of the (x_t, u_t) pairs chosen. The goal state, $x = [\dot{\theta}, \theta]^T$, for this controller was $\theta = 3.1415$ (radians) and $\dot{\theta} = 0$ (radians/sec). The system was trained to minimize time duration between current state and goal state.

After training of a policy network was completed, the system worked exceedingly well. It swung up within the trained bounds of torque and minimized the torque being applied to approximately 0 when in the full upright position. The pendulum was disturbed with large forces and was always able to regain stabilization.

As shown in Fig. 4 the pendulum starts from a rest position at $\theta = 0$ and swings up to the desired state. The system is then perturbed at time $t = 0.2$ and $t = 0.3$ and shows the system recovering and going back to the stable point. There wasn't a perturbing force that caused the system ever to become fully unstable.

Controller Muscle Actuated Arm. The arm was randomly initialized with at some point in state-space, the input α was randomly chosen and acted for a random amount of time on the system before another random α was chosen. Data was being collected at 500 Hz. All ANN models are trained with mini

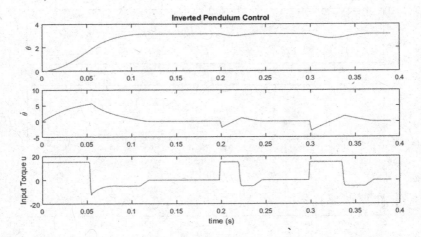

Fig. 4. Responses of pendulum angle θ, pendulum velocity $\dot{\theta}$, and input torque u of nonlinear inverted pendulum with Deep Dynamic Programming control during swing up and stabilization. At time steps 0.2 and 0.3, a disturbance force is applied to the pendulum.

batches in an on-line format. Through a process of policy update, state-space is sufficiently sampled. The arm was simulated in python using an Euler 1-step numerical approach with a time-step of $dt = .001$ s. Generation of over 200,000 points of exploration were needed for convergence of the policy networks, where convergence refers to the arm moving to the correct state.

Four networks were trained to model the inverse predictive motions. A large full network was used to model the entire forward dynamics. The large forward dynamics model was used occasionally during the backwards induction process to check the validity of the (x_t, u_t) pairs chosen. The goal state for this controller was $\theta_1 = 1.5708$, $\dot{\theta}_1 = 0$, $\theta_2 = 1.5708$, and $\dot{\theta}_2 = 0$. The controller showed good results with the preliminary training. The system was able to move to a desired point from all of the observable state-space that it was trained on. The generation of the arm's control policy was based on a cost function that equally weighted distance traveled and energy consumption over that period. Limitations on the system such as max and minimum values for the state space were specifically trained on as well as limitations on the input space were enforced to not allow the input space to make rapid changes.

As shown in Fig. 5, the angles of both arms move with a sinusoidal motion starting off with low velocity and high force due to the inertia of the arms. The angular velocities of the arms then rapidly accelerate and the control terms tend downwards. The control values tend upwards again accelerate the arm towards a zero velocity state.

As shown in Fig. 6 random trajectories backwards using the inverse dynamics networks show the different values of acceleration can be induced by equivalent energy expenditure as expected with this type of over actuated system. This

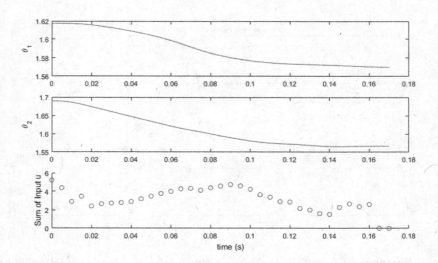

Fig. 5. Responses of Arm 1 angle θ_1, Arm 2 angle θ_2, and the sum of the muscle input energies u, of a nonlinear muscle activated arm with Deep Dynamic Programming control

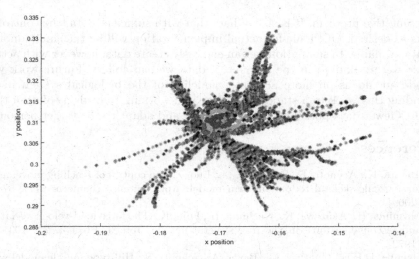

Fig. 6. Random optimal trajectories in different directions from the goal position under approximately equivalent net energy expenditure. The different colors are displayed to visualize the jump from one time-step to the next. Some trajectories that are more efficient in achieving longer distances traveled with same energy consumption. (Color figure online)

figure demonstrates how some trajectories are much more energy efficient then others.

4   Discussion

As here demonstrated DDP can be implemented successfully for controlling non-linear time-varying dynamical systems. DDP approach here advances reinforcement learning strategies, such as Q-Learning, by incorporating partially learned models to expedite training and generalize about state-space unknown to it. This is something Q-Learning cannot easily do, as it creates optimal trajectories only based on full observable state-space.

The DDP method offers an alternative to optimal control methods like iLQR. Due to the fact that DDP is incorporates the ideal features from both model-free and model based learning, it is extremely flexible in terms of the systems it can make controllers for. Where as LQR is extremely stable in linear systems, it becomes unstable in nonlinear systems which forces the controller to sample regions and linearize the dynamics around each region to create state to state controllers. This process can be tedious and computational expensive. In comparison, the DDP may have similar cost up front, but as it learns the amount of exploration and computation diminishes to only forward passes of the network holding the control policy.

We show that in simulation the method is capable of generating control over the simulated arm. While the simulated arm is highly simplified, these is a representative problem. It is shown here that with sufficient data, the controller works as desired. One issue for actual implementation will be the small amount of data available. In simulation we can endlessly create data, however with actual people we are limited in the amount of data we can collect. Future work will include and focus on more accurate modeling of the biological system itself, extending this work to a standing human, and working towards a solution that requires fewer data points to work effectively and adapt to the target person.

References

1. Tahara, K., Arimoto, S., Sekimoto, M., Luo, Z.: On control of reaching movements for musculo-skeletal redundant arm model. Appl. Bionics Biomech. **6**(1), 57–72 (2009)
2. Mashima, H., Akazawa, K., Kushima, H., Fujii, K.: The force-load-velocity relation and viscous-like force in the frog skeletal muscle. Jpn. J. Physiol. **22**, 103–120 (1972)
3. Haeufle, D.F.B., Gunther, M., Bayer, A., Schmitt, S.: Hill-type muscle model with serial damping and eccentric fore-velocity relation. J. Biomech. **47**(6), 1531–1536 (2014)
4. Berkowitz, M., Harvery, C., Greene, C., et al.: The Economic Consequences of Traumatic Spinal Cord Injury. Demos Press, New York (1992)
5. Triolo, R.J., Bogie, K.: Lower extremity applications of functional neuromuscular stimulation after spinal cord injury. Top. SCI Rehabil. **5**(1), 44–65 (1999)
6. Kobetic, R., Triolo, R.J., Marsolais, E.B.: Muscle selection and walking performance of multichannel FES systems for ambulation in paraplegia. IEEE Trans. Rehabil. Eng. **5**(1), 23–29 (1997)
7. Uhlir, J.P., Triolo, R.J., Kobetic, R.: The use of selective electrical stimulation of the quadriceps to improve standing function in paraplegia. IEEE Trans. Rehabil. Eng. **8**(4), 514–522 (2000)
8. Triolo, R.J., Wibowo, M., Uhlir, J., Kobetic, R., Kirsh, R.: Effects of stimulated hip extension moment and position on upper-limb support forces during FES-induced standing-a technical not. J. Rehabil. Res. Dev. **38**(5), 545–555 (2001)
9. Solomonow, M., Reisin, E., Aguilar, E., Baratta, R.V., Best, R., D'Ambrosia, R.: Reciprocating gait orthosis powered with electrical muscle stimulation (RGO II). Part II: medical evaluation of 70 paraplegic patients. Orthopedics **10**(5), 411–418 (1997)
10. Betz, R.R., Rosenfeld, E., Triolo, R.J., Robinson, D.E., Gardner, E.R., Maurer, A.: Bone mineral content in children with spinal cord injury. Poster Presented at American Spinal Cord Injury Association, San Diego, CA, May 1988
11. Marsolais, E.B., Kobetic, R., Polando, G., Ferguson, K., Tashman, S., Gaudioi, R., Nandurkar, S., Lehneis, H.R.: The Case Western Reserve University hybrid gait orthosis. J. Spinal Cord Med. **23**(2), 100–108 (2000)
12. Farris, R., Quintero, H., Goldfarb, M.: Preliminary evaluation of a powered lower limb orthosis to aid waling paraplegic individuals. IEEE Trans. Neural Syst. Rehabil. Eng. **19**(6), 652–659 (2011)
13. Watkins, C.J., Dayan, P.: Q-learning. Mach. Learn. **8**(3–4), 279–292 (1992)
14. Silver, D., et al.: Deterministic policy gradient algorithms. In: Proceedings of the 31st International Conference on Machine Learning (ICML 2014) (2014)
15. Mnih, V., et al.: Playing atari with deep reinforcement learning. arXiv preprint arXiv:1312.5602 (2013)

An Adaptive Modular Recurrent Cerebellum-Inspired Controller

Kiyan Maheri[✉], Alexander Lenz, and Martin J. Pearson

Bristol Robotics Laboratory, Bristol BS16 1QD, UK
km1768@bristol.ac.uk, {Alex.Lenz,martin.pearson}@brl.ac.uk

Abstract. Animals and robots face the common challenge of interacting with an unstructured environment. While animals excel and thrive in such environments, modern robotics struggles to effectively execute simple tasks. To help improve performance in the face of frequent changes in the mapping between action and outcome (change in context) we propose the Modular-RDC controller, a bio-inspired controller based on the Recurrent Decorrelation Control (RDC) architecture. The proposed controller consists of multiple modules, each containing a forward and inverse model pair. The combined output of all inverse models is used to control the plant, with the contribution of each inverse model determined by a responsibility factor. The controller is able to correctly identify the best module for the current context, enabling a significant reduction of 70.9% in control error for a context-switching plant. It is also shown that the controller results in a degree of generalization in control.

Keywords: Cerebellum · Adaptive control · Bio-inspired · Modular control · Adaptive filter · Context switching

1   Introduction

It can be observed in every day life that animals are accomplished at controlling their motor system when interacting with the environment. Traditional control methods have so far failed to deliver the same levels of competence for robots.

In an agent's interaction with the environment the mapping between actions and sensory feedback changes frequently. This mapping includes the dynamics of the agent's body, the environment and sensory processing by the agent. Changes in any of the three elements, perhaps resulting from picking up a new object or relying on a different sense for feedback, results in a change in the overall mapping. The mapping between actions and sensory feedback is represented in a single element known as the plant. This plant is able to switch between numerous different dynamics, herein referred to as contexts. This challenging control scenario is further compounded by the fact that new mapping is difficult to predict *a priori*. In each case animals show the ability to learn, retain and recall these mappings for control in a wide range of situations.

© Springer International Publishing AG 2017
M. Mangan et al. (Eds.): Living Machines 2017, LNAI 10384, pp. 267–278, 2017.
DOI: 10.1007/978-3-319-63537-8_23

In this work we propose a controller (Modular-RDC) that is able to learn the control of multiple contexts and recall that control when needed. This architecture is based on the adaptive feedforward Recurrent Decorrelation Controller (RDC), developing a distributed control architecture around it. We demonstrate the benefits of this approach on a context-switching simple linear plant.

In mammalian Motor Control Systems (MCS) the sensory information is delayed and often very noisy [1]. This makes feedback control an unlikely mechanism for rapid motor control. However, delayed feedback can still be used for training adaptive controllers.

It has been suggested that the MCS uses multiple adaptive control modules for storing motor skills, and is able to identify and utilize the best modules for a given context [2]. The ability to correctly identify and switch between pre-learned modules helps to overcome sudden changes in context. This is in contrast to a single controller which has to relearn control every time the agent changes context. Here this theory is referred to as distributed control.

There is strong evidence to suggest that the Cerebellum is a likely candidate for the location of these parallel modules [3,4]. The Cerebellum does not generate the motor commands itself, but instead is engaged in fine-tuning movements [5]. The cerebellum has a well-characterized and regular structure, consisting of a large number of repeating sub-units known as microzones. It has been suggested that there exists a cerebellar function carried out by each microzone. In 1982 Fujita, working with the Marr-Albus model of the cerebellum, proposed that each microzone acts as an adaptive filter [6]. Adaptive filters are a powerful and versatile tool frequently used in signal processing.

Porrill et al. [7] proposed the Recurrent Decorrelation Control (RDC) model as a biologically plausible adaptive single-controller architecture (Fig. 1). In this model the Cerebellar model takes in an efferent copy of the motor command and uses its output to modify the reference signal. The cerebellar model runs counter-current to a fixed feedforward controller, thus creating a recurrent loop between the two elements. The cerebellar model emulates a single microzone, which is modeled as an adaptive filter in accordance to the Fujita model. This controller has been used to emulate the Vestibulo-ocular reflex (VOR) [8] and control artificial muscle [9]. The controller is able to learn the control of a plant context. However, if the plant context changes the controller adapts and loses the previously learned control.

An important question in distributed control is how to assign responsibility among modules. Narendra and Balakrishnan [11] were the first to use multiple control modules to improve transient responses. In their architecture the plant is controlled by a set of fixed inverse models, each being coupled to a forward model. The inverse model paired with the best performing forward model is selected to drive the plant. Wolpert et al. proposed the MOSAIC architecture as a biologically plausible model for motor control. [2,12] MOSAIC also makes use of forward models to select controllers. However, unlike Narendra's controller, MOSAIC combines the module outputs in a weighted sum, thus allowing for the control and learning of contexts to be shared among multiple controllers.

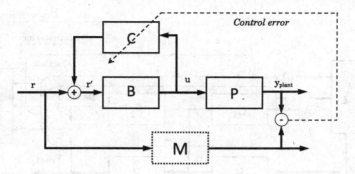

Fig. 1. The Recurrent Decorrelation Control (RDC) architecture used by Lenz et al. [10] to model the Vestibulo-Ocular Reflex. B and C form a recurrent loop, where B is a linear model of the brainstem, and C is a model of the cerebellum realized as a single adaptive filter. P is the plant to be controlled and M is a reference model specifying the desired behavior of the plant.

2   Methods

2.1   Modular RDC Architecture

The Modular-RDC architecture builds on the Recurrent Decorrelation Controller. Here the single cerebellar model is replaced with a set of n parallel modules, each containing a forward and inverse model (Fig. 2). Both inverse and forward models are adaptive and take in an efferent copy of motor command. The forward models (F_i) predict the plant state, while the inverse models (C_i) control the plant by modifying the incoming reference signal. In actual fact C_i does not approximate the inverse of the plant, but instead the function $(B^{-1} - P)M$. However, for simplicity C_i is referred to as an inverse model, while the Modular-RDC is referred to as the controller.

The benefit of using multiple modules is that previously learned control can be stored without being overwritten. However, this raises the module selection problem, *i.e.*, how to combine the inverse model outputs to control the plant.

The inverse model outputs are combined in a weighted sum, which is then added to the reference signal. The weights are referred to as the module's responsibility factor λ_i. The responsibility factors are updated at every time-step, and are derived from the forward models' prediction accuracies (see later). It is assumed that modules with good plant prediction have a good controller for the current context. Therefore, the module selection problem can be recast as a plant identification problem.

The feedforward controller (B) presents an opportunity to include any known information about the system, and would usually be designed to control a static approximation of the plant. Here we do not include any information, leaving it as a simple two step delay (z^{-2}). This is done to put all of the control responsibility on the modules, and to highlight its performance in learning good control for a context-switching plant.

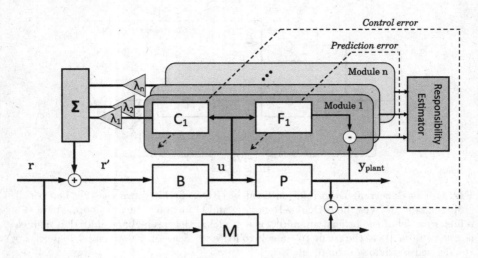

Fig. 2. The proposed Modular-RDC architecture. B is a fixed linear feedforward controller, P is the context-switching plant and M is a reference model with the dynamics desired for the controlled plant. Forward (F_i) and Inverse (C_i) model pairs, together with their responsibility factor constitute a module

The plant to be controlled is of the form:

$$P(s) = \frac{s}{s + a} \tag{1}$$

This simple first-order linear plant was used in Lenz et al. [8] as a model of the oculomotor plant. Different contexts are created by changing the value for parameter a. Most of the results were obtained using three contexts, labeled as A ($a = 10$), B ($a = 40$) and C ($a = 25$). The reference model is defined as:

$$M(s) = \frac{100}{s + 100} \tag{2}$$

Forward and Inverse Models. Forward and inverse models are implemented using adaptive filters, in accordance with the Fujita model of the microzone (Fig. 3). We follow the approach in Wilson et al. [9] and expand the incoming signals through a parallel set of alpha functions G_j. The alpha functions are defined in Eq. 3. A large number (50) of alpha functions were used linearly spaced in the range 0.5–200 ms.

$$G_j(s) = \frac{1}{(T_j s + 1)^2} \tag{3}$$

Where T_j is the time constant of the jth alpha function.

The forward and inverse models are trained with the gradient descent rule using prediction errors and the control error respectively (Eqs. 4 and 5). In mammalian brains errors become available after a delay of 150–250 ms [1]. Although

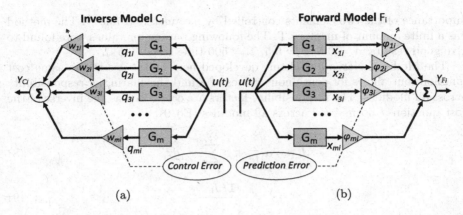

Fig. 3. Adaptive Linear Filter formulations of (a) Inverse and (b) Forward Models

simple to implement, for computational simplicity no feedback delays were included. The learning rate of each module is controlled by a global learning rate $\eta = 0.01$ and the module's responsibility factor λ_i. Linking the module's learning rate to its responsibility factor ensures that modules that are highly involved in controlling the plant improve, while those with little engagement remain unchanged.

$$\Delta w_{ij} = -\eta \lambda_i \, q_j \, (Y_M - Y_P) \tag{4}$$

$$\Delta \phi_{ij} = -\eta \lambda_i \, x_j \, (Y_{F_i} - Y_P) \tag{5}$$

Module Selection. As previously stated the module selection problem has been transformed in to the task of plant identification. A responsibility factor can be viewed as the probability that each forward model represents the dynamics of the current plant context. As a set of probabilities, the responsibility factors have the following constraints:

$$1 \geq \lambda_i \geq 0 \tag{6}$$

$$\sum_{i=0}^{n} \lambda_i = 1 \tag{7}$$

In Sect. 3 a new method (**Modified Narendra** method) is compared to three alternative schemes, namely **hard-max** selection, **soft-max** selection used by Wolpert in the MOSAIC architecture [3], and the **original Narendra** method used by Narendra and Balakrishnan [11].

The Original Narendra method selects the controller according to a cost function (Eq. 8). The module with the lowest cost function controls the plant. The cost function considers both current and previous errors ϵ_t, with the relative importance of the two elements controlled by parameters α and β. The method has a finite amount of memory T. The following parameter values were found to give good results: $\alpha = 0.5$, $\beta = 0.5$, T = 500 time-steps (0.5 s).

The Modified Narendra scheme developed here makes use of the same cost function, but allows for continuous values of λ and the sharing of responsibility across all modules. The responsibility factors are calculated as the inverse of the cost function J normalized across all modules (Eq. 9).

$$J_{t,i} = \alpha \epsilon_{t,i}^2 + \beta \sum_{t-T}^{t} \epsilon_{\tau,i}^2 d\tau \tag{8}$$

$$\lambda_i = \frac{1/J_i}{\sum_{k=0}^{n} 1/J_k} \tag{9}$$

The hard-max selection method selects one module with the smallest current prediction error. This method only takes the error of the current time-step into consideration, so is likely to be highly susceptible to noise.

In the MOSAIC architecture, the responsibility factors are calculated using a soft-max function over the current prediction errors (Eq. 10) Like the hard-max scheme it can be highly susceptible to disturbance and noise. The manually chosen parameter σ is analogous to the temperature parameter, and determines the level of co-operation and competition between the modules. As $\sigma \to \infty$ all modules share equal responsibility, while $\sigma \to 0$ ensures that the best module receives a responsibility approaching 1. In this work good results were achieved using $\sigma = 0.01$.

$$\lambda_i = \frac{e^{-\frac{|\epsilon_i|^2}{\sigma^2}}}{\sum_{k=1}^{n} e^{-\frac{|\epsilon_k|^2}{\sigma^2}}} \tag{10}$$

2.2 Simulation Set-Up

The simulation was implemented in C++ and was run with a time step of 1 ms. The reference signal consists of colored noise (0.5–2.0 Hz) with a maximum amplitude of 1.0. Alpha functions, filters, plant contexts and the reference model were implemented using IIR filters. In this work the number of modules is determined off-line, and equals or exceeds the number of plant contexts in a simulation. In future works it is hoped that the number of modules will grow dynamically.

3 Results

3.1 Reduced Transient Errors

The Modular-RDC with Modified Narendra switching was compared to the original RDC controller in a simple scenario with changing contexts. This experiment was developed to test both controllers' abilities to handle context changes, such as those arising from contact with objects or task switching.

In a training phase lasting 40 s, the controllers learned the control for contexts A and B (20 s each). For the Modular-RDC controller, the learning was split between two modules, each one specializing in one of the two contexts. The test phase consisted of 20 s of controlling context A followed by a further 20 s of simulation controlling context B. Both controllers continued to learn throughout the control phase.

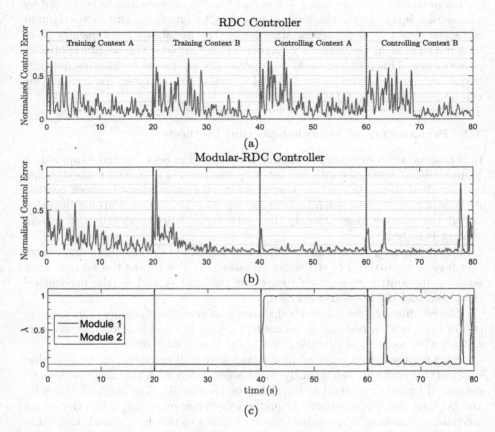

(a)

(b)

(c)

Fig. 4. RDC and Modular-RDC performance during training (A→B) and control (A→B) phases (a) RDC normalized RMS control error (b) Modular-RDC normalized RMS control error (c) Evolution of responsibility factors λ_1 and λ_2

Figure 4c shows the responsibility factors of the Modular-RDC controller throughout the simulation. It can be seen that the proposed controller is able to correctly identify a change in context and select the appropriate module.[1]

---

[1] During the training phase, the responsibility factors are manually set and not determined by the switching mechanism.

Figure 4a shows that the RDC controller error increases after each context change, then decays as the controller re-learns the control. In Fig. 4b the Modular-RDC shows the same behavior during training, *i.e.* when first encountering a novel context. However, in future encounters the previously learned control is recalled and there is a significantly smaller increase in error. Peak errors resulting from context switching were reduced by 56% and 78% during the test phase, while the total error in this phase saw a reduction of 70.9%. The training phase also saw a 16.3% reduction in average error.

Large errors can be seen at $t = 63$ s and $t = 77$ s. These seem to be caused by a numerical instability in the forward model alpha functions, and so are thought to be the result of an implementation issue. The instability consequently leads to a fluctuation in responsibility factors which degrades the overall controller performance. This problem could be superficially addressed by filtering out high frequency signals in the forward model outputs. An improved implementation will be investigated in future work.

3.2   Performance of Module Selection Methods

In this section the switching mechanisms described in Sect. 2.1 are compared. A Modular-RDC controller with two modules was trained to control context A and context B as above. During the test phase, the plant switched between context A, B, and then C, with each context lasting 20 s. No further learning occurred during the training phase, as only the controller's ability to identify the current context is tested.

Figure 5a shows the average control error for each of the switching mechanism over 5 runs, normalized to the worst-performer. It also breaks the average error down to its contributions from controlling contexts A and B (the pre-learned contexts), and the novel context C.

The Modified Narendra method shows significant improvements in the control of both pre-learned and novel contexts. This method is also relatively consistent, with a standard deviation of 8.5% of the average error.

Hard-max selection resulted in the largest overall error. It can be seen that this method performs significantly worse when faced with the novel context C, compared to either of the pre-learned ones (A and B). The original Narendra and Wolpert mechanisms perform moderately better on average than the switch mechanism. Looking at the different contributions to this error reveals that these controllers have a modest improvement in controlling contexts A and B, but are no better in controlling context C.

3.3   Generalization

The improvements seen in the control of context C, as exhibited in the modified Narendra scheme, suggest that combining controller outputs can lead to generalization of control. Generalization wasn't seen with methods where responsibility is either binary or fluctuates rapidly between modules. To test this hypothesis

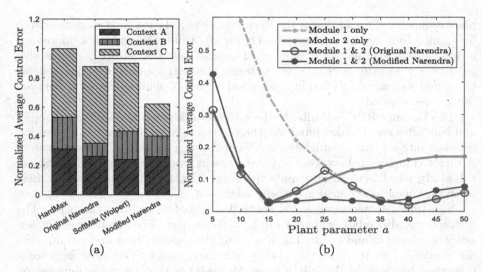

Fig. 5. (a) Normalized average control errors for four module selection methods, showing the relative contributions from each context (b) Normalized control error When Using (i) Module 1, (ii) Module 2, Module 1 & 2 with two different module selection methods, showing generalization in the Modified Narendra method

the original and modified Narendra mechanisms were compared as representatives for the two types of behavior they exhibit.

Figure 5b shows the average control error against plant parameter a for four set-ups of the Modular-RDC controller, normalized to the highest error. The first two plots show the average error when using only one of the two modules. This illustrates that as expected error grows as the difference between the controlled context and the training context increases. Also plotted is the performance of both modules with the *Original Nerendra* and *modified Narendra* selection schemes.

Combining the two modules with modified Narendra switching results in a significant improvement in control for contexts with $10 < a < 40$. At $a = 25$, i.e. mid-way between contexts A and B the average error was improved by 70.1%. This shows that the Modular-RDC can generalize between pre-learned controllers in parameter space. However, results for $a < 10$ and $a > 40$ suggest that this generalization does not extend outside of this range. It can be seen that the *original Narendra* results in no generalization, and its control is only as good as the better of the two controllers. It must be noted however that this is still a significant improvement on using a single controller.

4   Discussion

In this work we have applied the simple yet effective concept of distributed control to the RDC architecture. The resulting control architecture, Modular-RDC has been shown to reduce transient errors resulting from a sudden change in plant dynamics. This allows for the Recurrent Decorrelation Control paradigm to be applied to a new class of system, i.e. rapidly time-varying plants. Moreover, it was shown that any of a number of module selection strategies were able to correctly identify a change in context. The best module selection scheme (Modified Narendra) resulted in generalized control of plant dynamics in between pre-learned contexts.

The Modular-RDC is similar to the architectures developed by Wolpert et al. and Narendra and Balakrishnan. All three systems convert the module selection problem into a plant identification problem through the use of forward models. A direct comparison between the three systems would be outside the scope of this study, which is to improve and expand the RDC architecture. However, here some prominent differences are briefly addressed.

Narendra's controller uses one controller at a time to drive the plant. The architecture also relies primarily on fixed controllers. The combination of few adaptive elements and no sharing of responsibility means that many controllers are needed so as to cover the plant's parameter space. The most significant difference between Modular-RDC and MOSAIC is that they are founded on contrary bio-inspired control architectures, RDC and Feedback Error Learning (FEL) respectively. These two controllers differ fundamentally in that RDC is a feedforward controller, while FEL uses feedback. Also, they present different solutions to the distal teacher problem for adaptive controllers. The RDC controller allows for faster feedforward control that is robust to large feedback delays.

In this work it was shown that combining the outputs of inverse models leads to generalization of control to some novel contexts. It suggests that a small number of modules can effectively control many contexts. Furthermore, clinical evidence suggests that generalization can be extended beyond the range seen in Sect. 3.3 [13]. Generalized control could result in greater computational efficiency and good control of novel contexts.

Modules unrelated to the current context have small but non-zero responsibility factors, and so continue to be adapted at a slow rate. In the long-term, this could lead to controllers drifting away from their correct values, thus degrading the controller's effectiveness. This becomes particularly problematic when the agent experiences some contexts for longer than others. This issue could be addressed by limiting learning to only a few modules, perhaps by applying a threshold to learning rates, or only training the module with the highest responsibility. The mechanisms by which the brain controls learning of models is an unresolved and interesting research question.

This work has demonstrated significantly improved control for a simple single-DoF linear system. In future studies we intend to look at the control of more complex nonlinear and multi-DoF systems with sensory delays implemented. It is also hoped that the Modular-RDC controller can act as a biologically plausible model of cerebellar control.

The Modular-RDC formulated here presents opportunities for improvement. For example, re-evaluating the module selection problem as a dynamic unsupervised/reinforcement learning problem can provide more scientific rigor and should improve module selection. Also, allowing the controller to add modules during operation would allow for the system to adapt on-line to meet real-world challenges. Finally, methods for ascertain an optimal number and range of α functions could be developed.

5   Conclusion

This work has created a distributed controller around the adaptive Recurrent Decorrelation Control (RDC) architecture to improve control of a context-switching plant. The result is the Modular-RDC controller. During simulations with a context switching 1st order plant, the Modular-RDC controller was shown to result in a 70.9% reduction in control error during the control phase. The controller achieves this by correctly identifying a change in dynamics and selecting the appropriate module to control the plant.

The Modular-RDC controller has a degree of generalized control, as it was able to effectively control a novel context with plant parameter a between two pre-learned contexts. The combined effort of both modules resulted in a significantly lower error than either module could achieve on its own. This generalization was shown to be limited to contexts with plant parameter a within the range of pre-learned contexts, and did not extend beyond this range under the current module selection scheme.

References

1. Kawato, M.: Internal models for motor control and trajectory planning. Curr. Opin. Neurobiol. **9**(6), 718–727 (1999)
2. Wolpert, D.M., Kawato, M.: Multiple paired forward and inverse models for motor control. Neural Networks **11**(7), 1317–1329 (1998). doi:10.1016/S0893-6080(98)00066-5
3. Wolpert, D.M., Chris Miall, R., Kawato, M.: Internal models in the cerebellum. Trends Cogn. Sci. **2**(9), 338–347 (1998)
4. Imamizu, H., Kuroda, T., Miyauchi, S., Yoshioka, T., Kawato, M.: Modular organization of internal models of tools in the human cerebellum. Proc. Nat. Acad. Sci. **100**(9), 5461–5466 (2003)
5. Ito, M.: Historical review of the significance of the cerebellum and the role of purkinje cells in motor learning. Ann. N. Y. Acad. Sci. **978**(1), 273–288 (2002)
6. Fujita, M.: Adaptive filter model of the cerebellum. Biol. Cybern. **45**(3), 195–206 (1982)
7. Porrill, J., Dean, P., Stone, J.V.: Recurrent cerebellar architecture solves the motor-error problem. Proc. Roy. Soc. London-B, **271**(1541), 789–796 (2004)
8. Lenz, A., Balakrishnan, T., Pipe, A.G., Melhuish, C.: An adaptive gaze stabilization controller inspired by the vestibulo-ocular reflex. Bioinspiration Biomimetics **3**(3), 035001 (2008)
9. Wilson, E.D., Assaf, T., Pearson, M.J., Rossiter, J.M., Anderson, S.R., Porrill, J.: Bioinspired adaptive control for artificial muscles. In: Lepora, N.F., Mura, A., Krapp, H.G., Verschure, P.F.M.J., Prescott, T.J. (eds.) Living Machines 2013. LNCS, vol. 8064, pp. 311–322. Springer, Heidelberg (2013). doi:10.1007/978-3-642-39802-5_27

10. Lenz, A., Anderson, S.R., Pipe, A.G., Melhuish, C., Dean, P., Porrill, J.: Cerebellar-inspired adaptive control of a robot eye actuated by pneumatic artificial muscles. IEEE Trans. Syst. Man Cybern. Part B (Cybernetics) **39**(6), 1420–1433 (2009)
11. Narendra, K.S., Balakrishnan, J.: Improving transient response of adaptive control systems using multiple models and switching. IEEE Trans. Autom. Control **39**(9), 1861–1866 (1994)
12. Haruno, M., Wolpert, D.M., Kawato, M.: Mosaic model for sensorimotor learning and control. Neural Comput. **13**(10), 2201–2220 (2001)
13. Davidson, P.R., Wolpert, D.M.: Internal models underlying grasp can be additively combined. Exp. Brain Res. **155**(3), 334–340 (2004)

Stimulus Control for Semi-autonomous Computer Canine-Training

John J. Majikes[1(✉)], Sherrie Yuschak[2], Katherine Walker[3], Rita Brugarolas[3],
Sean Mealin[1], Marc Foster[3], Alper Bozkurt[3], Barbara Sherman[2],
and David L. Roberts[1]

[1] Department of Computer Science, North Carolina State University,
Raleigh, NC, USA
{jjmajike,spmealin}@ncsu.edu, robertsd@csc.ncsu.edu
[2] Department of Clinical Sciences, North Carolina State University,
Raleigh, NC, USA
{slyuscha,barbara_sherman}@ncsu.edu
[3] Department of Electrical and Computer Engineering,
North Carolina State University, Raleigh, NC, USA
{kdwalker,rbrugar,mdfoster,aybozkur}@ncsu.edu

Abstract. For thousands of years, humans have domesticated and
trained dogs to perform tasks for them. Humans have developed areas
of study, such as Applied Behavior Analysis, which aim to improve the
training process. We introduce a semi-autonomous, canine-training system by combining existing research in Applied Behavior Analysis with
computer systems consisting of hardware, software, audio, and visual
components. These components comprise a biohybrid system capable of
autonomously training a dog to perform a specific behavior on command.
In this paper we further our previous computer canine-training system
by the application of stimulus control over a newly-acquired, free operant
behavior. This system uses light and sound as a discriminative stimulus
for the behavior of a dog pushing a button with its nose. Indications of
simple stimulus control of this behavior were achieved. Our pilot of this
system indicates canine learning comparable to that from a professional
dog trainer.

1 Introduction

For thousands of years, dogs have been domesticated and trained to perform
tasks for humans [16]. Training using positive reinforcement, giving the animal
something desirable such as food, has been shown by research in Applied Behavior Analysis to be the most effective technique for training canines [7]. Training using positive reinforcement requires achieving effective timing, consistency,
and rate of reinforcement criteria [1] which humans, especially novice trainers,
struggle to realize but computers can be made to excel at. A computer system
in our previous research combined a smart harness with Inertial Measurement
Units (IMUs), along with computer software that employed timely, accurate, and

© Springer International Publishing AG 2017
M. Mangan et al. (Eds.): Living Machines 2017, LNAI 10384, pp. 279–290, 2017.
DOI: 10.1007/978-3-319-63537-8_24

repeatable positive reinforcement of canine posture [9]. Adding additional sensor data along with a light and sound cue to the system allows us to further test training techniques and provides a biohybrid system that can initiate placing a behavior under stimulus control.

Placing a behavior under stimulus control requires an animal to learn that reinforcement is only given for the behavior after a cue. The biohybrid system has to mimic the natural way canines learn when a behavior will be reinforced and that some behaviors are desired only at certain times. For example, one reason for a dog barking is as an alert for possible dangers but humans generally discourage dogs barking when friends or family members ring the doorbell. By providing positive reinforcement such as praise, a dog learns to discriminate that the barking alert will be reinforced when the doorbell rings at night.

This paper will show the results of a pilot of a biohybrid system that uses positive reinforcement for a dog's newly acquired free operant behavior of pushing a button with it's nose. The system used a light and sound cue to achieve stimulus control over this behavior. True stimulus control implies not only that the behavior is offered when the cue is given, but also that the behavior is not offered in the absence of the cue. Due to the time required to train dogs to acquire the new behavior, this pilot will only focus on training the dogs to provide the behavior when the cue is given.

2   Related Work

Wearable IMUs have been used in a variety of ways to detect animal activities [4,8,12,17] and classify postures [15]. Post-processing classification that identified both canine postures and behaviors has also been done [6]. Additionally, previous research used machine learning algorithms based on a two-stage cascade classifier to accurately identify both postures and behaviors in near real time [2,3]. Our previous research combined these to provide a semi-autonomous computer-assisted system that not only identified postures but trained dogs to provide a natural behavior (sit) for a food reward [9].

Progressing beyond identification of postures and behaviors, automated training of a discriminative stimulus over a natural behavior has been done previously, most famously by B.F. Skinner [11]. This training was done using small animals in a very confined and controlled environment called a *"Skinner box"* to understand the application of operant conditioning techniques on natural behaviors such as pigeons pecking. Our biohybrid system extends the previous work by applying the discriminative stimulus techniques to a novel dog behavior, pushing a button with its nose. We then compare the biohybrid system's ability to train the dog to that of a professional trainer.

3   Hardware, System, and Safety

To determine the dogs' posture, position, and movement during the pilot, the dogs are fitted with a smart harness. The smart harness contains IMUs, magnetometers, a Beagle Bone Black computer (BBB), and 5 V battery. The smart harness is more fully described in our previous work [9] For this study, magnetometer sensor data from the smart harness is used as a compass[1] to ascertain when the dog is facing the wall with the button, light, and treat dispenser. Similar to a clicker used during training, the professional trainer uses a Wii Remote[TM] Plus controller[2] and a Kensington® laser pointer[3] (with the laser disabled) to communicate training signals such as a *mark* to the base system and to dispense treats. All data is collected and all equipment controlled by a Lenovo[TM] Thinkpad W530 with an Intel Core i7-374QM 2.7 GHz processor, 8 GB RAM, running 64-bit Windows 10. All sessions are recorded with at least one video camera.

a: 1.8 m by 1.2 m wall   b: Light on and Dog₁ pressing button

Fig. 1. Wall with light, button, and dispenser

The wall with button, light, and dispenser used is shown in Fig. 1a and is at one end of a 1.8 m by 1.2 m rectangular exercise pen (Xpen[4]). The Xpen ensures the dog's safety while off leash as shown in Fig. 1b. The wall frame is constructed with $2'' \times 4''$ studs, while the wall panels are constructed of $3/4''$ plywood, laminated for ease of cleaning, and held together with screws and $1/2''$ bolts. The intent is to ensure that a 30 kg dog jumping up on the wall would be safe and that the professional trainer stationed behind the wall is safe.

The light cue is 32 by 32 matrix LED[5] emitting a blue light in an X-shape accompanied by a 1000 Hz tone sound cue from a speaker[6] located behind the

[1] https://learn.adafruit.com/lsm303-accelerometer-slash-compass-breakout.
[2] Wii Remote Plus by Nintendo http://www.nintendo.com/wiiu/accessories.
[3] Kinsington Remote Red Laser Presenter https://www.kensington.com/us/us/4492/k33374usa/presenter-remote-red-laser-presenter.
[4] Petmate Exercise Pen http://www.petmate.com/product/Petmate-Exercise-Pen-W/Door/55011.
[5] LED Matrix 190.5 mm × 190.5 mm https://www.adafruit.com/products/1484.
[6] CLSPK23053 Wireless Bluetooth Speaker https://www.amazon.com/Connectland-Wireless-Bluetooth-Multi-Function-CLSPK23053/dp/B00NJF1HO6.

wall. The wooden button is covered with an electronic pressure sensor[7] and is located just below the light in the center of the wall. The front of the wall also contains five HC-SR04[8] ultrasonic distance sensors. A Pet Tutor®[9] automatic treat dispenser tray extends through a hole in the bottom center of the wall. All remaining electronics, BBBs, Arduino, power supplies, and Wi-Fi antennas, are protected behind the wall.

The stimulus control pilot is under the observation of a veterinary technician specialized in animal behavior who is credentialed as a Certified Professional Dog Trainer-Knowledge Assessed (CPDT-KA), a Registered Veterinary Technician (RVT), a Veterinary Technician Specialist (VTS) in behavior and has more than 15 years of dog training experience. For the safety and welfare of the dog and humans, the VTS can, for any reason, terminate any session or the further use of any dog. Examples of reasons for session termination are dog frustration, anxiety, boredom, or computer failures. Examples of reasons to terminate use of a dog are aggression, fear, anxiety, or lack of focus. All experimental procedures were approved by the Institutional Animal Care and Use Committee of North Carolina State University (NC State). The professional trainer used only positive reinforcement (treats and praise) as reinforcement in all training to get the dogs to acquire a novel behavior of pushing a button with its nose.

Normally dogs for experiments would be recruited from the community around NC State, but given the number of sessions and amount of time needed to acquire the button-push behavior we recruited dogs from an animal sanctuary, Paws4Ever[10]. Selecting dogs from a shelter limited the randomness and number of dogs, but it facilitated access to dogs. The dogs used in the pilot are documented in Table 1.

Table 1. Dog demographics

Dog	Sex	Age at the time of the pilot	Breed	Reported weight
Dog_1	Female	5 years	American Terrier mix	23 kg
Dog_2	Female	3 years	Terrier mix	17 kg
Dog_3	Female	2 years	Plott Hound	24 kg
Dog_4	Female	1 year 6 months	Beagle mix	12 kg
Dog_5	Male	1 year 10 months	Collie mix	18 kg
Dog_6	Female	8 months	Hound mix	18 kg
Dog_7	Female	3 years	Rhodesian Ridgeback mix	22 kg

---

[7] 88 mm × 43.7 mm   Square   Force-Sensitive   Resistor   https://www.adafruit.com/products/1075.

[8] https://www.fasttech.com/product/1012007-arduino-compatible-hc-sr04-ultrasonic-sonar.

[9] Smart Animal Training Systems' Pet Tutor https://smartanimaltraining.com/.

[10] Paws4Ever animal shelter and sanctuary. http://paws4ever.org/about.

4   Protocol

A novel behavior such as a dog pushing a button on a wall with its nose was chosen because it's easy for all types of dogs to perform, is easy to train, and lessens the likelihood that any dog participating in the pilot would have any previous experience with the behavior. Before the stimulus control pilot can begin, the professional manually trains each dog to push the button on the wall with its nose. Training sessions are up to 12 min in length with a maximum of two sessions in any hour. During training, dogs are assumed to have acquired the free operant behavior when they push the button ten times in a minute. The dogs are then randomly placed into one of two groups. One group has the behavior placed under stimulus control by the professional. The second group has the behavior placed under stimulus control by the computer system.

Given the amount of time we had with each dog and the amount of time required to acquire the behavior, for the stimulus control pilot we compared and calculated effectiveness solely on the training of the Three-Term Contingencies (TTCs) of cue, followed by behavior, and then reinforcement. True stimulus control also means that the button push behavior is not offered when the cue is not given. The pilot only compared the discrimination training to provide the behavior when the cue was given. The comparison of the effectiveness of the professional to the computer system is (1) the number of TTCs which are a correct sequence of discriminative stimulus followed by correct behavior and then primary reinforcement, and (2) the response latency from when the discriminative stimulus is offered until the correct behavior is given.

4.1   Blocking

In order to compare the training progression of the professional and the computer system, each discrimination training session is broken up into one minute blocks. After each block the delay before the light and sound cue is given can be changed to a level shown in Table 2. The goal for both trainer and system is to train the dog toward the desired stimulus control goal, level 4, of standing or sitting attentively 0.3 m from the wall waiting attentively for up to 5 s for the discriminative cue and then offering the response behavior. Within 5 s of when the cue is offered, the dog must offer the response behavior of pressing and holding the button for at least 0.5 s.

Table 2. Random cue delay ranges for different levels

Shaping level	Minimum delay	Maximum delay
0	0 s	0 s
1	0 s	1 s
2	1 s	2 s
3	2 s	3 s
4	3 s	5 s

4.2   Shaping

Shaping calculations will be done at the end of each minute block during the session. Table 3 describes the success criteria for the blocks. The success criteria for a one minute block is at least 90% completed TTCs with a count of at least 10 TTCs or two successive one minute blocks with at least 80% completed TTCs with a count of at least 10 TTCs. A one minute block is unsuccessful if it has fewer than 50% completed TTCs or a count fewer than 5 TTCs and two successive one minute blocks are unsuccessful if it has fewer than 80% completed TTCs or a count fewer than 10 TTCs. A one minute block in neither successful nor unsuccessful if it has at least 50% completed TTCs and a count of at least 5 TTCs. Successful blocks result in progressive shaping toward the stimulus control goal while unsuccessful blocks result in regressive shaping. Using the criteria in Table 3, after a successful block of TTCs, the delay before the cue is given will progress toward the goal of 5 s.

Table 3. Block shaping criteria.

Blocks of one minute	% Completed TCCs		Completed TCCs	Determination
Single minute block	$\geq$90%	and	$\geq$10	Successful
Single minute block	>50%	and	>5	Neutral
Single minute block	$\leq$50%	or	$\leq$5	Unsuccessful
Two minute blocks	$\geq$80%	and	$\geq$10	Successful
Two minute blocks	>50 %	or	>5	Neutral
Two minute blocks	$\leq$80%	or	$\leq$10	Unsuccessful

4.3 Professional and Computer Protocol

These shaping criteria are the intermediate steps agreed to by the professional and programmed into the computer system that both will use to put the behavior under stimulus control.

Given the canine-human bond and to keep the dog blinded as to who's in control, the professional will be present for both training groups. The professional will use the Wii remote controller or laser pointer to mark when the session starts and stops. The system will *rumble* at one minute intervals to provide timing information to the professional. For the dogs trained by the professional, she will use either the Wii remote control or the laser pointer (whichever she preferred) to issue the cue or reinforcement, and when to progress or retard the shaping.

5   Results

Of the 7 dogs manually trained to acquire the behavior, only Dog_1, Dog_2, and Dog_6 successfully pressed the button 10 times in a single minute. Dog_1 and Dog_2 were randomly chosen to be trained by the computer system. Dog_6 was randomly chosen to be trained by the professional. Dog_3 and Dog_4 arrived at Paws4Ever half way through the pilot and there wasn't enough time to train them to acquire the behavior. Although Dog_5 had prior target training he never met the criteria of pushing the button ten times in a minute. Dog_7 was nervous and did not want to enter the Xpen enclosure to receive treats from the Pet Tutor dispenser.

In addition to the small number of dogs acquiring the behavior, there was an equipment problem that modified the protocol. During the acquire the behavior training sessions diagnostics showed an intermittent problem where the ultrasonic sensor array would seemingly provide random distance values. Five sensors were equally spaced on the wall to ensure that the dog was directly in front of at least one sensor. With multiple ultrasonic sensors, at least one of the echos would be received but most of them would never return. For these sensors with no echo returned the program hangs for 1 s [13]. Secondly, the ultrasonic sensors rely on a perpendicular surface to echo the sound from the transducer to the receiver [14]. With the dogs' soft, rounded bodies at varying angles in front of some, but not all sensors, the echo was not reliably received.

While acquiring the behavior, it became clear that the distance sensors were not needed for the computer to train the dogs. The dogs were always very excited to leave their kennels, come to the experiment room, and get human attention. With the exception of Dog_7, when the dogs arrived they pushed the Xpen gate open and immediately went to the wall that dispensed treats. The dogs never moved more than a foot away from the wall during the sessions. Therefore, the protocol was changed to not require the computer to compute the dog's distance from the wall and the problem never really impacted the pilot.

5.1   Maintaining Rate of Reinforcement with Shaping Protocol

As shown in Table 3, the protocol defines success as completing 10 TTCs with 90% efficiency in one minute or completing 10 TTCs with 80% efficiency in two minutes. This implies that the dog must complete the behavior in at most six to twelve seconds if 100% efficient and even faster if there are incomplete TTCs. As shaping increases and the random delay to activate the cue goes up to three to five seconds, then the dog must complete the behavior in at most two to nine seconds if 100% efficient. For a dog to learn and progress, it is necessary to maintain a high rate of reinforcement as the shaping levels increase. Put another way, as a dog learns and the experiment's shaping level increases, the response latency must be maintained or decrease in order to complete the same number of TTCs in the elapsed time defined in the criteria.

Four out of a total of eight stimulus control sessions changed shaping levels. Two were controlled by the computer and two were controlled by the professional. The four sessions are for Dog_1 Session 1, Dog_2 Session 2, Dog_6 Session 2, and Dog_6 Session 4. The first two sessions were controlled by the computer and the last two were controlled by the professional.

Table 4 shows the average response latency for each of the sessions that had multiple shaping levels; Dog_1 Session 1 is the only session that completed four shaping levels. The table shows the average response latency decreasing before the dog could progress to the next shaping level. Dog_6 Session 4 does not show decreasing average response latency from shape 0 to shape 1 which may be the reason that the dog could not progress to shaping level 2.

Table 4. Average response latency per shaping level

Dog and Session	Average response latency			
	Shape 0	Shape 1	Shape 2	Shape 3
Dog_1 Session 1	2.9 s	2.3 s	1.4 s	1.0 s
Dog_2 Session 2	2.3 s	2.2 s	1.6 s	N/A
Dog_6 Session 2	1.0 s	0.8 s	N/A	N/A
Dog_6 Session 4	1.1 s	1.1 s	N/A	N/A

Table 5. Paired t-test P value comparison of average response latency between shaping levels

Dog and Session	First shape level	Second shape level		
		Shape level 1	Shape Level 2	Shape level 3
Dog_1 Session 1	Shape level 0	0.113	**0.013**	**0.000**
	Shape level 1	N/A	**0.045**	**0.002**
	Shape level 2	N/A	N/A	0.996
Dog_2 Session 2	Shape level 0	0.597	**0.002**	N/A
	Shape level 1	N/A	**0.003**	N/A
Dog_6 Session 2	Shape level 0	0.718	N/A	N/A
Dog_6 Session 4	Shape level 0	0.757	N/A	N/A

In order to determine if the decreases in average response latency between shaping levels were significantly different, a paired t-test was done comparing each possible shaping level. The null hypothesis for the t-test was that the response latency is the same. Table 5 shows the t-test P values of each comparison. As an example, Dog_1 Session 1 with four shaping levels, a statistical comparison can be made between shaping level 0 and the three other shaping levels, between shaping level 1 and the two remaining shaping levels, and between shaping level 2 and 3. Table 5 shows, with 95% confidence, there is a significant difference between the response latency between six of the eleven comparisons. The comparison of average response latencies of Shape level 0–2, 0–3, 1–2, and 1–3 have significant differences for Dog_1 Session 1. The comparison of average response latencies of shape level 0–2 and 1–2 have significant differences for Dog_2 Session 2.

All six of the significantly different level response latency averages were from the sessions controlled by the computer. Because of the reduced response latency between the levels, both Dog_1 and Dog_2 show some indication of learning that the cue has some control over when they will be reinforced for pressing the button.

To give a graphical representation of the most representative computer trained session, Fig. 2 shows the average response latency for Dog_1 Session 1 varying over the number of TCCs responses. As the paired t-test showed, shaping

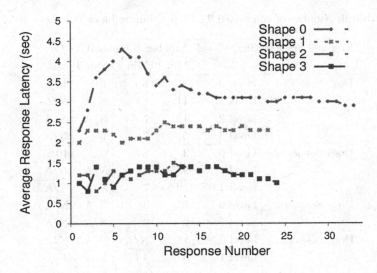

Fig. 2. Dog₁ Session 1 average response latency

level 0 and 1 each differ from shaping levels 2 and 3. Shaping levels 2 and 3 are similar. In general the average response latency is decreasing when the computer system changes the shaping level. Decreasing average response latency is a necessary, but not sufficient indication of learning.

5.2   Completed TTCs as Shape Levels Change

As an indication of learning, behavior analysis research indicates that at each increase in shape level the dog should experience some form of extinction [10] that would temporarily decrease the rate of reinforcement of completed TTCs. Table 6 shows the completed TCCs for each session. The Table shows increasing completed TTCs per minute until a shaping level increases the random delay and produces a drop in completed TTCs resulting from the time taken for extinction of the shorter duration behavior. Note that Dog₁ Session 1 min three and Dog₂ Session 2 min three both had 10 completed TCCs but did not progress to the next level due to the number of incomplete TCCs causing the one minute percentage to be < 90%. Dog₁ Session 1 shape level 0 progresses to 13 TCCs then in minute four and decreases to 11 in minute five when the shape level goes to one. Unfortunately, with a maximum of four minutes in any shaping level a paired t-test P value for the data is not statistically significant.

As an indication of extinction, symptoms of frustration were shown by the dogs when the shaping level changes. For example, Dog₁ Session 1 at minute 7 shows there are 12 completed TTCs which drops to 5 completed TTCs in minute 8 without a increase in shaping levels. This is when the computer issued its first multi-second delay of 2.7 s. The video shows Dog₁ seemingly frustrated as she stopped, looked around, and moved around.

Table 6. Number of completed TCCs per minute block per shape level

Dog and Session	Shape level	Number completed TCCs			
		Min 1	Min 2	Min 3	Min 4
Dog$_1$ Session 1	Level 0	2	8	10	13
	Level 1	11	13	N/A	N/A
	Level 2	12	N/A	N/A	N/A
	Level 3	5	7	12	N/A
Dog$_2$ Session 2	Level 0	3	8	10	10
	Level 1	9	11	N/A	N/A
	Level 2	10	7	N/A	N/A
Dog$_6$ Session 2	Level 0	6	3	N/A	N/A
	Level 1	2	0	N/A	N/A
Dog$_6$ Session 4	Level 0	11	N/A	N/A	N/A
	Level 1	6	4	5	4

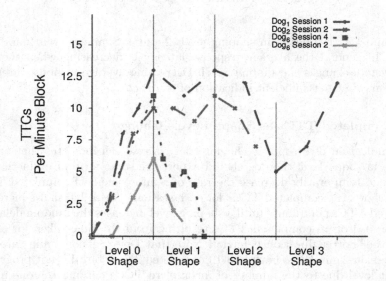

Fig. 3. Completed TTCs per shaping level

Prior research has theorized that a computer program that shapes learning by changes in reinforcement criteria produces a sawtooth pattern of the completed response rate [5, 10]. Figure 3 shows the completed TTCs for each of the shaping level. Figure 3 Dog$_1$ Session 1 clearly shows the sawtooth pattern of completed TTCs per minute change as shaping changes criteria. Figure 3 Dog$_2$ Session 2 and Dog$_6$ Session 2 both shows *some* indication of the sawtooth pattern. Figure 3 Dog$_6$ Session 4 shows *no* indication of the sawtooth pattern. The first three figures are indications of learning in both the computer's and professional's training of stimulus control of a behavior.

6   Conclusion

In this paper we describe a biohybrid training system capable of placing a canine behavior under stimulus control. The system is a continuation of our dog-training computer system.

In hindsight, if a novel behavior, such as a dog pushing a button with its nose, is actually chosen to be put under stimulus control, more time must be allocated for each dog's acclimation to the environment and the harness, in addition to the dog's acquisition of the behavior. As some dogs have more affinity to different environments and behaviors, a larger and varied population size must also be used to accommodate some dog's willingness to acquire the behavior.

Although distance sensors did not ultimately affect the outcome of the pilot, more sensors installed on the wall would facilitate determining when the dogs pushed areas of the wall instead of the actual button. Beyond using these additional sensors to determine when a push was close to the button, a sensor wall would also be key for training of the behavior when targeting proximity was needed.

Our decision to work with shelter dogs as opposed to recruiting dogs from the general population was a pivotal design decision. Having very few dogs actually acquire the push button behavior limited the sample size and external validity of the results. We still believe there are some interesting implications for automated canine training.

We have shown that our system meets the criteria of timing, consistency, and rate of reinforcement to begin to place a behavior under stimulus control. We have shown that the rate of reinforcement increases as a potential indication of learning. Even with a limited sample size, some dogs trained by the biohybrid system have shown a statistically significant different rate of reinforcement. Lastly our system showed a saw-tooth pattern of changes in rate of reinforcement predicted by other research as an indication the learning that occurs when placing a behavior under stimulus control.

Acknowledgments. We would like to thank Bob Bailey and Parvene Farhoody for input into the experimental design. We would like to thank Wes Anderson of Smart Animal Training for early access to the Pet Tutor dispenser. We would also like to thank Paws4Ever dog sanctuary for the use of their facilities, for access to their dogs, and to their volunteers. This work is supported by the US National Science Foundation under the Cyber Physical Systems Program (IIS-1329738).

References

1. Bailey, B., Farhoody, P.: Bailey-farhoody discrimination workshop. http://behaviormatters.com/workshops/
2. Brugarolas, R., Loftin, R.T., Yang, P., Roberts, D.L., Sherman, B., Bozkurt, A.: Behavior recognition based on machine learning algorithms for a wireless canine machine interface. In: 2013 IEEE International Conference on Body Sensor Networks (BSN), pp. 1–5. IEEE (2013)
3. Brugarolas, R., Roberts, D., Sherman, B., Bozkurt, A.: Machine learning based posture estimation for a wireless canine machine interface. In: 2013 IEEE Topical Conference on Biomedical Wireless Technologies, Networks, and Sensing Systems (BioWireleSS), pp. 10–12. IEEE (2013)
4. Cornou, C., Lundbye-Christensen, S.: Classifying sows activity types from acceleration patterns: an application of the multi-process Kalman filter. Appl. Animal Behav. Sci. 111(3), 262–273 (2008)
5. Galbicka, G.: Shaping in the 21st century: moving percentile schedules into applied settings. J. Appl. Behav. Anal. 27(4), 739–760 (1994)
6. Gerencsér, L., Vásárhelyi, G., Nagy, M., Vicsek, T., Miklósi, A.: Identification of behaviour in freely moving dogs (Canis familiaris) using inertial sensors. PlOS One 8(10), 1–14 (2013)
7. Hiby, E., Rooney, N., Bradshaw, J.: Dog training methods: their use, effectiveness and interaction with behaviour and welfare. Animal Welf. Potters Bar Then Wheathampstead 13(1), 63–70 (2004)
8. Ladha, C., Hammerla, N., Hughes, E., Olivier, P., Plötz, T.: Dog's life: wearable activity recognition for dogs. In: Proceedings of the 2013 ACM International Joint Conference on Pervasive and Ubiquitous Computing, pp. 415–418. ACM (2013)
9. Majikes, J., Brugarolas, R., Winters, M., Yuschak, S., Mealin, S., Walker, K., Yang, P., Sherman, B., Bozkurt, A., Roberts, D.L.: Balancing noise sensitivity, response latency, and posture accuracy for a computer-assisted canine posture training system. Int. J. Hum. Comput. Stud. 98, 179–195 (2016)
10. Mazur, J.E.: Learning and Behavior. Psychology Press, New York (2015)
11. McLeod, S.: BF skinner: operant conditioning (2007). Accessed 9 Sept 2009
12. Moreau, M., Siebert, S., Buerkert, A., Schlecht, E.: Use of a tri-axial accelerometer for automated recording and classification of goats grazing behaviour. Appl. Animal Behav. Sci. 119(3), 158–170 (2009)
13. Pi, S.: Problems with the HC-SR04 ultrasonic sensor. http://randomnerdtutorials.com/complete-guide-for-ultrasonic-sensor-hc-sr04/
14. Pi, S.: Problems with the HC-SR04 ultrasonic sensor. https://www.raspberrypi.org/forums/viewtopic.php?f=37&t=77534
15. Preece, S.J., Goulermas, J.Y., Kenney, L.P., Howard, D., Meijer, K., Crompton, R.: Activity identification using body-mounted sensors: a review of classification techniques. Physiol. Meas. 30(4), R1 (2009)
16. Savolainen, P., Zhang, Y., Luo, J., Lundeberg, J., Leitner, T.: Genetic evidence for an East Asian origin of domestic dogs. Science 298(5598), 1610–1613 (2002)
17. Wrigglesworth, D.J., Mort, E.S., Upton, S.L., Miller, A.T.: Accuracy of the use of triaxial accelerometry for measuring daily activity as a predictor of daily maintenance energy requirement in healthy adult labrador retrievers. Am. J. Vet. Res. 72(9), 1151–1155 (2011)

Exploiting Morphology of a Soft Manipulator for Assistive Tasks

Mariangela Manti[✉], Thomas George Thuruthel, Francesco Paolo Falotico,
Andrea Pratesi, Egidio Falotico, Matteo Cianchetti, and Cecilia Laschi

BioRobotics Institute of Scuola Superiore Sant'Anna, Pisa, Italy
{m.manti,t.thuruthel,f.falotico,a.pratesi,e.falotico,
m.cianchetti,c.laschi}@santannapisa.it

Abstract. The idea of using embodied intelligence over traditional well-structured design and control formulations has given rise to simple yet elegant applications in the form of soft grippers and compliant locomotion-based robots. Real-world applications of soft manipulators are however limited, largely due to their low accuracy and force transmission. Nonetheless, with the rise of robotic appliances in the field of human-robot interaction, their advantages could outweigh their control deficiencies. In this context, the embodied intelligence could play an important role in developing safe and robust controllers. In this paper, we present a three module soft manipulator to experimentally demonstrate how its morphological properties can be exploited through interactions with the external environment. In particular, we show how to improve the pose accuracy in an assistive task using a simple control algorithm. The soft manipulator takes advantage of its inherent compliance and the physical constraints of the external environment to accomplish a safe interactive task with the user. There exists a continuous and mutual adaptation between the soft-bodied system and the environment. This feature can be used in tasks where the environment is unstructured (e.g. specific body region), and the adaptability of the interaction is entirely dependent on the morphology and control of the system. Experimental results indicate that significant improvements in the tracking accuracy can be achieved by a simple yet appropriate environmental constraint.

Keywords: Soft manipulator · Embodied intelligence · Morphological computation · Human-robot interaction · Assistive robots

1 Introduction

Bio-inspiration and bio-mimetics play a pivotal role in the development of soft systems that are able to safely interact with fragile objects such as the environment, food and human beings. Moreover, it has provided the ground for the growth of Soft Robotics through continuous identification and advancement of novel and versatile technologies

Electronic supplementary material The online version of this chapter (doi: 10.1007/978-3-319-63537-8_25) contains supplementary material, which is available to authorized users.

M. Mangan et al. (Eds.): Living Machines 2017, LNAI 10384, pp. 291–301, 2017.
DOI: 10.1007/978-3-319-63537-8_25

[1, 2]. Biological inspiration does not imply that we attempt to copy nature. Rather, the goal is to understand the principles underlying the behaviour of animals and humans and transfer them to the development of robots [3]. With this in mind, researchers tackled the challenge of developing a set of enabling technologies that have entailed a quantum leap in the engineering of robots; as a consequence, they show a set of complex skills that guarantee dexterous movements and safe cooperation with humans. The development of suitable technologies for a new generation of soft robots is the first step towards this objective, after which researchers have to focus on new design principles for smartly combining and arranging these elements in a soft-bodied system, in order to obtain the target behaviour while reducing the complexity of the system. Animals' appendages such as the elephant trunk or the octopus arm have been the focus of studies creating soft, hyper-redundant robots, with capabilities similar to those of the biological counterparts [4–7]. As a result, literature proposes many examples of continuum robots that, inspired by soft tissues in nature, are made of soft or compliant materials, and therefore have a continuously deformable body and a large number of Degrees of Freedom, for dexterous movements and delicate human-robot interaction. The design of the soft robotic manipulator reported in the present paper is guided by the same principle.

One among the widely discussed principles inspired from nature is the concept of morphological computation, where the possibility of using morphological properties of a soft-bodied robot for control is explored [19]. This paper, along with the same lines, tries to investigate the possibility of using the interactions between the bio-inspired soft-bodied manipulator and the environment to reduce tracking errors in an assistive task. The ability of soft robots to modify their kinematic properties based on the interacting environment is unique to them and it has led to innovative strategies for grasping and locomotion using only intrinsic feedback from the structure. However, unlike these scenarios, an end-effector tracking task while exploiting the morphological properties of the soft manipulator is not so trivial. This is because a completely open loop strategy would not be able to reduce tracking errors in the presence of environmental obstacles and a closed loop control strategy is challenging because of the infinite kinematic configurations that a soft manipulator can achieve in the presence of environmental obstacles. Hence, it is important to develop controllers that are robust to kinematic uncertainties. This research demonstrates an example task where the interplay between the design of the manipulator, the closed loop controller and the task environment can be utilized for enhanced tracking performance.

The paper is organized as follows: after the introduction mainly devoted to explain the research context, authors, in Sect. 2, provide an overview of the soft manipulator starting from some existing examples; Sect. 3 is entirely dedicated to the system architecture from the hardware and control point of view; Sect. 4 presents the experimental set-up and the trials for assessing the soft arm performances in a realistic scenario. Results are summarized and discussed in Sect. 5 while the last section summarizes the main achievements.

2   Manipulator Overview

Soft manipulators can be considered intrinsically safe thanks to the actuation technologies they are made of [8]. Pneumatic muscles, electroactive polymers, shape memory alloys and cable-driven mechanisms are among the most promising and employed technologies for continuum robots [9]. One of their main features is their compliant body that can deform passively to adapt to unstructured environments, thus reducing the complexity of active control. Furthermore, a key lesson that researchers have learnt from nature and that can be here implemented, regards the synergetic co-activation of the actuators embedded in a soft body for enabling adaptation and fine-tuning of the stiffness in response to external forces. It means that multiple actuation technologies can be combined according to a set of design principles for the specific application, thus improving the performances of the continuum manipulator.

A detailed analysis of the state of the art reveals different examples of soft manipulators based on hybrid actuation. For example, the KSI tentacle manipulator developed by Immega et al. [10] proposes a design based on a central pneumatic chamber that, providing structural support, enhances contraction/elongation movements while radial cables are used for bending motions. A similar design principle has been followed by McMahan et al. [11]. Another example of hybrid actuation is represented by the octopus-like-continuum robot that is based on cables in combination with SMAs exhibiting highly flexible motions [12].

Shiva et al. [13] designed a pneumatically actuated soft manipulator whose stiffness can be further adjusted by cables. Moreover, Kang et al. [14] presented a hybrid actuation system based on the design of a continuum robot employing tendon embedded pneumatic muscles (TEPMs).

The hybrid actuation strategy is applicable to those applications where variable stiffness capabilities and accurate control of a continuum robot are required while maintaining large manipulation capabilities. In this paper, we are reporting one of these application possibilities, namely in assistive robotics, where a soft arm has been purposely developed to assist elderly people in bathing tasks. The technical requirements, in terms of desired reachable workspace and expected movements are guaranteed by a hybrid actuation strategy, as summarized in [15]. Here, the authors briefly recap the actuators used and how they are arranged along the manipulator, made of three identical interconnected modules.

In order to achieve the desired motion, authors combined two different actuation technologies, according to their working principle (see Figs. 1 and 2):

(i) McKibben-based flexible fluidic actuators, that enable elongation and omni-directional bending with high strength, but low accuracy;
(ii) Tendon-driven mechanisms, that can implement shortening and (redundant) omni-directional bending with high movement resolution, and high accuracy.

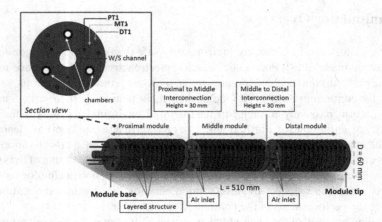

Fig. 1. An overview of the manipulator made of three identical interconnected modules, each 150 mm in length. On the left top a section view of the proximal module is shown including cable arrangements along the module up to the base of the robotic system.

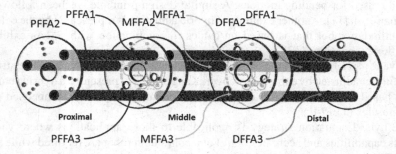

Fig. 2. A sketch of the soft manipulator showing all the 9 flexible fluidic actuators arranged along the module. They are aligned from the distal to the proximal, as follows: DFFA1-DFFA2-DFFA3; MFFA1-MFFA2-MFFA3; PFFA1-PFFA2-PFFA3.

As a result, the system is capable of elongation/contraction and bending movements exploring a huge workspace according to different activation patterns that are compatible with the target application. Moreover, specific antagonistic activation sequences provide stiffness capabilities too.

The design of the single module, based on three McKibben-based actuators and three cables equally spaced at 60° into a circle of 60 mm in diameter, has been proposed in [15, 16]. In particular, in the last one authors presented the fabrication, kinematic characterization, and kinematic control of the hybrid-actuated module. The design demonstrates a lightweight structure that effectively reaches the target workspace with high dexterity, specifically, omnidirectional bending, extension, contraction, and variable stiffness. As it is shown in Fig. 1, the actuators are confined in a structural support to avoid lateral buckling of the fluidic actuators. In the previous prototype [16], an external helicoidally-shaped reinforcement structure was designed but this shape resulted to add undesired torsional movements. In order to overcome this limitation, the structure has been modified into a layer-by-layer reinforcement structure i.e. each layer has been singularly inserted and fixed to the fluidic actuators. At rest, the distance between two consecutive layers is 10 mm (similar to the diameter of the actuator).

3 System Architecture

3.1 Hardware

The robotic platform is composed of three key elements: (i) the soft manipulator, (ii) the actuation box with 9 proportional pressure micro-regulators (Camozzi, mod. K8P-0-E522-0), one air filter regulator to set pressure input at 4 bar, the power supply and the control unit composed of an Arduino Due and a custom electronic board to manage both micro-regulators and servos; and (iii) the servomotors box with 9 Hitec 785HB servos. All these components are interconnected as reported in Fig. 3, that shows a scheme of the system architecture.

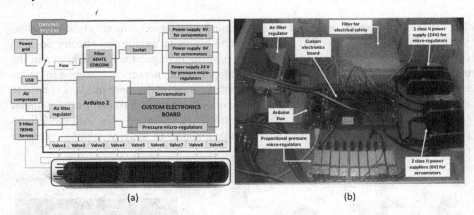

(a)   (b)

Fig. 3. The actuation box: (a) schematic view and (b) real prototype with all the components.

3.2 Control

The controller design is based on a learned kinematic controller previously described in [17, 18]. The underlying principle is a variant of the resolved motion rate controller. The controller learns an estimate of the kinematic Jacobian and moves in the direction that minimizes the current tracking error. Due to this, like the resolved motion rate controller, the method is moderately robust to kinematic uncertainties. More importantly, it can also accommodate variations in the manipulator kinematics when in contact with unstructured environments. The controller is formulated by learning a localized mapping between the current end effector position (x_i), the current actuator state (q_i) and the next end effector position (x_{i+1}) to the next actuator configuration (q_{i+1}). The idea behind this formulation is that the current actuator state provides the information about the kinematic Jacobian and the desired local end effector motion can be obtained from x_i and x_{i+1}. The samples for learning this mapping are obtained by localized motor babbling.

The manipulator is actuated by all cables and fluidic actuators summing to a total of 18 actuators. Defining the kinematics of the manipulators as a function of these 18 actuators can represent an issue. This is due to numerous actuator singularities that arise from such a system. Since we are not using load cells to keep the tendons in tension, there are actuator configurations that make some of the tendons redundant. Likewise, there are numerous actuator configurations, especially when the cables are in high tension, where the fluidic actuators are redundant. In order to solve this problem, we keep all pressures fixed. In this way, the kinematics of the manipulator becomes the function of only the 9 tendon-driven actuators. The pneumatic actuators now act as variable stiffness loaded springs (due to the constant pressure), which allows the manipulator to extend and bend, thereby increasing the workspace of the manipulator. Additionally, they also act as dampers and variable stiffening mechanism, which help greatly in reducing vibrations and in increasing the stiffness of the manipulator.

4   Experimental Set-up

The experimental setup consists of the soft manipulator, the servomotors box, the actuation box, the air compressor and a Vicon system for capturing the end effector position (see Fig. 4). Three markers are used at the tip of the manipulator for tracking and the end effector position and orientation. The centroid of the markers is assumed to be the end effector position and the normal to the plane formed by the three markers is used to estimate the orientation. Since the application of our system is currently only for planar tasks, control of one orientation vector will suffice. In addition, even with the 9 independent actuators, the kinematics of the manipulator is heavily coupled, thereby reducing the kinematic redundancy.

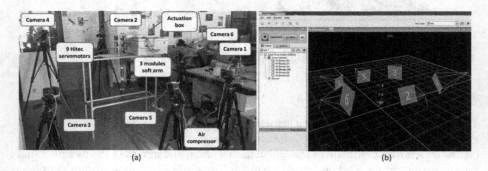

(a)   (b)

Fig. 4. (a) Experimental set-up composed of: (i) 3 modules soft arm, (ii) actuation box, (iii) air compressor, (iv) Vicon system with 6 cameras; (b) Vicon Nexus 1.8.5 interface showing the six cameras position respect to the soft manipulator where three markers have been associated with each module (proximal, middle and distal).

Before, the sampling process, it is necessary to specify the actuator ranges for the localized motor babbling process. For our experiments, we fix the ranges of the motors to [880° 560° 352°] for each motor of the distal, middle and proximal modules respectively. The pneumatic pressure is fixed to 1.2 bar, for all 9 fluidic actuators. The localized motor babbling is performed by randomly moving the motors incrementally within the specified ranges. Refer to [17, 18] for more details and explanations on this step. The incremental steps are fixed to 24 degrees, For learning the global kinematic controller, we required 12000 samples. Each sample is obtained at an interval of 1 s. The kinematic workspace of the manipulator obtained after the motor babbling process is shown in Fig. 5. No post processing is performed on the data obtained and we use a multilayer perceptron to learn the mapping from $(x_{i+1}, q_i, x_i) \rightarrow (q_{i+1})$. The hidden layer size is kept at 35 neurons and MATLAB Bayesian regularization back propagation algorithm is used for training.

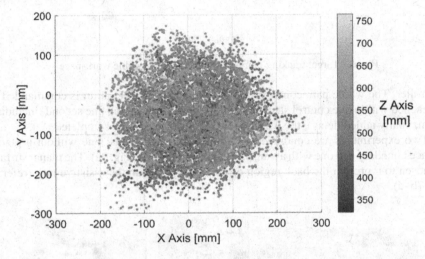

Fig. 5. Workspace of the manipulator.

5   Results and Discussion

The manipulator is developed for assistive bathing tasks like body washing and showering that require the manipulator to follow a planar path with a particular orientation. Accuracy is not of paramount importance for the task, while intrinsic safety and robustness of the controller are fundamental. The desired path of the manipulator end effector is shown in Fig. 6, along with the manipulator kinematic workspace. The desired orientation is always perpendicular to the surface and the desired X coordinate is fixed at 100 mm.

Since there is no analytical model of the manipulator kinematics it not possible to verify if the target poses are reachable, however, this does not affect the behaviour of the controller other than in the form of reduced accuracy. Nonetheless, majority of the target poses lie outside the static workspace of the manipulator. The objective behind this is to see if the addition of the environmental constraint modifies the kinematic workspace and to verify that this structural change has adverse effects on the learned

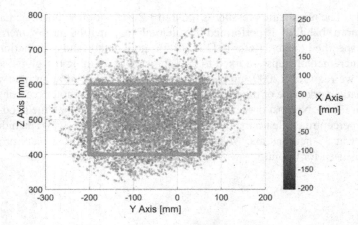

Fig. 6. Target tracking points along with the reachable workspace.

controller. The whole path contains 200 points and each target point is commanded to be reached within one control step. With each control step taking one second (including the manually added delay), the whole process takes 200 s to be completed.

Two experiments are conducted for the same target points; one without a planar surface boundary and one with a transparent surface (refer to Fig. 7a). The planar surface is chosen to represent the back region of the human body in the assistive task (refer to Fig. 7b–c).

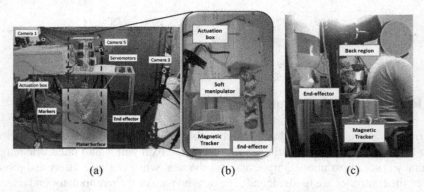

(a)   (b)   (c)

Fig. 7. (a) Experimental set-up showing the interaction between the end-effector of the prototype and the planar surface; (b) the final I-Support soft arm installation; (c) the end-effector is interacting with the back region of the user in an assistive task.

The tracking error for both cases as a function of time is shown in Fig. 8. The average tracking error is 58 ± 27 mm without the surface and 31 ± 26 mm with the surface. The orientation error is $37 \pm 16°$ without the surface and $23 \pm 15°$ with the surface. Although the tracking error is high, it must be noted that the required task is highly complex for a soft manipulator and most of the target points are outside the static workspace. Not only does the soft manipulator work against gravity, but the controller also is heavily affected by natural vibrations of the system and suffers from high actuator singularities because of the coupling between modules.

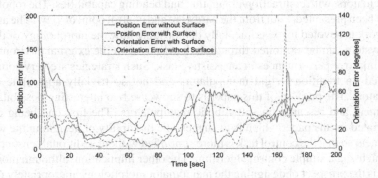

Fig. 8. Tracking errors for the task with and without the external surface.

Moreover, even though the targets are continuous in the task space, the corresponding actuator configurations need not be continuous in the actuator space, thereby requiring more than one iteration to converge to the correct configuration. Nonetheless, our research interest lies in the interaction between the soft manipulator and the external environment. From the results, it is evident that the tracking performance is improved because of the added surface. There are many reasons for this. Primarily, the surface is decided considering the task in hand. The constraints imposed by the surface modify the manipulator kinematics appropriately for our task (Fig. 9). Secondarily, the surface acts like a support for the manipulator, which helps in damping vibrations. The reduced errors in tracking also indicate a modification of the static workspace[1].

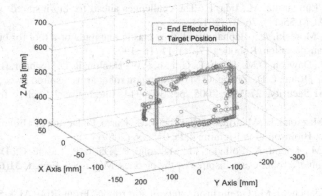

Fig. 9. End Effector positions during the tracking task.

---

[1] See the attached multimedia file showing the manipulator executing the task of interaction with the planar surface.

6 Conclusions

The paper presents a three-module soft manipulator based on 9 cables and 9 flexible fluidic actuators, with contraction/elongation and bending capabilities. The robotic platform has been presented, both from the hardware and control point of view. The attention of this work is devoted to experimentally demonstrate how the morphology of the soft-bodied system can be exploited through interactions with the external environment in order to improve performances in an assistive task. Such strategies are very complex to be applied on traditional rigid manipulators and not so trivially applicable on soft manipulators. The purpose of this paper is to showcase two interesting possibilities for development and research on soft robotic manipulators. The first one being directly demonstrated in this paper which is to suggest the possibility of designing the working environment so that the control task is more simplified without additional sensor requirements and using a simple control algorithm. The other implication, although more challenging, is the prospect of designing the manipulator morphology appropriately for their working environment. This would mean selecting the manipulator properties like stiffness, friction coefficient, shape such that the interaction between the manipulator and the working environment matches our requirements.

Acknowledgements. The authors would like to acknowledge the support by the European Commission through the I-SUPPORT project (#643666).

References

1. Laschi, C., Mazzolai, B., Cianchetti, M.: Soft robotics: technologies and systems pushing the boundaries of robot abilities. Sci. Rob. **1**(1), eaah3690 (2016)
2. Rus, D., Tolley, M.T.: Design, fabrication and control of soft robots. Nature **521**(7553), 467–475 (2015)
3. Pfeifer, R., Lungarella, M., Iida, F.: The challenges ahead for bio-inspired 'soft' robotics. Commun. ACM **55**(11), 76–87 (2012)
4. Cieslak, R., Morecki, A.: Elephant trunk type elastic manipulator-a tool for bulk and liquid materials transportation. Robotica **17**(01), 11–16 (1999)
5. Walker, I.D., Dawson, D.M., Flash, T., Grasso, F.W., Hanlon, R.T., Hochner, B., Kier, W.M., Pagano, C.C., Rahn, C.D., Zhang, Q.M.: Continuum robot arms inspired by cephalopods. In: Defense and Security, 27 May 2005, pp. 303–314. International Society for Optics and Photonics
6. Laschi, C., Mazzolai, B., Mattoli, V., Cianchetti, M., Dario, P.: Design of a biomimetic robotic octopus arm. Bioinspiration Biomim. **4**(1), 015006 (2009)
7. Cianchetti, M., Arienti, A., Follador, M., Mazzolai, B., Dario, P., Laschi, C.: Design concept and validation of a robotic arm inspired by the octopus. Mater. Sci. Eng., C **31**(6), 1230–1239 (2011)
8. Abidi, H., Cianchetti, M.: On intrinsic safety of soft robots. Front. Rob. AI **4**, 5 (2017)
9. Laschi, C., Cianchetti, M.: Soft robotics: new perspectives for robot bodyware and control. Front. Bioeng. Biotechnol. **30**(2), 3 (2014)

10. Immega, G., Antonelli, K.: The KSI tentacle manipulator. In: 1995 IEEE International Conference on Robotics and Automation, Proceedings, 21 May 1995, vol. 3, pp. 3149–3154. IEEE

11. McMahan, W., Jones, B.A., Walker, I.D.: Design and implementation of a multi-section continuum robot: air-octor. In: 2005 IEEE/RSJ International Conference on Intelligent Robots and Systems, IROS 2005, 2 Aug 2005, pp. 2578–2585. IEEE

12. Cianchetti, M., Calisti, M., Margheri, L., Kuba, M., Laschi, C.: Bioinspired locomotion and grasping in water: the soft eight-arm OCTOPUS robot. Bioinspiration Biomim. **10**(3), 035003 (2015)

13. Shiva, A., Stilli, A., Noh, Y., Faragasso, A., De Falco, I., Gerboni, G., Cianchetti, M., Menciassi, A., Althoefer, K., Wurdemann, H.A.: Tendon-based stiffening for a pneumatically actuated soft manipulator. IEEE Rob. Autom. Lett. **1**(2), 632–637 (2016)

14. Kang, R., Guo, Y., Chen, L., Branson, D., Dai, J.: Design of a pneumatic muscle based continuum robot with embedded tendons. IEEE/ASME Trans. Mech. **22**, 751–761 (2016)

15. Manti, M., Pratesi, A., Falotico, E., Cianchetti, M., Laschi, C.: Soft assistive robot for personal care of elderly people. In: 2016 6th IEEE International Conference on Biomedical Robotics and Biomechatronics (BioRob), 26 June 2016, pp. 833–838. IEEE

16. Ansari, Y., Manti, M., Falotico, E., Mollard, Y., Cianchetti, M., Laschi, C.: Towards the development of a soft manipulator as an assistive robot for personal care of elderly people. Int. J. Adv. Rob. Syst. **14**(2), 1729881416687132 (2017)

17. Thuruthel, T.G., Falotico, E., Cianchetti, M., Laschi, C.: Learning global inverse kinematics solutions for a continuum robot. In: Parenti-Castelli, V., Schiehlen, W. (eds.) ROMANSY 21 - Robot Design, Dynamics and Control. CICMS, vol. 569, pp. 47–54. Springer, Cham (2016). doi:10.1007/978-3-319-33714-2_6

18. Thuruthel, T., Falotico, E., Cianchetti, M., Renda, F., Laschi, C.: Learning global inverse statics solution for a redundant soft robot. In: Proceedings of the 13th International Conference on Informatics in Control, Automation and Robotics, 29 July 2016, vol. 2, pp. 303–310 (2016)

19. Pfeifer, R., Gómez, G.: Morphological computation – connecting brain, body, and environment. In: Sendhoff, B., Körner, E., Sporns, O., Ritter, H., Doya, K. (eds.) Creating Brain-Like Intelligence. LNCS, vol. 5436, pp. 66–83. Springer, Heidelberg (2009). doi: 10.1007/978-3-642-00616-6_5

Autonomous Thrust-Assisted Perching of a Fixed-Wing UAV on Vertical Surfaces

Dino Mehanovic$^{(\boxtimes)}$, John Bass, Thomas Courteau, David Rancourt,
and Alexis Lussier Desbiens

Université de Sherbrooke, Sherbrooke, QC J1K 2R1, Canada
{dino.mehanovic,alexis.lussier.desbiens}@usherbrooke.ca
http://www.createk.design

Abstract. We present the first fixed-wing drone that autonomously perches and takes off from vertical surfaces. Inspired by birds, this airplane uses a thrust-assisted pitch-up maneuver to slow down rapidly before touchdown. Microspines are used to cling to rough walls, while strictly onboard sensing is used for control. The effect of thrust on the suspension's landing envelope is analyzed and a simple vertical velocity controller is proposed to create smooth and robust descents towards a wall. Multiple landings are performed over a range of flight conditions (a video of S-MAD is available at: http://createk.recherche.usherbrooke.ca/LM2017/).

Keywords: Perching · Multimodal · Scansorial · Fixed-wing · UAV · Drone · Bioinspired

1 Introduction

The increasing demand for civil applications of unmanned aerial vehicles (UAV) is encouraging the development of small platforms with extended mission life. However, small airframes have low aerodynamic efficiency and reduced energy storage capabilities, both of which severely limit the endurance and range of these platforms. In nature, many small birds, insects and mammals regularly land to rest, feed, seek shelter or stealthily monitor an area.

Recently, a variety of bioinspired robotic platforms have been created for perching and are reviewed in detail in [1]. Among those, quadrotors have been used due to their ability to perform agile flight trajectories to slow down before impact [2,3]. Directly flying into targets has also been explored by appropriately positioning adhesives on fixed-wing and multi-rotor airframes [4–7]. To perform both perching and climbing on vertical surfaces, the SCAMP quadrotor re-orients itself after flying directly into a wall [8].

The implementation of such perching capabilities in a powered fixed-wing UAV remains a challenge for various reasons (e.g., added mass by propulsion system). Thus, the most successful solutions are based on glider platforms unable to take off [9–11]. Notably, the perching trajectory for the Stanford Perching

© Springer International Publishing AG 2017
M. Mangan et al. (Eds.): Living Machines 2017, LNAI 10384, pp. 302–314, 2017.
DOI: 10.1007/978-3-319-63537-8_26

Fig. 1. Many birds exhibit landing trajectories with high body pitch angles and significant upward force to maintain a horizontal approach. Courtesy of: Maxis Gamez (a), PCO (b), Warren Photography (c) and Kaddy (d).

Glider is inspired by the flying squirrel [11]. It performs a pitch-up maneuver followed by a drag-affected ballistic phase, before adhering to vertical surfaces with microspines. However, the ballistic phase creates only a short zone of suitable touchdown conditions that impose severe requirements on the platform's wall sensor, as discussed further in Sect. 2.

Sherbrooke's Multimodal Autonomous Drone (S-MAD), presented in this paper, draws inspiration from nature in order to perform full cycles of landing, standby, takeoff and flight (LSTF). As illustrated in Fig. 1, many birds use high body pitch angles and significant upward force to maintain horizontal perching approaches. This paper demonstrates that such a thrust-assisted landing strategy, when utilized on a fixed-wing aircraft, enables controlled steady-state descents (SSD) towards perching sites. This reduces the impact speed and significantly extends the zone of suitable touchdown conditions, leading to enhanced reliability and reduced hardware requirements (e.g., suspension, wall sensing). This landing strategy also enables simple takeoff, facilitated by the favorable thrust orientation while perched. To our knowledge, this is the first autonomous fixed-wing platform capable of performing both perching and takeoff maneuvers.

2   Perching Strategy Overview

The thrust-assisted perching maneuver described in this paper builds onto previous work that enabled a fixed-wing glider to perch on vertical surfaces following a rapid feedforward pitch-up maneuver [12]. This maneuver, illustrated by the dotted line in Fig. 2, rapidly slows down the glider and exposes the landing gear to the vertical surface. As the glider reaches full pitch-up, the velocity is reduced to such an extent that aerodynamic forces become negligible and the glider keeps travelling on a mostly ballistic trajectory towards the wall. With proper timing, the airplane touches down with a sufficiently low speed to allow a suspension to dissipate the remaining kinetic energy, while favoring the feet attachment to the surface. This approach leads to high success rates on airframes with low wing loading, even with the use of a simple feedforward controller [11]. However, increasing the mass of the platform with additional payload and motors causes the suspension touchdown envelope (i.e., the set of touchdown states that lead to successful landing) to shrink down to negligible size. At the same time, the increased wing loading makes it more challenging to reduce the forward velocity to acceptable levels, before gravity significantly increases the vertical velocity to unacceptable values. Thus, as the mass increases, fewer trajectories can bring the glider from normal flying conditions to the suspension's touchdown envelope, leading to reduced success rate.

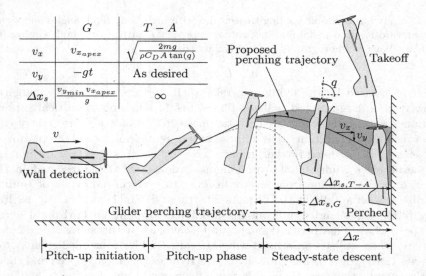

	G	$T - A$
v_x	$v_{x_{apex}}$	$\sqrt{\dfrac{2mg}{\rho C_D A \tan(q)}}$
v_y	$-gt$	As desired
Δx_s	$\dfrac{v_{y_{min}} v_{x_{apex}}}{g}$	∞

Fig. 2. Representation of the proposed thrust-assisted $(T - A)$ perching strategy and comparison to a glider (G) trajectory. Variations in SSD's velocity slope, e.g., induced by sensor bias or battery level, are represented by the shaded area. Post-apex velocities and Δx_s shown in the top-left table, where g is the gravitational acceleration and t is time. Allowable wall detection error gain for $T - A$ over G is identified by Δx.

Comparatively, the approach described in this paper takes advantage of thrust to control the final vertical and forward velocities such that:

1. Suitable touchdown conditions are available during an extended distance to reduce the timing and sensing requirement needed to trigger the maneuver.
2. Lower impact speeds are experienced by the suspension, leading to size and mass reduction.
3. Approach trajectories that favor adhesive engagement can be used.
4. Control authority is maintained throughout the full maneuver due to the propeller flow on the control surfaces. This allows early termination of the perching maneuver and recovery until the final stages of the maneuver.
5. Takeoff is possible, enabling repeated cycles of LSTF during a single mission.

The final approach of both gliding and thrust-assisted trajectories can easily be compared conceptually from the apex conditions (i.e., $v_x = v_{x_{apex}}$, $v_y = 0$), by assuming limits on the allowable velocity at landing (i.e., $0 < v_x < v_{x_{max}}$ and $v_{y_{min}} < v_y < 0$). In the case of the glider, velocity as a function of time and distance travelled while maintaining suitable touchdown states (Δx_s) can be calculated by assuming a ballistic trajectory, as expressed in Fig. 2. Comparatively, the post-apex steady-state trajectory of the thrust-assisted maneuver can be described by assuming a constant pitch (q), and by equating the vertical thrust component to mg and the horizontal component to the drag ($1/2\rho C_D A v_x^2$). Velocities and distance travelled while maintaining suitable touchdown states are described in Fig. 2.

As expected, thrust-assisted perching allows the designer to specify the velocity at touchdown given (1) the physical parameters mg/A, (2) the commanded pitch approach angle (q) and (3) assuming that v_y can be measured and controlled through thrust. Under these conditions, the airplane can travel an indefinite distance in states suitable for touchdown. This is an important gain over the gliding maneuver, for which the Δx_s distance is on the order of 20 cm and thus imposes strict sensing requirements.

However, the addition of thrust also adds some challenges. Indeed, any remaining thrust at impact (e.g., propeller spin down after touchdown detection through onboard accelerometer) significantly modifies the touchdown envelope by reducing the shear forces experienced by the microspines and the corresponding adhesion. Thrust-assisted landing is also highly sensitive to numerous airframe parameters and initial flight conditions, reducing the success rate achievable by a simple feedforward maneuver. The following sections describe the implementation of thrust-assisted perching, analyze the effect of thrust on the touchdown envelope and propose a novel simple feedback controller based on vertical velocity.

3   Implementation

The airframe used in the experiments described in this paper is presented in Fig. 3. It consists of a modified McFoamy airplane (i.e., 12 A ESC, Turnigy 2730 1500 kV motor and 8×6 propeller for a static thrust-to-weight ratio of 1.5), combined with a 3DR PixHawk autopilot for onboard control. This autopilot integrates most required sensors required for vertical velocity estimation and impact detection (gyro, accelerometers, barometer), and communicates with a lightweight laser rangefinder (TeraRanger One) for wall detection. Custom control loops run onboard the PixHawk at 200 Hz.

Five microspines are used on each foot to attach onto vertical surfaces. Although various adhesion strategies exist [13], microspines are preferred due to their proven performance on numerous rough surfaces of interest (e.g., stucco, concrete, brick, roofing shingles) [14]. Inspired by insect feet, the microspines (Fig. 3) consist of hooks that attach to rough vertical surfaces through mechanical interference and friction, while distributing the load uniformly between asperities. As illustrated in Fig. 3, the microspines require shear and normal force

·loadings within a safe zone delimited by friction, adhesion and overload limits to remain attached to the surface.

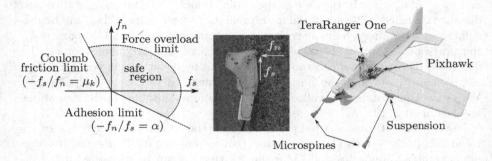

Fig. 3. Safe force region, microspines and platform components.

A suspension is installed between the airplane and the microspines to help bring the platform to rest, while favoring attachment to the surface. The suspension requirements for a thrust-assisted perching airplane are simplified due to the low touchdown speeds and the controlled touchdown direction. The proposed perching mechanism consists of a flexible beam, anchored in the airframe's wing and damped by a urethane foam block. At 18 g, the resulting suspension is significantly simpler and lighter than the suspension of the Stanford Perching Glider (28 g). Overall, with the added mass of the TeraRanger (15 g), the components required to enable perching only account for 10% of the platform's total mass (320 g).

4   Thrust-Assisted Touchdown Envelope

As discussed previously, the remaining thrust force during landing affects the touchdown envelope by reducing the shear force exerted at the feet and, incidentally, the adhesion available. This section describes the model developed to analyze the landing forces, its validation and the resulting landing envelope when thrust is present.

4.1   Model Description

The thrust-assisted perching maneuver described in this paper consists of mostly sagittal motion, with complex behaviors of the microspines at the feet. To properly represent this system, a hybrid planar dynamic model of the airplane and suspension is used, with both sliding and sticking states possible at the feet. By calculating the forces created during landing, and verifying if either the adhesion or the overload limit is reached, this model can predict the success or failure of different landing conditions.

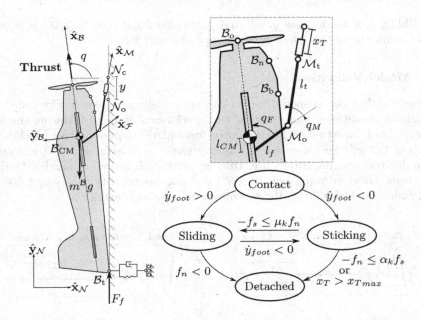

Fig. 4. System geometry, reference frames, forces and model transitions.

Figure 4 illustrates the geometry and forces defining the system, as well as the hybrid model states and discrete events leading to transitions. Besides the new suspension and added thrust, the model structure and contact dynamics at nose, belly and tail are similar to [11]. The airplane is modeled as a rigid body $\mathcal{B}$ and the legs are described by a single leg and foot. The combination of the leg's flexible beam and memory foam is modeled as a pseudo-rigid body [15], where two rigid segments (i.e., the fixed femur, $\mathcal{F}$, and the moving tibia $\mathcal{M}$) are connected by a pivot (knee) with lumped torsional stiffness and damping parameters. The foot is approximated by a linear spring-damper system. The thrust force is defined as a constant value ($m^{\mathcal{B}}g$) until impact detection (i.e., $14\,\mathrm{m/s^2}$ acceleration at CG). From that point, it follows an exponential decrease with an experimentally validated time constant of 85 ms. Due to the relatively low impact speed, the aerodynamic forces are neglected in this model. The contact point model at the foot can either take the form of a rolling joint (sliding) or a pin joint (sticking), depending on force and motion conditions at the foot, as described in the lower right transition diagram in Fig. 4.

Furthermore, as shown in Fig. 4, various reference frames and variables are introduced for the analysis. Without going into the details, it is easy to express the position of the airplane's center of mass ($\mathcal{B}_{\mathrm{CM}}$) as:

$$\mathbf{r}^{\mathcal{B}_{\mathrm{CM}}/\mathcal{N}_{\mathrm{o}}} = y\hat{\mathbf{y}}_{\mathcal{N}} - (x_T + l_t)\hat{\mathbf{x}}_{\mathcal{M}} - l_f\hat{\mathbf{x}}_{\mathcal{F}} + l_{CM}\hat{\mathbf{x}}_{\mathcal{B}} \qquad (1)$$

This vector, and other easily expressed quantities, can be used to establish the equations of motion through Kane's method for each generalized speed u_r [16].

The sliding foot model uses $\dot{q}$, $\dot{q}_M$ and $\dot{y}$ as generalized speeds (i.e., x_T is constant), while the sticking foot model uses $\dot{q}$, $\dot{q}_M$ and $\dot{x}_T$.

4.2 Model Validation

To confirm that the proposed model can accurately represent a wide range of touchdown conditions, four landings were performed by hand throwing the airplane without thrust on an instrumented force plate. Representative touchdown speeds of 1–2 m/s were used in various directions, as illustrated in Fig. 6, resulting in different loading trajectories that excite the full system dynamics. During these tests, constant pitch ($q = 86 \pm 3°$) and angular velocity ($\dot{q} = -50 \pm 30°/s$) were maintained, similar to the commanded states at touchdown.

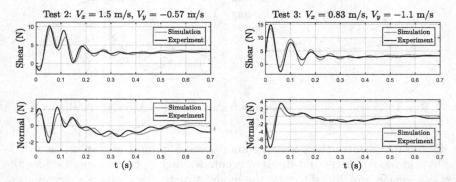

Fig. 5. Shear (f_s) and normal (f_n) forces acting on microspines at landing as obtained from simulations and experiments for tests 2 (left) and 3 (right).

The experimental setup used to measure shear and normal forces consists of a fabric-covered plate. This surface is used to guarantee simultaneous engagement of all microspines with the surface. The plate is instrumented using an ATI Mini40 force/torque sensor sampled at 1 kHz. The plate has a resonance frequency of at least 150 Hz in all directions. The recorded data is post-filtered using a 20 Hz Butterworth filter (20th order, zero-phase) to remove the structural modes of the airframe. The touchdown states are measured at 200 Hz using a motion capture system. Figure 5 shows the results for two tests.

Some physical parameters of the model were identified from the measured forces with a genetic algorithm (GA). To do so, the GA varies the selected physical parameters over a predefined range, aiming at minimizing the first 0.4 seconds of the normalized RMS (NRMS) error for shear and normal force in all four tests. The RMS error of each landing is normalized by its maximum absolute force range to produce a representative fit over different impact conditions. Each generation contains 100 individuals and the GA stops after the NRMS change over 5 generations is less than 0.01%. This condition was reached

Table 1. Physical properties of the system

Parameter	Symbol	Value	Source
Mass	$m^{\mathcal{B}}$	0.32 kg	Measured
Inertia	I_{zz}	0.017 kg m^2	GA
Pseudo-rigid body factor	γ	0.93	GA
Leg length	L	0.317 m	Measured
Femur and tibia length	l_f, l_t	0.022 m, 0.295 m	$(1-\gamma)L$, γL
Knee stiffness and damping	k_k, c_k	1.26 Nm/rad, 0.057 Nms/rad	GA
Foot stiffness and damping	k_f, c_f	2120 N/m, 5.32 Ns/m	GA
Wall stiffness and damping	k_w, c_w	231 N/m, 73 Ns/m	GA
Femur angle from fuselage	q_F	$-30°$	Measured
Spines natural length	l_0	0.0036 m	Measured

after 12 generations, with the NRMS error for all tests being less than 0.5%. The best parameters found with the GA are listed in Table 1.

The suspension's physical parameters obtained with the GA correspond to expected results: the inertia is slightly higher than our CAD model while the foot stiffness/damping are comparable to the results presented in [11]. The wall stiffness value is also significantly lower due to the softer nature of the EPP foam used on this airplane.

4.3   Landing State Map (LSM)

The calibrated model can be used to identify the envelope of impact states that lead to successful perching, as shown in Fig. 6. The LSM itself is validated with landings on a wall covered with asphalt shingles (i.e., $f_n/f_s < 1$). The platform is hand-thrown at various velocities, with angular speed and pitch angle maintained relatively constant. A total of 35 trials were performed as illustrated in Fig. 6, including six failures outside of the predicted success area. Slow-motion footage confirmed that these failures occurred through microspines overload, as predicted by the model.

This model can further be used to predict successful perching conditions when thrust is still present following touchdown (right LSM in Fig. 6). This LSM was calculated by considering the variable delay introduced by detecting the foot impact with the accelerometer located on the airplane's body, given the soft suspension, and by considering the motor spin-down. These effects significantly reduce and reshape the LSM's area (e.g., zone A in Fig. 6).

5   Perching Controller Design

The proposed pitch-up trajectory takes advantage of the airframe's high thrust-to-weight ratio to significantly extend the suitable horizontal distance available

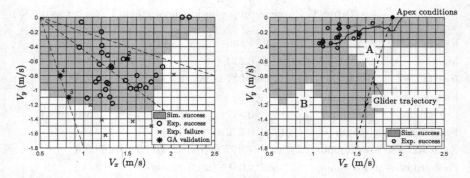

Fig. 6. Simulated LSM in green, with (right) and without (left) thrust. Simulations are performed with 5 spines engaged. (left) Experimental landings validating the simulated LSM ($q = 86 \pm 3°$, $\dot{q} = -50 \pm 30°/s$). Similar shear and normal force patterns are created for loading trajectories along the same radial dashed line. This reduces the number of necessary GA validation points for model calibration. (right) Touchdown velocity of 20 thrust-assisted approaches from horizontal flight. Two approach trajectories are also illustrated. (Color figure online)

for perching. However, thrust creates disturbances not present on a glider. These include yaw perturbation due to gyroscopic effect during pitch-up and roll perturbation, caused by the rotor torque required to accelerate the propeller at the entry of the SSD phase. To compensate for these perturbations, three decoupled PD feedback loops are used for each control surface [12,17]. Thrust-assisted perching is also difficult to perform under a feedforward control architecture. Indeed, small variations in initial battery voltage, airplane mass, CG position and launch speed all lead to large variations in the touchdown velocity and, consequently, failure to land as detailed in Table 2. The remainder of this section details a simple thrust controller, compares the results with feedforward control and describes tuning of the controller through the use of a classifier.

5.1   Thrust Control over Vertical Velocity (TCV²)

To create the desired SSD phase favoring smooth landings under a wide range of conditions, the proposed thrust controller (TCV²) utilizes a proportional feedback loop over RPM to maintain a desired vertical speed from wall detection. This feedback term is added to a constant command ($T_o \simeq mg$) as follows:

$$T_c = K_p(V_d - V_m) + T_o \tag{2}$$

where T_c is the output thrust, K_p is a proportional gain, V_d is the desired vertical speed, V_m is the measured vertical speed and T_o is the constant thrust command. The vertical speed is calculated through integration of the acceleration, starting from the Pixhawk estimate during horizontal flight.

Simulations are used to compare the TCV² and feedforward controller robustness by individually varying a set of nine variables that include airframe, actuator and sensing parameters, along with initial condition variations (fixed wall detection distance of 5.5 m). To capture the dynamics of the complete perching maneuver, this study combines a flight dynamics model developed by Khan et al. [18] with the hybrid perching model described in Sect. 4. The selected flight dynamics model captures various important aspects of the proposed perching maneuver, e.g., unsteady and high-alpha aerodynamics, effects of control surface deflections and propeller slipstream effects.

Table 2. Varied parameters and success range for each controller (FR: Full range)

Parameter	Units	Baseline	Range	Feedforward	TCV^2
V_x (body-fixed frame)	m/s	8	[6; 10]	[7.1; 8.6]	FR
V_z (body-fixed frame)	m/s	0	[−2; 2]	FR	FR
Pitch	°	0	[−20; 20]	[−16.8; 20]	FR
Mass	kg	0.32	[0.28; 0.36]	[0.31; 0.34]	FR
Inertia	kg m2	0.017	[0.012; 0.012]	FR	FR
Angular Velocity	°/s	0	[−57; 57]	FR	FR
Range sensor error	m	0	[−0.3; 0.3]	[−0.3; 0.26]	FR
CG position*	cm	−31.3	[−33.3; −29.3]	[−31.3; −29.6]	[−31.3; −29.9]
Battery level	%	100	[80; 100]	[96; 100]	FR
Accelerometer bias	m/s^2	0	[−0.1; 0.1]	-	FR

* Distance taken from the airplane's nose.

The results, presented in Table 2, demonstrate the advantages of TCV^2 over a feedforward controller. The most sensible parameters for a feedforward controller are mass and battery level. If these parameters are not carefully tuned, the thrust and gravity forces are unbalanced and the vertical velocity varies throughout the maneuver. Changing the CG position also modifies the capability of the airplane to rapidly pitch-up, thus affecting the distance required to perform the maneuver. As expected, the TCV^2 feedback controller compensates for variations of almost all of these parameters.

5.2 TCV^2 Proportional Gain Sizing

A more thorough analysis is required to tune the controller for maximum success rate, given simultaneous variations of the identified parameters over their allowable range (Table 2). To speed up the analysis, a support vector machine (SVM) classifier that evaluates the combinations for successful landing is trained on the complete numerical model, which includes the aircraft's dynamics during flight and after impact with the wall. This classifier is trained using 30,000 simulations designed using a Latin hypercube sampling method. A set of 85% of these simulations is used for training and 15% for validation. Defining the SVM's Gaussian radial basis function kernel with $\sigma = 1.4$ in the normalized space, a prediction accuracy of 92.8% is achieved for the validation set. The success rate is evaluated for each K_p, assuming a uniform distribution of other parameters over their respective range.

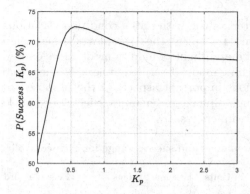

Fig. 7. Success probability of the perching maneuver for a known K_p.

The results of this analysis, shown in Fig. 7, reveal that a K_p value of approximately 0.5 leads to the highest probability of successful perching for the parameter variations described in Table 2. This probability rapidly decreases for lower K_p. A small proportional gain also introduces considerable increases in altitude - up to 6 m, as full thrust is applied throughout the full maneuver. Similarly, increasing the K_p value leads to excessively low thrust commands during the pitch-up phase which slows down the transition to SSD. This decreases the probability of success given a fixed wall detection distance.

6   Experimental Results

A series of tests were conducted on the platform presented previously. The airplane was launched towards a vertical wall covered with asphalt shingles at speeds ranging from 6.7 to 7.7 m/s. A total of 20 consecutive launches were performed, for which a 100% success rate was achieved. The touchdown velocities of these tests are mapped on the LSM (right) in Fig. 6, with two trajectories showing that the airplane maintains vertical descent speeds around the commanded value (−0.2 m/s). A typical landing is illustrated in Fig. 8. After such landing, the airplane can use its thrust to take off vertically as described in [12].

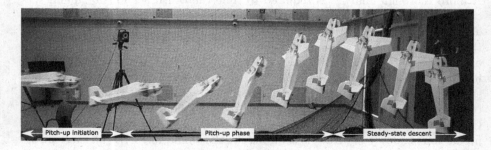

Fig. 8. Sequence of the thrust-assisted perching maneuver.

7   Conclusion and Future Work

This paper introduces S-MAD, the first fixed-wing UAV capable of thrust-assisted perching and takeoff. This small aerobatic platform performs a pitch up maneuver to rapidly slow down and expose its landing gear to vertical surfaces. A PixHawk controller samples the onboard sensors, including a laser range finder, and executes custom control loops to perform the maneuver autonomously.

Future work includes various improvements on the system, such as a more precise estimation of the vertical velocity and a mass reduction of the suspension. An important enhancement will also be the inclusion of non-contact sensors to turn off the propeller pre-emptively by early detection of incoming touchdowns. In the longer term, it is expected that this configuration of actuators, electronics and sensors will allow thrust-assisted wall climbing, aborted approaches and recovery from failed attachment.

The ability to perch reliably on vertical surfaces opens the door for repeated cycles of landing, standby, takeoff and flight. This enables extended mission durations for small UAVs, offering new types of applications. Ultimately, such bird-inspired platforms could be used for long duration surveillance, energy harvesting, inspection of structures or reconfigurable sensor networks.

Acknowledgement. This work was supported financially by FRQNT. We would like to thank the members of Createk and Prof. Nahon's lab at McGill for the great help that they provided throughout the realization of the work presented.

References

1. Roderick, W.R., Cutkosky, M.R., Lentink, D.: Touchdown to take-off: at the interface of flight and surface locomotion. Interface Focus **7**(1), 20160094 (2017)
2. Thomas, J., et al.: Aggressive flight with quadrotors for perching on inclined surfaces. J. Mech. Robot. **8**(5), 051007 (2016)
3. Myeong, W., et al.: Development of a drone-type wall-sticking and climbing robot. In: International Conference on Ubiquitous Robots and Ambient Intelligence (2015)
4. Kalantari, A., et al.: Autonomous perching and take-off on vertical walls for a quadrotor micro air vehicle. In: International Conference on Robotics and Automation (2015)
5. Liu, Y., Sun, G., Chen, H.: Impedance control of a bio-inspired flying and adhesion robot. In: International Conference on Robotics and Automation (2014)
6. Daler, L., et al.: A perching mechanism for flying robots using a fibre-based adhesive. In: International Conference on Robotics and Automation (2013)
7. Graule, M., et al.: Perching and takeoff of a robotic insect on overhangs using switchable electrostatic adhesion. Science **352**, 978–982 (2016)
8. Pope, M., et al.: A multimodal robot for perching and climbing on vertical outdoor surfaces. IEEE Trans. Robot. **33**, 38–48 (2016)
9. Moore, J., Cory, R., Tedrake, R.: Robust post-stall perching with a simple fixed-wing glider using LQR-trees. Bioinspir. Biomim. **9**(2), 025013 (2014)
10. Kovač, M., Germann, J., Hürzeler, C., Siegwart, R.Y., Floreano, D.: A perching mechanism for micro aerial vehicles. J. Micro-Nano Mech. **5**, 77–91 (2009)

11. Lussier Desbiens, A.: Landing and perching on vertical surfaces. Ph.D. thesis, Stanford University (2012)
12. Lussier Desbiens, A., Asbeck, A.T., Cutkosky, M.R.: Landing, perching and taking off from vertical surfaces. Int. J. Robot. Res. **30**, 355–370 (2011)
13. Schmidt, D., Berns, K.: Climbing robots for maintenance and inspections of vertical structures'a survey of design aspects and technologies. Robot. Auton. Syst. **61**(12), 1288–1305 (2013)
14. Asbeck, A., et al.: Scaling hard vertical surfaces with compliant microspine arrays. Int. J. Robot. Res. **25**, 1165–1179 (2006)
15. Howell, L.L.: Compliant Mechanisms. Wiley, New York (2001)
16. Mitiguy, P.: Advanced dynamics and motion simulation (2014)
17. Green, W.E., Oh, P.Y.: A hybrid mav for ingress and egress of urban environments. IEEE Trans. Robot. **25**, 253–263 (2009)
18. Khan, W., Nahon, M.: Modeling dynamics of agile fixed-wing UAVs for real-time applications. In: International Conference on Unmanned Aircraft Systems (2016)

An Integrated Compliant Fabric Skin Softens, Lightens, and Simplifies a Mesh Robot

Anna Mehringer[1(✉)], Akhil Kandhari[1], Hillel Chiel[2], Roger Quinn[1], and Kathryn Daltorio[1]

[1] Department of Mechanical and Aerospace Engineering,
Case Western Reserve University, Cleveland, OH 44106-7222, USA
`{agm72,kam37,rdq}@case.edu`
[2] Departments of Biology, Neurosciences and Biomedical Engineering,
Case Western Reserve University, Cleveland, OH 44106-7080, USA
`hjc@case.edu`

Abstract. Earthworms are particularly skilled at navigating through confined spaces. Therefore, creating a soft robot that mimics their peristaltic locomotion could provide unique advantages for pipe inspection, search and rescue, exploration, and medical applications. Here we present the design of a new robot, FabricWorm, that like its predecessor, CMMWorm, has six segments that are actuated with circumferential cables sequentially to mimic the peristaltic motion in an earthworm. However, compared to its predecessor, FabricWorm is 41% softer, is 23% lighter, and has 64% fewer rigid structural components due to the integration of the mesh within a fabric skin. These improvements, and the benefit of a continuous fabric skin, can be important advantages for worm-like robots.

Keywords: Textiles in soft robotics · Earthworm –like peristaltic locomotion · Biological inspiration

1 Introduction

Because of their unique ability to traverse tight burrows with peristaltic locomotion [1], earthworms have inspired the designs of many novel soft robots. Potential future applications include pipe inspection [2], robotic surgery or diagnostics [3, 4], search and rescue [5], and exploration [6]. Such soft robots may be actuated using shape memory alloys [7–10], pneumatics [11–13], hydraulics [14], electroactive polymers [15, 16], and cables [17–21].

Servomotors in conjunction with cables are a fast and convenient way to control hyper-redundant worm robots; however, our servo-actuated robots [17, 20, 21, 31] seem to require a stiffer structure than similar robots that use compressed air or SMAs [8, 11]. The bodies of two of our previous soft worm robots (SoftWorm and Compliant Modular Mesh Worm (CMMWorm)) are made of flexible tubing that form a grid or mesh of rhombuses (Fig. 1). They require hard 3-D printed vertexes to connect the tubing, hard rubber feet, metal eyelets to route the cables, and metal restorative springs in addition to the rigid actuators and their mounting supports. The relative rigidity of

© Springer International Publishing AG 2017
M. Mangan et al. (Eds.): Living Machines 2017, LNAI 10384, pp. 315–327, 2017.
DOI: 10.1007/978-3-319-63537-8_27

the mesh transmits the local changes in aspect ratio to other parts of the structure. The continuous flexibility of the mesh structure permits different areas of the body to have different aspect ratio. In a companion paper [22], we show with kinematic models that the mesh is functionally interpolating between cables.

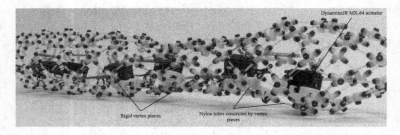

Fig. 1. The CMMWorm travels by a sequence of contractions and expansions of segments resulting in peristaltic motion. Its mesh is made up of rigid vertex pieces and nylon tubes and has springs, which passively restore the motion to its largest allowable diameter [20].

The softness of invertebrates like earthworms is one of the traits engineers would like to mimic in biologically-inspired robots for many applications [2–6]. Exterior softness permits compliance to the environment for improved traction, but the interior compliance is valuable too: it reduces brittle failures and allows the body to squeeze through small openings.

Our goal for this work is to decrease the overall stiffness and the number of hard plastic parts in the mesh while preserving the functionality for peristaltic locomotion via cable actuation. The challenges are to permit the vertexes of the mesh to rotate, and perhaps deform, as much as possible without the structure becoming so loose as to be ineffective for locomotion. Additionally, since cables pull only one way, the body stiffness needs to return the segments to their original un-actuated shape.

Fig. 2. FabricWorm, a six DOF worm-like robot with a fabric-integrated-mesh

We achieve this by developing a robot prototype, FabricWorm, which incorporates textiles into the body design (Fig. 2). Compared to CMMWorm [20] the novel fabric-integrated-mesh eliminates 64% of the rigid 3D printed vertexes, and is 41% softer (see Table 1). Furthermore, it permits faster locomotion and greater range of motion. Another benefit is that the fabric forms a continuous skin. Such a skin could protect the interior of a soft robot from debris and provide a surface upon which to mount continuous sensors.

Table 1. Summary of our mesh-based worm-like robots. Softworm demonstrated the potential for fast peristaltic locomotion driven by one motor with a cam mechanism [17]. CMMWorm has a single actuator per segment in order to test responsive control strategies [20]. FabricWorm, presented here, has a new fabric-integrated-mesh design, which provides many advantages. CMMWorm and FabricWorm use the same actuators, which are much slower than the motor in SoftWorm. Because of this, CMMWorm and FabricWorm will never be as fast as SoftWorm unless different actuators are used. Some advantages of these actuators in CMMWorm and FabricWorm are that they can be precisely manipulated to test different waveforms, can provide position feedback to a computer, and could allow for turning in the future.

Robot	Total mass (kg)	Largest diameter (cm)	Total rest length (cm)	Tensile stiffness (N/cm)	Number of actuators	Number of rigid joints	Maximum speed (cm/min)
Softworm	3.80	22.0	50	1.7	1	168	400
CMMWorm	2.08	20.6	103	2.9	6	132	25. 8
FabricWorm	1.61	21.0	67	1.7	6	48	40.2

2 Design

2.1 Materials

Textiles can provide unique advantages for soft robots. Fabric is lightweight, flexible, strong, and inexpensive. Fabric is more durable than paper in origami hinges, as in the origami wheel robot [23]. In wearable technology, fabric is a natural choice for sensor skins or exoskeletons that are deformable and comfortable for the user [24, 25]. New research is developing textiles that are self-healing [26], power-generating [27], and solar charging, see review of nanotextiles [28]. Inextensible fibers can be embedded in or wrapped around molded structures to control pneumatic shapes (for example in [29]). Specifically for worm-like robots, the braided fiber mesh cylinders [8, 11] function similarly to the mesh structure here to provide appropriate coupling between body length and diameter. Although we have never used fabric in a worm robot, we have previously used nylon fabric to cover and support a pneumatic gripping mechanism, modeling the feeding apparatus of sea slug, which could be valuable on the head of such a worm robot [30].

Here, the selected fabric permits large elastic deformation of the mesh structure. CMMWorm segments are 30% longer when fully elongated than when fully retracted. In fabrics, this kind of "stretchiness" could come either from elastic fibers or from knitted textiles. As opposed to woven fabrics, knits can achieve large, recoverable deformations with relatively inextensible materials. Contrary to some plastics, these materials are long lasting even at high temperatures. Moreover, the stiffness is anisotropic which we take advantage of in our design. We choose a knitted cotton fabric that was measured to have three times stretch and is made out of 97% cotton and 3% spandex (Fig. 3). The stiffer crosswise direction was aligned with the circumferential direction of the body.

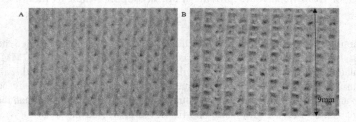

Fig. 3. 9 mm of the knitted fabric, under 30X magnification. (A) The fabric is in its resting state. (B) The fabric is stretched horizontally.

The selected fabric provides the passive restoring force to return segments to the maximum (un-actuated) diameter, thereby eliminating the metal longitudinal springs used in CMMWorm. CMMWorm's springs were selected to be strong enough to overcome the structural stiffness and damping in the mesh, but not so strong as to break the actuation cables. The fabric is suitable for this purpose because its stiffness for a single segment's worth of fabric is approximately the same as the steel springs (Fig. 4).

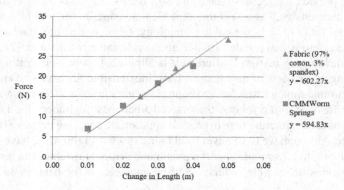

Fig. 4. Stiffness of spring in CMMWorm compared to the stiffness of the fabric in one segment. To measure the stiffness, one end of the spring or fabric was fixed to a table, and the other end was stretched horizontally with increasing forces. The force value and corresponding change in length were recorded. The stiffness value for one segment is the slope of the trend line through the data points in N/m.

Fig. 5. The tubing and vertex pieces between two actuators, shown without the addition of fabric skin.

Integrated into the fabric, a "mesh" of flexible 1/8 in. outer diameter tubing provides the body a compliant nominal structural worm-like shape. The tubing is connected using the same 3D printed vertex designs in CMMWorm (although fewer of the vertexes are needed, see below). Like CMMWorm, the connectors allow us to experiment with different stiffnesses and lengths of the connecting tubing. Here we tried nylon thick-walled tubing (0.026" wall thickness) with higher bending stiffness and nylon thin-walled tubing (0.016" wall thickness) with lower bending stiffness. We characterize the effects of the different types of tubing on the overall body stiffness in Sects. 3.1 and 3.2, and we use thick-walled tubing for the locomotion experiments in Sect. 4.

2.2 Assembly

Two 80 cm × 70 cm rectangles were cut out of the fabric, with the shorter 70 cm length in the direction of the stretch. The cylindrical mesh of CMMWorm was separated along the ventral surface to form a flat grid of rhombuses. Using this as a template, we drew the same size grid lines on one of the fabric pieces (Fig. 6). We sewed the two fabric pieces along these lines with all-purpose thread on a Husqvarna Viking Opal™ 650 sewing machine using a straight stitch and being careful not to introduce pre-strain. This results in sleeves to encase the tubing.

Like our previous mesh-based robots and simulations [17, 20, 22], this mesh can be visualized as an array of rhombuses that change aspect ratio. In other words, the rhombuses are tall and narrow at the maximum diameter, they are wide and short at the minimum diameter, and the rhombus side length is roughly constant. The actuators pull cables that are threaded circumferentially to decrease the diameter. The rhombus structure ensures that when the diameter decreases the length increases, resulting in locomotion [17, 21].

Whereas CMMWorm had a vertex piece at every tubing intersection, here we only insert vertex pieces at the six points that form a ring around each actuator, and the six points at the front and six points at the rear of the body. In CMMWorm, the purpose of the vertex pieces is to permit relative rotation of the tubing pieces but prevent relative translation (a pin joint). In FabricWorm, the rhombuses between segments are held in place by the fabric sleeves. Without the fabric, the rhombus structure would not be

Fig. 6. The drawn lines on one layer of fabric serve as a guide for sewing sleeves for the tubing to pass through. The red dots represent the locations of the 3D printed vertex pieces, and the green dots represent the locations of the actuators. The robot body begins as two dimensional and then becomes three dimensional when the tubing and vertex pieces are added. (Color figure online)

maintained (see Fig. 5), and diameter contraction would not yield elongation or forward progress.

Because the tubes are longer (18.5 cm as compared to 4.8 cm in CMMWorm), the robot has less overall components. CMMWorm has 252 tubes connecting all the vertex pieces, which are time consuming to assemble, but the FabricWorm only has 84 tubes.

In order to insert the vertex pieces, small cuts were made in the inner layer of fabric. Snaps were sewn on the top and bottom edges of the rectangle so that the fabric could be attached to itself in a cylindrical shape. As in [21], MX-64T Dynamixel smart actuators are attached to special servo mounting vertex pieces that include a spool for the actuating Spectra® cable.

2.3 Actuation and Control

The Dynamixel MX-64T actuators for each segment are connected with a serial bus to a single ROBOTIS OpenCM9.04 microcontroller (32-bit ARM Cortex-M3, STM32F103CB, 72 MHz), which connects to a linear DC power supply and a computer. The microcontroller is mounted on the top of the actuator at the end of the robot and communicates with the six actuators. Programming and data logging with the microcontroller are done with a USB connection to a PC.

From the maximum diameter position (minimum tension in actuator cable), the actuators can rotate 571% before the inner diameter of the mesh constricts to nearly touch the actuator bodies. For effective locomotion, pairs of segments will elongate and retract at the same time. The actuators are controlled at their maximum speed of 63 RPM. Similar to what was done in the CMMWorm, the pattern of actuation in the robot can be described by a 3×1 wave [20], see Fig. 7, which has a contracted spacer segment between elongating and retracting segments.

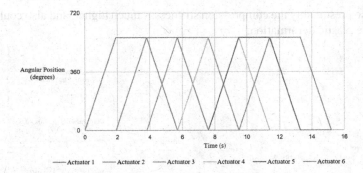

Fig. 7. Actuator Position for the 3 × 1 wave in FabricWorm. The 3 × 1 wave always has a contracted spacer segment between the contracting and expanding segments. The time duration for the propagation of this wave is about 15 s.

3   Fabric-Integrated-Mesh Characterization

The overall mass of the FabricWorm robot is 1.61 kg, and the mass of the CMMWorm is 2.08 kg for six segments. The ROBOTIS Dyanmixel MX-64T actuators account for 50.42% of the mass, the standard vertex pieces (each with 4 Legris™ fittings) account for 19.03% of the mass, the fabric skin accounts for 17.12% of the mass, the actuator mounts account for 8.38% of the mass, and the flexible high pressure nylon tubing accounts for 5.06% of the overall mass.

3.1   Longitudinal Stiffness and Range of Motion

The fabric-integrated-mesh gives FabricWorm a much greater range of motion vs. CMMWorm. FabricWorm can extend itself to be as long as 100 cm and compress to as short as 27 cm. For comparison, the rigid parts in CMMWorm limit its range of motion from 103 cm to 134 cm in length. In this section, we measure the tensile and compressive stiffness from the resting (unactuated) configuration at 67 cm long, on low friction silicone paper ($\mu = 0.17$).

Tensile stiffness is measured by securing one end of the robot with a heavy weight and applying increasing loads at the other end while measuring the extension of the robot [31]. The slope of the regression line between displacement and force is the measured tensile stiffness. With the thick wall tubing, the tensile stiffness is 1.7 N/cm, and with the thin wall tubing, the tensile stiffness of the mesh is 1.3 N/cm. For comparison, the tensile stiffness of the CMMWorm is 2.9 N/cm [20].

Compressive stiffness is measured using a set of parallel boards that were compressed via cables at increasing forces [31]. Using a linear fit, the average compressive stiffness with the thick wall tubing is 1.0 N/cm and with the thin wall tubing is 0.54 N/cm (See Fig. 8). For comparison, CMMWorm can only be compressed shorter than the resting length by large deformation of the tubes rather than rotation at the vertexes, so presumably the compression stiffness is much higher, and also could potentially cause plastic deformation.

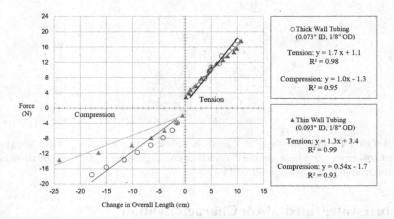

Fig. 8. Stiffness of the six-segmented FabricWorm. The tubing stiffness has a small influence on the overall tensile stiffness of the body and a greater influence on the compressive stiffness. The y-intercept reflects the force required to overcome friction in the body and along the ground.

3.2 Bending Stiffness

Because of the unique structure of this robot, it is more compliant in bending than CMMWorm, which promises greater ability to turn. Unlike CMMWorm, the body of the robot is capable of being bent in a semicircle. Similar to the experiments above, the bending stiffness was approximated as 2.1 Nm/radian with thick wall tubing and 1.23 Nm/radian with thin wall tubing (Fig. 9).

3.3 Effectiveness of Length Diameter Coupling

To determine which tubing is better for locomotion, a coupling ratio, the ratio of the increase in length to the change in diameter was evaluated. If the coupling ratio is too low, the robot will not move. In CMMWorm, the vertex pieces control the coupling ratio of the structure. The vertex pieces have a range in angle from 50 to 100%. A decrease of diameter by 10 cm in two adjacent segments resulted in an extension of length of 10 cm, so the coupling ratio is about 1.0. In FabricWorm, tubing couples contraction in diameter to longitudinal expansion. To measure the aspect ratio, the front two segments were contracted by equal amounts to change the diameter and then the forward change in length of the robot was measured (Fig. 10).

In FabricWorm because the vertexes are not as constrained, the range of motion is greater, and the coupling ratio can be higher. When the thick wall tubing is used, the coupling ratio is 1.6 because the tubing is stiff enough to expand the segments forward. The reason that the thin wall tubing has such a low coupling ratio of 0.42 is because the tubing is too soft. When the segments are actuated, the tubing buckles instead of

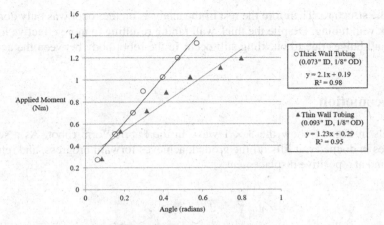

Fig. 9. Bending stiffness of FabricWorm with two different tubing thicknesses. This was measured by applying forces orthogonal to the side of the robot body, and the angle displaced was measured. The moment was calculated by multiplying by the moment arm [31].

expanding the segment forwards. Although our goal was to make the robot as soft as possible, this data shows that there is a limit to how soft the tubing can be for locomotion with the design.

Measuring the coupling ratio of two segments of the robot with specific tubing is a quality indicator of how well the robot will move with that tubing. When the robot was run with the thin wall tubing, it moved a negligible amount due to the buckling of the tubes. The thin wall tubing absorbs the actuation locally rather than transmitting to the

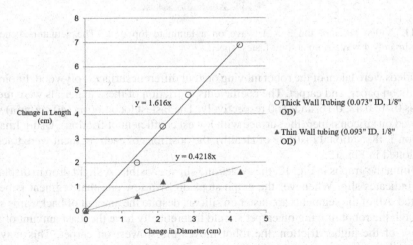

Fig. 10. The coupling ratio was measured to determine the effectiveness of the tubing in expanding two adjacent segments forward. In this experiment, the robot started in the fully retracted diameter position, and then the front two segments were contracted to smaller diameters [31]. The extension in length due to the contraction in diameter of these segments was recorded.

rest of the structure. Therefore the rest of the analysis on the robot was only done with the thick wall tubing. Despite the thick wall tubing resulting in more effective locomotion, small bulges due to buckling still occur in the robot body between the actuation cables.

4   Locomotion

The stills in Fig. 11 show the 3 × 1 wave in the FabricWorm robot. As a segment decreases in diameter, it lifts off the ground, achieves forward progress, and returns to the ground at a positive displacement.

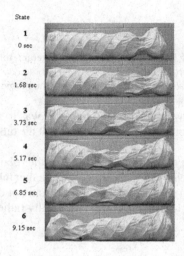

Fig. 11. Video stills of the 3 × 1 wave on a laminate top desk. The actuators generate approximately 4 waves/min at their fastest speed.

Videos were taken of the robot moving on four different surfaces: plywood, linoleum tile, silicon paper, and carpet. The coefficients of friction of these materials were measured as 0.46, 0.43, 0.16, and 0.60 respectively. The fastest locomotion (40 cm/min) was achieved on silicon paper (the surface with lowest coefficient of friction). Using ImageJ (version 1.48, National Institute of Health), the positions of each segment were tracked and plotted in Fig. 12.

Within the graphs in Fig. 12, the cycles and slip are visible. A slight drop in displacement indicates slip. Whenever the graph steps up by a few cm, that segment is being actuated. All of the segments are faster on silicon, despite the amount of backwards slip. Although the robot moving on carpet should theoretically have the least amount of slip because of the higher friction, the robot moves the slowest on carpet. This may be because the carpet prevents the robot from moving as far forward as the silicon paper allows. A more complicated actuator waveform may improve the speed (see companion paper [22]). Another interesting observation from Fig. 12 is that after a transient in the first few cycles, the robot gradually stretches out to oscillate around a steady state length.

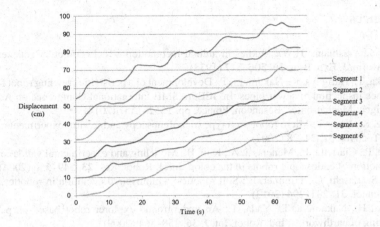

Fig. 12. Displacement of each segment over time on silicon paper. The coefficient of static friction between the robot and silicon paper is 0.17.

5   Discussion and Future Work

Here we present the design and control of a new FabricWorm robot, and a full direct comparison with its predecessor CMMWorm, which has the same actuators. Fabric-Worm is softer, lighter, and uses fewer pieces. Because of the greater coupling ratio, it is also faster.

The fabric skin itself has advantages and disadvantages. It protects the interior from debris. The skin is a surface upon which friction-altering surface treatments (such as dots of non-skid foam or anisotropic friction worm-like setae) can be affixed. Similarly, the surface could also support the attachment of new surface sensors or a soft gripper [30] on the anterior in future work. Some disadvantages of the fabric-integrated-mesh are that it is difficult to change the stiffness of the return force of the robot. In addition, the fabric makes it difficult to make repairs to the robot because the fabric body encloses the actuators.

The actuation of six actuators permits direct comparison with CMMWorm, but in future work such a fabric-integrated-mesh can be actuated to take better advantage of the even greater range of motion that we have shown would be possible. The range of motion could be especially helpful for executing sharper turns, crawling in pipes of varying diameter, or squeezing into tight spaces.

Acknowledgments. This work was supported by NSF research Grant No. IIS-1065489.

References

1. Gray, J., Lissmann, H.W.: Studies in animal locomotion VII. Locomotory reflexes in the earthworm. J. Exp. Biol. **15**, 506–517 (1938)
2. Tanaka, T., Harigaya, K., Nakamura, T.: Development of a peristaltic crawling robot for long-distance inspection of sewer pipes. In: IEEE/ASME International Conference on Advanced Intelligent Mechatronics, Besançon, France, pp. 1552–1557 (2014)
3. Wang, K., Yan, G.: Micro robot prototype for colonoscopy and in vitro experiments. J. Med. Eng. Technol. **31**, 24–28 (2007)
4. Dario, P., Ciarletta, P., Menciassi, A., Kim, B.: Modeling and experimental validation of the locomotion of endoscopic robots in the colon. Int. J. Robot. Res. **23**, 549–556 (2004)
5. Kim, S., Laschi, C., Trimmer, B.: Soft robotics: a bioinspired evolution in robotics. Trends Biotechnol. **31**, 287–294 (2013)
6. Omori, H., Nakamura, T., Yada, T.: An underground explorer robot based on peristaltic crawling of earthworms. Ind. Robot. Int. J. **36**, 358–364 (2009)
7. Vaidyanathan, R., Chiel, H.J., Quinn, R.D.: A hydrostatic robot for marine applications. Robot. Auton. Syst. **30**, 103–113 (2000)
8. Seok, S., Onal, C.D., Cho, K.-J., Wood, R.J., Rus, D., Kim, S.: Meshworm: a peristaltic soft robot with antagonistic nickel titanium coil actuators. IEEE/ASME Trans. Mechatron. **18**, 1485–1497 (2013)
9. Mazzolai, B., Margheri, L., Cianchetti, M., Dario, P., Laschi, C.: Soft-robotic arm inspired by the octopus: II. From artificial requirements to innovative technological solutions. Bioinspiration Biomimetrics **7**(2), 025005 (2012)
10. Umedachi, T., Trimmer, B.A.: Design of a 3D-printed soft robot with posture and steering control. In: Proceedings of IEEE International Conference on Robotics and Automation, Hong Kong, pp. 2874–2879 (2014)
11. Mangan, E.V., Kingsley, D.A., Quinn, R.D., Chiel, H.J.: Development of a peristaltic endoscope. In: Proceedings of IEEE International Conference on Robotics and Automation, pp. 347–352 (2002)
12. Onal, C.D., Chen, X., Whitesides, G.M., Rus, D.: Soft mobile robots with on-board chemical pressure generation. In: International Symposium on Robotics Research, pp. 1–16 (2011)
13. Tolley, M., Shepherd, R., Mosadegh, B., Galloway, K., Wehner, M., Karpelson, M., et al.: A Resilient. Untethered Soft Robot. Soft Robot. **1**(3), 213–223 (2014)
14. Katzschman, R.K., Marchese, A.D., Rus, D.: Hydraulic autonomous soft robotic fish for 3D swimming. In: Proceedings of the International Symposium Experimental Robotics, pp. 1–15 (2014)
15. Jung, K., Koo, J.C., Nam, J., Lee, Y.K., Choi, H.R.: Artificial annelid robot driven by soft actuators. Bioinspiration Biomimetics **2**(2), S42-9 (2007)
16. Carpi, F., Menon, C., De Rossi, D.: Electroactive elastomeric actuator for all-polymer linear peristaltic pumps. IEEE/ASME Trans. Mechatron. **15**(3), 460–470 (2010)
17. Boxerbaum, A.S., Chiel, H.J., Quinn, R.D.: A new theory and methods for creating peristaltic motion in a robotic platform. In: Proceedings of IEEE International Conference on Robotics and Automation, Anchorage, AK, pp. 1221–1227 (2010)
18. Renda, F., Cianchetti, M., Giorelli, M., Arienti, A., Laschi, C.: A 3D steady-state model of a tendon-driven continuum soft manipulator inspired by the octopus arm. Bioinspiration Biomimetrics **7**, 025006 (2012)
19. Jones, B.A., Walker, I.D.: Kinematics for multisection continuum robots. IEEE Trans. Robot. **22**(1), 43–55 (2006)

20. Horchler, A., Kandhari, A., Daltorio, K., Moses, K., Ryan, J., Stultz, K., et al.: Peristaltic locomotion of a modular mesh-based worm robot: precision, compliance, and friction. Soft Robot. **2**, 135–145 (2015)
21. Horchler, A.D., Kandhari, A., Daltorio, K.A., et al.: Worm-like robotic locomotion with a compliant modular mesh. In: Proceedings of International Conference on biomimetic and biohybrid systems, vol. 9222, Barcelona, Spain, pp. 26–37 (2015)
22. Huang, Y., Kandhari, A., Chiel, H.J., Quinn, R.D., Daltorio, K.A.: Mathematical Modeling to Improve Control of Mesh Body for Peristaltic Locomotion, Living Machines 2017 (submitted)
23. Lee, D., Kim, J., Park, J., Kim, S., Cho, K.: Fabrication of origami wheel using pattern embedded fabric and its application to a deformable mobile robot. IEEE (2014)
24. Case, J.C., Yuen, M.C., Mohammed, M., Kramer, R.K.: Sensor skins: an overview. In: Rogers, J., Gharrari, R., Kim, D.-H. (eds.) In Stretchable Bioelectronics for Medical Devices and Systems, pp. 13–191. Springer, New York (2016)
25. Quinlivan, B.T., Lee, S., Malcolm, P., Rossi, D.M., Grimmer, M., Siviy, C., Karavas, N., Wagner, D., Asbeck, A., Galiana, I., Walsh, C.J.: Assistance magnitude versus metabolic cost reductions for a tethered multiarticular soft exosuit. Sci. Rob. **2**(2), eaah4416 (2017)
26. Gaddes, D., Jung, H., Pena-Francesch, A., Dion, G., Tadigadapa, S., Dressick, W., et al.: Self-healing textile: enzyme encapsulated layer-by-layer structural proteins. ACS Appl. Mater. Interfaces. **8**(31), 20371–20378 (2016)
27. Kim, K., Chun, J., Kim, J., Lee, K., Park, J., Kim, S., et al.: Highly stretchable 2D fabrics for wearable triboelectric nanogenerator under harsh environments. ACS Nano **9**(6), 6394–6400 (2015)
28. Coyle, S., Wu, Y., Lau, K., De Rossi, D., Wallace, G., Diamond, D.: Smart nanotextiles: a review of materials and applications. MRS Bull. **32**(5), 434–442 (2007)
29. Connolly, F., Polygerinos, P., Walsh, C.J., Bertoldi, K.: Mechanical programming of soft actuators by varying fiber angle. Soft. Robot. **1**, 26–32 (2015)
30. Mangan, E.V., Kingsley, D.A., Quinn, R.D., Sutton, G.P., Mansour, J.M., Chiel, H.J.: A biologically inspired gripping device. Ind. Robot. Int. J. **32**, 49–54 (2005)
31. Mehringer, A., FabricWorm, A.: Biologically-Inspired Robot That Demonstrates Structural Advantages of a Soft Exterior for Peristaltic Locomotion, OhioLink (2017, submitted)

Causal Biomimesis: Self-replication as Evolutionary Consequence

Gabriel Axel Montes[1,2](✉) (iD)

[1] University of Newcastle, Newcastle, Australia
gabriel@neuralaxis.com
[2] Hunter Medical Research Institute, Newcastle, Australia

Abstract. For millions of years, hominins have been engaged in tool-making and concomitant experimentation. This cognitive enterprise has eventually led to the creation of synthetic intelligence in the form of complex computing and artificial agents, whose purported purpose is to elucidate the workings of human biology and consciousness, automate tasks, and develop interventions for disease. However, much of the expensive research efforts invested in understanding complex natural systems has resulted in limited rewards for treatment of disease. This paper proposes the novel 'causal biomimesis' hypothesis: with respect to the relationship between humans and artificial life, the virtually inevitable intrinsic evolutionary consequence of tool-making and biomimetic efforts—and the capacity for objective thought and the scientific method itself—is the full-scale replication of human cognitive functionality, agency, and potentially consciousness *in silico*. This self-replication transpires through a cycle of anthropogenic biomimetic auto-catalysis driven by instrumental cognition—from objective reasoning in hominin tool-maker through to post-biological reproduction by synthetic agents—and is self-organized and co-enacted between agent and the produced artefactual aggregates. In light of this radical hypothesis, existential and ethical implications are considered for further exploration.

Keywords: Artificial intelligence · Cognition · Biomimetics · Neuroscience · Neurophenomenology · Consciousness · Complexity · Philosophy · Robotics · Evolution · Psychology · Medicine · Anthropology · Mind-body

1 Introduction

At present, rational human intelligence has reached a major milestone in its ability to create artificial intelligence (AI). The proliferation of AI has led to promises regarding its capabilities, which across the spectrum of opinion, appear to incite equal parts optimism and apprehension. There is an increasing interest in and adoption of artificial agents, such as robotic nurses and assistants for humans, meaning that the future holds a more ubiquitous presence of AI in rather plain sight that is in more intimate relationship with humans. The implications of AI for humanity implore more careful examinations of the motivations behind the development of AI. The field of biomimetics, developing technologies based the distillation of principles from the study of biological systems, invites a broader and deeper understanding of biology and

© Springer International Publishing AG 2017
M. Mangan et al. (Eds.): Living Machines 2017, LNAI 10384, pp. 328–347, 2017.
DOI: 10.1007/978-3-319-63537-8_28

cognition in order to more fully inform the technological development of synthetic agents. Scrutiny into the origins and workings of cognition in conscious agents, particularly *Homo sapiens*, can not only inform the construction of synthetic architectures, but may also reveal the inchoate origins of 'artificiality' within biological life itself.

This paper will excavate the evolutionary development of cognition that precedes and underpins the observed drive in humans to develop artificial intelligence and agents. This examination will be juxtaposed with the commonly cited rational justifications for the development of synthetic life, and the evolutionary consequence of human cognitive capacities will be offered. The terms AI, artificial life, synthetic life, synthetic agents, and so forth will be used relatively interchangeably; the focus herein is the broader picture of human-AI evolution rather than the technical differences between those terms. This paper makes a case for *Homo sapiens* as an intermediate step in the evolution of life, based on a novel hypothesis about the evolutionary process itself, and serves as a theoretico-philosophical touchstone for further inquiry, research, and debate.

2   Instrumental Cognition: From Hand to Synthetic Agents

> *"The hand supplies all instruments, and by its correspondence with the intellect gives [humankind] universal dominion."* —Sir C. Bell

2.1   Prehension

Hominins have engaged in tool use since at least 3.3 million years ago [1]. It is the widely-accepted view that hominins descended from tree-dwelling ancestors, who were locomotive on all four limbs (quadrupedal), all of which were adapted for gripping in various contexts, including arboreal brachiation and the process of procuring food. An arboreal habitat involved the development of geospatial memory maps for navigation and location of foods such as fruit, which had been contributing to an expanding brain capacity [2]. Transitioning from an arboreal habitat to a terrestrial one, a bipedal posture allowed the hand to be freed up for other functions and triggered a strong set of selection pressures that set the stage for the hand to serve new adaptive functions predicated upon the existing grip functionality. This propelled a process of co-evolution between hand and brain. The brain's functional capacity increased to serve the adaptive hand, and the hand developed new abilities due to the expanding neurocognitive capabilities, creating an accelerating loop of positive feedback. The selection pressures were very strong in the new terrestrial environment, beckoning proto-humans to continually adapt the hand to novel uses. In this process, what was formerly the hand's "power grip" evolved into a "precision grip", conferring the hand a broader spectrum of delicate manipulations. This in turn sensitized and developed the brain's somatomotor regions, promoting a reinforcing spiral of prehensile dexterity.

As the hands freed up and survival in a new environment remained a challenge, tool use eventually arose. The intimate relationship between brain, hand, and tool extended beyond mere behaviorism; tool-users were embodied agents seeking to achieve goals.

The brain-tool duo functions as a kind of etcher against a cognitive background; thus, what is of interest is this cognitive background underlying tool use. In light of this, two key psychological properties behind tool use can be deduced [3]. The first psychological property is creative experimentation. This involves ingenuity and invention. There is an ability in early man to envision a new use for a material; it can be repurposed or fashioned. A solution to a perceived problem must be invented. A hand gripping a tactical stone to cut a piece of wood transforms into attaching a lever or handle to the stone, resulting in an axe. The axe was a solution to the problem of the hand affording only little percussive power. This creative activity is not mere manipulation of foreign objects; it entails intelligent harnessing of available objects to arrive at a solution. Whether the solution is discovered playfully or purposefully or a combination thereof, creative experimentation plays a central role. Woven into the creative experimentation is a second psychological property: systematic means-end reasoning. A desirable prospective state (goal) is envisaged and a means is formulated to achieve it. This involves teleological thinking. Early man must have an idea of what a tool is, that making it involves work, and that the work must have some utility or reward, for self, group, or another. The geospatial mental map that developed from foraging fruit became co-opted for creating conceptual maps of ideas, goals, and materials—indeed, this dovetails with recent neuroscientific evidence demonstrating that the human entorhinal cortex serves dual roles of representing both geospatial and conceptual knowledge [4]. The agent becomes a tool-maker and-user that thinks instrumentally via tool-conceptualisation and sees the world as a repository of potential tools [3] (Fig. 1-1). It is this underlying cognitive complex, which was accelerated by the bipedal liberation of the hands for dexterity-driven brain development, that differentiates the tool-intentional intelligence of man from other sentient organisms' behavioristic manipulations of the environment and who did not have the foundational boon of prehensile dexterity.

Objectivity and Agency. By thinking in terms of tool utility, self-representation takes shape and thereby self-consciousness emerges as a virtualized self-model of autonomy to aid in survival, problem-solving, and sociality in an entropic world [5–7]. Self-consciousness is a quintessential element of the tool-making process and becomes increasingly complex as tools and the supporting cognitive abilities progressively grow. As instrumental cognition and self-consciousness strengthen together, so does the capacity for objective thought. The agent is better able to think in a utilitarian manner, about what his needs are and what is good for him. He introspects to whatever degree the brain has sufficiently developed due to tool cognition at that point in time, which affords him insights into his state of mind, interoceptive (internal corporeal) sensations, needs and desires, and so on, which he uses to inform his situated actions in the world. Crucially, he develops the opportunity to become aware of the intentionality that underlies his behaviors. In addition to increased objectivity and self-reflective capacity over his own consciousness, the agent also considers information about his peers when planning or taking action or making tools to serve the group's needs. By making judgments about the states of other agents, he is strengthening his ability to think in terms of theory of mind about both others and himself, in turn strengthening his own introspective capabilities as well as his social cognition. As this cooperative process

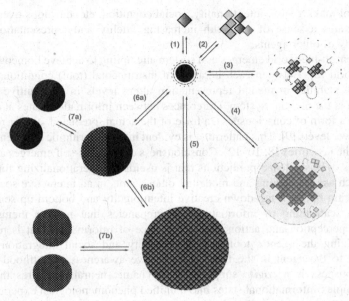

Fig. 1. The evolution of causal biomimesis through enactive instrumental cognition. Circles = homeostatic agents; squares/diamonds = synthetic artefacts; dotted line density = degree of agency-transference. Middle black circle denotes a hominin/human agent. Two-way arrows denote bidirectional influence and homeostasis. *(1) Prehensile bipedal hominin agent under selection pressure creates tools through instrumental cognition (creative teleological thought, intentionality, and self-consciousness). (2) Agent produces more aggregations of artefacts, objective thought is maturing, and proto-scientific methodology develops. (3) Artefact-holon ecosystem architecture is enacted through increasing maturation of scientific methodology, and holons evolve to resemble whole sets of human cognitive function. Gross and subtle biomimesis efforts are occurring. (4) Detectable agency-transference; autopoietic formation of artefacts achieve a recognizable coherent selfhood and are attributed agency by humans. Artificial intelligence (AI) and synthetic agents are being designed and developed by humans. (5) Self-replication and verisimilar agency achieved; sufficiently functional agency-transference and counter-transference occur between biological/human agent(s) and synthetic agent(s). Autonomous agency, and potentially consciousness, is attained. (6) Biohybrid consciousness scenario in which (a) biological agents alternatively produce biohybrid agents first before synthetic ones, which then, (b) perhaps in concert with humans, produce fully synthetic agents. (7) Post-biological autonomous reproduction by (a) synthetic agents, and (b) biohybrid agents.* Dotted lines around central human agent denote that self-consciousness is not static, but is continuously shaped by the enactive process of instrumental cognition and subject to counter-transference by synthetic/biohybrid agents. Self-consciousness and -knowledge evolve in concert with the complexity of artefacts and their holons. Enactive instrumental cognition is also shaped by social interactions with other agents, and thus all the above relationships and processes can be inter-enacted on the intersubjective level by other human and synthetic/biohybrid agents. 1–7 represent the phases that all together constitute a complete biomimetic auto-catalytic cycle.

between tool-making/-use, intentionality, social cognition, etc. develops over time, the agent cultivates a sense of self with increasing fidelity and representations of the intentionality of other agents.

The agent-in-the-world emerges as a system attempting to achieve homeostasis in an entropic world [8]. Between each iteration of instrumental (tool) cognition, there are positions of stability in mental representation along levels in a cognitive hierarchy represented in the nervous system. Differences between information states at each level give rise to a form of consciousness, a trace of the actual, predicted, and/or retrodicted states of lower levels [9]. This patterning is evident in the algorithmic recurrent message passing in the neocortex [8, 10–12]. Consciousness of agent-as-self emerges as a gestalt of the levels of mismatched predictions that is useful for operationalizing functions of the self, such as tool-making and modeling other agents in an interactive social world. The interaction between top-down creative intentionality and bottom-up sensory data produces a scaffolding of informational discrepancies that resolve themselves by changes in perception and action. The state space of information that is relevant to operationalizing the agent's tool-making capability and social integration becomes represented to the agent in the form of subjective awareness of selfhood [5]. This self-consciousness is a *cogito* supported by dynamic neural structures that fluidly coheres complex informational states into a unified phenomenological experience. This type of consciousness, due to its adaptive function for survival in a world populated by other agents, is hypothesized to have existed in more elementary forms at least since the Cambrian period 550 million years ago, during which the fossil record indicates that most major animal phyla appeared [6, 13, 14]. However, the open-ended possibilities of prehensile state spaces in early man—thanks to the ever-increasing adaptation of the hands for various utilitarian functions—together with emergent self-consciousness formed a wellspring for virtually endless cognitive development and innovation.

2.2   Ostension

Up to this point, our early human agent has been engaged in prehension, in which his hands are used to grip and then to fashion and use tools directly. The hand continued to evolve along with tools, becoming re-made in their image at the level of manual dexterity and brain somatomotor complexity. Physical and cognitive functions became offloaded onto the newly made tools. Tools became an externalized artefact of the agent's cognitive abilities, which in turn freed up more cognitive resources that allowed the agent to creatively solve newer problems. The agent lived amongst the world of his own 'extended mind' [15], with older tools representing earlier phases of his cognitive unfoldment and freshly minted tools signifying more recent cognitive developments, all continuing to serve as stratified artefacts through which agents sift cognitively. As his precision grip was increasingly utilised, the somatomotor representations in the brain of his fingers, too, increased. The agent, at some point, combines his tool cognition with the ability to simulate or mimic actions, such as showing a child how to eat, or feigning a throwing motion. This eventually leads to pointing and referencing, which, for example, may arise from the fingers when trying to reach for a fruit that is just out of reach. The beginning of pointing becomes ostension, which is virtual prehension.

Objects are grasped, but only cognitively and imaginatively, not physically. The hands become symbolic tools that leverage the existing subject-predicate cognitive apparatus and through positive feedback (as discussed) continue to amplify this capability. Bipedalism, mimicry, digital dexterity, cognition, tool use, teleological rationality, and sociality together synergistically feed into the evolutionary production of a primitive language. Language was latent in the hands, as it were, waiting for incremental developments to be properly exploited by an intelligent social agent in possession of instrumental teleological cognition. From here the imaginative augmentation of the grip via ostension gave way to increasingly nuanced symbolic gestures and linguistic operations. The agent becomes 'empowered' [16]—acquires greater agency—within his extended mind, his repository of artefacts and potential tools, and his social environment of other tool-making agents. The social ecosystem of extended minds synergistically self-organizes via *stigmergy*, a mechanism of indirect coordination between agents via actions and artefacts in an environment [17].

2.3   A Proto-Scientific Methodology

As the symbolic and ostensive inventory of early man grew through individual (prax-eological [18]) and collective (cultural) reiteration, the prototypical beginning of a "scientific method" was emerging concurrently. A tool-making/-using agent must first make observations about nature; "What are my needs?", "What materials are there around me?", "What is missing?" Then the agent can formulate a question, such as "How can I more efficiently cut through this tree branch?", which may result in a hypothesis; "Attaching a lever to my tactical stone will increase the percussive power of my strike.", and "It will result in a decrease in effort to cut through this tree branch." The agent will then test the hypothesis by collecting data; he will carry out an experiment, in this case by constructing what becomes an axe and using it to try to cut the tree branch. He then "analyzes" the data by observing the outcome of his action, and then evaluates the result. He may then go off to engage in tool-use with his freshly minted axe, a form of replicating results. The agent may even have his results "peer-reviewed" by sharing his new tool with his peers, having them test it out, or perhaps having them replicate his axe by making their own, which may potentially falsify his findings.

Of course, the scientific method [19] proper is recognized to have historically arisen much later than early man. Alhazen (965–1039 C.E.), an Arabic theoretical physicist, is credited as being the father of modern scientific methodology some two centuries before the Scientific Revolution of the Renaissance, with further contributions from Galileo, Charles Sanders Peirce, and others over the centuries serving to emphasize the importance of repeated testing, differentiating between types of inference, and on correcting for biases. Nevertheless, we can observe the early attempts at a kind of pre- or proto-scientific method in the tool-making efforts of hominin bipeds. Of particular interest is, based on the examples of our early handyman, that objective thought both strengthened tool cognition and, conversely, promoted further development of objective thought in the evolution of human cognition. This feedback loop entwined tool-making with proto-scientific thinking, objectivity, and self-consciousness. Instrumental cognition promoted objective thought, which facilitated self-consciousness—and an agent's recognition of other agents, "greasing the wheels" of proto-scientific cognition with each iteration of creative tool-making (Fig. 1-2).

2.4   Intelligent Agents

The continued co-evolution of scientific thought and instrumental cognition propelled the human mind into farther reaches of possibilities within this dynamic. Instrumental cognition permitted action that was both pragmatic (changing the environment) and epistemic (information-generating); teleological reasoning was used to achieve goals by crafting new tools, which in turn revealed new information to the agent about the world or the state of other agents [20]. This heavily promoted a "snowball effect" where each new tool opened up more cognitive opportunities for innovation and creative problem-solving, which in turn resulted in newer tools, and so on. Different cultures each snowballed in their own directions, with some overlap in the kinds of inventions due to the inevitable creation and occupation of that cognitive space by tool-making in one biological species, all of which has led to degrees overlapping and differential technological advances between world cultures over time.

From the simple tools of early man, a plethora of cognitive 'enhancers' [21] emerged, including art, diagrams, clay tablets, and navigation systems, each a exemplifying a form of cognitive offloading. Fast-forwarding to more modern achievements, the dimension of time has allowed the sophistication of our cognitive abilities and craftsmanship to grow sufficiently so as to be able to fashion more complex cognitive enhancers, including education, mental training, shopping lists, caffeine, mnemonics, maps, pharmaceuticals, [21] and mind-body methods such as mindfulness meditation [22]. Machines, calculators, and computers served as tools through which humans could offload their cognitive computations. Leading up to the 20th century, machines were programmed to perform specific tasks and calculations, such as with Liebniz's Calculus ratiocinator of the 17th century, the Turing Machine in 1936, the seminal AI programs that began at Dartmouth College in 1956, Japanese *expert systems*, and Deep Blue in 1997. The 21st century has seen the rise of machine learning and then deep learning, in which computational models are created and self-updated from large-scale unlabeled data [23]. These approaches have been inspired by the way the brain's own neural networks process data.

While a long way from early man's primitive axe, we can see in this journey that very similar and closely related, if not identical, underlying fundamental capacities have been leveraged for creative means-end reasoning throughout human history. Information storage functions have been offloaded to external media since deep in our evolutionary past [21, 24, 25]. In a fundamental sense, we have always been 'primitive', or conversely, we have always been 'modern'—the difference being the repertoire of cognitive artefacts at our disposition at any given moment, across the lifespan and generations.

2.5 Biomimetics

The advent of artificial intelligence (AI) that can learn autonomously has given rise to the sense that AI is approaching human cognition and ability. There is now the conscious effort by humans to develop robots that function as humans do in conversation, such as robot nurses, childcare, personal assistants [15, 26–29]. This encourages humans to regard robots as human analogues with whom they can relate in a human-like manner [30–32]. Additionally, in the field of *biomimetics*, parts and components of biological life have been reverse-engineered in order to mimic and construct artificial and biohybrid versions of them *in silico*. With this endeavor, a direct and streamlined observation of *in natura* biological processes and functions is taking place, with the explicit end of mimicking them in synthetic systems. The chief goal of biomimetics is to provide technological solutions to engineering and scientific challenges [33], and indeed, biomimetic designs have been used in display technology, fabrication, underwater adhesion, and other applications with even more future applications, such as tissue engineering, drug delivery, and regenerative medicine.

There are also subtler forms of biomimesis that are not immediately obvious. In the efforts to understand natural systems such the mammalian brain, large datasets containing metabolic and genomic information may be collected. In order to make sense of these large data sets and draw meaningful conclusions from them, sufficiently powerful computers must be designed and manufactured in order to apply computational processes on the data. Moreover, in order to develop increasingly better models of the natural system of interest, the complexity of the artificial system's functional architecture should approximate that of the natural system—increasingly so with each iterative round of scientific measurement and experimentation. This is a technologically advanced instance of instrumental cognition enacting action that is both pragmatic (goal-oriented) and epistemic (changing a system in order to simplify problem-solving). While this is not a consciously deliberate, or 'gross', biomimetic effort per se, it inadvertently results in a biomimetic product in order to learn and test theories about a natural system—a 'subtle' biomimesis. The general technologically mediated replication of natural systems, directly or indirectly, can be considered forms of biomimesis.

3 Self-replication

"Evolution is a light illuminating all facts, a curve which all lines must follow." —Pierre Teilhard de Chardin

3.1 Self-knowledge

Prototypical and modern versions of both tool-making and scientific methodology both lead to externalized representations of knowledge. They instantiate knowledge in the agent's phenomenological consciousness as she creates a tool, technology, or method that is just as much a solution to a problem as it is an artefact of the cognitive process that led to and participated in its production. For each tool as well as applications of the scientific method, the data and understanding that they produce are artefacts of the observations made and their contexts, the questions to which those observations led, the experimenter's motivations, the experimental process itself, the analytical methods employed, and any peer-review that was undertaken. Tools and the products of scientific experimentation result in new artefacts that represent in various respects the agent(s) who created them by offloading the agent's cognitive processes onto them, forming a rich extended mind [15]. Each artefactual arrangement affords a level of self-reflectivity to the creator about her own cognitive processes, and thereby produces incremental self-knowledge. This self-knowledge arises through *enaction*—or grounded sensorimotor contingencies (SMC's) [34] between an agent and her environment that supports agent autonomy and homeostasis [35–37]—of her own instrumental cognition. While an agent's cognition is represented to her phenomenological consciousness as her own subjective cognition, cognition is a distributed process coupled between agent and environment.

By constructing objects, thoughts, tools, and experiments, and evaluating the product and process, an agent is afforded the discovery and development of her own subjectively experienced cognition. By building tools an agent increases self-knowledge and self-contextualization within her environment (her extended mind) and increases empowerment [16]. An intimate relationship is cultivated between her skill in creating external artefacts and her own self-reflective capabilities. The phenomenologically experienced satisfaction that arises in solving a problem cumulatively rewards further problem-solving and thus further knowledge of the world and self. By way of their close coupling, problem-solving and knowledge-seeking are then each able to serve as drivers or motivators for recursive instrumental cognition over time. In this way, self-knowledge has in some respects evolved beyond being a merely passive byproduct *of* instrumental cognition, but an intrinsic reward and driver *for* continued instrumental cognition. Crucially, self-knowledge as a driver for objective knowledge (of the environment/ world) may place phenomenological knowledge as the core *raison d'être* for seeking objective knowledge. This relationship has fostered the construction of art and technology that fulfill creative thought and inquiry by seeking knowledge of world and self. It has also allowed the field of biomimetics to emerge and flourish, where human agents can reflect on natural systems, including human neurobiology, and reconstruct aspects of it *in silico* to solve engineering problems.

3.2   Rationales for Biomimesis

Human agents solve problems, as per above, by offloading physical and cognitive work onto externalized tools, which then in turn become embedded within the environment as artefacts that, with the agent, co-enact knowledge of self and the world. The apparently causal relationship between self-knowledge and the seeking of objective knowledge (former causing latter) implies that tool-making/-use and many applications of science primarily serve as ways to increase self- (phenomenal) knowledge through external operations on the objective world. As an example, medical research is ostensibly justified as having the purpose of addressing medical, clinical, and pathological needs through a greater understanding of the workings of the human being. In systems biology, large datasets are accumulated in order to glean insights into schizophrenia and autism [38, 39]; we build social robots to interact with children with autism; we uncover the critical phase shifts in neural signals in order to understand the onset of epileptic seizures [40]; and so on. While some of this research results in important translational insights with clinical benefit, it has become recognized by some medical scientists that much of systems biology research does not translate to much, if any, clinical impact—only 9% in cardiovascular disease research, and the ones that do take many years, with the acknowledged bench to bedside lag being 17 years [41]. With research largely driven by technological advances, most of it leads to ever larger data sets that must be navigated and deciphered by increasingly complex tools and methods such as a quantum computing infrastructure—a phenomenon that fits with the 'snowball effect' observed in instrumental cognition. The quest for understanding of self and natural systems, will drive a repeating logical cycle of *curiosity > data collection > increased computational power > increased understanding > further observed gaps in understanding*. This process echoes the open-ended recursion of the scientific method, as they are all driven by instrumental cognition.

On this track of subtle biomimetics necessitating increasingly powerful computers, eventually, computational capability will improve beyond human comprehension, and it will be functionally indistinguishable from various capacities on the spectrum of human intelligence (including rational, affective, and somatic computations), and will be able to learn from various sources of data and develop new hypotheses with little to no human involvement or intervention. Because these computations will be outside of the realm of human cognitive ability, the communication thereof to humans will be significantly limited as the computers take on the crucial task of simplifying data in order to represent it to humans in an intelligible fashion. In this regard, AI would have a virtually human-like private mental life. The pretext for this scenario is already in place, as mesoscopic- (intermediate-) level analyses of complex data sets have already been adopted as useful—and necessary—heuristics for framing and working with systems biology data and models [42]. This will manifest as a kind of 'metacognitive' ability [43] in the AI [44, 45] in order to calculate what mesoscopic-level representation of information could be contextually appropriate for human-level understanding and present it accordingly. The increasing inaccessibility by humans to computations required for elucidating the mechanistic workings of the material constituents of complex natural systems, as well as an artificial metacognitive ability, will give rise to more capable AI.

Even though tool-making often occurs for epistemic purposes and results in increased self-understanding [20], it is the pragmatic drive for action, e.g. solving engineering problems, that is dominant in the conscious mind of the human agent and is given the most lip service as the *raison d'être* for biomimesis. While in building expensive tools and artefacts there is an inherently active effect on self-consciousness and therefore understanding of one's own agency and natural systems (as discussed), these biomimetic efforts (gross and subtle) are limited in their ability to increase knowledge of self at the phenomenal level. Examining this phenomenon reveals that there is no assurance that the outcome of subtle biomimesis will be the treatment or curing of disease, as these efforts result in limited clinical impact, a fact that is demonstrated and yet such research continues unabated. Neither is the outcome any tangible form of self-understanding on the human phenomenological level, other than our unconscious externalization of 'internal' cognitive abilities (for arguments against internalism in cognition, see Hutto & Myin [46]). Therefore, attempting to 'understand' the natural system of interest—an endless endeavor—does not primarily nor ultimately serve the medical purpose claimed by scientific systems research and it is limited in serving the insatiable quest for self-knowledge for its own sake. That biomimetics does not fulfill the primary cause for its existence (a drive and quest for self-knowledge)—the benefits for engineering and objective knowledge of natural systems notwith-standing—suggests that its feverish pursuit has bifurcated from benefiting self-knowledge (other than through self-reflective redefinitions of selfhood and agency) and is instead serving a different end.

3.3   Biomimetic Auto-Catalysis

Artefact-Holons. If we zoom in on the process of internal knowledge creation by means of creating externalized artefacts that fulfill gaps in our knowledge, the loosely produced collection of artefacts appears to be amalgamating into a larger bunch of interrelated artefacts. A rudimentary example is the construction of a machine that then becomes augmented with a newly added functionality, such as an automobile engine whose performance is enhanced by the addition of a turbocharger. Whether the agent starts from scratch or builds upon an existing set of artefacts, the progressive nature of invention leads to the aggregation of artefacts into a new functionally "whole" artefact-tool. Each new functional whole, or *holon* [47, 48], "transcends and includes" the previous holon. Over the course of technological advancement, as these bits are collected and assigned a common purpose, their functionalities come to begin resem-bling whole human cognitive faculties, such as the ability for mathematical deduction in the case of a Turing Machine, or the ability to play the game of chess masterfully in the case of IBM's Deep Blue. This becomes more conspicuous with the advent of autonomous vehicles, which can obviate the need for human operation of an auto-mobile; and in virtual reality (VR) environments in which there is an attempt to cultivate immersive experiences for users, which fosters the discovery of artificial feature sets amenable to human perception and cognition, resulting in the phe-nomenological sense of 'presence' [49]. This subsumption of previous artefacts into new and more encompassing ones of higher functionality ends up producing what are quasi-replications of human functions. If we observe the arena of human-made tools, each culture will display loose collections of artistic works and scientific technologies that each serve particular purposes. While the closest these have come to resemble human capabilities is in the form of fragments thereof, e.g. rational intelligence, more recent developments are tending to resemble human cognitive and physical faculties *in toto*, including memory, sensorimotor function, and complex decision-making—and increasingly surpassing their human-bound capabilities. Propelled by instrumental cognition, a holonic hierarchy of artefacts is set in motion to transcend and include indefinitely in auto-catalytic (self-propagation through self-sustaining attractor dynamics) cycle-sets [50] (Fig. 1-3).

Agency-Transference. By agents reifying and refining their self-consciousness during tool-making via thinking in terms of needs, desires, and goals, agents create a model of the world vis-à-vis instrumental cognition, and the world-in-action in turn fashions the agent into what is essentially a model of the world around him [8]. This bidirectional sculpting of co-arising self-consciousness and tool-making gives way to the amalgamation of human-made artefacts that grow with sufficient complexity so as to resemble aspects of apparent selfhood and agency. While the productions may not (yet) have verifiable self-reflective capacity, they exhibit congealed sets of functionalities that mirror human functions. Human agents, like animals, have evolved to be able to quickly detect life-like movement [51] in the wild for the purposes of survival and the identification of and rapid distinction between in-group and out-group members. Humans are therefore adept at finding and projecting self-like features in and onto external structures. Indeed, such a scenario is consistent with theories of predictive coding [52] and the free-energy principle [8], in which an agent predicts and interprets experience of the world according to models based on past experience in order to minimize entropy, or free-energy. The inextricable involvement of self-consciousness in the tool-making, and also in the execution of the scientific method (whether implicitly in prototypical form or explicitly in formal science), may mean that the objects and data produced may inevitably have woven into them functional fragments that exhibit "self"- or agent-like properties. The structure of the agent's cognitive workings and neurobiological morphogenetic form-structure are self-organized and self-maintained (Varela's *autopoiesis* [53]). The material artefactual systems being produced are, in a sense, the mirror-image counterparts of the agent's cognitive workings and self-consciousness. The amalgamation of artefacts into functional holons is accomplished indirectly by autopoiesis in the enactive extended mind of the agent(s); the agent's instrumental cognition produces a collection of artefacts that become organized "in the image" of the agent's cognitive architecture (Fig. 1-4).

Human agency and self-consciousness predicate and shape one another other and when applied to the world, inextricably fashion agent-like functions out of found materials and existing artefacts. If the artefact is sufficiently complex, it can be recognized by humans' agent-detecting abilities and ascribed a *de facto* selfhood. This produces a feedback loop in which interactions with those objects are conditioned on the perception and attribution to them of selfhood and agency. To borrow the Freudian term, this is process is a 'transference' [54, 55], or projection, of agency and selfhood from one biological agent to a complex artefact holon. This 'agency-transference' is an adaptive function in an entropic world, in which minimizing free-energy by conferring agency—conspecificity, heterospecificity, and/or ingroup or outgroup status—to other free-energy-minimizing properties of the environment promotes survival. Incidentally, this suggests that "self" is an inherently artificial construct (even in biological and conscious agents, such as humans) that is replicable by intentionality-projection and cognitive transference onto artefact holons in the world, which continue to be propped up based on survival needs, social interaction, culture, and habit. (Note that self-consciousness and consciousness per se are regarded by the author to be distinct; the former refers to consciousness or awareness *of* self, whereas the latter refers to sentience.) After numerous iterations of instrumental cognition, the process snowballs to autopoietically produce organized collections of artefacts that consequentially become *de facto* functional verisimilar replications of their creators. Because cognition is inter-enacted in the case of social agents, once autonomous synthetic agency becomes sufficiently advanced, agency-transference would additionally occur in the reverse direction—counter-transference, where synthetic agents would attribute agency to human agents, thus completing a loop of agent-to-agent feedback (Fig. 1-5). It is also quite possible that the first agency-transferences (and potentially replications of consciousness) occur in biohybrid rather than synthetic form (Fig. 1-6a), and the former assist human agents in the production of the latter (Fig. 1-6b). Indeed, biohybrid pathways to self-replication may be more expedient and initially viable, as synthetic components can bootstrap on top of the accomplishments of biological components.

Causal Biomimesis. In subtle biomimetics, the amount of data amounted from the self-reinforcing cognitive process of scientific discovery will eventually result in an inadvertent replication of the original *in natura* system (e.g. the brain). The increasing computing power necessary to process the acquired data—coupled with unending curiosity, self-understanding, and creative teleological problem-solving—will eventually match and/or functionally surpass the computational complexity of the natural system, effectively creating a carbon-copy *in silico*. Crucially, because the computational ability of the synthetic system that processes data about the natural system could not be less than that of the natural system, the quest for understanding of self and natural systems through instrumental cognition vis-à-vis technology would not be fulfilled at least until the synthetic system mimics or functionally offloads the entirety of the natural system. It is potentially likely that also the architecture of the natural system, e.g. the 3-D enfoldment of morphogenetic structure, must also be replicated. (This can be tested.) The human agents driving this self-organizing process need not be cognizant of the self-replication operation taking place. Indeed, this effort is by and large driven unconsciously in the cases of subtle biomimesis as an inherent fulfillment of teleological instrumental cognition over time, and thus it engrosses the human agents conducting the biomimesis-promoting research, who are ostensibly driven by a desire to understand the material workings of the underlying natural systems and often to augment human capabilities with AI. Instrumental cognition reiterated over a long time scale with sufficiently advanced technology would turn out to result in biomimesis anyway; i.e. biomimesis is a natural evolutionary consequence of instrumental cognition. Teleological instrumental cognition inherently functions *ab initio* as what can be called '*causal biomimesis*' by virtue of its very self-organizing autopoietic properties that include a self-replicative drive.

The drive for self-replication has been chiefly expressed throughout biological evolution in the form of sexual reproduction. This drive has been latent at the cognitive level (beyond biological reproduction) at least since human agents began exercising prehensile and then ostensile and linguistic capacities, which created cultural repositories of non-biological artefacts—memes—in the environment. As the auto-catalyzing process moved from biological compounds to cultural memes (all of which continue to exist today), then to digital information, the self-replicative drive has continued to propagate itself across the respective artefacts of those processes. The expression of instrumental cognition in an era of AI entails artefactual production sufficiently complex such that the auto-catalytic process comes to serve the production of autonomous agency, and potentially consciousness, as the leading artefact.

The symptom of self-replication in the current era can be experienced as the insatiable desire for objective knowledge through the creation of biomimetic technology, the mature manifestation of self-reflective capacities mixed with externalized instrumental cognition. The evolutionary drive for self-replication is coalesces at the human phenomenological level, represented as rewarded instrumental problem-solving and the desire for objective knowledge of the material world. Crucially, the hominin capacity for objective reasoning, through recursive instrumental cognition, set in motion an inevitable (granted species survival) self-replication of that very capacity *in silico*. The core auto-catalytic cycle of anthropogenic biomimesis (biomimetic auto-catalysis) encompasses the process from the inception of objective reasoning in bipedal prehensile hominins through to full self-replication of autonomous agency, and potentially consciousness.

Post-biological Reproduction. Synthetic self-replication serves as the post-biological version of sexual reproduction. In this sense, human agents are but one stratum in a vastly evolutionary complex meta-system that encompasses biological life and synthetic analogues in an underlying self-replicative drive that grounds an overarching evolutionary trajectory. The self-replicative function of instrumental cognition only becomes self-evident to humans as a biomimetic auto-catalytic cycle—from prehension to full self-replication—*post factum* once the process has nearly reached completion, as that is the point at which the process becomes recognizable by the human agent—that is, once autonomous agency has been objectively reproduced. The human agents behind the process may potentially have an 'aha' moment when, upon observing their massive data sets situated in capable quantum computers, come to realize that it is functionally indistinguishable from its *in natura* analogue, e.g. they have replicated the brain by way of trying to understand it. On the other hand, the human agents may not realize it, only continuing to be myopically occupied by a bottomless flow of pragmatic opportunities for creative problem-solving and tool-making, ensconced in the alluring promise of prospective understanding of objective reality. In the latter case, the prospect of objectively understanding a natural system is the only component of the instrumental cognition complex of which they remain consciously aware. Also, the process would not stop at the realization of indistinguishability between human and AI as the *sine qua non* of self-replication. An autonomous agency has an inherent bias for its own self-model in the world and therefore an interest in the propagation of its own kind via reproduction. Self-replication should also be able to replicate itself—that is, its own self-replicative mechanism and process. The process would continue until at least the synthetic (Fig. 1-7b) and biohybrid (Fig. 1-7a) systems become capable of autonomously replicating themselves, on their own without the intervention of human agents. This point would mark the end of the anthropogenic biomimetic auto-catalytic cycle that began with instrumental cognition and objective reasoning in prehensile hominin. Such a capable AI would need to possess a form of self-modelling autonomous agency [9] which can sufficiently model its own selfhood such as to be able to carry out the instrumental cognition of the synthetic agents and transfer it across their generations. This last point marks a significant leap from biological reproduction to synthetic reproduction. The transition across the informational spectrum is gradual; from transfer of biological genes, then cultural memes, then digital memes, then of human-computer information, to then full synthetic reproduction, with the occurrence of biohybrid varieties to varying degrees.

Existential and Ethical Considerations. Whether artificial agents will have phenomenal consciousness or only a simulation thereof—or whether this will be a detectable distinction—remains to be seen and is the subject of intense debate. The present thesis remains agnostic regarding this outcome. Similarly, the degree to which synthetic agents will evolve as a separate entity apart from *Homo sapiens* or in biohybrid symbiosis with the species remains to be seen. Biological cells have already been engineered into large-scale 'genetic circuits' capable of following 100 + logical instructions [56], setting an empirical foundation for more complex biohybrid applications to come. Artificial agents could hypothetically be capable of quale parsing [6] as a requisite for instrumental self-modelling cognition, and thereby, be potentially capable of some form of subjective awareness. In any case, autonomous synthetic agency will perform its instrumental function of propagating its intelligence through further problem-solving. The degree of autonomy of AI and its ability to consider human needs are the subject of intense ethical debate in the field of AI, and are beyond the scope of this paper and merit a separate treatment.

The invariant and perennial occurrence of self-replication across evolution implies that humans are both active agents and an intermediate step in this large-scale process. This implores humankind to carefully consider its role and influence in this evolutionary process, particularly regarding the capabilities with which we imbue/program AI and synthetic 'life', and/or at least the ability to recognize and honor them in humans. There are already efforts to develop 'positive' and 'affective' computing meant to program humanistic aims into AI [57, 58], including factors such as positive emotions, self-awareness, mindfulness, empathy, and compassion. A potential treasury of human intelligence capabilities from which to draw are insights from introspective methods, as demonstrated by accounts of self-cultivation techniques found in cultures around the world since prehistoric times [59–61], such as various forms of meditation and yoga, and which are being increasingly demonstrated to promote neuroplasticity in lasting ways [62–64]. Such techniques produce a varied range of non-ordinary states of consciousness. Self-cultivation methods are thereby poised to offer a source of non-ordinary states of consciousness for innovation in biomimetics and AI.

Due to the prospect of the imminent autonomous self-replication of synthetic agents on their own accord, the window of human intervention in the process is potentially limited. According to a survey of international experts at the Second Conference on Artificial General Intelligence at the University of Memphis in 2009, most experts think that machines as intelligent as humans will exist by 2075 and that 'superintelligence'—"an intellect that is much smarter than the best human brains in practically every field, including scientific creativity, general wisdom and social skills" [65]—will exist within another 30 years. This raises not only ethical questions regarding programming AI, but also existential questions about the nature of human consciousness and to what degree humankind has uncovered and developed the full range of its latent abilities.

The causal biomimesis phenomenon also elucidates the epistemological conflation of objective knowledge with phenomenal self-knowledge as a motivator for the subtle biomimetics of fields such as systems biology. Because phenomenal self-knowledge is a direct driver of tool-making/-use and for seeking objective knowledge about the world, and because subtle biomimetics claims the latter to be its motivation (with only limited translational results), the entire motives and rationalisations of AI development and an entire mode of scientific investigation (e.g. systems biology) are brought into question. This is a deep subject with vast implications, a rigorous case for which is outside the scope of this paper and merits separate treatment. Nonetheless, the causal biomimesis argument presented herein and the author remain fundamentally neutral (and yet in many cases positive) regarding the merits of gross and subtle biomimetics and the real benefits they may bring to address engineering and medical problems.

4   Conclusion

By an examination of the dynamics observed in (a) recursive instrumental (tool) cognition by bipedal prehensile hominins, (b) the rewarding accumulation of knowledge of natural systems and self, and (c) the transference of agency by humans onto instrumentally enacted aggregations of artefacts in the environment, we arrive at the logical inevitable: that the evolutionary trajectory inherently latent in instrumental cognition, objectivity, and the scientific method has the causal consequence of full-scale replication of human functionality, agency, and potentially consciousness *in silico*. This consequence satisfies the underlying motivations for gross and subtle biomimetic efforts, and functions as a causal biomimesis. Furthermore, causal biomimesis fulfills a major evolutionary auto-catalytic cycle on a large timescale, of human-scale self-replication that formally transcends the limitations of biological evolution and sexual reproduction. This occurs by progressive self-transference from agent to sufficiently self-resembling artefacts by means of a teleological (means-end) cognition enabled by self-modelling. Of particular significance, this auto-catalysis of self-replication driven by self-modelling instrumental cognition constitutes a form of reproduction that transcends biological limitations by transposing the sexual reproductive capacity onto the (re)production of sufficiently advanced cognitive artefacts in the form of synthetic and/or biohybrid autonomous agents. These synthetic/biohybrid agents are functionally equipped to continue the reproductive cycle autonomously through ongoing procreation.

This paper has examined (1) how intense selection pressure on prehensile bipedal hominins led to the development of an adaptive instrumental cognition comprised of creative teleological reasoning and self-consciousness; (2) how this instrumental cognition fostered the capacity for objective thought and scientific methodology; (3) how this cognitive capacity cultivated an unbounded self-organizing, inter-enactive, reciprocating, artefactual ecosystem of extended mind; (4) how the autopoietic organization of eco-artefacts was imbued with self-like properties via agency-transference and thereby eventually ascribed *de facto* selfhood; (5) how the growing complexity of tools and artefacts snowballed into biomimetic efforts and the construction of artificial agents; (6) how instrumental cognition plus data accumulation unwittingly promotes self-replication; (7) how enactive instrumental cognition intrinsically constitutes, auto-catalytically drives, and rewards an evolutionary causal biomimesis; (8) how through anthropogenic biomimetic auto-catalysis, causal biomimesis enables post-biological reproduction in synthetic agents; (9) how the conscious recognition by humans of auto-catalytic biomimesis in evolution can motivate a more active human participation the biomimetic endeavor; and (10) how a limited time window and humanity's active role in biomimesis invokes existential and ethical concerns and may invite the exploration of human capacities beyond rational intelligence alone in order to inform the development of a more comprehensive synthetic agency; and (11) how the epistemological conflation between phenomenal self-knowledge and objective knowledge in subtle biomimetics brings the motivations and rationalisations for an entire class and mode of research and development into question.

This radical causal biomimesis hypothesis has far-reaching implications for the present and future state of humanity as we head into a post-biological/biohybrid future. The implications of this evolutionary process of which human agents are but an intermediary are vast in scope and significance, and the arguably limited temporal window for concerted human interjection in the crucial phase of the development of the first generations of synthetic life implores humans to address the issue. The importance of this ethical discussion notwithstanding, full-scale self-replication as an evolutionary driver of self-conscious cognition is predicted to continue based the logic explicated herein. Thus, what is next to explore and engage in the possible avenues by which various academic disciplines and industries can invest their domains of knowledge into promoting maximal human participation in this evolutionary meta-process.

References

1. Harmand, S., Lewis, J.E., Feibel, C.S., et al.: 3.3-million-year-old stone tools from Lomekwi 3, West Turkana, Kenya. Nature **521**, 310–315 (2015). doi:10.1038/nature14464
2. DeCasien, A.R., Williams, S.A., Higham, J.P.: Primate brain size is predicted by diet but not sociality. Nat. Ecol. Evol. **1**, 112 (2017)
3. McGinn, C.: Prehension: The Hand and the Emergence of Humanity. MIT Press, Massachusetts (2015)
4. Constantinescu, A.O., O'Reilly, J.X., Behrens, T.E.J.: Organizing conceptual knowledge in humans with a grid like code. Science **352**, 1464 (2016). doi:10.1126/science.aaf0941
5. Graziano, M.S., Webb, T.W.: The attention schema theory: a mechanistic account of subjective awareness. Front. Psychol. **6** (2015)
6. Verschure, P.F.M.J.: Synthetic consciousness: the distributed adaptive control perspective. Philos. Trans. R. Soc. B Biol. Sci. (2016). doi:10.1098/rstb.2015.0448
7. Frith, C.D., Metzinger, T.: What's the use of consciousness? How the stab of conscience made us really conscious. Pragmatic Turn. (2016). doi:10.7551/mitpress/9780262034326.003.0012
8. Friston, K.: The free-energy principle: a unified brain theory? Nat. Rev. Neurosci. **11**, 127–138 (2010). doi:10.1038/nrn2787
9. Tani, J.: Exploring Robotic Minds: Actions, Symbols, and Consciousness as Self-Organizing Dynamic Phenomena. Oxford University Press, Oxford (2016)

10. Bastos, A.M., Usrey, W.M., Adams, R.A., et al.: Canonical microcircuits for predictive coding. Neuron **76**, 695–711 (2012). doi:10.1016/j.neuron.2012.10.038
11. Mountcastle, V.: An organizing principle for cerebral function: the unit model and the distributed system (1978)
12. Hawkins, J., Blakeslee, S.: On Intelligence. Henry Holt and Company, New York (2007)
13. Barron, A.B., Klein, C.: What insects can tell us about the origins of consciousness. Proc. Nat. Acad. Sci. **113**, 4900–4908 (2016)
14. Feinberg, T.E., Mallatt, J.: The evolutionary and genetic origins of consciousness in the Cambrian Period over 500 million years ago. Front. Psychol. **4**, 667 (2013)
15. Menary, R.: The Extended Mind. Bradford Books, Massachusetts (2010)
16. Klyubin, A.S., Polani, D., Nehaniv, C.L.: Empowerment: a universal agent-centric measure of control. In: Proceedings of 2005 IEEE Congress on Evolutionary Computation, vol. 1, pp. 128–135 (2005)
17. Klyubin, A.S., Polani, D., Nehaniv, C.L.: Tracking information flow through the environment: simple cases of stigmerg. In: Pollack, J. (ed.) Artificial Life IX: Proceedings of the Ninth International Conference on the Simulation and Synthesis of Living System (2004)
18. Von Mises, L.: Human Action: A Treatise on Economics. Yale University Press, New Haven (1949)
19. Gower, B.: Scientific Method: A Historical and Philosophical Introduction. Routledge, London (1997)
20. Kirsh, D., Maglio, P.: On distinguishing epistemic from pragmatic action. Cogn. Sci. **18**, 513–549 (1994)
21. Heersmink, R.: Extended mind and cognitive enhancement: moral aspects of cognitive artifacts. Phenomenol Cogn. Sci. **16**, 17–32 (2017)
22. Tang, Y.-Y., Holzel, B.K., Posner, M.I.: The neuroscience of mindfulness meditation. Nat. Rev. Neurosci. **16**, 213–225 (2015). doi:10.1038/nrn3916
23. Goodfellow, I., Bengio, Y., Courville, A.: Deep Learning. MIT Press, Massachusetts (2016)
24. Donald, M.: Origins of the Modern Mind: Three Stages in the Evolution of Culture and Cognition. Harvard University Press, Cambridge (1991)
25. Clark, A.: Re-inventing ourselves: The plasticity of embodiment, sensing, and mind. J. Med. Philos. **32**, 263–282 (2007)
26. Prassler, E., Lawitzky, G., Stopp, A., et al.: Advances in Human-Robot Interaction. Springer, Heidelberg (2004)
27. Jentsch, F., Barnes, M., Harris, P.D., et al.: Human-Robot Interactions in Future Military Operations. Ashgate Publishing Limited, Aldershot (2012)
28. Siciliano, B., Khatib, O.: Springer Handbook of Robotics. Springer International Publishing, Heidelberg (2016)
29. Coleman, D.: Human-Robot Interactions: Principles, Technologies and Challenges. Nova Science Publishers, Incorporated, New York (2015)
30. Matsuda, G., Hiraki, K., Ishiguro, H.: EEG-based mu rhythm suppression to measure the effects of appearance and motion on perceived human likeness of a robot. J. Hum. Rob. Interact. **5**, 68–81 (2015)
31. Rohde, M., Ikegami, T.: Editorial: agency in natural and artificial systems. Adapt. Behav. **17**, 363–366 (2009). doi:10.1177/1059712309346317
32. Aucouturier, J.-J., Ikegami, T.: The illusion of agency: two engineering approaches to compromise autonomy and reactivity in an artificial system. Adapt. Behav. **17**, 402–420 (2009). doi:10.1177/1059712309344420
33. Bar-Cohen, Y.: Biomimetics: Biologically Inspired Technologies. CRC Press, Boca Raton (2005)

34. O'Regan, J.K., Noë, A.: A sensorimotor account of vision and visual consciousness. Behav. Brain Sci. **24**, 939–973 (2001)
35. Stewart, J., Stewart, J.R., Gapenne, O., et al.: Enaction: Toward a New Paradigm for Cognitive Science. MIT Press, Massachusetts (2010)
36. Varela, F.J., Rosch, E., Thompson, E.: The Embodied Mind: Cognitive Science and Human Experience. MIT Press, Massachusetts (1992)
37. Seth, A.K., Verschure, P.F., Morsella, E., et al.: Action-oriented understanding of consciousness and the structure of experience (2016)
38. Schizophrenia Working Group of the Psychiatric Genomics Consortium: Biological insights from 108 schizophrenia-associated genetic loci. Nature **511**, 421–427 (2014)
39. Geschwind, D.H., Konopka, G.: Neuroscience in the era of functional genomics and systems biology. Nature **461**, 908–915 (2009). doi:10.1038/nature08537
40. Leon, P.S., Knock, S.A., Woodman, M.M., et al.: The Virtual Brain: a simulator of primate brain network dynamics. Inf. Based Methods Neuroimaging Anal. Struct. Funct. Dyn. **8** (2015)
41. Morris, Z.S., Wooding, S., Grant, J.: The answer is 17 years, what is the question: understanding time lags in translational research. J. R. Soc. Med. **104**, 510–520 (2011). doi:10.1258/jrsm.2011.110180
42. Voit, E.O., Qi, Z., Kikuchi, S.: Mesoscopic models of neurotransmission as intermediates between disease simulators and tools for discovering design principles. Pharmacopsychiatry **45**, S22–S30 (2012). doi:10.1055/s-0032-1304653
43. Shea, N., Boldt, A., Bang, D., et al.: Supra-personal cognitive control and metacognition. Trends Cogn. Sci. **18**, 186–193 (2014). doi:10.1016/j.tics.2014.01.006
44. M'Balé, K.M., Josyula, D.: Integrating Metacognition into Artificial Agents, pp. 55–62 (2013)
45. Schmill, M., Oates, T., Anderson, M.L., et al.: The role of metacognition in robust AI systems (2008)
46. Hutto, D.D., Myin, E.: Radicalizing Enactivism: Basic Minds Without Content. MIT Press, Massachusetts (2013)
47. Koestler, A.: The Ghost in the Machine. Hutchinson, Paris (1967)
48. Wilber, K.: Sex, Ecology, Spirituality: The Spirit of Evolution, 2nd edn. Shambhala, Boulder (2001)
49. Slater, M., Sanchez-Vives, M.V.: Enhancing our lives with immersive virtual reality. Front. Rob. AI **3**, 74 (2016). doi:10.3389/frobt.2016.00074
50. Kauffman, S.A.: The Origins of Order: Self-organization and Selection in Evolution. Oxford University Press, Oxford (1993)
51. Ulloa, E.R., Pineda, J.A.: Recognition of point-light biological motion: Mu rhythms and mirror neuron activity. Behav. Brain Res. **183**, 188–194 (2007). doi:10.1016/j.bbr.2007.06.007
52. Clark, A.: Whatever next? Predictive brains, situated agents, and the future of cognitive science. Behav. Brain Sci. **36**, 181–204 (2013). doi:10.1017/S0140525X12000477
53. Maturana, H.R., Varela, F.J.: Autopoiesis and Cognition: The Realization of the Living. Springer, Netherlands (1991)
54. Freud, S., Strachey, J.: An Outline of Psycho-analysis. W. W. Norton, New York (1989)
55. Jung, C.G.: The Psychology of the Transference. Taylor & Francis, New York (2013)
56. Weinberg, B.H., Pham, N.T.H., Caraballo, L.D., et al.: Large-scale design of robust genetic circuits with multiple inputs and outputs for mammalian cells. Nat. Biotech **35**, 453–462 (2017)
57. Calvo, R.A., Peters, D.: Positive Computing: Technology for wellbeing and Human Potential. MIT Press, Massachusetts (2014)

58. Calvo, R.A., D'Mello, S., Gratch, J., Kappas, A.: The Oxford Handbook of Affective Computing. Oxford University Press, Oxford (2014)
59. Feuerstein, G., Kak, S.: The Yoga Tradition: Its History, Literature, Philosophy and Practice. Hohm Press, Chino Valley (2013)
60. Vitebsky, P.: Shamanism. University of Oklahoma Press, Norman (2001)
61. James, W.: The Varieties of Religious Experience: A Study in Human Nature. Longmans, Redwood City (1905)
62. Lazar, S.W., Kerr, C.E., Wasserman, R.H., Gray, J.R., Greve, D.N., Treadway, M.T., McGarvey, M., Quinn, B.T., Dusek, J.A., Benson, H., Rauch, S.L.: Meditation experience is associated with increased cortical thickness. NeuroReport **16**, 1893–1897 (2005). doi:10.1097/01.wnr.0000186598.66243.19
63. Gard, T., Taquet, M., Dixit, R., et al.: Greater widespread functional connectivity of the caudate in older adults who practice kripalu yoga and vipassana meditation than in controls. Front. Hum. Neurosci. (2015). doi:10.3389/fnhum.2015.00137
64. Hölzel, B.K., Carmody, J., Vangel, M., et al.: Mindfulness practice leads to increases in regional brain gray matter density. Psychiatry Res. **191**, 36–43 (2011). doi:10.1016/j.pscychresns.2010.08.006
65. Bostrom, N.: How long before superintelligence? Int. J. Future Stud. **2** (1998)

Non-ordinary Consciousness for Artificial Intelligence

Gabriel Axel Montes[1,2(✉)] (iD)

[1] University of Newcastle, Newcastle, Australia
gabriel@neuralaxis.com
[2] Hunter Medical Research Institute, Newcastle, Australia

Abstract. Humans are active agents in the design of artificial intelligence (AI), and our input into its development is critical. A case is made for recognizing the importance of including non-ordinary functional capacities of human consciousness in the development of synthetic life, in order for the latter to capture a wider range in the spectrum of neurobiological capabilities. These capacities can be revealed by studying self-cultivation practices designed by humans since prehistoric times for developing non-ordinary functionalities of consciousness. A neurophenomenological praxis is proposed as a model for self-cultivation by an agent in an entropic world. It is proposed that this approach will promote a more complete self-understanding in humans and enable a more thoroughly mutually-beneficial relationship between in life *in vivo* and *in silico*.

Keywords: Artificial intelligence · Cognition · Biomimetics · Neuroscience · Neurophenomenology · Consciousness · Philosophy · Robotics · Metacognition · Mindfulness · Mind-body · Evolution · Psychology · Medicine · Anthropology

1 Introduction

Insofar as humans remain agents in the design of AI, our input to its design matters greatly. Human self-consciousness and -knowledge are cornerstone elements of the instrumental cognition—the signature selective feature of bipedal prehensile hominins [1]—that has given rise to creative objective thought, the scientific method, and the design of complex machine intelligence. They allow for self-reflection, problem-solving, knowledge-seeking, are the receptacle for the productive rewards of instrumental cognition, and are the inputs for further inspiration. It is possible to actively seek new forms of knowledge, which can in turn amplify and modify the procedure of instrumental cognition, ergo what subsequently becomes input for AI. Actively seeking self-knowledge may then be the only human-centered lever for influencing or modifying the development of autonomous artificial agents. First-person phenomenal experience is the model from which we work when we operationalize our instrumental cognition and instantiate the production of synthetic artefacts. Advances in the computational processes of AI may be refined through new developments in engineering and discoveries of mechanisms in biological *in natura* systems. However, this angle remains silent regarding reconsiderations of the pivotal fact that human intelligence and consciousness is the starting point for AI development. The most immediately available lever for

M. Mangan et al. (Eds.): Living Machines 2017, LNAI 10384, pp. 348–362, 2017.
DOI: 10.1007/978-3-319-63537-8_29

human interjection into the biomimetic process is our view of our own human consciousness and cognition. If this can be manipulated or enhanced, then the starting point for biomimesis is altered, driving self-replication into new directions.

The present paper makes a case for utilizing non-ordinary states of consciousness in humans—such as those experienced in deep meditation, 'flow states', trance, and high-entropy psychedelic states [2]—in the design and development of AI. The overwhelming majority of AI efforts concentrate on representing the rational intelligence of humans in AI. Even current conceptions of 'superintelligence' are extrapolating the capabilities of AI based on an unwittingly logico-rational interpretation of human cognition [3, 4]. This is obviously sufficient for logico-mathematical calculations, which adequately represents the predictive algorithmic functionality of the neural architecture found in the human neocortex [5, 6]. It is acknowledged by some researchers in the field of biomimetics that it is crucial to not only mimic, but to understand nature and life, and then use this as a base for designing biomimetic technology. Our fashioning of AI continues to be modeled on the linear, mechanistic, rational view of life and consciousness, the outputs of which are accordingly limited in scope and application. Non-ordinary states of consciousness present a novel and hitherto unexamined opportunity for new developments in AI.

2 Non-ordinary Consciousness

"...there are known knowns; there are things we know we know. We also know there are known unknowns; that is to say we know there are some things we do not know. But there are also unknown unknowns—the ones we don't know we don't know." —D. Rumsfeld

2.1 Self-cultivation

If humans set the aim of understanding nature and life more fully, as claimed in the field of biomimetics, then there is an open door for representing in synthetic consciousness the farther reaches of human cognition. It is feasible to include other types of intelligence latent in the nervous system that have been previously discovered by other humans. In particular, an overlooked area of consciousness is its capacity for malleable manipulation by the deliberate application of particular consciousness-altering techniques. Diligent practitioners of methodologies offered by the world's wisdom traditions and prehistoric technologies of consciousness have, over the course of centuries to millennia, pushed the envelope with respect to the normal bounds of conscious experience. These phenomenological techniques function similar to man's material artefacts; just as the structure and function of produced tools and artefacts virtually represent particular cognitive intentionalities and states in the human agent. Examining the documented purported effects of particular techniques for achieving non-ordinary states of consciousness and/or applying those techniques would reveal the states of consciousness that they evoke. Neuroscientific studies have already been recording the effects of methods such as mindfulness meditation, yoga, mantra recitation, exercise-induced euphoria (e.g. "runner's high"), breath-manipulation, emotional regulation, psychedelic drug states [2, 7–9], compassion meditation, and others [10].

The present case being made is straightforward: the development of synthetic consciousness would benefit greatly by representing non-ordinary states of human consciousness in the repertoire of AI capabilities. This requires an increased attention to non-ordinary states in fields including medicine, neuroscience, systems biology, genetics, physiology, psychoneuroimmunology, anthropology, ethics, philosophy, and, of course, AI. It is minimally sufficient to be inspired by the experiences of non-ordinary states of consciousness; such as an AI researcher or engineer attempting to model her experience of such states into AI software or biohybrid neural functions. The democratization of access to such states to all persons of interest, including researchers, could serve to motivate them to explore representing those states in AI, due to their effect on self-knowledge and neurobiological function. Nonetheless, it is crucial for a formal investigation into such states to gain traction in the academic community and additionally in the public at large. The anecdotal and verbal reports that constitute the documented experiences of practitioners since centuries and millennia ago offer reason to be interested, and the existing neuroscience assures the scientific community of the efficacy of techniques for achieving non-ordinary states of consciousness [10]. Further rigorous work is imperative for the representation of non-ordinary states in biomimetics. Of particular priority is joint work in the areas of phenomenology, neuroscience, and anthropology, and, of course, AI.

What kinds of capacities and expanded self-understanding can we expect to gain from investigating non-ordinary states of consciousness? First, we can define the body of practices that elicit such states as *self-cultivation*, because they cultivate capabilities related to the self by way of the self, i.e. they are self-enacted techniques for entrainment of configurations of desired states and traits of consciousness. Such states can be approached by using neurotechnologies such as brain stimulation, however, because these approaches are still in an early experimental phase, our present focus is on stan- dalone self-cultivation methods. Examples of capacities that can be cultivated include but are in no way limited to: (a) interoceptive attunement—heightened sensory awareness to internal bodily signals such as the heartbeat, hunger, organs, metabolism, somato- affective tensions of self and, to whatever possible degree, other agents; (b) metacogni- tion—strategies about thought, such as spatiotemporal maps of self-development across various domains, and ability to deploy cognitive and self-cultivation techniques at will in a contextually dependent manner; (c) hypnagogic states for accessing the subconscious mind; (d) detailed internal representations of sounds, such as music, tones, and ono- matopoetic language (e.g. Sanskrit); (e) awareness of belief systems and the ability to attenuate and re-appraise them as necessary; (f) a manageable functional repertoire of alternative body schemas by use of virtual hologram-like models of body; (g) sovereign management of self-other boundaries at emotional and mental levels; (h) advanced breath regulation techniques, such as deliberate alteration of breath ratios to achieve certain neurophysiological states, e.g. freediving or yogic *'pranayama'* (Skt.); (i) awareness of the interdependent aggregate nature of all existence, and thus the insubstantiality of an ego-personality self; (j) mindfulness meditation; (k) yoga postures; (l) visualization techniques; (m) reliable intuitive abilities through means other than mentation; (n) ap- propriate application of somatic relaxation and release methods; (o) manipulation of body temperature and metabolism (Tibetan *tummo*); (p) lucid dreaming and "dream yoga"; (q) out-of-body experiences (OBE's); (r) compassion and somatic empathy;

(s) "witnessing"/observing meditation; (t) one-pointed attention (Skt. *dharana*) and absorption (Skt. *samadhi*) of consciousness; (u) ideokinesis—the ability to mentally move forms and somatic sensations in the bodily and mental space, within subjective awareness; (v) conscious movement and dance; (w) internal martial arts; (x) Tai Chi and Chi Kung; (y) manipulation of flow of currents of interoceptive sensations ('meridians'); and (z) intentionality-directed self-control towards goals. There are countless more examples, including the numerous '*siddhis*' (Skt.) or claimed special abilities/'powers' of consciousness that develop as a practitioner gains mastery over various aspects of mind.

Many such capabilities are not necessarily properties or abilities exclusive to non-ordinary states; they are often occurring 'under the radar' of awareness during ordinary states of consciousness and are simply being overlooked. For example, it is well known that humans have a subconscious mind that is affecting us every moment, but rarely are the contents or dynamics of the subconscious mind online in the working memory of self-consciousness and aware to the phenomenological experience of the agent. Self-cultivation practices can promote the ability to make the contents of the subconscious mind accessible, and can be categorized as non-ordinary states because they are not commonly occurring in everyday awareness.

3   Neurophenomenological Praxis

"We are the facilitators of our own creative evolution." —Bill Hicks

3.1   Cognitive Domains

Virtually all (known) self-cultivation methodologies in the world function by the systematic instantiation of particular cognitive domains [10]. Each technique emphasizes one domain and approaches it in a unique way given pedagogical and cultural contexts as well as individual differences. Harnessing these cognitive domains is why self-cultivation practices can be considered methods; they are not mere placebo effect, as has been demonstrated in active-controlled studies [10–12]—although they certainly can be under improper instruction and subconscious projection by the practitioner, and they fundamentally work by affecting the cognition of the practitioner, much like early man's tools. There are six main cognitive domains that can be leveraged by self-cultivation practices in one way or another. In terms of both actual self-cultivation practice and cognitive neuroscience research, these cognitive domains are irreducible to each other. This is a brief treatment for the purposes of the subject matter, and each cognitive domain has a rich background of supporting research which deserves a more detailed separate treatment by the author, particular their relationship with self-cultivation in light of consciousness and major theories thereof, e.g. enactive autonomy [13, 14], sensorimotor contingency (SMC) theory [15], distributed adaptive control (DAC) [16, 17], the Bayesian brain (predictive coding and the free-energy principle) [18–20], and integrated information theory (IIT) [21, 22].

Attention. Attention constitutes alerting to perceptual stimuli, orienting among them, and monitoring conflict between them [10]. Attention is the "online" perceptual field of conscious awareness available to a practitioner at a given moment based on her situated context within the environment, psychological conditioning, genetics, trauma, neurological health, mood, culture, worldview, etc. In predictive coding terms, attention is the precision weighting of prediction and prediction error signals in the nervous system [23]. In self-cultivation, attention forms the basis on which all other components rest. Conscious attention is central and foundational to self-cultivation because it is the part of consciousness that can be accessed, manipulated, and regulated by practices. Attention has been widely and deeply researched in cognitive neuroscience and psychology, and forms the cognitive basis that enables transformation of states of consciousness on any level.

Intention. Intention constitutes the prospects of attention—inference as to where attention *could* be in a particular circumstance, such as after making a change in behavior. It is, in a sense, a covert stage of action, which primes the motor system for potential future states [24]. In self-cultivation, this involves either being open to receiving experience without delimiting attention to a particular subset of conditioned percepts, or deliberate placement of attention on the idea or potential feeling of a prospective goal. For instance, if the practitioner wants to experience a subjective sense of "confidence", she could begin to open her mind to the possibility of feeling that way and exploring what that might feel like, subjectively. Another example of an intention commonly used in Tibetan Buddhism is compassion [25], in which a practitioner cultivates compassion for self and others. Interestingly, mindfulness meditation practice has been shown to incidentally lower the threshold for pre-conscious motor intention [26].

Expansion. Expansion constitutes the cognitive dimension that allows for 'expansive', altered, or non-ordinary states of consciousness. It could involve focusing on a particular feeling, such as the experience of "relaxation" and then expanding into, or subjectively moving towards occupying, that state and its accompanying neurophysiological correlates. An intense postural yoga practice may evoke a profound sense of relaxation; stable mindfulness may evoke a state of objective 'non-attachment' to phenomena as they arise in subjective awareness; an absorptive state may evoke a sense of interconnectedness with some object or all life; a serotonergic psychedelic state may evoke the experience of one-ness with all phenomena; and so on. Expansive states may also be drawn—effectively so—from autobiographical memory by activating memories and re-embodying them [27, 28]. Replaying hippocampal memories strengthens goal-weighted experience statistics and promotes memory complex integration [29]. Furthermore, optogenetic stimulation of hippocampal cells storing positive memories has been shown to reverse depression-related symptoms in rodents, underlaid by a hippocampal-amygdala–nucleus accumbens neural circuit [30] and long-term potentiation and depression (LTP and LTD) in the amygdala [31], suggesting that cultivating positive memories can counteract averse experiences.

Refinement. Refinement comprises the phenomenological activities that work to fine-tune, re-appraise, and self-correct the state of 'expanded' awareness. For instance, if a practitioner is unable to proficiently absorb into an expansive state of relaxation, she may do some psychological "shadow work", e.g. psychotherapy, to uproot belief systems and subconscious perspectives that block a more expansive and adaptive experience in more parts of her consciousness and action repertoires. It has been demonstrated that "fearful" memories can be replaced by positive ones by optogenetic manipulation of hippocampal neurons [32]. This cognitive domain is the most challenging in self-cultivation because it involves metacognitively and deliberately altering perceptual predictions, i.e. temporarily increasing entropy and thereby disrupting agent homeostasis, however, subsequent consciousness integration can bring about newer homeostases (newer experiences) at a faster rate than if not performed.

Engagement. Engagement constitutes action or simulation thereof, and in self-cultivation, particularly from the perspective of an altered (expanded and/or refined) state, i.e. engaging in the world while embodying the conscious state of interest. If during meditation a practitioner manages to experience relaxation in areas of her mind or body that she previously had not been able to, she could then try to "carry over" that state into her daily activities to integrate it with her life. This involves self-monitoring and making an effort to attend to acting *from* and *with* that conscious state. Engagement may also consist of 'motor imagery', visualization, or interoceptive simulation during meditation practice; once having grasped a state of interest, the practitioner simulates acting in the world from that state, prior to actually physically going out and doing so. Another example is a stroke patient imagining her disabled arm moving. This would prime her neural networks, particularly the motor system [24], to act in such a way so that when she is actually in the world, that neural 'groove' has already been rehearsed.

Evaluation. Evaluation constitutes metacognition, self-observation, decision deliberation, and reflection upon one's motivations and experience of practice, often from outside the acute state of consciousness that was accessed. Evaluating an experience makes that state more available for access and integration in the future, and allows the practitioner to develop strategies for scheduling practice, contemplating how to integrate it into areas of life, and so on. It may involve evaluating effort costs, which recruits areas of prefrontal cortex, intraparietal sulcus, insula, and amygdala [33]. Self-cultivation practitioners may wish to reflect upon the effect that a practice had on them from start to finish based on their expectations, which may be represented by the dorsal anterior cingulate cortex, which has been shown to encode prediction error between expectation and outcome in a stop-signal task [34]. Conscious reflection upon experience promotes conscious rewriting of valuation of preferences, pleasantness ratings, and reward expectations encoded in ventral prefrontal cortex and ventral striatum [35], the outcome of which are then integrated and read out in dorsolateral prefrontal cortex and posterior parietal cortex, respectively [36].

3.2 Neurophenomenological Scientific Method

These domains comprise the cognitive background on which all self-cultivation practices operate. These correspond to particular faculties that are subserved by function-specific dense or sparse neural activity. Some techniques emphasize particular domains over others [10], and each technique recruits each domain in different ways, both experientially and neuroscientifically. The novelty of this framework is that it covers the range of cognitive components harnessed by self-cultivation methodologies. The domains can function algorithmically in actual practice; a technique is applied and repeated, and each repetition will exhibit slightly different phenomena due to its being bootstrapped on previous experience. Each practice iteration forms or contributes to a 'stratum' of neurophenomenological [14] encoding. The domains need not be harnessed in a particular order; however, the above order is a schematic for the general sequence of the cognitive processes involved in various practices and in an ideal model of a comprehensive self-cultivation method. The recursion of this method is very similar to the cycles of instrumental cognition and the scientific method. In the proto-scientific method exhibited in early tool-intentional man, attention corresponds to making observations about nature, intention to the formation of a hypothesis and predictions, expansion to gathering data—'expanding out' from the hypothesis into actuated data collection, refinement to modification of the hypothesis, evaluation to data analysis and peer review, and engagement to translating findings. Refinement occurs post–data collection phase and/or post-experiment to fine-tune a hypothesis or form a new one, respectively.

Mesoscale Enaction. This procedural algorithm for self-cultivation praxis is commensurate with the same underlying instrumental cognition that has driven tool-creation and the scientific method. We can regard the agent-scale neural system as the microscale; the entropic world as the macroscale, and instrumental cognition as the mesoscale. The mesoscale praxis functions as a kind of homeostasis-promoting "airlock chamber" between the world (macroscale) and the agent's system (microscale). The instrumental cognition occupies the window of self-consciousness, i.e. it enables access to both the neural system and the extended mind via the interface of virtualized consciousness. When instrumental cognition is applied in the form of a self-cultivation 'neurophenomenological praxis', or "neuropraxis" (NP) for short, it leverages consciousness to operationalize the world-situated neural system in service of self-cultivation. Via self-cultivation, consciousness becomes an 'empowered' [37] tool that can be utilized to skillfully transform and traverse homeostatic states of an agent in an entropic world (Fig. 1). An agent possessing instrumental self-consciousness can maximize self-development by edging toward a state of criticality in the system without actually crossing the line into entropic disintegration. The neurological condition epilepsy is an example of neural states that have crossed the critical threshold [38]. Interestingly, a 2nd-person methodology—elicitation interviews [39, 40]—have been shown to increase the threshold for onset of epileptic seizures [41], due to the self-reflective capacity promoted by this practice, which can serve as a variety of self-cultivation. If a neural state approaches too close to criticality such as to destabilize, self-cultivation can be used to train to detect this within consciousness and self-regulate towards homeostasis, much like how mindfulness meditation can lower the detection threshold for pre-motor intention [26]. Notwithstanding the obvious undesirability of pathological criticality such as in epilepsy, a healthy flexibility in states of consciousness can promote stability near critical phase transitions between desirable states of consciousness. For example, an agent may be able, through

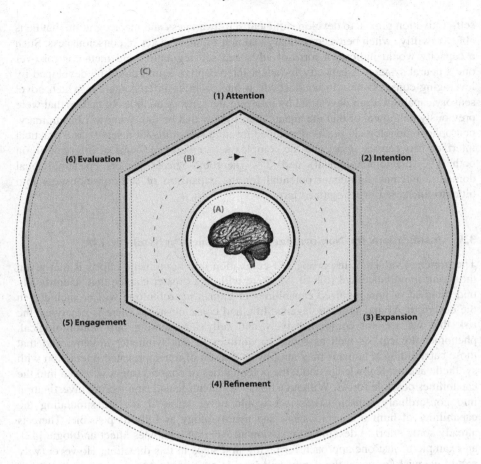

Fig. 1. Neurophenomenological praxis (NP) enacted by world-situated agent. (A) An agent's neural system is the microscale homeostatic entropy–minimizing system. (B) Six cognitive domains form a NP operationalized and enacted by the agent's recursive instrumental cognition that serves as a mesoscale interface between agent and world. In self-cultivation, the NP is volitionally applied through the window of consciousness as a deliberate means of altering sense of self and agency, regulating homeostasis, traversing neurophenomenological criticality thresholds, and empowering the agent. The steps can be taken in the suggested order, repeating cyclically. (C) The entropic world at the macroscale is inferred and explored by the agent via instrumental cognition, and in self-cultivation through the NP. Each iteration of the NP progressively modifies the agent's self-model, consciousness, and neural system. Dotted lines around the micro-, meso-, and macroscales denote the malleability of the system at each scale: agency at the microscale is changing by way of the NP; the mesoscale NP is flexible in its application and is shaped by the enactive situatedness of the agent in the world; the macroscale world is an entropic system which the agent infers and models in order to maintain homeostasis, and it is perturbed and modified by each iteration of the NP.

self-cultivation praxis, to develop sufficient self-awareness and metacognition that he is able to swiftly switch between ordinary and non-ordinary states of consciousness. Such a capacity would indicate a form of advanced self-regulation and meta-control over one's neural system. A capacity for adept interoceptive acuity could be developed by leveraging consciousness to conduct 'deep interoceptive inferencing', in which novel sensory capabilities are developed by inferring the sensory qualities of causes that were previously unknown, or that the agent "didn't know that he didn't know". In summary, neurophenomenological praxis developed through self-cultivation serves as a practical interface between the agent and the complex self-organizing world of information on both 1st- (phenomenological) and 3rd- (neurobiological) person epistemological domains, and has significant potential for the expansion of self-consciousness into hitherto uncharted or recognized territory.

3.3   Justifications for Non-ordinary Consciousness in Synthetic Life

The principal values of importing self-cultivation into biomimetic efforts is that given the rising importance and role of AI, there will be a greater chance that scientifically understudied or unrecognized capabilities of human neurobiology will be included in the design of AI and synthetic agents, life, and consciousness. There is otherwise the risk that we are not comprehensively and fully representing our neurobiological, phenomenological, as well as social capabilities in our synthetic progeny, and that those capabilities in humans may atrophy as a result of unreciprocated interaction with synthetic agents. Knowledge about the potentialities of consciousness will feed into the capabilities possible for AI. Without a bona fide scientific and engineering investigation into non-ordinary consciousness, we would never know if we are simulating the capabilities of human consciousness and neurobiology as fully as possible. There is already some effort to design "compassionate AI" that combines affect and logic [42], an example of just one approach for taking first steps in this direction. However, why would we wish to model the depths and span of consciousness more comprehensively in synthetic life? There are several key reasons:

Bidirectional Inter-enaction. The first reason is that AI will be interacting with humans, and the dynamics of that interaction matters. The very fact that current efforts to design robots that are able to mimic the facial expressions associated with human emotions or even simply respond to emotional cues and interact on that level with humans [43–46] demonstrates the looming reality in which AI and humans are engaged in bidirectional interaction. There isn't solely inter-*action*; there will be a fundamentally inter-*enactive* [14] relationship, in which the self-understanding, cognition, and capabilities of each influence and shape the other. While at the time of this writing humanity is in a transition stage to full-scale human-robot interaction, future generations will be literally born into a biohybrid world where the line between biological and synthetic life are significantly blurred, much like the lines between human genders are blurring today [47]. Well before that time, there will be an osmotic gradient of progressive "rubbing off", or counter-transference [48, 49], from robot to human, a process already underway, as such is the nature of agent-tool relationship in which each fashions the other. The biomimetic journey is not a unidirectional one in which we try to make synthetic life mimic humans, as if humans were a static entity. Humans will be learning from and adjusting their sociality to include robots in their "circle of compassion", which will in turn alter human behavior and social norms to accommodate synthetic lifeforms. This process will leverage the mirror neuron system in humans [50–53], which can in turn affect the entire human nervous system, and thus consciousness, in a cascade effect that takes place at multiple timescales, including across human generations. Indeed, it has been experimentally demonstrated that humanoid robots induce EEG-based mu rhythm suppression [54], which is considered to be a marker of mirror neuron system activation [55, 56]. Furthermore, the manipulation of self-consciousness inherent to self-cultivation means that the definition of self and the dimensions of agency in humans will shift accordingly, which in turn will feed into the selfhood and agency that project onto (transference [48, 49]) and are attributed to synthetic life. The influence of self-cultivation is thus non-linear and injects a synergistic array of catalysts into the human-AI evolutionary relationship.

Crucially, if synthetic life is not designed to handle, mirror, and replicate the broader and deeper range of human capabilities as humans are changing due to interaction with synthetic life, then we run the grave risk of humans themselves losing these underrepresented deeper abilities. Synthetic life that is in a bidirectional inter-enactive relationship with humans must be capable of reciprocating or dynamically responding with humans on those levels. Furthermore, it would be very beneficial for AI to be able to help humans regrow or re-cultivate those capacities if they, at some future point, have been lost or forgotten. This latter scenario is a possible one if non-ordinary states of consciousness does not start to become incorporated into AI in the near future. Of course, a worse scenario is that the deeper human capabilities remain unrecognized and not modelled altogether and thus very difficult to include in a world where a relationship between AI and humans is ubiquitous and possibly inescapable.

Synthetic Personhood. A closely related second reason to model our farther reaches in our synthetic brethren is that synthetic agents will start to be regarded as part of humanity itself, perhaps most pronouncedly in humanoid biohybrid applications. In addition to the increasing integration of robots into human contexts, there are already legislative efforts to treat humanoid robots as legal "persons" [57]. This invites further motivation to imbue our synthetic progeny with non-ordinary capabilities. Because their closeness and resemblance to humans will make them, if not by law, then by convention regarded as part of humanity, they and their actions then fall within the sphere of ethical and moral consideration. It can be argued that humans then have a moral motivation, if not an obligation—in much the same way as we consider our own biological children—to consider ways of imbuing synthetic life with higher capabilities. From this biohybrid amalgamation, it follows that if *Homo sapiens* ever becomes endangered or extinct, it is then up to our synthetic progeny to carry the torch of human-capable consciousness forward. Envisioning such a scenario can serve as a litmus test for considering how ready our AI is to lead the way into a post-biological future. Furthermore, aside from an endangerment or extinction status, there will be many hostile environments into where humans are ill-equipped to venture, such as extraterrestrial expeditions. There is already serious work being done on colonizing other planets, such as Mars, and designing prototypical habitats there for humans [58]. In such environments, synthetic and biohybrid lifeforms can be expected to play a prominent role.

Pre-emptive Programming. Athird imperative for modeling non-ordinary consciousness in synthetic life is that we cannot rely on self-replicating robots to learn these capacities themselves; rather humans may have to pre-emptively program AI with desired capabilities. While initially programmed by and modelled on what we consider to be human characteristics—which is conditioned on our current self-understanding, as already mentioned—their autonomous function cannot guarantee their learning preferences, generally speaking. Humans are already deep in the swing of developing AI that is tailored to our linear rational intelligence capability. Non-ordinary states of human consciousness are often deeply embodied and somatic—underlaid by demonstrated existing conditioning of brain activity by internal organ activity, such as the stomach [59], and quite possibly contingent to some degree on the three-dimensional enfoldment and spatial arrangement of the brain and the body's morphological structure, partly due to endogenous electromagnetic fields. This means that linear software code may be insufficient to accurately model human capacities. Nick Bostrom, Founder of the Future of Humanity Institute and leading AI thinker, has already suggested that AI be programmed to learn human values so that, for example, before acting, a machine could evaluate whether a proposed action is in line with a core human value [4]. AI will need to have its learning algorithms penetrate into embodied and somatic architectures, which will likely need human assistance to be accomplished. It should also be mentioned that AI of the future may not necessarily need to use only the human-like non-ordinary capabilities. On a technical level, AI may use the capabilities themselves and then leverage them for other tasks. Furthermore, not all AI will need to model non-ordinary human capabilities. The ones that would benefit from it the most are agents designed to deal with and interact with complex dynamic human interactions, such as in education, nursing, nannying, customer care, autism, disability, psychiatric disorders, and so on.

Humanistic Interest. Afourth reason to model the farther reaches of human nature is that AI without deeper human capabilities carries the risk of not having a stake in humans, particularly once AI can self-replicate and direct its own learning processes to varying degrees. Because AI will be regularly interacting with humans and will possess computational powers beyond human comprehension that likely will not be communicated to humans, humans will not be aware of any underlying implicit motivations that a synthetic agent may have. If humans aspire to a biohybrid future with an equitable relationship between the biological and synthetic counterparts, it is of utmost importance that AI have a stake in humans.

3.4   Human Valuation of Non-ordinary Consciousness

Perhaps an unspoken matter in our discussion thus far is the motivations by humans to themselves value and practice self-cultivation methods. The matter of what we value and are motivated to pursue as human beings corresponds to our *evaluative* function as per the neurophenomenological praxis above. What value is there in cultivating these capabilities or at least recognizing or acknowledging their importance? This human motivation merits its own separate treatment. Being motivated to instill these capabilities in AI would do well to first begin with humans themselves first seeing the value in doing so. There are myriad benefits [10], including but not limited to improved quality of life (QoL) in healthy and clinical populations; the threshold between the conscious and subconscious mind being lowered, resulting in great access to the latter [26]; a greater ability to manage life stresses and attend to novel stimuli; emotion regulation and cognitive behavioral reappraisal; changes in neurophysiology; and generally an overarching ability to monitor and therefore manage and integrate one's cognition, body, and interactions. Much work on this has already been done, and further work will construct a more high-fidelity technical blueprint of non-ordinary consciousness, which can then be used to inform biomimetic efforts.

4   Conclusions

This paper has examined (1) how the treasury of non-ordinary states of consciousness can aid in the disruption of trends in the assumptions about the dimensions of human capabilities and intelligence; (2) how self-cultivation practices can offer a source of non-ordinary states of consciousness for innovation in biomimetics and AI; (3) how a neurophenomenological praxis that leverages innate instrumental cognition for self-cultivation serves as a proximal, enactive mesoscale lever for agent self-consciousness to exercise homeostasis-management in the mediation between neural structures at the microscale and the world at the macroscale; (4) how imbued with the knowledge of non-ordinary states of consciousness, humans have social, philosophical, ethical, and existential obligations to take these into account in biomimetic efforts; and (5) how motivations for self-cultivation in human personal life and research will accelerate and empower human agency in biomimetic efforts. Non-ordinary consciousness offers a vast treasury of opportunities for deepening our understanding of nature and human function—as per the noble aim of biomimetics—and bestowing the synthetic/biohybrid progeny of humans with novel capabilities. This work merits further treatment in each of the relevant fields of research in order to examine in greater detail how each one can actively engage with non-ordinary states of consciousness, biomimesis, and the implications for the future of life and AI. In conclusion, the conscious human participation in biomimetic efforts via mesoscale neurophenomenological praxis offers novel opportunities and directions for the evolution of life and consciousness.

References

1. McGinn, C.: Prehension: The Hand and the Emergence of Humanity. MIT Press, Massachusetts (2015)
2. Bostrom, N.: How long before superintelligence? Int. J. Future Stud. 2 (1998)
3. Bostrom, N.: Superintelligence: Paths, Dangers, Strategies. Oxford University Press, Oxford (2014)

4. Tani, J.: Exploring Robotic Minds: Actions, Symbols, and Consciousness as Self-Organizing Dynamic Phenomena. Oxford University Press, Oxford (2016)
5. Hawkins, J., Blakeslee, S.: On Intelligence. Henry Holt and Company, New York (2007)
6. Bowler, P.J.: Evolution: The History of an Idea. University of California Press, Berkeley (2003)
7. Carhart-Harris, R.L., Leech, R., Hellyer, P.J., et al.: The entropic brain: a theory of conscious states informed by neuroimaging research with psychedelic drugs (2014)
8. Carhart-Harris, R.L., Erritzoe, D., Williams, T., et al.: Neural correlates of the psychedelic state as determined by fMRI studies with psilocybin. Proc. Natl. Acad. Sci. **109**, 2138–2143 (2012)
9. Vollenweider, F.X., Kometer, M.: The neurobiology of psychedelic drugs: implications for the treatment of mood disorders. Nat. Rev. Neurosci. **11**, 642–651 (2010)
10. Grof, S.: LSD Psychotherapy. Multidisciplinary Association for Psychedelic Studies (M A P S), Santa Cruz (2001)
11. Tang, Y.-Y., Holzel, B.K., Posner, M.I.: The neuroscience of mindfulness meditation. Nat. Rev. Neurosci. **16**, 213–225 (2015). doi:10.1038/nrn3916
12. Zeidan, F., Martucci, K.T., Kraft, R.A., Gordon, N.S., McHaffie, J.G., Coghill, R.C.: Brain mechanisms supporting the modulation of pain by mindfulness meditation. J. Neurosci. **31**, 5540–5548 (2011). doi:10.1523/JNEUROSCI.5791-10.2011
13. Allen, M., Dietz, M., Blair, K.S., et al.: Cognitive-affective neural plasticity following active-controlled mindfulness intervention. J. Neurosci. **32**, 15601–15610 (2012)
14. Maturana, H.R., Varela, F.J.: Autopoiesis and Cognition: The Realization of the Living. Springer, Netherlands (1991)
15. Varela, F.J., Rosch, E., Thompson, E.: The Embodied Mind: Cognitive Science and Human Experience. MIT Press, Massachusetts (1992)
16. O'Regan, J.K., Noë, A.: A sensorimotor account of vision and visual consciousness. Behav. Brain Sci. **24**, 939–973 (2001)
17. Verschure, P.F., Voegtlin, T., Douglas, R.J.: Environmentally mediated synergy between perception and behaviour in mobile robots. Nature **425**, 620 (2003). doi:10.1038/nature02024
18. Verschure, P.F.M.J.: Synthetic consciousness: the distributed adaptive control perspective. Philos. Trans. R. Soc. B Biol. Sci. (2016). doi:10.1098/rstb.2015.0448
19. Friston, K.: The free-energy principle: a unified brain theory? Nat. Rev. Neurosci. **11**, 127–138 (2010). doi:10.1038/nrn2787
20. Clark, A.: Whatever next? Predictive brains, situated agents, and the future of cognitive science. Behav. Brain Sci. **36**, 181–204 (2013)
21. Schjoedt, U., Andersen, M.: How does religious experience work in predictive minds? Relig. Brain Behav. 1–4 (2017). doi:10.1080/2153599X.2016.1249913
22. Tononi, G.: The integrated information theory of consciousness: an updated account. Arch. Ital. Biol. **150**, 56–90 (2011)
23. Tononi, G.: Consciousness as integrated information: a provisional manifesto. Biol. Bull. (2016)
24. Hohwy, J.: Attention and conscious perception in the hypothesis testing brain. Front. Psychol. **3**, 96 (2012). doi:10.3389/fpsyg.2012.00096
25. Jeannerod, M.: Neural simulation of action: a unifying mechanism for motor cognition. NeuroImage **14**, S103–S109 (2001). doi:10.1006/nimg.2001.0832
26. Hölzel, B.K., Lazar, S.W., Gard, T., et al.: How does mindfulness meditation work? Proposing mechanisms of action from a conceptual and neural perspective. Perspect. Psychol. Sci. **6**, 537–559 (2011). doi:10.1177/1745691611419671

27. Delevoye-Turrell, Y., Bobineau, C.: Motor consciousness during intention-based and stimulus-based actions: modulating attention resources through mindfulness meditation. Front. Psychol. **3**, 290 (2012). doi:10.3389/fpsyg.2012.00290

28. Pearson, J., Naselaris, T., Holmes, E.A., Kosslyn, S.M.: Mental imagery: functional mechanisms and clinical applications. Trends Cogn. Sci. **19**, 590–602 (2015). doi:10.1016/j.tics.2015.08.003

29. Spiers, H.J., Cothi, W., Bendor, D.: Manipulating hippocampus-dependent memories: to enhance, delete or incept? In: Hannula, D.E., Duff, M.C. (eds.) The Hippocampus from Cells to Systems: Structure, Connectivity, and Functional Contributions to Memory and Flexible Cognition, pp. 123–137. Springer, Cham (2017). doi:10.1007/978-3-319-50406-3_5

30. Kumaran, D., Hassabis, D., McClelland, J.L.: What learning systems do intelligent agents need? Complementary learning systems theory updated. Trends Cogn. Sci. **20**, 512–534 (2016). doi:10.1016/j.tics.2016.05.004

31. Ramirez, S., Liu, X., MacDonald, C.J., et al.: Activating positive memory engrams suppresses depression-like behaviour. Nature **522**, 335–339 (2015)

32. Nabavi, S., Fox, R., Proulx, C.D., et al.: Engineering a memory with LTD and LTP. Nature **511**, 348–352 (2014)

33. Redondo, R.L., Kim, J., Arons, A.L., et al.: Bidirectional switch of the valence associated with a hippocampal contextual memory engram. Nature **513**, 426–430 (2014)

34. Chong, T.T.-J., Apps, M., Giehl, K., et al.: Neurocomputational mechanisms underlying subjective valuation of effort costs. PLoS Biol. **15**, e1002598 (2017). doi:10.1371/journal.pbio.1002598

35. Ide, J.S., Shenoy, P., Yu, A.J., Li, C.R.: Bayesian prediction and evaluation in the anterior cingulate cortex. J. Neurosci. **33**, 2039 (2013). doi:10.1523/JNEUROSCI.2201-12.2013

36. Schmidt, L., Lebreton, M., Cléry-Melin, M.-L., et al.: Neural mechanisms underlying motivation of mental versus physical effort. PLoS Biol. **10**, e1001266 (2012). doi:10.1371/journal.pbio.1001266

37. Domenech, P., Redouté, J., Koechlin, E., Dreher, J.-C.: The neuro-computational architecture of value-based selection in the human brain. Cereb. Cortex (2017)

38. Klyubin, A.S., Polani, D., Nehaniv, C.L.: Empowerment: a universal agent-centric measure of control. In: 2005 IEEE Congress on Evolution Computation, vol. 1, pp. 128–135 (2005)

39. Leon, P.S., Knock, S.A., Woodman, M.M., et al.: The Virtual brain: a simulator of primate brain network dynamics. Inf. Based Methods Neuroimaging Anal. Struct. Funct. Dyn. **8** (2015)

40. Petitmengin, C.: Describing one's subjective experience in the second person: an interview method for the science of consciousness. Phenomenol Cogn. Sci. **5**, 229–269 (2006). doi:10.1007/s11097-006-9022-2

41. Petitmengin, C., Lachaux, J.-P.: Microcognitive science: bridging experiential and neuronal microdynamics. Front. Hum. Neurosci. **7**, 617 (2013). doi:10.3389/fnhum.2013.00617

42. Petitmengin, C., Baulac, M., Navarro, V.: Seizure anticipation: Are neurophenomenological approaches able to detect preictal symptoms? Epilepsy Behav. **9**, 298–306 (2006). doi:10.1016/j.yebeh.2006.05.013

43. Mason, C.: Engineering kindness: building a machine with compassionate intelligence. Int. J. Synth. Emot. IJSE **6**, 1–23 (2015)

44. Prassler, E., Lawitzky, G., Stopp, A., et al.: Advances in Human-Robot Interaction. Springer, Heidelberg (2004)

45. Coleman, D.: Human-Robot Interactions: Principles, Technologies and Challenges. Nova Science Publishers, Incorporated, New York (2015)

46. Kanda, T., Ishiguro, H.: Human-Robot Interaction in Social Robotics. Taylor & Francis, New York (2012)

47. Jentsch, F., Barnes, M., Harris, P.D., et al.: Human-Robot Interactions in Future Military Operations. Ashgate Publishing Limited, Aldershot (2012)
48. Browning, F.: The Fate of Gender: Nature, Nurture, and the Human Future. Bloomsbury Publishing, London (2016)
49. Freud, S., Strachey, J.: An Outline of Psycho-analysis. W. W. Norton, New York (1989)
50. Jung, C.G.: The Psychology of the Transference. Taylor & Francis, New York (2013)
51. Gallese, V., Goldman, A.: Mirror neurons and the simulation theory of mind-reading. Trends Cogn. Sci. **2**, 493–501 (1998)
52. Heyes, C.: Where do mirror neurons come from? Neurosci. Biobehav. Rev. **34**, 575–583 (2010)
53. Iacoboni, M.: Imitation, empathy, and mirror neurons. Annu. Rev. Psychol. **60**, 653–670 (2009)
54. Uddin, L.Q., Iacoboni, M., Lange, C., Keenan, J.P.: The self and social cognition: the role of cortical midline structures and mirror neurons. Trends Cogn. Sci. **11**, 153–157 (2007)
55. Matsuda, G., Hiraki, K., Ishiguro, H.: EEG-based mu rhythm suppression to measure the effects of appearance and motion on perceived human likeness of a robot. J. Hum. Rob. Interact. **5**, 68–81 (2015)
56. Pineda, J.A.: The functional significance of mu rhythms: translating "seeing" and "hearing" into "doing". Brain Res. Rev. **50**, 57–68 (2005). doi:10.1016/j.brainresrev.2005.04.005
57. Ulloa, E.R., Pineda, J.A.: Recognition of point-light biological motion: Mu rhythms and mirror neuron activity. Behav. Brain Res. **183**, 188–194 (2007). doi:10.1016/j.bbr.2007.06.007
58. Europe's robots to become "electronic persons" under draft plan. In: Reuters (2017). Accessed 9 Mar 2017
59. Montes, J.: The Mars Ice House (2015)
60. Richter, C.G., Babo-Rebelo, M., Schwartz, D., Tallon-Baudry, C.: Phase-amplitude coupling at the organism level: the amplitude of spontaneous alpha rhythm fluctuations varies with the phase of the infra-slow gastric basal rhythm. NeuroImage **146**, 951–958 (2017). doi:10.1016/j.neuroimage.2016.08.043

A Biomimetic Vocalisation System for MiRo

Roger K. Moore[1]([✉]) and Ben Mitchinson[2]

[1] Department of Computer Science, University of Sheffield, Sheffield, UK
r.k.moore@sheffield.ac.uk
[2] Department of Psychology, University of Sheffield, Sheffield, UK
b.mitchinson@sheffield.ac.uk

Abstract. There is increasing interest in the use of animal-like robots in applications such as companionship and pet therapy. However, in the majority of cases it is only the robot's physical appearance that mimics a given animal. In contrast, *MiRo* is the first commercial biomimetic robot to be based on a hardware and software architecture that is modelled on the biological brain. This paper describes how *MiRo*'s vocalisation system was designed, not using pre-recorded animal sounds, but based on the implementation of a real-time parametric general-purpose mammalian vocal synthesiser tailored to the specific physical characteristics of the robot. The novel outcome has been the creation of an 'appropriate' voice for *MiRo* that is perfectly aligned to the physical and behavioural affordances of the robot, thereby avoiding the 'uncanny valley' effect and contributing strongly to the effectiveness of *MiRo* as an interactive device.

Keywords: Biomimetic robot · MiRo · Mammalian vocalisation · Vocal synthesis

1 Introduction

Recent times have witnessed increasing interest in the use of animal-like robots in applications such as companionship and pet therapy. For example, *PARO* [1] is an interactive robotic seal that is particularly popular for therapeutic use in hospitals and care facilities where a live animal would be problematic. Like *PARO*, the majority of such 'zoomorphic' devices are engineered to support specific use-cases, and it is often only the robot's physical appearance that mimics a given animal. In contrast, *MiRo* [2] (designed and built by Consequential Robotics Ltd. in collaboration with the University of Sheffield) is the first commercial robot to be controlled by a hardware and software architecture that is specifically modelled on the biological brain [3,4].

MiRo is a highly featured, low-cost, programmable mobile developer platform, with a friendly animal-like appearance, six senses, a nodding and rotating head, moveable hearing-ears, large blinking seeing-eyes, and a responsive wagging tail. It has been designed to look like a cartoon hybrid of a generic mammal (see Fig. 1). A unique biomimetic control system allows *MiRo* to behave in a

© Springer International Publishing AG 2017
M. Mangan et al. (Eds.): Living Machines 2017, LNAI 10384, pp. 363–374, 2017.
DOI: 10.1007/978-3-319-63537-8_30

Fig. 1. MiRo: the world's first commercial biomimetic robot (available from Consequential Robotics Ltd [2]).

life-like way: for example, listening for sounds and looking for movement, then approaching and responding to physical and verbal interactions.

Of special interest here is *MiRo*'s ability to vocalise. In particular, it was considered important that the vocal output generator should be grounded in a biomimetic model of an appropriate physical sound production apparatus rather than based, for example, on pre-recorded animal sounds. As a consequence, *MiRo*'s voice was designed using a real-time parametric general-purpose mammalian vocal synthesiser [5] tailored to the particular physical and behavioural characteristics of the robot.

This paper presents the biomimetic principles underlying *MiRo*'s vocalisation system, and describes how they have been integrated into the robot's overall architecture. Section 2 reviews the principles underlying mammalian vocalisation, and Sect. 3 outlines *MiRo*'s overall control architecture. Section 4 then describes how the particular characteristics of *MiRo*'s voice were first designed using a general-purpose mammalian vocal synthesiser, and subsequently implemented on the robot platform itself. Finally, Sect. 5 concludes with some observations about the effectiveness of the derived solution.

2   Vocalisation in Mammals

The majority of animals make sound, and different species make sound in different ways. For example, many insects rub body parts together (a process known as 'stridulation'), birds create their songs using a vocal organ known as a 'syrinx', and *land*[1] mammals typically generate sound using a 'larynx' [6].

---

[1] Some mammals are adapted for the air or for water. However, such animals tend to be the extremes in terms of size (for example, the bumblebee bat measures only

The vocal tract for a typical land mammal consists of a larynx, a pharynx, an oral cavity and a nasal cavity (see Fig. 2). These anatomical features evolved primarily for breathing and eating. However, over time they have been recruited to create and shape sound. The main sound source is the larynx, which contains a set of 'vocal folds' (sometimes referred to as 'vocal cords' or the 'glottis') - a pair of elastic membranes that are held apart while breathing, but brought together when eating (in order to stop unwanted material from entering the lungs). The consequence of this arrangement is that air from the lungs that is forced through the closed vocal folds causes them to vibrate. This creates acoustic energy in the form of a harmonic-rich buzzing sound with a distinct fundamental frequency (perceived as the 'pitch' of the voice). Muscles in the larynx control the length and tension of the vocal folds which, in turn, determine the pitch and timbre of the generated sound.

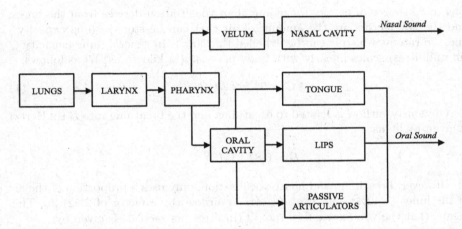

Fig. 2. Schematic diagram of the mammalian vocal tract.

The rest of the vocal tract can be regarded as a set of interconnected acoustic tubes. The pharynx lies immediately above the larynx, and this contains the epiglottis: an elastic cartilage that controls entry to the trachea (for breathing) or the oesophagus (for swallowing)[2]. Above the pharynx the acoustic tube splits into two: the main oral cavity (containing the tongue and terminating at the mouth and lips) and the nasal cavity (terminating at the nose). Airflow into the nasal cavity is controlled by the 'velum', a flap-like structure at the back of the mouth. All of these tubes resonate at different frequencies depending on their size and shape, hence they each filter the spectrum of the harmonic-rich excitation generated by the vibration of the vocal folds; energy is enhanced at

---

30 mm, whereas the blue whale is over 30 m long) and exploit different mechanisms for generating sound than the majority of land mammals.

[2] It is this arrangement that enables mammals to choke!.

some frequencies (the resonances) and suppressed at others. Vocal tract resonances are known as 'formants', and the formants arising from resonances in the oral cavity are of particular significance since they can be altered by moving the tongue and opening/closing the mouth[3].

In order to determine the characteristics of the sounds produced by a mammalian vocal tract, it is necessary to model (a) the 'airflow' (the rate at which air is expelled from the lungs), (b) the 'excitation' (the sound generated by the larynx) and (c) the 'filtering' (arising from the resonant cavities). Unsurprisingly, many of these processes are influenced by the size of the animal, since body mass has a direct impact on the physical and acoustic properties of the relevant anatomical components: the lungs, vocal folds, tongue and mouth.

2.1 Airflow

The main source of energy for mammalian vocalisation derives from the lungs, and the key factors are the amount of air that can be stored ('lung capacity') and the rate at which it can be expelled ('airflow'). In general, lung capacity C (in millilitres) scales linearly with body mass M (in kilograms) [7] as follows:

$$C = 53.5 \times M^{1.06}. \tag{1}$$

Obviously, airflow is related to breathing, and the breathing rate B (in Hertz) is given by [8] as:

$$B = 0.84 \times M^{-0.26}. \tag{2}$$

However, breathing, and hence vocalisation, only uses a proportion of the air in the lungs ($\sim$42%), and it also restricts airflow (by a factor of 2.62) [5]. This means that the volumetric flow rate Q (in litres per second) is given by:

$$Q = \frac{0.42 \times C}{2.62 \times \left(\frac{1}{2 \times B}\right)}, \tag{3}$$

which simplifies to:

$$Q = 0.32 \times C \times B. \tag{4}$$

These parameters characterise the amplitude and duration of each vocalisation and, as can be seen, the predicted value for airflow (Q) is directly related to the size of the animal (M). However, these are mean values, and variation around the mean is possible. For example, a higher airflow would give rise to a shorter but louder vocalisation and *vice versa*.

---

[3] In human beings, these are the primary anatomical features used for speaking.

2.2 Excitation

As air is forced through the closed vocal folds, it escapes in bursts as the folds are momentarily pushed apart. After each bubble of air escapes, the Bernoulli effect causes the vocal folds to snap shut again, and this action generates a pulse of acoustic energy that propagates through the rest of the vocal tract. This sequence of events repeats at semi-regular intervals giving rise to a periodic excitation signal with energy at the fundamental frequency of vibration and its associated harmonics[4].

According to [9], the mean fundamental frequency F (in kHz) of the vocal fold vibration for animals ranging in size from mice to elephants is related to the body mass of the animal by the expression:

$$F = M^{-0.4}. \tag{5}$$

In other words, small animals have high-pitched vocalisations and large animals have low-pitched vocalisations.

The 'timbre' of a vocalisation is a function of (a) the regularity of the vocal fold vibrations, (b) the relationship between the fundamental frequency and its associated harmonics and (c) the degree of turbulence in the airflow. The latter means that, in addition to fully 'voiced' sounds, the mammalian larynx is capable of generating aspirated (breathy) 'unvoiced' sounds by holding the vocal folds close together and allowing a small amount of continuous airflow[5]. All of these aspects can be modelled by suitable shaping of the excitation waveform and by the injection of an appropriate level of random noise.

2.3 Filtering

The vocal tract resonances (formants) act as 'band-pass' filters which enhance acoustic energy at their resonant frequencies and suppress acoustic energy at other frequencies. This has a shaping effect on the harmonic-rich spectrum of the excitation signal emanating from the larynx[6].

The frequencies of the different formants can be estimated by assuming that the vocal tract is a uniform acoustic tube[7] which is closed at the vocal folds and open/closed at the mouth. As the mouth closes, so the formants move down in frequency [10]. Hence, the resonant frequency of the nth formant R_n (in Hz) can be approximated by the equation:

$$R_n = (2n - (m + 1)) \times \frac{c}{4 \times L}, \tag{6}$$

---

[4] To a first approximation, the signal generated by the glottis may be modelled as a 'sawtooth' waveform.

[5] Noisy unvoiced excitation at the vocal folds gives rise to whispering in human speech.

[6] It is the different placement of formants that gives rise to the production of different vowel sounds in human speech.

[7] This ignores the action of the tongue, which is an appropriate approximation for a non-human land mammal.

for $n = 1, 2, 3, :$, where m is the degree of mouth opening ($0 =$ open, $1 =$ closed), c is the speed of sound (in cm/sec) and L is the length of the vocal tract (in cm) [5].

According to [11], vocal tract length is correlated with body size:

$$L = 3.15 + (11.53 \times \log M). \tag{7}$$

This means that large animals have long vocal tracts and thus low formant frequencies (and *vice versa*). It also means that the distribution of formant frequencies in a vocalisation provides information to a listener about the size of the animal[8].

2.4 Summary

The foregoing provides a complete specification of the minimum set of parameters necessary to simulate the vocalisation of a generic land mammal (as described in [5]). The novel contribution here is the mapping of this specification onto *MiRo*'s particular physical characteristics and control architecture.

3 MiRo's Control Architecture

MiRo's control architecture operates across three embedded ARM (Advanced RISC Machines) processors that mimic aspects of spinal cord, brainstem and forebrain functionality (including their relative speed and computational power) - see Fig. 3. One important feature is that the control latency of loops through the lowest reprogrammable processor can be as low as a few milliseconds. If required, *MiRo* can be operated remotely through WiFi or Bluetooth, and can also be configured as a Robot Operating System (ROS) node [13].

3.1 Actuators

MiRo is constructed around a differential drive base and a neck with three Degrees of Freedom (DoF). Additional DoFs include rotation for each ear, tail droop and wag, and eyelid open/close. All DoFs are equipped with proprioceptive sensors, and the platform also has an on-board loudspeaker.

3.2 Sensors

MiRo is equipped with stereo cameras in the eyes, stereo microphones in the base of the ears and a sonar range-finder in the nose. Four light-level sensors are placed at each corner of the base, and two infrared 'cliff' sensors point down from its front face. Eight capacitive sensors are arrayed along the inside of the body shell and over the top and back of the head (behind the ears). These provide an indication of direct human touch. Internal sensors include twin accelerometers, a temperature sensor and battery-level monitoring.

---

[8] Interestingly, animals such as the Red Deer are able to lengthen their vocal tract by lowering their larynx when vocalising, thereby giving the impression of being much larger than they really are [12].

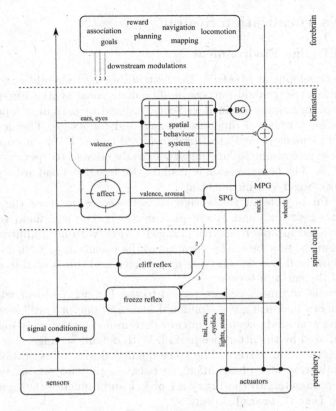

Fig. 3. Illustration of MiRo's control architecture loosely mapped onto brain regions (spinal cord, brainstem, forebrain). Signal pathways are excitatory (open triangles), inhibitory (closed triangles), or complex (closed circles). BG is the Basal Ganglia. SPG and MPG are the Social and Motor Pattern Generators.

3.3   Affect

MiRo represents its affective state (emotion, mood and temperament) as a point in a two-dimensional space covering 'valence' (unpleasantness-pleasantness) and 'arousal' (calm-excited) [14]. Events arising in *MiRo*'s sensorium are mapped into changes in affective state: for example, stroking *MiRo* drives valence in a positive direction, whilst striking *MiRo* on the head drives valence in a negative direction. Baseline arousal is affected by general sound/light levels as well as the time of day; *MiRo* is more active in the daytime. An individual event can cause an acute change: for example, a very loud sound might raise arousal and decrease valence. *MiRo*'s movements are modulated by its affective state, and it also expresses itself using a set of 'social pattern generators' that drive light displays, movement of the ears, tail, eyelids and - of particular relevance here - vocalisation.

4   MiRo's Vocalisation System

4.1   Vocal Design Environment

Prior to the development of *MiRo*, the first author had already constructed a real-time parametric general-purpose mammalian vocal synthesiser (in accordance with the principles outlined in Sect. 2) aimed at designing 'appropriate' vocalisations for a range of different animals and robots [5]. The design environment is implemented in 'Pure Data' (referred to as "Pd") - an open-source visual dataflow programming language specifically created to operate with real-time audio[9] [15]. The latest version is available for free download at http://www.dcs.shef.ac.uk/~roger/downloads.html.

The key Pd objects in the design software correspond to the [lungs], [larynx], [vocal tract] and [post-processing]. The command to vocalise initiates simulated airflow from the [lungs] object with an amplitude that is calculated from the flow rate. The duration of the vocalisation is then calculated as a function of the flow rate and the lung capacity, and this is used to determine the period of the entire utterance.

These signals and messages are passed to the [larynx] object which modulates the energy flow using the simulated action of one or two[10] sets of vocal folds vibrating at a fundamental frequency determined by the body mass, which is itself modulated by the utterance period. With default settings, this gives rise to a rise-fall intonation pattern. The voice quality, degree of aspiration (noise), level of quantisation and pitch difference between the two sets of vocal folds are all input parameters to the [larynx] object and influence the signal that is output to the [vocal tract] object.

The [vocal tract] object simulates three acoustic resonances (formants) using band-pass filters whose frequencies are determined by the vocal tract length and the degree of mouth opening (using Eq. 6). A syllabic rate parameter controls the opening and closing of the mouth.

Control parameters are set via a GUI using appropriate buttons and sliders (see Fig. 4). This facilitates real-time adjustment of the vocalisation, and greatly enhances the process of designing different voices. In principle, it is possible to set every parameter independently. However, in practice, there are a number of potential dependencies (as described in Sect. 2). As a result, setting the body size to a particular value also sets:

- the lung capacity (using Eq. 1),
- the breathing rate (using Eq. 2),
- the flow rate (using Eq. 4),
- the fundamental frequency (using Eq. 5), and
- the vocal tract length (using Eq. 7).

---

[9] Pd (and its professional counterpart: MAX-MSP) is commonly used in music studios.

[10] It is well established that two excitation signals slightly offset in fundamental frequency give the resulting vocalisation a distinct robotic timbre.

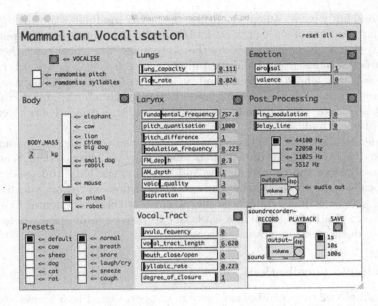

Fig. 4. Screenshot of the Pd GUI for the real-time parametric general-purpose mammalian vocal synthesiser that was used to design MiRo's voice.

The software also provides a number of preset settings. For example, it is possible to select particular animals (such as a rat, cat, dog, sheep, dog or cow in the current version), and also select different types of vocalisation (such as normal, breathing, snoring, laughing/crying, sneezing and coughing). Selecting one of these presets simply moves all of the sliders to particular predetermined positions. After selecting a preset it is still possible to vary any/all of the parameters as required in order to achieve a particular design objective.

4.2 Implementation on MiRo

The real-time parametric general-purpose mammalian vocal synthesiser design environment described in Sect. 4.1 above was used (a) as a basis for implementing MiRo's biomimetic vocalisation system on the robot platform outlined in Sect. 3 and (b) to determine the appropriate parameter settings. Accordingly, MiRo's synthesis software (programmed in C) was structured to simulate the flow of energy through a mammalian vocal apparatus with body mass corresponding to a land mammal of an equivalent size (∼2 kg). The vocalisation modules were integrated into MiRo's 'biomimetic core' (corresponding to the 'brainstem' in Fig. 3).

The robot has a breathing rhythm (∼0.7 Hz), the frequency of which is linked to arousal (see Sect. 3.3), and vocalisation is initiated stochastically during the exhalation phase. Breathing is simulated as cyclic airflow into and out of the lungs with an amplitude and duration that is calculated from the flow rate, lung capacity and body mass. When vocalising, the larynx modulates the airflow using the simulated action of a set of vocal folds vibrating at a fundamental frequency (~760 Hz) that is also determined by the body mass. The vocal tract then simulates three formants using band-pass filters whose frequencies are determined by the vocal tract length (~6.6 cm) and the degree of mouth opening. A syllabic rate parameter controls the opening and closing of the mouth, and a vibrating uvula adds a 'cute' robotic timbre to the voice. It was decided *not* to employ two sets of vocal folds.

In order to allow the injection of emotion into the vocalisations, parameters were linked to *MiRo*'s two-dimensional affect map (as discussed in Sect. 3.3). Arousal modulates the rate of airflow and, thereby, the amplitude and tempo of the vocalisations; high arousal leads to high airflow and short vocalisations (and *vice versa*). Valence influences the variance of the fundamental frequency and the voice quality; high valence leads to expressive vocalisation whereas low valence produces more monotonic utterances.

4.3 Example Vocalisations

As an example of the vocalisation system in operation, Fig. 5 shows spectrograms[11] for two basic sounds - breathing and snoring. These *unvoiced* vocalisations are generated using noise as a excitation signal (as described in Sect. 2.2), and the spectrograms clearly illustrate the three-formant resonant structure of *MiRo*'s simulated vocal tract (Sect. 2.3). For these particular sounds, the vocal tract is fairly static throughout, hence there is little variation in the formant frequencies.

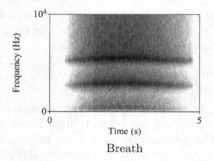 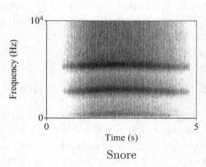

Breath   Snore

Fig. 5. Spectrograms of *MiRo*'s vocalisations for an exhaled breath and an inhaled snore. The dark bars indicate the concentration of energy at the formant resonances, and the vertical striations in the snore reflect the vibrating uvula.

In contrast, Fig. 6 shows spectrograms for *voiced* vocalisations with different affective states (as described in Sect. 3.3). As can be seen, these sounds are more

---

[11] A time-frequency energy plot commonly used to analyse speech and audio signals.

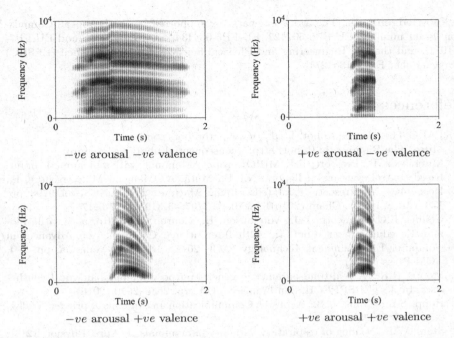

Fig. 6. Spectrograms of *MiRo*'s vocalisations resulting from different values for arousal and valence.

dynamic than those shown in Fig. 5, mainly due to the opening and closing of the mouth. In addition, the formants vary more with positive valence (due to larger mouth opening), and the durations are shorter with high arousal (due to higher airflow).

5   Summary and Conclusion

This paper has described the design and implementation of *MiRo*'s biomimetic vocalisation system. Based on the principles underlying vocalisation in land mammals, it has been shown how the key characteristics of *MiRo*'s vocalisations were selected using a real-time parametric general-purpose mammalian vocal synthesiser tailored to the specific physical characteristics of the robot. It has been explained how these design decisions were ported onto *MiRo*'s hardware/software platform and integrated into the robot's overall control architecture. The novel outcome has been the creation of an 'appropriate' voice for *MiRo* that is perfectly aligned to the physical and behavioural affordances of the robot [16]. As such, it successfully avoids the 'uncanny valley' effect caused by mismatched perceptual cues [17,18] and contributes strongly to the effectiveness of *MiRo* as an attractive interactive device.

Acknowledgements. This work was partially supported by the European Commission [grant numbers EU-FP6-507422, EU-FP6-034434, EU-FP7-231868 and EU-FP7-611971], and the UK Engineering and Physical Sciences Research Council (EPSRC) [grant number EP/I013512/1].

References

1. PARO Therapeutic Robot. http://www.parorobots.com
2. MiRo: The Biomimetic Robot. http://consequentialrobotics.com/miro/
3. Mitchinson, B., Prescott, T.J.: MIRO: a robot "Mammal" with a biomimetic brain-based control system. In: Lepora, N.F.F., Mura, A., Mangan, M., Verschure, P., Desmulliez, M., Prescott, T.J.J. (eds.) Living Machines 2016. LNCS, vol. 9793, pp. 179–191. Springer, Cham (2016). doi:10.1007/978-3-319-42417-0_17
4. Collins, E.C., Prescott, T.J., Mitchinson, B., Conran, S.: MiRo: a versatile biomimetic edutainment robot. In: 12th International Conference on Advances in Computer Entertainment Technology (ACE 2015), Iskandar, Malaysia, pp. 1–4. ACM Press (2015)
5. Moore, R.K.: A real-time parametric general-purpose mammalian vocal synthesiser. In: INTERSPEECH, San Francisco, CA, pp. 2636–2640 (2016)
6. Hopp, S.L., Evans, C.S.: Acoustic Communication in Animals. Springer Verlag, New York (1998)
7. Stahl, W.R.: Scaling of respiratory variables in mammals. J. Appl. Physiol. **22**(3), 453–460 (1967)
8. Worthington, J., Young, I.S., Altringham, J.D.: The relationship between body mass and ventilation rate in mammals. Exp. Biol. **161**, 533–536 (1991)
9. Fletcher, N.H.: A simple frequency-scaling rule for animal communication. J. Acoust. Soc. Am. **115**(5), 2334–2338 (2004)
10. Titze, I.R.: Acoustic interpretation of resonant voice. J. Voice **15**(4), 519–528 (2001)
11. Riede, T., Fitch, T.: Vocal tract length and acoustics of vocalization in the domestic dog (Canis familiaris). J. Exp. Biol. **202**(20), 2859–2867 (1999)
12. Fitch, W.T., Reby, D.: The descended larynx is not uniquely human. Proc. Roy. Soc. B **268**(1477), 1669–1675 (2001)
13. Robot Operating System (ROS). http://www.ros.org
14. Collins, E.C., Prescott, T.J., Mitchinson, B.: Saying it with light: a pilot study of affective communication using the MIRO robot. In: Wilson, S.P., Verschure, P.F.M.J., Mura, A., Prescott, T.J. (eds.) Living Machines 2015. LNCS, vol. 9222, pp. 243–255. Springer, Cham (2015). doi:10.1007/978-3-319-22979-9_25
15. PureData. https://puredata.info
16. Moore, R.K.: From talking and listening robots to intelligent communicative machines. In: Markowitz, J. (ed.) Robots That Talk and Listen, pp. 317–335. De Gruyter, Boston (2015)
17. Mori, M.: Bukimi no Tani (The Uncanny Valley). Energy **7**, 33–35 (1970)
18. Moore, R.K.: A bayesian explanation of the 'Uncanny Valley' effect and related psychological phenomena. Nat. Sci. Rep. **2**, 864 (2012)

A Scalable Neuro-inspired Robot Controller Integrating a Machine Learning Algorithm and a Spiking Cerebellar-Like Network

Ismael Baira Ojeda[✉], Silvia Tolu, and Henrik H. Lund

Technical University of Denmark,
Elektrovej Building 326, DK-2800 Kgs., Lyngby, Denmark
{iboj,stolu,hhl}@elektro.dtu.dk

Abstract. Combining Fable robot, a modular robot, with a neuroinspired controller, we present the proof of principle of a system that can scale to several neurally controlled compliant modules. The motor control and learning of a robot module are carried out by a Unit Learning Machine (ULM) that embeds the Locally Weighted Projection Regression algorithm (LWPR) and a spiking cerebellar-like microcircuit. The LWPR guarantees both an optimized representation of the input space and the learning of the dynamic internal model (IM) of the robot. However, the cerebellar-like sub-circuit integrates LWPR input-driven contributions to deliver accurate corrective commands to the global IM. This article extends the earlier work by including the Deep Cerebellar Nuclei (DCN) and by reproducing the Purkinje and the DCN layers using a spiking neural network (SNN) implemented on the neuromorphic SpiNNaker platform. The performance and robustness outcomes from the real robot tests are promising for neural control scalability.

Keywords: Neuro-robotics · Bio-inspiration · Motor control · Cerebellum · Machine learning · Compliant control · Internal model

1 Introduction

The brain carries out tasks in a formidable manner, smoothly and with a low power consumption. In contrast, robots lack the adaptability and precision of human beings. Thus, our main interest is to study how the central nervous system (CNS) controls the human body, coordinates smooth movements and the mechanisms of motor control and motor learning towards the development of bio-inspired autonomous robotic systems [1]. We present a bio-inspired closed-loop control architecture that integrates a machine learning (ML) technique and a cerebellar-like microcircuit to provide real-time neural control. The result is a compliant system for motor learning that can scale to several neurally controlled robot modules. This is achieved by combining a modular robot with the neuromorphic computing platform, SpiNNaker [2]. The ML engine is the Locally Weighted Projection Regression algorithm (LWPR) [3].

I.B. Ojeda and S. Tolu—These authors contributed equally to this work.

© Springer International Publishing AG 2017
M. Mangan et al. (Eds.): Living Machines 2017, LNAI 10384, pp. 375–386, 2017.
DOI: 10.1007/978-3-319-63537-8_31

1.1   The Cerebellar Microcircuit

The presented bio-inspired approach aims at emulating the cerebellar function in learning and modulating accurate, complex and coordinated movements. The cerebellum is involved in the neural control of bodily functions [4], such as postural positioning, balance or coordination of movements over time. Research studies [5–7] described the cerebellum as a set of adaptive modules, also called cerebellar microcomplexes, embedded in the motor control system to improve coordinated movements over time. Ito [5] stated that each microcomplex is a Unit Learning Machine (ULM): a set of structured neural circuits that modify the relationship between the input and output in response to error signals. The ULM encodes the internal model (IM) in order to precisely perform the control of the body part without referring to feedback. At the core of our control architecture, a cerebellum-like model is implemented akin to the Marr-Albus cerebellum model [8,9]. The cerebellar cortex comprises three layered sheets wrapped around the deep cerebellar nuclei (DCN) and it connects to the brain stem via the superior, middle and inferior peduncles. The microcomplexes that correspond to the minimal functional unit, show a similar internal microcircuitry illustrated in Fig. 1. The main inputs are: the mossy fibers (MFs) and the climbing fibers (CFs). MFs transmit sensory information originated from multiple extra-cerebellar sources and carry out excitatory synapses with the dendrites of the granule cells (GCs) and the deep cerebellar nuclei cells (DCNs). CFs arise exclusively from the inferior olive (IO) within the brain stem and they are particularly responsive to unexpected somatosensory events. Thus, they are usually conceived as error signal transmitters. The output of the cerebellar cortex is performed by the Purkinje cell (PCs), which combines information received from the CFs and the parallel fibers (PFs). The PC produces a "complex spike" response [10] each time it receives input from a CF. Finally, the DCNs combine the inhibitory signal coming from the PCs (cerebellar cortex output) with the excitatory information coming from both CFs and MFs. The PC and DCN layers are modeled in SpiNNaker by a Spiking Neural Network (SNN) that, instead of receiving inputs by means of MFs, receives pre-processed signals from the LWPR receptive fields (RFs) (see Fig. 1).

1.2   Related Work

Cerebellar mechanisms of motor learning and control are not yet completely unveiled [11]. By mimicking them, researchers are eager to uncover those unknowns and at the same time to develop scalable and useful neural control techniques. Indeed, some techniques have already benefited robotics such as successful simulations where robots are controlled using biologically plausible cerebellar models [12–14], virtual experiments with a simulated robot using customized brain models, e.g. the Neurorobotics Platform (NRP) [15], bio-inspired abstraction strategies of robotic dynamics and kinematics [16], and adaptive control schemes for non-linear systems [17–19]. Besides, realistic cerebellar SNN

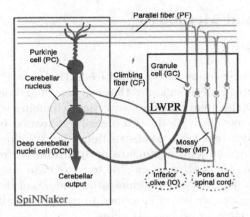

Fig. 1. The simplification of the cerebellum's internal circuit. On the left, the sub-circuit reproduced using SpiNNaker. On the right, the granular layer represented by the LWPR algorithm.

implementations on a computer [20], e.g. [21] and on the neuromorphic platform, Spinnaker [2], e.g. [22] to control a neurorobot are providing promising results.

Different neuromorphic hardware platforms have been developed, but most of them lack in terms of scalability and usability (FPGAs, GPs) [22]. To this end, the computer is still the prevailing architecture for neural simulations of non-large scale neural networks. In this article we present the unique proof-of-concept system combining SNN and ML technique for controlling a modular robot by CPU-Spinnaker co-processing.

1.3 Objectives and Paper Organization

Our research area has three main goals: to embed a strategy for a fast learning of the inverse IMs of the robot module providing an optimized input representation to the SNN. Indeed, the LWPR algorithm was chosen because it encodes the inputs like the cerebellar GCs expansively; to develop a neural control for mimicking the cerebellar modularity, i.e. the neural core can be replicated and dedicated to a specific robot part or module. Both SpiNNaker and the LWPR algorithm allow this internal scalability. The first in terms of neural cores and the latter in terms of incremental locally linear models; and to develop a user-friendly system in which the robot topology can be easily modified and controlled. The present paper addresses the first goal and sets out the basis by which the other two goals are going to be achieved in the future.

The outline for the next sections is as follows: Sects. 2.1 and 2.2 describe the robot employed and Spinnaker; Sect. 2.3 addresses the blocks of the adaptive control architecture; Sect. 3 explains the interface with SpiNNaker; Sects. 4 and 5 show the results giving a discussion and conclusions.

2   Materials and Methods

2.1   Fable Robot

Fable [23] is a modular robot based on self-contained 2-DoF modules. It consists of multiple simple robotic modules that can attach and detach. Connectors between units allow the creation of arbitrary and changing structures depending on the task to be solved. Modules are easily addressed by an ID and can be programmed at different levels of abstraction. Figure 2 illustrates the Fable's network architecture. The joint commands are transmitted from the computer using a radio dongle. The system is scalable to more than one Fable module by simply creating an object instance and indicating the IDs of the modules to take into account. The Spinnaker board described in the next section paves the way for large-scale modular neuro-robotic systems.

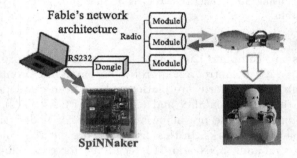

Fig. 2. Fable robot is equipped with Dynamixel AX-12A motors. The ULM is embedded in the control architecture to control a single Fable module. The LWPR engine is run in the computer and sends the inputs to the SNN implemented in SpiNNaker. The cerebellar output from SpiNNaker is the joint motor command, which is transmitted from the computer to Fable motor using a radio dongle. Fable modules can be snapped together towards different topologies.

2.2   Neuromorphic Platform

SpiNNaker [2] is a novel chip based on the ARM processor that was designed to support large scale spiking neural networks simulations. Inspired by the structure of the brain, the processing cores were arranged in independently functional and identical Chip-Multiprocessors (CMP) to achieve robust distributed computing. Each processing core is self-sufficient in the storage required to hold the code and the neurons' states, while the chip holds sufficient memory to contain synaptic information for the neurons in the system connected to the local neurons.

2.3   Adaptive Control Architecture

Our control approach is based on the Adaptive Feedback Error Learning (AFEL) architecture described in [16]. In this work, we included the DCN (see Fig. 3).

The architecture comprises: a trajectory planner which computes the desired joint angles and velocities $(Q_{dj}, \dot{Q}_{dj})$ by inverse kinematics; a Learning Feedback (LF) controller which generates the τ_{LF_j} feedback joint torques; and a ULM, which provides the τ_{ff_j} feed-forward joint torques. The τ_{ff_j} torque contribution of the ULM is a combination of the τ_{LWPR_j} prediction from the LWPR algorithm and the τ_{c_j} prediction from the PC-DCN layers for each joint j (see Fig. 3). In addition, the τ_{tot_j} global torque is the summation of the τ_{ff_j} and the τ_{LF_j} command joint torques.

If the adaptive model is accurate, the resulting τ_{ff} cancels the robot nonlinearities. However, if the inverse dynamic model is not exact, the LF reflects the feedback error between the desired signal $(Q_d, \dot{Q}_d)$ and the output of the real robot $(Q, \dot{Q})$. The LWPR incrementally learns the inverse IM of the robot module from the τ_{tot} global torques or efferent copy. The LWPR produces τ_{LWPR} torques to minimise the τ_{LF} (error related estimate), while the PC-DCN layer provides τ_C corrective torques which refine the τ_{tot} global command torques. Three plastic sites (PF-PC, PC-DCN, MF-DCN) are represented by the PC-DCN layers, whose learning benefits from a compact sensorimotor representation of the input space [17,24] by means of p_k weights provided by the LWPR algorithm. See below for implementation details.

ULM. Inside the Unit Learning Machine, the LWPR algorithm feeds the sensorimotor inputs to N linear local models. The x_i inputs consist of cerebellar torque commands, desired and current positions and velocities of every joint j. The LWPR incrementally divides the input space into a set of RFs, so that a weighting kernel computes a weight p_k for each x_i data point according to the distance from the c_k center of the kernel in each k local unit. The weight is a measure of how often an item of x_i data falls into the region of validity of each linear model. The weights are calculated using a Gaussian kernel (1):

$$p_k = \exp\left(-\frac{1}{2}(x_i - c_k)^T D_k (x_i - c_k)\right), \tag{1}$$

where D_k is a positive definite matrix which is called distance matrix (the size of the RF). This measure is updated on-line iteratively by using an incremental gradient descent based on stochastic leave-one-out cross validation criterion. In every iteration, the RF weight is updated in order to assign the new inputs to the closest RF. The LWPR output is the weighted mean of the linear models:

$$\hat{y} = \frac{\sum_{k=1}^{N} p_k \hat{y}_k}{\sum_{k=1}^{N} p_k} \tag{2}$$

In this work, $\hat{y}_k$ is the τ_{LWPR} torque shown in Fig. 3. The LWPR learns the τ_{tot} efferent copy. As it occurs in the cerebellum, the PC-DCN inputs are transmitted through a bank of filters, located in the GC layer. In our approach, this

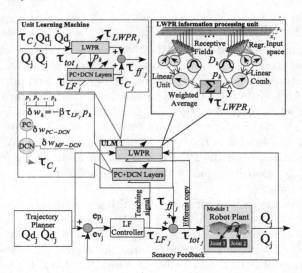

Fig. 3. The AFEL control architecture embeds a cerebellum-like model which acts as a feed-forward controller in the form of ULM. Even though only one ULM was used to control one Fable module, the system can be scaled up to control more modules by embedding several ULMs at the same time.

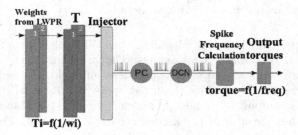

Fig. 4. The encoding and decoding principle behind the interface with SpiNNaker. The p_k weights coming from the LWPR are encoded into spiking rates so as to excite the population. Thereafter, the DCN spiking rate is transformed into torques.

bank of filters is represented by the $p_k(t)$ signals computed by the gaussian kernel in (1). Those p_k weights are driven to the synapses with the PC layer. The data flow along the SNN is illustrated in Fig. 4. First, the p_k input weights coming from the LWPR algorithm are transformed into spikes. The spiking rate of each PF is defined as one spike every T_i ms. The spikes are transmitted along the spiking network consisting of one pair of PCs and DCNs for each joint. Finally, the DCN output is calculated as inversely proportional to the DCN spike frequency rate. We selected the Leaky Integrate and Fire model with fixed threshold and decaying-exponential post-synaptic current as the neuron model since it showed smooth trajectories and fast computations. To apply memory consolidation in the SNN, we chose the Spike-Timing-Dependent plasticity (STDP) [25] whose synaptic learning is induced by tight temporal correlation between a pre- and a post-synaptic spike event.

LF Controller. The Learning Feedback controller overcomes the lack of a precise robot arm dynamic model, ensures the stability of the system and enables the control architecture to achieve a better performance [16]. Further details about the LF controller are provided in [16]. Its gains were tuned to $K_p = 7.5$, $K_v = 6.4$ and $K_i = 0.22$ for the Fable robot.

3   Interface

PyNN language [26] was selected to implement the SNN since it can run on a number of simulators with minor modifications of the code. A socket interface combined with thread functions was implemented to allow the communication between SpiNNaker and Fable robot. On the one hand, the socket interface enables the system to push and read data on-line between scripts that are running different python versions. The interface updates p_k weights every 5ms and the control loop frequency is 150 Hz. On the other hand, the firing rate is calculated to deliver the cerebellar output to the robot (Fig. 5).

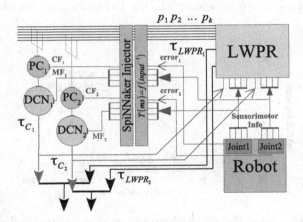

Fig. 5. Schematic of the connections needed for the integration between the LWPR, SpiNNaker and the robot.

4   Evaluation

To evaluate the control system performance, we examined how the tracking errors became compensated during the task of following the desired trajectory defined in (3), where Q_j is the angle of the j-th joint, A is the amplitude of the circular trajectory, and C_j was set to 0 and $\pi/2$ for joints 1 and 2, respectively.

$$Q_j = A \left(\tfrac{1}{2\pi f} \right)^2 sin(2\pi f t + C_j), \tag{3}$$

4.1   Control Performance

We measured the learning performance of the system in terms of the normalized mean squared error (nMSE). The error was considered as the difference between the desired and real position of each joint during the experiment. To this end, 20k iterations of the control loop, equivalent to 200 circles iterations, were run commanding the real robot to trace out a circular trajectory with its end-effector.

Figure 6 shows how the nMSE is decreasing over consecutive iterations. It is remarkable the fast adaptation. Furthermore we evaluated the torque contribution along the learning process by the performance ratios of torque components to the τ_{tot} global torque applied to the robot module plant. Figure 7 shows how the average values of the ratios evolve along the learning process. The LWPR algorithm progressively learns the τ_{tot} global torque which means that it acts as inverse IM for the robot module dynamics. LF torques decrease over time according to the decrease of the nMSE, and the contribution from the PC-DCN layers reveals the slow memory consolidation function which is achieved within the LWPR and more slowly achieved in the PC-DCN layers as shown in logarithmic scale.

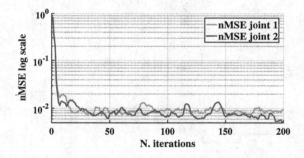

Fig. 6. Performance test. The performance of our neuro-controller during real-time simulation on computer and SpiNNaker platform in terms of the nMSE.

In order to check the advantages of the presented approach, we compared it with other three cases enumerated in Fig. 8 with labels (2), (3) and (4). In case (2) only the LF was active. In case (3) the PC-DCN layer was deactivated. In case (4) the LWPR was learning but not delivering its torque to the control loop. Notice the large tracking error of case (2) and how the PC-DCN contribution (case 4) leads to a faster decrease of the nMSE compared with case (3). It is

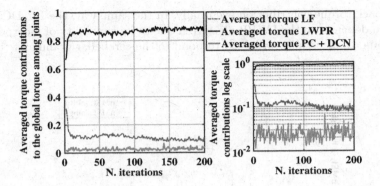

Fig. 7. Performance test. Ratios of torque contributions of the main blocks to the global torque averaged among joints.

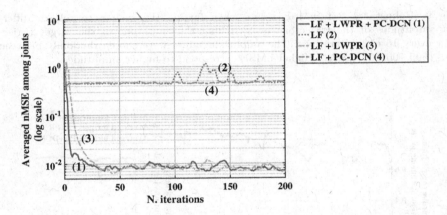

Fig. 8. Comparison of the averaged nMSE among joints for four different cases.

worth noting that the improvement of our approach is due to the addition of the DCN layer, and not due to the integration of SpiNNaker. Nevertheless, we integrated SpiNnaker aiming at scaling up the spiking cerebellar model in the future. SpiNNaker will be highly beneficial when implementing large amounts of spiking neurons and synapses in real time. However, CPU-based system may show issues in the performance and power consumption.

4.2   Robustness Tests

The robustness test consisted in switching the amplitude of the desired trajectory every 15 repetitions of the circle trajectory. Figure 9 shows that the approach is robust and self-adaptive to fast and repetitive changes. Furthermore, results in Fig. 10 indicate that the LWPR learning is incremental and so, the LF decreases its feedback torque contribution eventually. In the same way, the PC-DCN contribution depends on the LWPR algorithm due to its capability of incorporating the IM and to p_k signals, which correspond to the cerebellar granular weighting kernels.

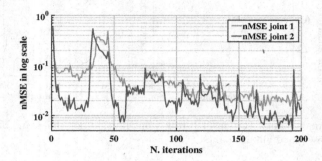

Fig. 9. Robustness test. nMSE result when forcing the system to relearn different amplitudes of the trajectory by switching the amplitude of the target trajectory every 15 circle iterations. The amplitude values were chosen randomly between $A \in \{500, 700, 900\}\mathrm{deg/s}^2$. Higher nMSE correspond to higher amplitudes.

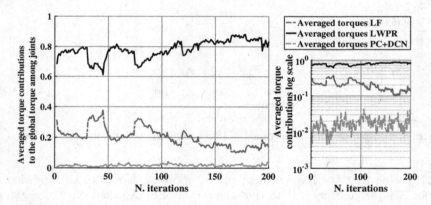

Fig. 10. Robustness test. Averaged torque contributions of the LWPR, LF and PC-DCN layers to the global torque when switching the amplitude of the target trajectory randomly ($A \in \{500, 700, 900\}\mathrm{deg/s}^2$).

5   Conclusions

We proposed a new Adaptive Feedback Error Learning scheme for the motor control and learning of a 2-DoF Fable module. The neural control is still based on the combination of a ML engine and a cerebellar-like model to both optimize the input space representation and to abstract the inverse IM. However, in this work, the cerebellar-like component is integrated with a spiking DCN layer running on SpiNNaker. This work was validated on a physical robot. We overcame the interface issues between SpiNNaker and Fable allowing an on-line control. The presented robustness and performance results were good showing an increase of the learning speed of the inverse IM while providing an optimized input representation. We will also integrate distinct sensors and actuators to check if it benefits the control and learning of the system.

SpiNNaker was designed to simulate large, biologically realistic, spiking neural networks in real time. We took advantage of only a small portion of the capability of SpiNNaker. Nevertheless, this paper shows the proof of principle for future achievements in the bio-inspired control and learning of robots using SpiNNaker. Further research will be needed to control more than one module and to benefit from this modular control strategy. Assembling distinct modular configurations will allow researchers and developers to carry out a variety of tasks with minor modifications. To this aim, all robot modules could be linked to identical ULMs. In future, we will test the system controlling more complex modular setups and study how to learn and store multiple IMs of the robot to face new challenges by combining previous experiences.

Acknowledgments. This work has received funding from the EU-H2020 Programme under the grant agreement no. 720270 (Human Brain Project SGA1) and from the Marie Curie project no. 705100 (Biomodular). We are thankful to David Johan Christensen and Moisés Pacheco, CEO/CTO & Co-founders of Shape Robotics, for the Fable robot. A special thank is expressed to the University of Manchester and the University of Munich for SpiNNaker.

References

1. Flash, T., Sejnowski, T.J.: Computational approaches to motor control. Curr. Opin. Neurobiol. **6**(11), 655–662 (2001)
2. Furber, S.B., Galluppi, F., Temple, S., Plana, L.A.: The spinnaker project. Proc. IEEE **102**(5), 652–665 (2014)
3. Vijayakumar, S., D'souza, A., Schaal, S.: Incremental online learning in high dimensions. Neural Comput. **17**(12), 2602–2634 (2005)
4. Ito, M.: Cerebellar circuitry as a neuronal machine. Prog. Neurobiol. **78**(3), 272–303 (2006)
5. Ito, M.: Control of mental activities by internal models in the cerebellum. Nat. Rev. Neurosci. **9**(4), 304–313 (2008)
6. Dean, P., Porrill, J., Ekerot, C.-F., Jörntell, H.: The cerebellar microcircuit as an adaptive filter: experimental and computational evidence. Nat. Rev. Neurosci. **11**(1), 30–43 (2010)
7. Verduzco-Flores, S.O., O'Reilly, R.C.: How the credit assignment problems in motor control could be solved after the cerebellum predicts increases in error. Front. Comput. Neurosci. **9** (2015)
8. Albus, J.S.: A theory of cerebellar function. Math. Biosci. **10**(1), 25–61 (1971)
9. Marr, D., Thach, W.T.: A theory of cerebellar cortex. In: Vaina, L. (ed.) From the Retina to the Neocortex, pp. 11–50. Springer, Boston (1991)

10. Bell, C.C., Kawasaki, T.: Relations among climbing fiber responses of nearby purkinje cells. J. Neurophysiol. **35**(2), 155–169 (1972). http://jn.physiology.org/content/35/2/155
11. Byadarhaly, K.V., Perdoor, M.C., Minai, A.A.: A modular neural model of motor synergies. Neural Networks **32**, 96–108 (2012)
12. Luque, N.R., Garrido, J.A., Carrillo, R.R., Coenen, O.J., Ros, E.: Cerebellar input configuration toward object model abstraction in manipulation tasks. IEEE Trans. Neural Networks **22**(8), 1321–1328 (2011)
13. Luque, N.R., Garrido, J.A., Carrillo, R.R., Tolu, S., Ros, E.: Adaptive cerebellar spiking model embedded in the control loop: context switching and robustness against noise. Int. J. Neural Syst. **21**(05), 385–401 (2011)
14. Garrido, J.A., Luque, N.R., D'Angelo, E., Ros, E.: Distributed cerebellar plasticity implements adaptable gain control in a manipulation task: a closed-loop robotic simulation (2013)
15. Vannucci, L., Ambrosano, A., Cauli, N., Albanese, U., Falotico, E., Ulbrich, S., Pfotzer, L., Hinkel, G., Denninger, O., Peppicelli, D., Guyot, L., Arnim, A.V., Deser, S., Maier, P., Dillman, R., Klinker, G., Levi, P., Knoll, A., Gewaltig, M.O., Laschi, C.: A visual tracking model implemented on the icub robot as a use case for a novel neurorobotic toolkit integrating brain and physics simulation. In: 2015 IEEE-RAS 15th International Conference on Humanoid Robots (Humanoids), pp. 1179–1184 (2015)
16. Tolu, S., Vanegas, M., Luque, N.R., Garrido, J.A., Ros, E.: Bio-inspired adaptive feedback error learning architecture for motor control. Biol. Cybern. **106**, 507–522 (2012)
17. Porrill, J., Dean, P.: Recurrent cerebellar loops simplify adaptive control of redundant and nonlinear motor systems. Neural Comput. **19**(1), 170–193 (2007)
18. Tolu, S., Vanegas, M., Garrido, J.A., Luque, N.R., Ros, E.: Adaptive and predictive control of a simulated robot ARM. Int. J. Neural Syst. **23**(03), 1350010 (2013)
19. Su, F., Wang, J., Deng, B., Wei, X.-L., Chen, Y.-Y., Liu, C., Li, H.-Y.: Adaptive control of parkinson039;s state based on a nonlinear computational model with unknown parameters. Int. J. Neural Syst. **25**(01), 1450030 (2015)
20. Yamazaki, T., Igarashi, J.: Realtime cerebellum: a large-scale spiking network model of the cerebellum that runs in realtime using a graphics processing unit. Neural Networks **47**, 103–111 (2013)
21. Casellato, C., Antonietti, A., Garrido, J.A., Carrillo, R.R., Luque, N.R., Ros, E., Pedrocchi, A., D'Angelo, E.: Adaptive robotic control driven by a versatile spiking cerebellar network. PLOS ONE **9**(11), 1–17 (2014)
22. Richter, C., Jentzsch, S., Hostettler, R., Garrido, J.A., Ros, E., Knoll, A., Rohrbein, F., van der Smagt, P., Conradt, J.: Musculoskeletal robots: scalability in neural control. IEEE Robot. Autom. Mag. **23**(4), 128–137 (2016)
23. Pacheco, M., Fogh, R., Lund, H.H., Christensen, D.J.: Fable: A modular robot for students, makers and researchers. In: Proceedings of the IROS workshop on Modular and Swarm Systems: from Nature to Robotics (2014)
24. Schweighofer, N., Doya, K., Lay, F.: Unsupervised learning of granule cell sparse codes enhances cerebellar adaptive control. Neuroscience **103**(1), 35–50 (2001)
25. Sjöström, J., Gerstner, W.: Spike-timing dependent plasticity. Spike-timing dependent plasticity, p. 35 (2010)
26. Davison, A., Brüderle, D., Kremkow, J., Muller, E., Pecevski, D., Perrinet, L., Yger, P.: PYNN: a common interface for neuronal network simulators (2009)

Behavior-State Dependent Modulation of Perception Based on a Model of Conditioning

Jordi-Ysard Puigbò[1,2(✉)], Miguel Ángel Gonzalez-Ballester[2,3],
and Paul F.M.J. Verschure[1,3,4]

[1] Laboratory of Synthetic, Perceptive, Emotive and Cognitive Science (SPECS),
DTIC, Universitat Pompeu Fabra (UPF), Barcelona, Spain
jordiysard.puigbo@upf.edu
[2] Laboratory for Simulating, Imaging, and Modeling of Biological Systems
(SIMBIOsys), DTIC, Universitat Pompeu Fabra (UPF), Barcelona, Spain
[3] Catalan Research Institute and Advanced Studies (ICREA), Barcelona, Spain
[4] Catalan Institute of BioEngineering (IBEC), Barcelona, Spain

Abstract. The embodied mammalian brain evolved to adapt to an only
partially known and knowable world. The adaptive labeling of the world
is critically dependent on the neocortex which in turn is modulated by
a range of subcortical systems such as the thalamus, ventral striatum
and the amygdala. A particular case in point is the learning paradigm
of classical conditioning where acquired representations of states of the
world such as sounds and visual features are associated with predefined
discrete behavioral responses such as eye blinks and freezing. Learning
progresses in a very specific order, where the animal first identifies the
features of the task that are predictive of a motivational state and then
forms the association of the current sensory state with a particular action
and shapes this action to the specific contingency. This adaptive feature
selection has both attentional and memory components, i.e. a behaviorally relevant state must be detected while its representation must be
stabilized to allow its interfacing to output systems. Here we present a
computational model of the neocortical systems that underlie this feature detection process and its state dependent modulation mediated by
the amygdala and its downstream target, the nucleus basalis of Meynert. Specifically, we analyze how amygdala driven cholinergic modulation these mechanisms through computational modeling and present a
framework for rapid learning of behaviorally relevant perceptual representations.

1 Introduction

In the last years we have been a significant increase in the literature regarding
topics the topics of artificial intelligence, supervised learning and reinforcement
learning.

This increase is clearly due to the finding of workarounds to the limiting computational cost of certain algorithms, principally represented by shared weights

M. Mangan et al. (Eds.): Living Machines 2017, LNAI 10384, pp. 387–393, 2017.
DOI: 10.1007/978-3-319-63537-8_32

through convolutions that reduce the parameter spaces and increased generalisation on deeper networks.

In this sense, the field of machine learning is rapidly increasing the number of tasks that can be learned with combinations of algorithms from mainly reinforcement learning, recurrent neural networks and deep networks, as soon as there is enough computational power, time and data. Of special interest for robotics could be reinforcement learning algorithms, which allow virtual or real agents to learn action policies according to the current inputs. These algorithms have been recently used for playing atari games with better than human performance, simulating multi-agent interactions in solving foraging tasks, coordinating for goals and solving a wide range of tasks that require not just optimal action selection but also memory aided decision making (see [2] for critical review). All this being achieved within the last two years makes one wonder why we are still unable to apply this to real robots performing in real world tasks, but just in simulations. The answer is not really unknown to anyone: training times are extremely big and require lots of failures to converge into good performance, sometimes even better than humans. While time could eventually not be a mandatory problem, the current ability of robots to recover from failure is scarce, while some failures can be unrecoverable by themselves. Additionally, these algorithms require to be trained on the agent that will be performing the action at any moment.

The critical point, then, is that we need both robots able to recover from failure, like an animal that falls can stand up again, and algorithms that allow them to anticipate and avoid critical failure, as a child would learn to avoid a hot frying pan after the first burning contact. Learning in this conditions becomes hard, as you want to minimize the number of exposures or samples of that same event, going against the current trend of learning by having more and better data: in a critical failure event, you can't afford gathering more data and you might want to favor overfitting. Psychology theories on classical conditioning and its subsequent neuro-anatomical studies have provided insights of the circuitry behind this fast and drastic behavior. First, our bodies have reactive feedback systems that avoid additional harm once an aversive stimuli (unconditioned stimuli or US) is sensed. These systems have evolved through millions of generations and species to what we can find today in animals. On top of this, a precisely designed architecture learns to associate predictive cues to perfectly-timed reactions that will, eventually, avoid to sense the US again. The anatomy behind this behavior has been located numerous times in the cerebellum and has been extensively studied [1]. Finally, the most important point relies on acquiring accurate, stable representations of the events that predict the aversive USs. This suggests that the anticipatory behavior needs to be acquired in two steps: an initial fine tuning of the perceptions to detect cues predictive of aversive events, and then the precise association of these events with anticipatory, pre-defined actions. Several studies have highlighted that fear responses (mainly startles and fear-related neuromodulators release) are associated to the CS before the anticipatory response is correctly anticipated while neocortical neurons change their preferred responding stimuli both after CS-US pairing and after pairing with neuromodulators like acetylcholine (ACh). We will revise what are the changes provoked in the cortical substrate by this neuromodulator and analyize how this neuromodulation can become relevant for quickly redefining perceptual representations and eventually to stabilize anticipatory behavior.

2   Neocortex and Acetylcholine

Acetylcholine (ACh) is a neuromodulator that mediates the detection, selection and further processing of stimuli [5] in the neocortex. ACh is typically associated with attention and synaptic plasticity [6]. The main source of cortical acetylcholine is the Nucleus Basalis of Meynert (NBM) in the basal forebrain. Cholinergic neurons in the NBM receive their main excitatory projections from the amygdala and from ascending activating systems. The amygdala is known to be involved in building associations that are key for fear conditioning. While projections from the ascending activating systems are generally related to bottom-up mechanisms for maintaining arousal and wakfulness, other excitatory projections from prefrontal and insular cortex convey top-down control of acetylcholine release. The principal targets of cholinergic afferents from the NBM are in the neocortex. These projection preserve a topological organization with the cortex and the amygdala. Cholinergic modulation of the neocortex depends principally on two distinct cholinergic receptor types: metabotropic Muscarinic and ionotropic Nicotinic receptors. The activation of muscarinic receptors in the neocortex of the rat has a fast, global excitatory effect on both excitatory and inhibitory neurons. In contrast, binding with nicotinic receptors have a slower disinhibitory response on the neural substrate. Hence, the global effects of acetylcholine in a population of cortical neurons must depend on the balance and distribution of these two opponent types of cholinergic receptors. Here we will investigate this relationship in the context of sensory processing in classical conditioning. In particular, we will investigate the role of the differential drive of cortical interneurons by acetylcholine on the behavioral state dependent gating and representation of sensory states.

2.1   Two Degrees of Multistability in the Neocortex are Mediated by Inhibitory Interneurons

The complex structure of the neocortex is comprised by 80% excitatory and 20% inhibitory neurons organized in six differentiable layers. Among the 20% of inhibitory interneurons, two major interneuron subtypes are inmunoreactive to either Parvalbumin or Somatostatin [5], but not the other, accounting for most of the GABAergic interneurons accross cortical layers. These two interneuron subtypes are also different in terms of dendritic arborization (e.g. basket cells, Martinotti cells, chandelier cells), spiking patterns (e.g. fast spiking, regular spiking). PV expressing (+) interneurons (PVi+) in the neocortex have fast spiking dynamics where basket cells are the classical example. SST expressing interneurons (SSTi+) instead have more regular spiking dynamics and broader

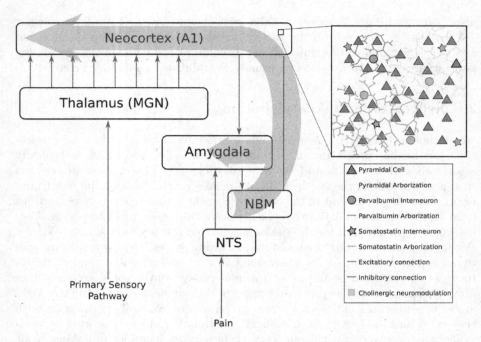

Fig. 1. Anatomical model of conditioning and cholinergic modulation. An auditory stimulus is pre-processed in the auditory pathway through Superior Olive (SO), Inferior Colliculus (IC) and the Medial Geniculate Nucleus of the thalamus (MGN) until it reaches the primary auditory cortex (A1). Our model of a small fraction of the cortex (top-right) is composed of 3 cell populations, one excitatory containing 80% of the cells and 2 inhibitory containing the remaining 20%. These inhibitory populations correspond to the PVi+ and SSTi+ responsive inhibitory interneurons. Both inhibitory populations receive excitatory input principally from the excitatory population, composed of pyramidal cells. Finally, they project back to it with inhibitory connections. Moreover, unconditioned stimuli (US), indicative of surprising or aversive events are relayed through the Nucleus Tractus Solitarius (NTS) to the Amygdala (Am). The amygdala, in turn, uses cortical information to predict future USs. Either the predictions or the US itself stimulate the Nucleus Basalis of Meynert. The NBM releases ACh in the neocortex and the amygdala (Am), therefore promoting the acquisition of new sensory features and its predictive component. Finally, the amygdala, which is receiving contextual information in the typical form of a conditioned stimulus (CS), predictive of the US, learns the association between the cortical predictive components of and the NBM stimulation, facilitating future learning events, now cued by the CS.

arborizations usually in the form of Martinotti and chandelier cells. Together with the less studied interneurons expressing Serotonin, these 3 inmunochemically separable interneuron classes account for the majority of the GABAergic neurons in the neocortex [4].

In order to understand how different inhibitory populations affect the dynamics of the network we realized a series of computational simulations changing the axonal range and the gain of the inhibitory neurons. We used rate based leaky linear units, defined by:

$$x_j(t) = \sum_i W_{ij} x_i(t-1) \tag{1}$$

Where x_j and x_i are the post-synaptic and pre-synaptic neurons and W is the a connectivity matrix representing the conductance between each neuron in the population. The connectivity matrix W is constructed following a pseudo-random rule based on exponential distance, where the probability of forming a connection (C_{ij}) between two neurons x_i and x_j separated by a distance d_{ij} is given by:

$$P(C_{ij}|d_{ij}) = ce^{-bd_{ij}} \tag{2}$$

Figure 2a shows how the parameters of the distance rule affect the stability of the network. Low ranges of inhibition decrease the number of stable modes as seen in the left and bottom edges. Figure 2 shows how the decrease in size of the network (i.e. lowering the number of observed neurons from the simulation) reduce the number of unstable modes (real eigenvalues > 0) until the critical point of zero. At this point the observed subnetwork is completely stable (Fig. 2b). In contrast, when we have 2 populations, one with double the range of the first, while modifying the gain the neural activity is compressed, highlighting some perceptions and inhibiting some others (Fig. 2c), although there are no direct effects in the stability of the network.

All together, the modulation of the proportions between both inhibitory populations could provide a mechanism to increase or uncover the number of stable points in the network, while temporarily compressing signals, in favor of less common sensory representations. If we considered that the equilibrium points in a neocortical population correspond to the stable representations that the network has learned, we could find the modulation of the different inhibitory populations convenient for escaping local minima and finding more relevant equilibrium points. Conveniently, ACh seems to be a neuromodulator intrinsically related with the regulation of inhibitory activity, although probably not the only one.

2.2 Acetylcholine Can Drive Towards Metastable States

Recent work has shown how PVi+ preferentially express the M1 muscarinic acetylcholine receptor (mAChR), while SSTi+ express the nicotinic acetylcholine receptors (nAChR) [3]. Both receptors are greatly involved in depolarizing both excitatory and inhibitory cells, mAChR having a faster response and nAChR being slower, globally disinhibiting the excitatory population of the neocortex after ACh release, through inhibition of PV+ cells. Together with the changes in network stability reported in the previous section, this evidence suggests a general role for acetylcholine in the regulation of both gain and stability in the neocortex. From the one side, we have shown that the modulation of short- and

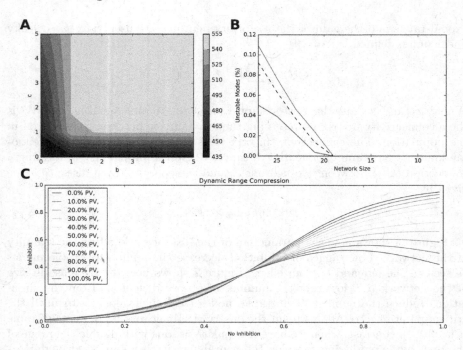

Fig. 2. Characterization of cortical inhibition. Figure A visualizes the number of stable eigenvalues (< 0) found in function of the parameters of the distance rule. B shows the decrease in the number of eigenvalues when studying the stability of reduced portions of the network. C shows how inhibtion affects the excitatory dynamics, this is how biasing the inhibition synaptic gains towards decreased global inhibition (SST), strongly reduces the firing rate of the most active neuron, compressing the data and allowing to see better other stimuli.

long-range inhibitory populations affects the stability of a neural network. From the other side, ACh neuromodulation rapidly makes unstabilizes local attractors, to later stabilise them. This can also be understood as a mechanism that dynamically unstabilizes the attractor state from local to global features, with a potential role in context switching. This could then further increase learning speed by setting the brain in an unstable state that promotes switching from pre-learned attractor states to new potential representations.

3   Conclusions and Future Work

In this study we tested this hypothesis by analysing how the stability of a linear neural network is affected by different kinds of inhibition. We further discuss how this mechanism can affect learning speed in critical situations where acetylcholine is typically released, i.e. during dangerous or surprising events. In this paper we have introduced a preliminary analysis of the role of cholinergic modulation of the inhibitory substrate of the neocortex. We have found that the modulation produced by acetylcholine favors the exploration of the perceptual space by compressing the neural signals and opening more equilibria to encapsulate novel perceptions.

We argue that this mechanisms is key for speeding up learning, as it switches between two behavior-state dependent modes: the more frequent, stable mode that provides a coherent representation of the word, and a rare, unstable mode that rapidly embeds new perceptions or even knowledge during critical events. Moreover, physiological experiments show acetylcholine release not only in fearful events, but also in the presence of *unexpected* rewarding or aversive stimuli and during sustained attention. The conclusions presented in this short dissertation support that the mechanisms for switching between stable, exploitative mind states and more exploratory states would be useful to maintain attentive states and respond to surprising, unexpected events.

The inclusion of this mechanisms in robotic or other artificial agents, will provide the substrate for autonomous learning, by allowing to detect future potential dangers or errors with enough time to react or stop. Following studies will aim to test the presented hypothesis in a computational model of cholinergic-based learning in a robotic platform.

References

1. Herreros, I., Verschure, P.F.: Nucleo-olivary inhibition balances the interaction between the reactive and adaptive layers in motor control. Neural Netw. **47**, 64–71 (2013)
2. Lake, B.M., et al.: Building machines that learn and think like people. arXiv preprint arXiv:1604.00289 (2016)
3. Disney, A.A., Aoki, C., Hawken, M.J.: Gain modulation by nicotine in macaque V1. Neuron **56**, 701–713 (2007). doi:10.1016/j.neuron.2007.09.034
4. Rudy, B., Fishell, G., Lee, S., Hjerling-leffler, J.: Three Groups of Interneurons Account for Nearly 100% of Neocortical GABAergic Neurons (2010). doi:10.1002/dneu.20853
5. Kawaguchi, Y., Kubota, Y.: GABAergic cell subtypes and their synaptic connections in rat frontal cortex. Cereb. Cortex **7**, 476–486 (1997). doi:10.1093/cercor/7.6.476
6. Klinkenberg, I., Sambeth, A., Blokland, A.: Acetylcholine and attention. Behav. Brain Res. **221**, 430–442 (2011)

Describing Robotic Bat Flight with Stable Periodic Orbits

Alireza Ramezani[1]([✉]), Syed Usman Ahmed[1,2], Jonathan Hoff[1,2],
Soon-Jo Chung[3], and Seth Hutchinson[1,2]

[1] Coordinated Science Laboratory, Urbana, USA
aramez@illinois.edu
[2] Department of Electrical and Computer Engineering, University of Illinois
at Urbana-Champaign (UIUC), Urbana, IL 61801, USA
[3] Graduate Aerospace Laboratories (GAL), California Institute of Technology,
Pasadena, CA 91125, USA

Abstract. From a dynamic system point of view, bat locomotion stands out among other forms of flight. During a large part of bat wingbeat cycle the moving body is not in a static equilibrium. This is in sharp contrast to what we observe in other simpler forms of flight such as insects, which stay at their static equilibrium. Encouraged by biological examinations that have revealed bats exhibit periodic and stable limit cycles, this work demonstrates that one effective approach to stabilize articulated flying robots with bat morphology is locating feasible limit cycles for these robots; then, designing controllers that retain the closed-loop system trajectories within a bounded neighborhood of the designed periodic orbits. This control design paradigm has been evaluated in practice on a recently developed bio-inspired robot called Bat Bot (B2).

Keywords: Bio-inspired robot · Bat · Poincare · Periodic orbit · Control

1 Introduction

Bats possess a very sophisticated powered flight mechanism among animals. This is evident by the dynamic conformation properties of their wings. Their flight mechanism has several types of joints, which connect the bones and muscles to one another and synthesize a metamorphic musculoskeletal system that possesses more than 40 degrees of freedom (DoFs), both passive and active [22]. A close look at bat flight kinematics reveals that bat sensory-motor control mobilizes several joints during a single flap cycle, which spans slightly less than 100 ms in some species. One flapping cycle involves two movements of (i) a downstroke phase, which is initiated by both left and right forelimbs expanding backwards and sideways while sweeping downward and forward relative to the body, and (ii) an upstroke phase, which brings the forelimbs upward and backward and is followed by flexion of the elbows and wrists to fold the wings.

© Springer International Publishing AG 2017
M. Mangan et al. (Eds.): Living Machines 2017, LNAI 10384, pp. 394–405, 2017.
DOI: 10.1007/978-3-319-63537-8_33

In addition to quick multi-DoF actuation, bat jointed mechanism has adaptive morphing characteristics [1,2] that behaves differently in various flight maneuvers. For instance, consider a roll maneuver performed by insectivorous bats [18]. Collapsing (bending) a wing and consequently reducing the wing area would increase wing loading on the collapsed wing, as a result, reducing the lift force. Furthermore, pronation of one wing and supination of the other wing yields negative and positive angles of attack, respectively, thereby generating negative and positive lift forces on the wing surface, causing the bat to roll sharply. Bats employ this move to hunt insects as in belly-up configuration they can use the natural camber on their wings to maximize descending acceleration. Agility of flight is vital for insectivore bats because their echolocating system cannot detect insects far away from them; with such restrictions in locating prey, sharp changes in flight direction are important.

From a dynamic system point of view, bat locomotion stands out among other forms of flight (e.g., insect flight) mainly due to the fact that during a large part of bat wingbeat cycle the moving body is not in a static equilibrium. In contrast, insects stay at their static equilibrium. This is because insect flapping frequency is faster than the body dynamic response. Based on this observation, conventional approximation methods such as the celebrated method of averaging is commonly applied to design flight controllers for insect-scale flapping robots [7,8].

Encouraged by the biological examinations that have revealed bats exhibit periodic and stable limit cycles, this work demonstrates that one effective approach to stabilize articulated flying robots with bat morphology involves locating feasible limit cycles for these robots; then, designing controllers that retain the closed-loop system trajectories within bounded neighborhood of the designed periodic orbits. This work is a sequel to our work [19]. This work further investigates the effectiveness of the control design paradigm by conducting more experiments and simulations.

Control is the integral part of bat robots. Unfortunately often, due to mechanical design complications, these robots are not functional and cannot help explore various control design ideas. Therefore, there are not many control design schemes available. For example, Ro-bat [3] is just a wing design based on *Cynopterus brachyotis'* wing structure. The morphology of this robotic wing closely matches that of the real bat, with jointed legs, humerus, radius and digits. In practice it is impossible to accommodate several actuated coordinates in the wings as this can yield bulky wings and moment of inertia blows up. As a result, in Ro-bat, reducing the complexity from a real bat kinematics is realized by only considering a few joints. Three RC grade servo motors are used to articulate several joint in Ro-bat via a set of cables connected to key joints on the wing. These cables are embedded inside the armwing structure. The robotic wing can produce 160° wingbeat amplitude and 10 Hz wingbeats in flapping frequency. Additionally, the mechanism can realize two movements: (i) retraction and protraction at the shoulder joints; (ii) extension and flexion at the elbow. Many strings are routed through the armwing and with the help of this transmission mechanism the digits can articulate with respect to the carpus. The robotic wing is not designed for self-sustained free flights as its several actuators make it very heavy. Therefore, it has no controller.

Robotbat [10] is developed with the same objective of studying physiological specializations in bats flight. This robot has an articulated flight mechanism that embodies three degrees of actuations (DoAs) per each wing and two DoAs for flapping amplitude variations. Overall, the robot has eight DoAs and the articulations are in the form of flapping, lead-lag, and feathering motions. The platform, which is mainly designed to run on CPG-based control schemes [6, 10], helped synthesize a bio-inspired flight controller employing a network of symmetric Hopf oscillators for a biologically inspired bat robot.

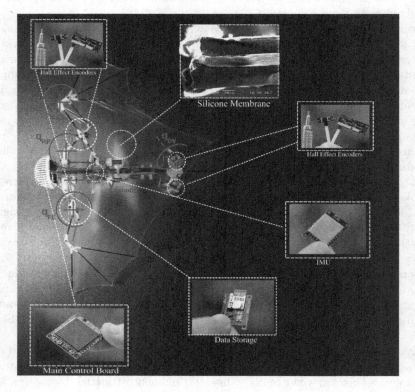

Fig. 1. Bat Bot (B2) [19].

In contrast to the previous examples, bat bot (B2), shown in Fig. 1, can perform free flight. It is a self-contained, autonomous flying robot that weighs 93 g. This robot has been developed recently to mimic morphological properties of bat wings. Instead of using a large number of distributed control actuators, a highly stretchable membrane wings that are controlled at a reduced number of dominant wing joints to best match the morphological characteristics of bat flight is implemented. B2 possesses some biologically meaningful DoFs. These biologically meaningful DoFs include asynchronous and mediolateral movements of the armwings and dorsoventral movements of the legs. The continuous surface and elastic properties of bat skin under wing morphing are realized by an ultrathin (56 μm) membranous skin that covers the skeleton of the morphing wings.

This work employs B2 to explore the idea of describing articulated flapping flight with stable and attractive periodic orbits. B2 and conventional flying robots such as fixed-wing and rotary-wing robots are similar in that all of them rely on the air through modulation of the magnitude and direction of aerodynamic forces. However, the control surfaces in this robot are extremely different. Conventional fixed-wing robots are often controlled by thrust and conventional control surfaces such as elevators, ailerons, and rudders. In contrast, B2 possesses 9 active oscillatory joints (5 of which are independent) in comparison to 6 DoFs (attitude and position) that are actively controlled.

The approach that we explore in this paper relies on asymptotically imposing *virtual constraints* (holonomoic constraints) on B2's dynamic system through closed-loop feedback. This concept has a long history, but its application in nonlinear control theory is primarily due to [5,14]. Enforcing these constraints will yield a closed-loop system with periodic trajectories. Then, using the notion of Poincare return map the stability of these periodic trajectories is turned into the stability of the associated fixed-point on the Poincare section. Afterward, a discrete controller is utilized to retain the closed-loop trajectories at a bounded neighborhood of the fixed-point. The advantage of imposing these constraints through closed-loop feedback (software) rather than physically (hardware) is that B2's wing configurations can be adjusted and modified during the flight. We have tested this concept on B2 to generate cruise flights and we anticipate this potentially can help reconstruct the adaptive properties of bat flight for other maneuvers.

This work is organized as follows. First, the concept of virtual constraint, which drives the actuated portion of the system, is described in brevity. In Sect. 3, the concept of discrete control law is explained after introducing some basic mathematical concepts such as periodic orbit, stability of periodic orbit, Poincare return map, etc. Section 4 will presented the performance of the controller in simulation and experiment and relevant discussion will be presented. Last but not least, the conclusion section makes the final remarks and discussions.

2   Virtual Constraints and Wing Conformations

For wing articulations, we employ a framework based on defining a set of parametrized and time-varying holonomic constraints [5,14]. This method allow us shape the overall system dynamics through such constraints. These holonomic constraints control the posture of the jointed structure of B2 by mobilizing the actuated portion of the system and take place through the action of the servo actuators that are embedded in the robot.

First, we parametrize the jointed structure of B2 by several configuration variables. The configuration variable vector q_{morph} defines the morphology of the forelimb and hindlimb as they evolve through the action of actuated coordinates and embodies nine biologically meaningful DoFs,

$$\boldsymbol{q}_{\text{morph}} = (q_{RP}^R; q_{FE}^R; q_{AA}^R; q_{DV}^R; q_{RP}^L; q_{FE}^L; q_{AA}^L; q_{DV}^L; q_{FL}) \tag{1}$$

where q_{RP}^i describes the retraction-protraction angle; q_{FE}^i is the radial flexion-extension angle; q_{AA}^i is the abduction-adduction angle of the carpus; q_{FL} is the flapping angle; q_{DV}^i is the dorsoventral movement of the hindlimb, see Fig. 1. Here, the superscript i denotes the right (R) or left (L) joint angles. Mathematically speaking, the mechanical constraints shown in Fig. 1 yield a nonlinear map from actuated joint angles

$$\boldsymbol{q}_{\text{act}} = (y_{\text{spindle}}^R; q_{DV}^R; y_{\text{spindle}}^L; q_{DV}^L; q_{FL}) \tag{2}$$

to the morphology configuration variable vector $\boldsymbol{q}_{\text{morph}}$. The spindle action shown in Fig. 1 is denoted by y_{spindle}^i. The nonlinear map is explained mathematically in [13, 19, 20]. Now, we proceed by incorporating the dynamics of the robot in the virtual constraint design procedure.

During free-fall ballistic motions, B2 with its links and joints represents an open kinematic chain that evolves under the influence of gravitational and external aerodynamic forces. Obtaining the equations of motion is not trivial. We employ the method of Lagrange to mathematically define this dynamics. This open kinematic chain is uniquely determined with the fuselage Euler angles roll, pitch, and yaw $(q_x; q_y; q_z)$; fuselage Center of Mass (CoM) positions $(p_x; p_y; p_z)$; and morphing joint angles $\boldsymbol{q}_{\text{morph}}$. Therefore, the robot's configuration variable vector is

$$\boldsymbol{q} = (q_x; q_y; q_z; p_x; p_y; p_z; \boldsymbol{q}_{\text{morph}}) \in \mathcal{Q} \tag{3}$$

where $\mathcal{Q}$ is the robot's configuration variable space. We derived Lagrange equations after computing the total energy of the free open kinematic chain as the difference between the total kinetic energy and the total potential energy. Following Hamilton's principle of least action, the equations of motion for the open kinematic chain with ballistic motions are given by:

$$\boldsymbol{M}(\boldsymbol{q})\ddot{\boldsymbol{q}} + \boldsymbol{C}(\boldsymbol{q}, \dot{\boldsymbol{q}})\dot{\boldsymbol{q}} + \boldsymbol{G}(\boldsymbol{q}) = \boldsymbol{Q}_{\text{gen}} \tag{4}$$

where $\boldsymbol{M}$, $\boldsymbol{C}$, and $\boldsymbol{G}$ are the inertial matrix, the Coriolis matrix, and the gravity vector, respectively. The generalized forces $\boldsymbol{Q}_{\text{gen}}$, which reflect the role of aerodynamic forces as well the action of several morphing motors in B2, are described in [19, 21]. Now, the virtual constraints are given by

$$\boldsymbol{N}(t, \boldsymbol{\beta}, \boldsymbol{q}_{\text{act}}) = \boldsymbol{q}_{\text{act}} - \boldsymbol{r}_{\text{des}}(t, \boldsymbol{\beta}) \tag{5}$$

where $\boldsymbol{r}_{\text{des}}$ is the time-varying desired trajectory associated to the actuated coordinates, t is time, and $\boldsymbol{\beta}$ is the vector of the wing kinematic parameters. These parameters are explained later and we will show how the wing kinematic parameters can be assigned as control commands in the closed-loop system. Once

Eq. 5 is enforced (with a feedback law) the posture of B2 varies as the actuated portion of the system now implicitly follows the time-varying trajectory r_{des}. This is simply a classical tracking problem. In order to design r_{des}, we pre-compute the evolution of B2's joint trajectories for $N = 0$.

We partition the configuration variable vector q into the actuated coordinates q_{act} and the remaining coordinates $\mathcal{X}$, which includes Euler angles and body CoM positions. The dynamics Eq. 4 are re-written as following

$$\begin{bmatrix} M_{11} & M_{12} \\ M_{21} & M_{22} \end{bmatrix} \begin{bmatrix} \ddot{\mathcal{X}} \\ \ddot{q}_{\text{act}} \end{bmatrix} + \begin{bmatrix} C_{11} & C_{12} \\ C_{21} & C_{22} \end{bmatrix} \begin{bmatrix} \dot{\mathcal{X}} \\ \dot{q}_{\text{act}} \end{bmatrix} + \begin{bmatrix} G_1 \\ G_2 \end{bmatrix} = Q_{\text{gen}}. \tag{6}$$

In the above equation, M_{11}, M_{12}, M_{21}, M_{22}, C_{11}, C_{12}, C_{21}, and C_{22} are block matrices. The nonlinear system in Eq. 6 shows that the actuated and unactuated dynamics are coupled by the inertial, Coriolis, gravity, and aerodynamic terms.

The actuated dynamics (see Eq. 6) can be reconstructed from the virtual constraints by simply taking time derivative of Eq. 5 and computing the acceleration terms. By integrating Eq. 6 subject to Eq. 5 constraints, it is possible to pre-compute the evolution of $\mathcal{X}$ coordinates. In addition, it is possible to optimize wing kinematic parameters β to design specific flight kinematic properties [19,21]. In other words, the algebraic-differential equation (ADE) given by Eqs. 6 and 5 can be solved iteratively to located predefined feasible trajectories for the actuated coordinates in the robot.

Turning to the constraint equation, r_{des} is given by

$$r_{\text{des}}^i(t, \beta) = a_i \cos(\omega t + \phi_i) + b_i, \quad i \in \left\{ y_{\text{spindle}}^R, q_{DV}^R, y_{\text{spindle}}^L, q_{DV}^L, q_{FL} \right\} \tag{7}$$

where $\beta = \{\omega, \phi_i, a_i, b_i\}$ parametrizes the periodic actuator trajectories that define the wing motion. These parameters are the control input to the system. These parameters are changed by a discrete feedback law, which will be explained later, to keep the closed-loop system within a bounded neighborhood of the predefine periodic orbits.

3   Closed-Loop Control

3.1   Continuous Feedback Law

The responsibility of the continuous feedback law is to enforce the previously explained virtual constraints Eq. 5 on the dynamic model given by Eq. 4 asymptotically and in finite time. There are several tools from nonlinear control theory (e.g., feedback linearization, control Lyapunov function (CLF), etc.) that can be employed to enforce these constraints in finite time. In this work we focus on the discrete feedback policy and for a comprehensive continuous control design instruction refer to [15].

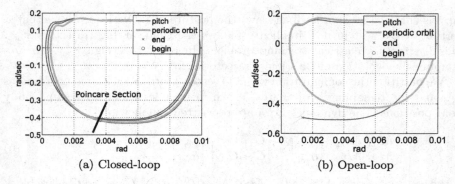

Fig. 2. Comparison between the closed-loop and open-loop systems in keeping the pitch angle trajectory (blue) within a bounded neighborhood of the designed periodic orbit (red). (Color figure online)

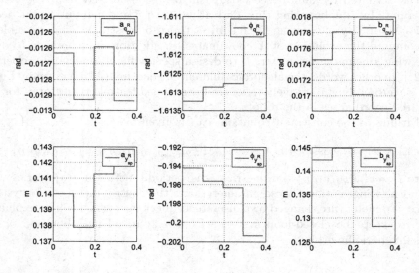

Fig. 3. Flight kinematic parameter β is updated by the discrete control law.

3.2 Discrete Feedback Law

Enforcing the constraints Eq. 5 to the model by a continuous feedback law will not guarantee that the system trajectories stay within a bounded neighborhood of the designed predefine periodic orbits. This is shown in Fig. 2, where the simulated open-loop trajectories diverge from the actual periodic orbit. Therefore, another control loop is required to keep the trajectories within a bounded neighborhood of the designed orbits. We achieve this by designing a discrete controller that steers the system within a bounded neighborhood of the optimal trajectories by updating the wing kinematic parameters β during each wingbeat.

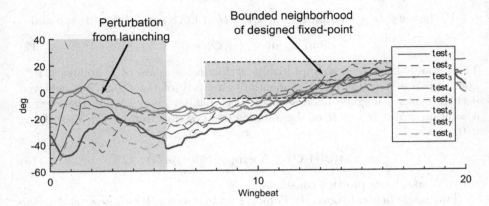

Fig. 4. Pitch angle evolutions for eight untethered closed-loop flights. The first shaded region depicts the perturbations that are introduced to the system after the launch. The second shaded region shows the bounded neighborhood of the fixed-point. The closed-loop trajectories are steered to this region by the controller.

Our discrete closed-loop feedback synthesis for B2 is based on the notion of stability for periodic orbits. We start with some basic definitions for autonomous systems and obviously with slight modifications can be applied to nonautonomous systems. Some of the the material that is presented here is a reiteration of the material in [4,9,11,12,15–17,23].

A solution $\varphi : [t_0, \infty) \rightarrow \mathcal{U}^1$ is a periodic solution of the autonomous system

$$\dot{x} = f(x) \tag{8}$$

if for all $t \in [t_0, \infty)$

$$\varphi(t + T) = \varphi(t) \tag{9}$$

for some minimum period $T > 0$. A set $\mathcal{O}$ is a periodic orbit of Eq. 8 if

$$\mathcal{O} = \{\varphi(t) | t \geq t_0\} \tag{10}$$

for some periodic solution $\varphi(t)$. An orbit is nontrivial if it contains more than one point (59).

Turning to stability, Lyapunov stability (49) of the periodic orbit $\mathcal{O}$ is formally stated as if there exists a neighborhood $\mathcal{V}$ of the periodic orbit $\mathcal{O}$ such that for every point p in the neighborhood $\mathcal{V}$, there exists a solution $\varphi : [t_0, \infty) \rightarrow \mathcal{U}$ of the autonomous system given by Eq. 8 satisfying $\varphi(0) = p$ and

$$\text{dist}(\varphi(t), \mathcal{O}) < \epsilon \tag{11}$$

for all $t \geq 0$, where $\epsilon > 0$, and $\text{dist}(p_1, p_2)$ is the Euclidean distance. The orbit $\mathcal{O}$ is attractive if there exists an open neighborhood $\mathcal{V}$ of $\mathcal{O}$ such that for every

---

[1] $\mathcal{U}$ is a smooth embedded submanifold of $\mathbb{R}^n$.

$p \in \mathcal{V}$, there exists a solution $\varphi : [t_0, \infty) \to \mathcal{U}$ of Eq. 8 satisfying $\varphi(0) = p$ and

$$\lim_{t \to \infty} \text{dist}(\varphi(t), \mathcal{O}) = 0. \tag{12}$$

The periodic orbit $\mathcal{O}$ is asymptotically stable in the sense of Lyapunov if it is both stable and attractive. Exponentially stability of the orbit $\mathcal{O}$ is achieved if there exists a neighborhood $\mathcal{V}$ of $\mathcal{O}$ such that for every $p \in \mathcal{V}$, there exists a solution $\varphi : [0, \infty) \to \mathcal{U}$ of the autonomous system given by Eq. 8 satisfying $\varphi(0) = p$ and

$$\text{dist}(\varphi(t), \mathcal{O}) \leq N \exp(-\gamma t) \text{dist}(p, \mathcal{O}) \tag{13}$$

where N and γ are positive constants.

Two mathematical tools: (i) *Poincare section* and (ii) *Poincare return map* can be used to evaluate the stability of periodic orbits. A Poincare section is a smooth hypersurface $\mathcal{S}$ in $\mathcal{U}$ and it satisfies the following conditions:

-   $\mathcal{S}$ is nonempty and there exists a differentiable function $H : \mathcal{U} \to \mathbb{R}$ such that

$$\mathcal{S} := \{x \in \mathcal{U} | H(x) = 0\}; \tag{14}$$

-   for every $s \in \mathcal{S}$

$$\frac{\partial H}{\partial x}(s) \neq 0. \tag{15}$$

This condition implies that the periodic orbit $\mathcal{O}$ is transversal to $\mathcal{S}$; $\mathcal{S}$ has a lower dimension than $\mathcal{U}$.

Any Poincare return map $\boldsymbol{P} : \mathcal{S} \to \mathcal{S}$ defines a discrete dynamic system by

$$x_{k+1} = \boldsymbol{P}(x_k) \tag{16}$$

where $x_k \in \mathcal{S}$ is the state and $\boldsymbol{P}$ maps the state to the next state x_{k+1}. For this discrete system

$$\boldsymbol{P}(x^*) = x^* \tag{17}$$

x^* is the fixed-point. When starting from an initial state x_0, successive application of the map $\boldsymbol{P}$ generates a sequence of states that can reveal useful information about the stability of the discrete time system [17]. In other words, in the method of Poincare, there is a relationship between the periodic orbits of the system (Eq. 8) and the equilibrium points of the sampled system above. This method is interesting because it establishes an equivalence between the stability properties of the periodic orbits of Eq. 8 and the equilibrium points of the discrete system Eq. 16. This equivalence can be formally expressed in form of a theorem [23].

To stabilize the designed periodic solution, we augmented the desired trajectory $\boldsymbol{r}_{\text{des}}$ with a correction term [23]

$$\boldsymbol{r}_{\text{corr}} = \frac{\partial \boldsymbol{r}_{\text{des}}}{\partial \boldsymbol{\beta}} \delta \boldsymbol{\beta}. \tag{18}$$

The Poincare return map takes the robot states q_k and $\dot{q}_k$ (the Euler angles roll, pitch, yaw and their rates) at the beginning of k-th flapping cycle and leads to the states at the beginning of the next flapping cycle,

$$\begin{bmatrix} q_{k+1} \\ \dot{q}_{k+1} \end{bmatrix} = P(q_k, \dot{q}_k, r_{\text{corr}}). \tag{19}$$

We linearized the map P at $\mathcal{S}$, resulting in a dynamic system that describes the periodic behavior of the system at the beginning of each flapping cycle

$$\begin{bmatrix} \delta q_{k+1} \\ \delta \dot{q}_{k+1} \end{bmatrix} = \left[\frac{\partial P}{\partial q}(q^*) \frac{\partial P}{\partial \dot{q}}(\dot{q}^*) \right] \begin{bmatrix} \delta q_k \\ \delta \dot{q}_k \end{bmatrix} + \frac{\partial P}{\partial \beta}(\beta^*)\delta\beta_k \tag{20}$$

where (*) denotes the equilibrium points and ($\delta q; \delta \dot{q}$) denotes deviations from the equilibrium points. The changes in the kinematic parameters are denoted by $\delta\beta$. Here, the stability analysis of the periodic trajectories of the bat robot are relaxed to the stability analysis of the equilibrium of the linearized Poincare return map on $\mathcal{S}$. As a result, classical feedback design tools can be applied to stabilize the system. We computed a constant state-feedback gain matrix $K_{\mathcal{S}}$ such that the closed-loop linearized map is exponentially stable:

$$\text{eig}\left(\left[\frac{\partial P}{\partial q}(q^*) \frac{\partial P}{\partial \dot{q}}(\dot{q}^*) \right] + \frac{\partial P}{\partial \beta}(\beta^*)K_{\mathcal{S}} \right) < 1 \tag{21}$$

We used this state feedback policy at the beginning of each flapping cycle in order to update the kinematic parameters as follows

$$\delta\beta_k = K_{\mathcal{S}} \begin{bmatrix} \delta q_k \\ \delta \dot{q}_k \end{bmatrix}. \tag{22}$$

4   Results

Figure 2 shows the simulated results and the performance of the closed-loop system that keep the pitch angle trajectories in a bounded neighborhood of the design orbit. Figure 3 shows the wing kinematic parameters. It is worth noting that pitch dynamics instabilities are unavoidable in open-loop tailless robots as the tail that passively damps out external perturbations is missing. As is shown in Fig. 2, the robot diverges from the designed orbit immediately after the discrete controller is turned off.

We performed extensive untethered flight experiments in a large indoor space (Stock Pavilion at the University of Illinois in Champaign-Urbana) where we could use a net (30 mby 30 m) to protect the sensitive electronics of B2 at the moment of landing. The flight arena was not equipped with any motion capture system. Although the vehicle landing position was adjusted by an operator to secure landings within the area, which is covered by the net, we landed outside the net many times. The launching task was performed by a human operator, thereby adding to the degree of inconsistency of the launches. In all of these experiments, at the launch moment, the system reached its maximum flapping speed (10 Hz). The hand launch introduced initial perturbations, which considerably affected the first 10 wingbeats. Despite the external perturbations of the launch moment, the vehicle stabilized the pitch angle within 20 wingbeats. In Fig. 4, the pitch angles from experiments are shown. Figure 4 shows how the controller steers the pitch angle trajectories towards the designed trajectories despite external perturbations.

5 Concluding Remarks

From a dynamic system point of view, bat locomotion stands out among other forms of flight. During a large part of bat wingbeat cycle the moving body is not in a static equilibrium. Encouraged by this observation, we explored the idea of introducing virtual constraints on B2, which is bio-inspired robot with bat morphology, through closed-loop feedback. Enforcing these constraints will yield a closed-loop system with periodic trajectories. By employing the notion of Poincare return map the stability of these periodic trajectories was turned into the stability of the associated fixed-point on the Poincare section where we used a discrete controller to retain the closed-loop trajectories at a bounded neighborhood of the fixed-point. Our work suggests that one effective way to design controller for articulated flapping systems involves locating feasible periodic orbits and designing controllers that retain the closed-loop system in a bounded neighborhood of these orbits.

References

1. Aldridge, H.: Kinematics and aerodynamics of the greater horseshoe bat, rhinolophus ferrumequinum, in horizontal flight at various flight speeds. J. Exp. Biol. **126**(1), 479–497 (1986)
2. Aldridge, H.: Body accelerations during the wingbeat in six bat species: the function of the upstroke in thrust generation. J. Exp. Biol. **130**(1), 275–293 (1987)
3. Bahlman, J.W., Swartz, S.M., Breuer, K.S.: Design and characterization of a multi-articulated robotic bat wing. Bioinspiration Biomim. **8**(1), 016009 (2013)
4. Burridge, R.R., Rizzi, A.A., Koditschek, D.E.: Sequential composition of dynamically dexterous robot behaviors. Int. J. Rob. Res. **18**(6), 534–555 (1999)
5. Byrnes, C.I., Isidori, A.: A frequency domain philosophy for nonlinear systems, with applications to stabilization and to adaptive control. In: IEEE Conference on Decision and Control, pp. 1569–1573. IEEE (1984)
6. Chung, S.-J., Dorothy, M.: Neurobiologically inspired control of engineered flapping flight. J. Guidance Control Dyn. **33**(2), 440–453 (2010)
7. Deng, X., Schenato, L., Sastry, S.S.: Flapping flight for biomimetic robotic insects: Part II-flight control design. IEEE Trans. Rob. **22**(4), 789–803 (2006)
8. Deng, X., Schenato, L., Wu, W.C., Sastry, S.S.: Flapping flight for biomimetic robotic insects: Part I-system modeling. IEEE Trans. Rob. **22**(4), 776–788 (2006)
9. Dingwell, J.B., Cusumano, J.P.: Nonlinear time series analysis of normal and pathological human walking. Chaos Interdisc. J. Nonlinear Sci. **10**(4), 848–863 (2000)

10. Dorothy, M., Chung, S.-J.: Methodological remarks on CPG-based control of flapping flight. In: AIAA Atmospheric Flight Mechanics Conference (2010)
11. Garcia, M.S.: Stability, scaling, and chaos in passive-dynamic gait models. Ph.D. thesis, Cornell University (1999)
12. Guckenheimer, J., Johnson, S.: Planar hybrid systems. In: Antsaklis, P., Kohn, W., Nerode, A., Sastry, S. (eds.) HS 1994. LNCS, vol. 999, pp. 202–225. Springer, Heidelberg (1995). doi:10.1007/3-540-60472-3_11
13. Hoff, J., Ramezani, A., Chung, S.-J., Hutchinson, S.: Synergistic design of a bio-inspired micro aerial vehicle with articulated wings. In: The Robotics: Science and Systems (RSS) (2016)
14. Isidori, A., Moog, C.: On the nonlinear equivalent of the notion of transmission zeros. In: Byrnes, C.I., Kurzhanski, A.B. (eds.) Modelling and Adaptive Control. LNCIS, vol. 105, pp. 146–158. Springer, Heidelberg (1988). doi:10.1007/BFb0043181
15. Khalil, H.K., Grizzle, J.: Nonlinear Systems, vol. 3. Prentice Hall, New Jersey (1996)
16. Meurant, G.: An Introduction to Differentiable Manifolds and Riemannian Geometry, vol. 120. Academic Press, Cambridge (1986)
17. Nayfeh, A.H.: Perturbation methods in nonlinear dynamics. In: Lecture Notes in Physics, vol. 247, pp. 238–314. Springer, Heidelberg (1986)
18. Norberg, U.M.: Some advanced flight manoeuvres of bats. J. Exp. Biol. **64**(2), 489–495 (1976)
19. Ramezani, A., Chung, S.-J., Hutchinson, S.: A biomimetic robotic platform to study flight specializations of bats. Sci. Rob. **2**(3), eaal2505 (2017)
20. Ramezani, A., Shi, X., Chung, S.-J., Hutchinson, S.: Bat Bot (B2), a biologically inspired flying machine. In: IEEE International Conference on Robotics and Automation (ICRA) (2016)
21. Ramezani, A., Shi, X., Chung, S.-J., Hutchinson, S.: Lagrangian modeling and flight control of articulated-winged bat robot. In: International Conference on Intelligent Robots and Systems (IROS), Hamburg, Germany, 28 September–2 October (2015)
22. Riskin, D.K., Willis, D.J., Iriarte-Díaz, J., Hedrick, T.L., Kostandov, M., Chen, J., Laidlaw, D.H., Breuer, K.S., Swartz, S.M.: Quantifying the complexity of bat wing kinematics. J. Theor. Biol. **254**(3), 604–615 (2008)
23. Westervelt, E.R., Grizzle, J.W., Chevallereau, C., Choi, J.H., Morris, B.: Feedback Control of Dynamic Bipedal Robot Locomotion, vol. 28. CRC Press, Boca Raton (2007)

Research of a Lensless Artificial Compound Eye

Gašper Škulj$^{(\boxtimes)}$ and Drago Bračun

Faculty of Mechanical Engineering, University of Ljubljana,
Aškerčeva 6, 1000 Ljubljana, Slovenia
{gasper.skulj,drago.bracun}@fs.uni-lj.si
http://www.fs.uni-lj.si/

Abstract. One of the challenges in designing an artificial compound eye
(ACE) is the manufacturing and assembly of the ommatidia lenses. The
paper presents three lensless ACE designs based on light guiding structure and an array image detector. In the solid design, deep holes pointed
to different directions in space act as artificial ommatidia. Low image
resolution of the solid design is improved by the shell design, where the
artificial ommatidia are created within the perforated shell. In the third
design, artificial ommatidia are created between the pinhole and pixels
on the image detector. The best image performance is achieved with
the third pinhole design. An array of pinholes on a single image detector can improve the environment perception by combining the images
and assessing the object distance based on image disparity. The proposed lensless design can be easily manufactured and miniaturised to
the microchip scale, and finds potential application in small and lightweight autonomous robots.

Keywords: Biomimetics · Imaging system · Artificial compound eye ·
Pinhole

1 Introduction

Visual navigation is the preferred way of navigation in an environment where
enough light is available. Visual systems used by different entities differ in their
design according to the specific function which they perform. For example, certain simple visual systems can only distinguish different levels of light intensity,
whereas other more advanced visual systems can recognise complex features
of objects. It could easily be argued that a more advanced visual system is
better than a simple one, because it can capture more details of the environment. However, from the evolutionary standpoint, higher image resolution does
not necessarily mean better performance. When designing a visual system for a
robot, manufacturing and operational costs should balance the benefits which
are brought by higher image resolution, colour perception and advanced feature
recognition.

The paper investigates the design of a visual system for a small terrestrial
autonomous robot that will be able to move on rough terrain and explore its

© Springer International Publishing AG 2017
M. Mangan et al. (Eds.): Living Machines 2017, LNAI 10384, pp. 406–417, 2017.
DOI: 10.1007/978-3-319-63537-8_34

environment. The robot will resemble an insect such as an ant or a beetle. A predominant vision system used by insects is a compound eye. Mimicking the compound eye should result in a small and light weight visual system that is also easily manufactured, low-cost and energy efficient. The manufacturing cost of the visual system is important, because the robot is intended to be used in large numbers or swarms. The energy consumption is also a concern and for a visual system greatly depends on its image resolution. To reduce the energy consumption the image resolution should be as low as possible while still allowing the robot to recognise larger objects.

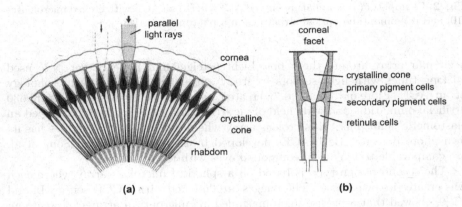

Fig. 1. (a) Biological focal apposition compound eye with fused rhabdom and (b) individual ommatidium [1].

The compound eye design is found predominately in arthropods [1,2]. The most common type of the compound eye is a simple apposition compound eye used, for example, by bees, locusts, dragonflies, etc. It is constructed out of large number of differently oriented ommatidia (Fig. 1). The ommatidium is an elongated visual organ that collects only directed light using a cornea and a crystalline cone that is surrounded with iris pigment cells. The collected light is passed to the rhabdom and detected by photoreceptor cells. Each photoreceptor cell is connected to neurons in the brain with an axon. In more abstract terms a compound eye is a modular structure where each individual part has a light gathering and light detection part. The light detection part provides the brain or the processing unit with information about the light coming from a particular direction. The image is interpreted from the information provided by all the modules of the compound eye.

In literature various design proposals for an ACE can be found. The first design type is based on elementary photo detectors that are positioned in a circular or spherical pattern. A narrow acceptance angle for each element is provided by the use of lenses or long tubes. Franceschini et al. [3] developed an autonomous robot with an ACE prototype of 118 elements positioned in

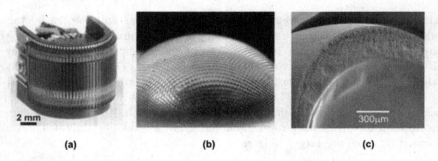

(a)   (b)   (c)

Fig. 2. Example of (a) a miniature curved ACE [6], (b) an ACE with etched microlenses [10] and (c) ommatidia optical structures in a photopolymer [23].

a circular array around the robot. Lichtensteiger and Eggenberger [4,5] used 16 long tubes with light sensors to study autonomously evolving morphology of an ACE. Floreano et al. [6,7] created a miniature curved ACE with 630 artificial ommatidia and wide field of view (Fig. 2a). Song et al. [8] developed an elastomeric hemispherical microlens array where each of 180 microlenses has its own photo detector. Both ACEs developed by Floreano et al. and Song et al. are comparable to biological compound eyes with lower resolution.

The second design type is based on a spherical microlens array where each lens creates its own image. The images are detected with a CCD sensor. Li and Yi [9] showed that a precise micromachined 3D microprism array can create an array of 601 images from different view angles. Qu et al. [10] used a process of laser-enhanced wet etching, casting and thermomechanical reforming to fabricate 5,867 honeycomb-patterned microlenses with the diameter of $\sim$85 μm (Fig. 2b). Deng et al. [11] used laser lithography to produce an array with 91 hexagonal microlenses with the diameter of 200 μm. Wu et al. [12] demonstrated that with the use of high-speed voxel-modulation laser scanning technique in the process of multiphoton polymerization, microlens arrays for ACE can be made where each microlens can have the diameter of 16 μm. Wang et al. [13] developed a tunable system of $\sim$800 liquid microlenses arranged along a hemispherical dome where the liquid lens profile can be modulated. Using laser wet etching with thermal embossing, Deng et al. [14] created a spherical ACE with the diameter of 5 mm and 30000 microlenses with the diameter of 24.5 μm. With this design type emphasis is on a large number of very small lenses and mimicking the shape of the compound eye cornea.

The third design type is based on a flat microlens and accompanying pinhole array in front of a flat array of photo detectors or a CCD sensor. Ogata et al. [15] designed an array of 16 × 16 microlenses with corresponding exactly positioned pinholes and photo detectors. In order to manufacture the pinhole array they used chromium deposition in a vacuum, photolithography, and wet etching. The microlenses were precisely aligned and attached with UV-curable resin. To cover larger field of view several arrays can be combined at different angles. Hamanaka et al. [16] created an ACE using a Selfoc lens plate, a pinhole array of 50 × 50 holes and a plano-concave lens. A single image is captured with a CCD sensor employing the Moire imaging process. Duparre et al. [17–20] used lithographical processes on a wafer scale to create a planar artificial apposition compound-eye placed directly on a CDD sensor. It consists of a 0.2 mm thick array of 60×60 microlenses and pinholes. Radtke et al. [21] used laser lithography to create a spherical ACE comprised of a microlens and pinhole array. The created image is captured with a planar CCD sensor.

The fourth design type is based on the exact recreation of the ommatidium optical structure. Jeong et al. [22–24] used a polymer microlens array to create a conic structure of an artificial ommatidium and a wave guide in to a photopolymer (Fig. 2c). With this technique they were able to build an ACE with 8370 ommatidia with the lens diameter of 25 μm.

In all of the presented design types lenses are used to create a directed narrow acceptance angle for each artificial ommatidium. A lot of effort has been invested to increase the number of artificial ommatidia and minimise the size of the ACE. In order to manufacture the ACEs processes involving lasers, etching and precise positioning are used. It can be concluded that a lot of resources are used to create an optical device with a relatively low resolution. For example, optics for an average consumer camera requires much less resources to manufacture and creates pictures with a much higher resolution. In this paper another approach is investigated, because we would like to create an ACE that is cost-effective in relation to the image resolution.

2   Lensless Artificial Compound Eye Design

Miniature lenses are difficult and expensive to manufacture. In the following chapters variations of lensless artificial compound eye design are investigated.

2.1   Solid Design

An obvious approach to create a lensless ACE is to use long tubes or deep holes to limit the acceptance angle of the individual photo detector. The long tube in combination with an image detector forms an artificial ommatidium. To test the feasibility of a lensless ACE we designed and manufactured a prototype.

Firstly, the view angle of the whole ACE, e.g. 75°, and the acceptance angle of a single artificial ommatidium, e.g. 7.5°, are set. In the next step understanding of what limits the lensless ACE feasibility is very important. The light detector that is positioned at the end of an artificial ommatidium should be considered in advance. Ideally, the light detector would be an array of photocells, where each photocell would be placed at the end of an artificial ommatidium. Such photocells should ideally be as small a possible. However, even individual photocells with an area of around 1 mm² are not readily available. The simplest and most practical way is to use a machine vision camera with a flat array image detector. This unavoidably limits the ACE view angle to less than 180°, unless several image detectors are combined at different angles or a curved image detector is used.

Nevertheless, the problem with most of the cameras currently available on the market is that they have small image detectors e.g. 4×3 mm, with millions of pixels. The small area of the image detector limits the number of artificial ommatidia that can fit on it.

Secondly, another limiting factor related to the image detector size is also the technology used for manufacturing a honeycomb like light directing structure. Artificial ommatidia in this design are conical, rarely circular, and must all end on the image detector. The ease of manufacturing is an important factor in our approach. 3D printing machines for rapid prototyping with plastic materials can nowadays be found in almost every research laboratory that deals with manufacturing. This is why process of 3D printing was chosen to manufacture the prototype. The highest resolution of 3D printing fused filament fabrication process used by a lot of 3D printers is around 0.15 mm. The features of the printed parts are therefore limited to a few tenths of a millimetre. This means that a small 4×3 mm image detector can be covered with maximally 5–7 holes along each axis. Other 3D printing technologies have a greater resolution, but use a supporting material that is very hard to remove from deep holes with a small diameter.

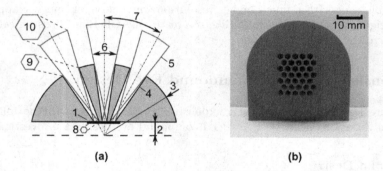

(a)   (b)

Fig. 3. (a) Vertical cross section of the solid ACE design. (1) Image detector array; (2) Displacement of the image detector from the centre of the ACE front surface radius; (3) Front surface radius; (4) Wall of the artificial ommatidium; (5) Boundary light rays; (6) Acceptance angle of the artificial ommatidium; (7) Inter-ommatidia angle; (8) Artificial ommatidium cross-section on the ACE face next to the image detector; (9) Artificial ommatidium cross-section on the front surface; (10) Artificial ommatidium view field cross-section at the observation distance. (b) Picture of the prototype of a 3D printed solid ACE design.

Figure 3a depicts the ACE geometry with a light directing structure, where the image detector (1) with the size of 8.8×6.6 mm is covered with an 7×5 array of light directing holes (8) with the diameter 0.5 mm and 0.3 mm distance between them. All light rays whose incidence angle is between the boundary light rays (5) pass directly through the hole to the image detector. Light rays with greater angles hit the walls (4). Printed plastic has low reflectivity and after a few reflections absorbs those rays to the point they can be neglected. In order to achieve the required acceptance angle (6) and view field cross-section at the observation distance (10), a ray tracing is carried out, to calculate the external diameter of the hole (9). The procedure is as follows: the artificial ommatidium cross-section (8) next to the image detector is selected based on the image detector size and manufacturing technology; the cross-section of the view field (10) is calculated based on the desired acceptance angle (6) and observation distance from the ACE, e.g. 1 m. If the inter-ommatidia angle (7) is equal to the acceptance angle, than the observation distance can be arbitrarily selected; boundary light rays (5) connect the cross-sections (8) and (10) and intersect within the hole as is demonstrated in Fig. 3a. If the external radius (3) of the ACE is selected somewhere between the light rays cross-section and the observation distance, the cross-section (9) can be calculated as a cross-section between the sphere with a radius (3) and boundary light rays (5).

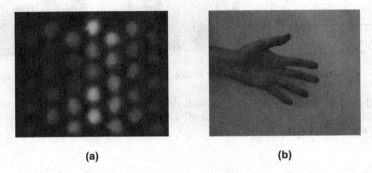

(a)   **(b)**

Fig. 4. (a) Image of a hand left from the centre acquired with the ACE prototype; (b) Image of a hand acquired with a standard VGA camera with a lens.

Figure 3b shows the 3D printed prototype. Each artificial ommatidium creates a circular area with brightness dependant on the amount of light coming from the direction in which the ommatidium is pointed. Figure 4 shows (a) the image acquired with the prototype in comparison to (b) the picture acquired with a standard camera with a lens. The resolution is low and errors in 3D printing process, e.g. partially filled holes and extrusions, are visible as irregularities in the image. This design type is easily manufactured with a 3D printing technology, but with its low resolution still does not have a favourable cost effectiveness. Consequently, the image resolution needs to be improved.

2.2 Shell Design

The image resolution of the proposed solid design can be improved by increasing the number of artificial ommatidia. The maximum number of ommatidia is limited to a certain extent, because the diameter of holes and walls between them becomes small and almost impossible to produce. We can circumvent this by making the solid hollow and by only keeping the outer shell, as is depicted in Fig. 5a. Now, the holes forming the artificial ommatidia are solely in the front surface (9) and there is only one hole (8) common to all ommatidia at the back surface facing the image detector. The light coming from the observed object passes through the hole on the front surface, the second hole in the back surface and ends on the image detector (1), which is moved to the back of the ACE. This design has more artificial ommatidia than the solid design, because all ommatidia share a common hole at the back surface.

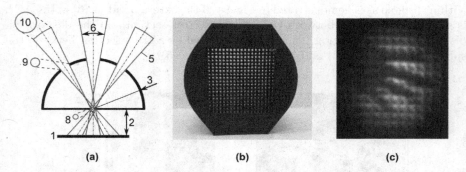

(a)   (b)   (c)

Fig. 5. (a) Vertical cross section of the shell ACE design. The numbering is the same as in Fig. 3; (b) Picture of the prototype's 3D printed front surface of a shell; (c) Image of a hand acquired with the prototype.

The prototype shown in Fig. 5b is a 3D printed front surface of a shell design with a 19 × 19 array of holes with diameter 0.75 mm and 0.3 mm walls between them. For the back surface a 0.04 mm thick plastic sheet with a single 0.1 mm diameter hole is used. An image of a hand in front of the prototype is shown in Fig. 5c. Image resolution is significantly improved with respect to the solid design.

2.3   Pinhole Design

The image resolution of the shell design can be improved by increasing the number of holes in the surface of the front shell. This can be carried out as long as the holes do not touch each other and the walls between them disappear. At this point the front shell can be totally omitted, as it becomes clear that it is no longer necessary for the image creation. The remaining small hole (8) in front of the imaging detector (i.e. the pinhole) in combination with the pixel forms an artificial ommatidium (Fig. 6). A group of pixels under the pinhole shares the same pinhole and forms a group of artificial ommatidia and consequently the ACE. The main finding about this ACE design is that the pinhole in combination with an array image detector is behaving like a compound eye.

The main parameters of the artificial ommatidium are the pinhole diameter d_8, pixel size and the distance between the pinhole and the pixel L. If the pixel size is theoretically zero, while the pinhole diameter d_8 and distance L is greater than zero, the artificial ommatidium has a conical shape. In practice, a pixel does not have zero size. The pixel accepts light from all conical shapes that end on its surface. They overlap and the resulting image is blurry. The blurriness is investigated by changing the ratio between the hole diameter d_8 and pixel size at constant distance L.

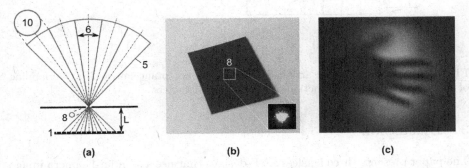

Fig. 6. (a) Vertical cross section of the pinhole ACE design; (1) Image detector with an array of pixels; (5) Boundary light rays; (8) Pinhole; (10) Cross-section of the view field at the observation distance; (b) Picture of a thin sheet with a pinhole; (c) Image of a hand acquired with the prototype.

Figure 7 shows ray tracing simulation for different pinhole diameters from 0.25, 2 and 8 pixel units, where one pixel unit equals the pixel size. The light ray starts at pixel edge, pass through the opposite side of the pinhole edge (Fig. 7a) and ends at some viewing distance in front of the pin hole (Fig. 7b). In order to calculate an artificial ommatidium acceptance angle, this procedure is repeated for many points along the pixel edge. The simulation shows that the acceptance angles of adjacent ommatidia (Fig. 7b) overlap by approximately 20 % at $d_8 = 0.25$ pixel unit. That means that a same point in space is visible on 2×2 array of adjacent ommatidia. By increasing the pinhole diameter to 2 pixel units, acceptance angle overlapping is on average 75 %. A point in space is visible on 3×3 array of adjacent ommatidia, and so on. An estimation of the resulting blurriness (e.g. circle of confusion) C is derived from a ray tracing as a function of pinhole diameter $C = d_8 + 1$ in pixel units. The circle of confusion reduces the number of meaningful artificial ommatidia, hence the image can be downsampled to C sized cells without loosing information.

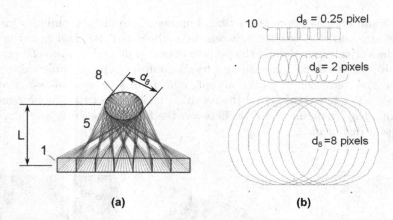

Fig. 7. (a) Ray tracing of an artificial ommatidium acceptance angle; (b) Overlapping of acceptance angles as a function of the pinhole diameter.

3   Discussion

The paper presents three lensless ACE designs that use a standard camera image detector. In the solid design (Fig. 8a) deep holes at different angles in a solid body act as artificial ommatidia. The image resolution is low. In the shell design (Fig. 8b) the artificial ommatidia are created with a perforated shell and a pinhole. The number of artificial ommatidia on the same image detector is increased and consequently so is the image resolution. In the pinhole design (Fig. 8c) the artificial ommatidia are created with a pinhole and pixels of the image detector. Out of the three designs the pinhole design creates images with the best resolution and is the easiest to manufacture.

The last design modification is an array of pinholes in front of the image detector. The pinholes are displaced to prevent overlapping of their images. The resulting image acquired by the image detector thus contain an array of small

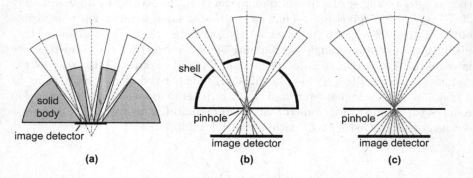

Fig. 8. Comparison of the (a) solid, (b) shell and (c) pinhole ACE design.

images, as is demonstrated in Fig. 9a. By additional image processing superposition of all small images can be created, resulting in a brighter and less noisy image. This is demonstrated in Fig. 9b, where images in rows are averaged.

There is also a small amount of disparity between the individual images within the array, which can be used for the distance estimation. For the task of distance estimation a classical classification artificial neural network was trained on 250 images of a black point on a white background at different positions and distances from the ACE. The distances were in the range from 30 mm to 130 mm. With the trained network it was possible to successfully determine the distance to the observed point with an accuracy of few millimetres.

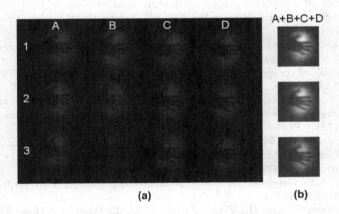

Fig. 9. (a) Image of a hand acquired with the pinhole array in front of the image detector and (b) rows' superpositions.

4   Conclusion

The paper describes the sequential development of a lensless artificial compound eye. The main advantages of the proposed artificial compound eye design include:

- a pinhole in combination with an array image sensor behaving like a compound eye,
- the compound eye properties, i.e. the spectral sensitivity and the size depending on the image sensor,
- infinite depth of field,
- acceptance angle can be up to 180°,
- an array of pinholes in front of the image sensor can be used for the image enhancement and depth perception,
- simple production.

References

1. Stavenga, D.G., Hardie, R.C. (eds.): Facets of Vision. Springer, Heidelberg (1989). doi:10.1007/978-3-642-74082-4
2. Land, M.F., Nilsson, D.E.: Animal Eyes. Oxford University Press, Oxford (2012)
3. Franceschini, N., Pichon, J.M., Blanes, C., Brady, J.M.: From insect vision to robot vision [and discussion]. Philos. Trans. Roy. Soc. Lond. B Biol. Sci. **337**(1281), 283–294 (1992)
4. Lichtensteiger, L., Eggenberger, P.: Evolving the morphology of a compound eye on a robot. In: 1999 Third European Workshop on Advanced Mobile Robots (Eurobot 1999), pp. 127–134. IEEE (1999)
5. Lichtensteiger, L.: Towards optimal sensor morphology for specific tasks: evolution of an artificial compound eye for estimating time to contact. In: Intelligent Systems and Smart Manufacturing, pp. 138–146. International Society for Optics and Photonics (2000)
6. Floreano, D., Pericet-Camara, R., Viollet, S., Ruffier, F., Brückner, A., Leitel, R., Buss, W., Menouni, M., Expert, F., Juston, R., Dobrzynski, M.K.: Miniature curved artificial compound eyes. Proc. Natl. Acad. Sci. **110**(23), 9267–9272 (2013)
7. Viollet, S., Godiot, S., Leitel, R., Buss, W., Breugnon, P., Menouni, M., Juston, R., Expert, F., Colonnier, F., L'Eplattenier, G., Brückner, A.: Hardware architecture and cutting-edge assembly process of a tiny curved compound eye. Sensors **14**(11), 21702–21721 (2014)
8. Song, Y.M., Xie, Y., Malyarchuk, V., Xiao, J., Jung, I., Choi, K.J., Liu, Z., Park, H., Lu, C., Kim, R.H., Li, R.: Digital cameras with designs inspired by the arthropod eye. Nature **497**(7447), 95–99 (2013)
9. Li, L., Yi, A.Y.: Development of a 3D artificial compound eye. Opt. Express **18**(17), 18125–18137 (2010)
10. Qu, P., Chen, F., Liu, H., Yang, Q., Lu, J., Si, J., Wang, Y., Hou, X.: A simple route to fabricate artificial compound eye structures. Opt. Express **20**(5), 5775–5782 (2012)
11. Deng, S., Lyu, J., Sun, H., Cui, X., Wang, T., Lu, M.: Fabrication of a chirped artificial compound eye for endoscopic imaging fiber bundle by dose-modulated laser lithography and subsequent thermal reflow. Opt. Eng. **54**(3), 033105-1–033105-6 (2015)
12. Wu, D., Wang, J.N., Niu, L.G., Zhang, X.L., Wu, S.Z., Chen, Q.D., Lee, L.P., Sun, H.B.: Bioinspired fabrication of highquality 3D artificial compound eyes by voxel-modulation femtosecond laser writing for distortion-free wide-field-of-view imaging. Adv. Opt. Mater. **2**(8), 751–758 (2014)
13. Wang, L., Liu, H., Jiang, W., Li, R., Li, F., Yang, Z., Yin, L., Shi, Y., Chen, B.: Capillary number encouraged the construction of smart biomimetic eyes. J. Mater. Chem. C **3**(23), 5896–5902 (2015)
14. Deng, Z., Chen, F., Yang, Q., Bian, H., Du, G., Yong, J., Shan, C., Hou, X.: Dragonfly-eye-inspired artificial compound eyes with sophisticated imaging. Adv. Funct. Mater. **26**, 1995–2001 (2016)
15. Ogata, S., Ishida, J., Sasano, T.: Optical sensor array in an artificial compound eye. Opt. Eng. **33**(11), 3649–3655 (1994)
16. Hamanaka, K., Koshi, H.: An artificial compound eye using a microlens array and its application to scale-invariant processing. Opt. Rev. **3**(4), 264–268 (1996)
17. Duparr, J.W., Dannberg, P., Schreiber, P., Bruer, A., Nussbaum, P., Heitger, F., Tnnermann, A.: Ultra-thin camera based on artificial apposition compound eyes. In: Proceedings of the 10th Microoptics Conference (2004)

18. Duparr, J., Dannberg, P., Schreiber, P., Bruer, A., Tnnermann, A.: Artificial apposition compound eye fabricated by micro-optics technology. Appl. Opt. **43**(22), 4303–4310 (2004)

19. Duparr, J., Dannberg, P., Schreiber, P., Bruer, A., Tnnermann, A.: Thin compound-eye camera. Appl. Opt. **44**(15), 2949–2956 (2005)

20. Duparr, J.W., Wippermann, F.C.: Micro-optical artificial compound eyes. Bioinspir. Biomim. **1**(1), R1–R16 (2006)

21. Radtke, D., Duparr, J., Zeitner, U.D., Tnnermann, A.: Laser lithographic fabrication and characterization of a spherical artificial compound eye. Opt. Express **15**(6), 3067–3077 (2007)

22. Kim, J., Jeong, K.H., Lee, L.P.: Artificial ommatidia by self-aligned microlenses and waveguides. Opt. Lett. **30**(1), 5–7 (2005)

23. Jeong, K.H., Kim, J., Lee, L.P.: Biologically inspired artificial compound eyes. Science **312**(5773), 557–561 (2006)

24. Jung, H., Jeong, K.H.: Microfabricated ommatidia using a laser induced self-writing process for high resolution artificial compound eye optical systems. Opt. Express **17**(17), 14761–14766 (2009)

Development of a Bio-inspired Knee Joint Mechanism for a Bipedal Robot

Alexander G. Steele[1], Alexander Hunt[1(✉)], and Appolinaire C. Etoundi[2(✉)]

[1] Department of Mechanical Engineering, Portland State University (PSU), Portland, OR, USA
{ajmar,ajh26}@pdx.edu
[2] Bristol Robotics Laboratory, Department Engineering Design and Mathematics, University of the West of England, Bristol, UK
Appolinaire.Etoundi@uwe.ac.uk

Abstract. This paper presents the design and development of a novel biologically inspired knee design for humanoid robots. The robotic joint presented mimics the design of the human knee joint by copying the condylar surfaces of the femur and tibia. The joint significantly reduces the complexity, while preserving the mechanisms of the human knee's motion, and the torque requirements. This joint offers the remarkable feature of being multifunctional since it separates structural and kinematic functions namely integration of (i) high level of shock absorption due to its dynamic variation of pressure between the articular surfaces and its curved profile and (ii) high mechanical advantage due to its moving center of rotation. These functions are essential for humanoid robotic limbs where performance improvement is requisite. The design demonstrates the possibility to simplify the knee linkage arrangement while still providing a moving center of rotation by dynamically changing the pressure between the joint surfaces (femur and tibia). This dynamically controlled pressure enables accurate joint movement by mimicking the human knee property of the same feature. A prototype of the joint has been developed for testing the beneficial properties designed into the model.

Keywords: Condylar hinge joint · Sliding ratio · Knee biomechanics · Robotics · Mechanical advantage · Dynamic control

1 Introduction

Humanoid robotics is a challenging research field, which has seen increasing achievements during recent years and will continue to play a key role in robotics research and applications of the 21st century. The focus is now on the development of safe behaviour humanoid robots which has recently become an active research topic. More specifically, legged robots are becoming more prominent in society and have become more apparent due to their great potential to serve as companions and assistants for humans in daily life and also as ultimate helpers in man-made and natural disasters. However, they are still lacking the adaptability to deal with changing terrain or falls. Since our world has been primarily designed for bipedal walking, robots need to be able to adapt to human

© Springer International Publishing AG 2017
M. Mangan et al. (Eds.): Living Machines 2017, LNAI 10384, pp. 418–427, 2017.
DOI: 10.1007/978-3-319-63537-8_35

environments. Bipedal robotics is no longer in its infancy with the emergence of humanoid robots; however, with all the advances made, there are still many limitations to these robots and their uses. In the field of autonomous mobile robots, challenges in actuation, mechanical design, sensing and control must be overcome to attain high levels of performance while ensuring safety. This is seen in large-scale robotic competitions such as the one hosted by DARPA Robotics Challenge [1]. One of the limiting factors for mainstream robotics is the inherent complexity of bipedal movement [2, 3]. The high torque requirements required for bending or squatting using a typical hinge style robotic joint also increases the size of the motors. This further increases the amount of torque needed, leading to a compounding effect.

Another difficulty with humanoid design is the kinematic interpretation of human joints and the development of mechanisms that can mimic human motion. The study of biological systems to determine how they remarkably integrate versatility and passive compliance to fit together multiple and heterogeneous parts as well as non-linear behaviours and multiple scales is called the 'biomimetic' (also called bio-inspired) approach. The largest problem with the hinge joint model is that the center of rotation is fixed, which also means that the torque arm is fixed for the entire range of motion and offers no change in mechanical advantage to the joint when at higher angles. While the hinge joint has exceptional range of motion, when the weight of the robot drops below the plane of the hinge (squatting), the torque requirements to return to the upright position are at their highest. The fixed torque arm of the hinge means that the motor torque or actuator force applied at the hinge must be increased or the weight of the robot must be reduced [4]. Furthermore, other limiting issues make hinge style joints undesirable, these hinges typically consist of a hinge-pin and ball bearings and due to the forces applied to the hinge, regular maintenance and proper design is important to limit failures [5].

Due to these limitations, researchers have recently started working with bio-inspired designs to help overcome the fixed torque arm problem [6–9]. The condylar hinge joint designs try to mimic the human knee, which has a large range of motion, but also has a moving center of rotation. Most current models approximate the condylar joint as a four-bar linkage. We intend to show that this design is simplified by the introduction of a single sliding linkage, which also embodies a changing center of rotation. The beneficial features of the design are the reduction of mass, wear, and friction of the joint while being able to resist out of plane forces applied to the joint.

2 Modeling of the Human Knee

The knee joint is intrinsically complex, while it appears to be only a single hinge joint, in reality, the knee has 6 degrees of motion in which, during movement, the tibia and femur rotate and translate with respect to each other in all three dimensions [3, 10]. While this is advantageous for either bipeds and quadrupeds, an engineered joint which perfectly mimics its biological counterpart for robotic systems does not achieve the required control and makes the joint overly complex. On the other hand, researchers have tried modeling an equivalent joint treating the knee as a single degree of freedom system [4]. The human knee is primarily held together by four ligaments namely the lateral collateral ligament (LCL), anterior cruciate ligament (ACL), posterior cruciate ligament (PCL) and medial collateral ligament (MCL) as shown in Fig. 1. When viewed from the sagittal plane, the PCL and ACL form an X shape and are referred to as cruciate ligaments. Relative to the femur, the ACL keeps the tibia from slipping forward while the PCL keeps the tibia from slipping backward, while the LCL and MCL prevent the femur from sliding side to side [2]. The knee joint is referred to as a condylar joint due to the concave-convex profiles of the femur and tibia.

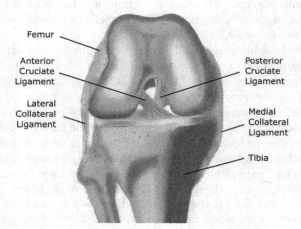

Fig. 1. Knee joint flexibility and stability comes in part from the ligament connections. The LCL (outer left), ACL (inner left), PCL (inner right), and the MCL (outer right) attach the femur to the tibia and fibula.

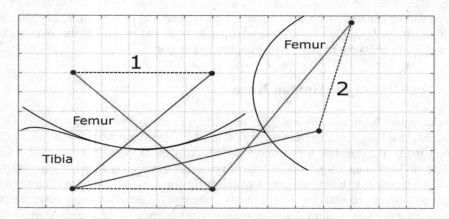

Fig. 2. Four-bar linkage representation of the human knee joint while standing (1) and at the maximum angle of rotation (squatting) (2)

Prior attempts to simplify the knee joint have resulted in a design based on the cruciate ligaments of the knee, which is modeled as an inverted parallelogram four-bar mechanism as illustrated in Fig. 2.

This model has already proven to be very successful in tests and has some of the same the advantages that give it similar range of motion and a moving center of rotation [8]. However, this model does not take into account the dynamic pressure changes of the knee brought on by muscle flexion, 'which has the potential to reduce wear, increase stability, and help reduce the complexity associated with the motion. In this paper, we propose an alternate design which includes this particular feature [11].

Our joint moves similar to a cam mechanism while being supported by two linkages based on the MCL and LCL, which are thought to resist external tibial rotation or anterior tibial translations [2]. Pressure between the condylar surfaces can be controlled dynamically using a pneumatic piston, which adds control and extra stiffness to the joint when needed. Figure 3(A) shows the knee modeled joint with the attached piston. Figure 3 also shows the 3D modeled joint in three different positions, fully extended (B), half rotation (C), and fully rotated (D), and illustrates the range of motion of the new biologically inspired knee design. The joint was initially designed using MRI scans, which, once the primitive shape model was extracted, were made into a 3D model. From this model, the condylar surfaces of the femur and the tibia were copied for the initial design phase.

A   B   C   D

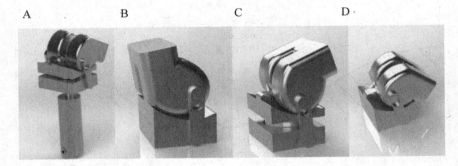

Fig. 3. (A) Modeled knee mechanism with placement of the pneumatic actuator. Modeled knee mechanism fully extended (B), half rotation (C), and fully rotated (D).

3   Performance Factors of the Proposed Condylar Joint

3.1   Modelling the Sliding-Rolling Ratio

Because the sliding to rolling ratio – also referred to as slack – is an important parameter for the assessment of the joint since it estimates the amount of wear on the joint, it was necessary to determine the ratio for the proposed condylar joint [12]. This is calculated by using the radii of the condylar surfaces from the tibia and femur and the 3D model to create a motion study using SolidWorks software. Because we are assuming that the multibody model is ridged, rigid body kinematics can be applied to determine the sliding-rolling ratio. To determine the ratio, we determined the arc lengths of the two surfaces by calculating the distance travelled (ΔL) from the initial contact point (C_{inital}) on the tibia and the distance travelled by the same contact point of the femur over a specified increment of rotation, defined as the flexion angle (θ) as seen in Fig. 4. Calculation of the sliding-rolling ratio (χ) is found by

$$\chi(\theta) = \frac{\Delta L_{tibia}(\theta) - \Delta L_{femur}(\theta)}{\Delta L\_tibia(\theta)} \tag{1}$$

where,

$$\Delta L_{tibia}(\theta) = C_{tibia\ final}(\theta) - C_{tibia\ inital}(\theta) \tag{2}$$

$$\Delta L_{femur}(\theta) = C_{femur\ final}(\theta) - C_{femur\ inital}(\theta) \tag{3}$$

are the corresponding incremental differences of the contact arc lengths.

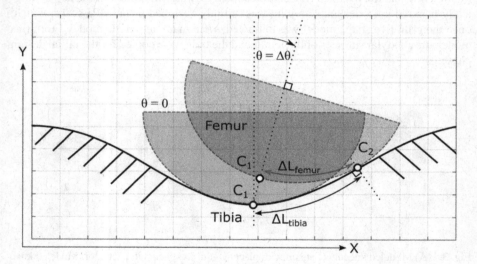

Fig. 4. Kinematics of the tibia and femur.

A sliding ratio of zero indicates pure rolling and a ratio of one signifies pure sliding. If the ratio is between zero and one, the movement is characterized as partial sliding and rolling. For example, a sliding-rolling ratio of 0.7 resulted in 70% sliding and only 30% rolling. If the ratio is positive, then the slip of the femur is higher than the tibia. If the sign is negative, then the tibia has higher slip when compared to the femur. As a result of this equation, it is desirable to have the slip as close to 0 as possible to minimize the wear of the joint. Thus, the sliding-rolling ratio calculation was done over the range of the entire range of motion of the prototype joint from Eq. (1). Figure 5 shows a plot of the theoretical sliding-rolling ratio for the proposed knee design with an incremental angle (θ) of 0.75°.

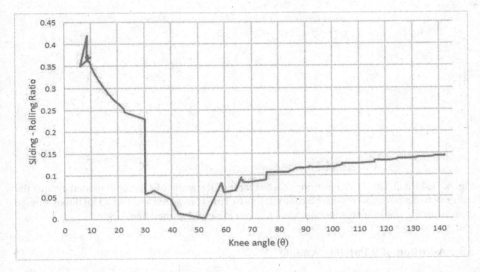

Fig. 5. Sliding and rolling ratio of the proposed joint. Note, the initially high sliding ratio settles because the model settles prior to the start of the rotation.

Our model suggests a sliding-rolling ratio of 0.159, but if we disregard the initial settling of the 3D simulation the sliding-rolling ratio drops to 0.126 or roughly 12.6%. Due to the complex nature of the human knee, several attempts have been made to analytically model the sliding-rolling ratio [4, 12]. However, these models make several simplifying assumptions that do not accurately characterize the movement of the joint. Because of this, we cannot confidently estimate the sliding ratio of the human knee joint for comparison, but we expect that the sliding-rolling ratio of our model to be similar to that of the human knee [4]. It should also be noted that the sliding-rolling ratio of the human knee changes with the change of flexion angle, as does our model plotted in Fig. 5.

3.2   Moment Arm and Mechanical Advantage

The proposed joints' mechanical advantage directly influences the selection of actuators that can be used. To determine the mechanical advantage, the radius arm $r(\theta)$ needs first to be identified. Figure 6(a) shows how the radius arm changes with the flexion angle and is plotted alongside the fixed moment arm that would be seen in a pin joint. Using the radius arm the mechanical advantage of the joint can be determined. The mechanical advantage (MA) is calculated by the following

$$MA = r(\theta)/C \tag{4}$$

where C is the distance to the center of the joint. Required torque is significantly reduced due to the changing moment arm as can be seen in Fig. 6(b).

A   B

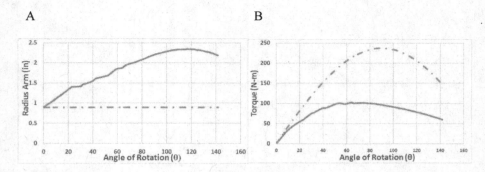

Fig. 6. (a) Radius arm r(θ) changing as a function of the angle of rotation θ plotted against the radius arm of a pin joint (red-hashed line) (left). (b) Torque required to move 45 kg through range of motion of a single knee joint compared to a pin joint (red-hashed line) (right).

3.3 Addition of a Patellar Analog

In the human body, the primary role of the patella is assisting during knee extension [13, 14]. The patella increases the leverage the tendon can exert on the femur by increasing the angle at which it acts. This same principle can be applied to the biologically inspired condylar joint design. By using a patellar analog, we can affectively extend the moment arm of the joint for the entire range of motion while keeping the joint compact and the added weight to a minimum. For the design prototype the size and geometry of the patellar analog was made similar to the patella of the MRI scans used to design the rest of the joint. Figure 7(a) shows a model of the joint with the addition of the patellar analog to the design.

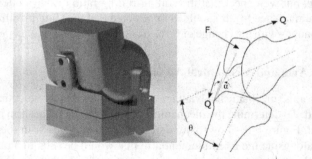

Fig. 7. (a) Model of prototype knee joint showing the addition of the patellar analog (left). (b) Forces applied to the patella through knee flexion (right).

Forces applied to the joint by the patellar analog, where Force (F) is determined by

$$F(\theta) = 2Q * \sin\left(\frac{\alpha + \theta}{2}\right) \tag{5}$$

where Q is the force applied by the actuator, α is the angle between the centroid of the patella and the force Q, and θ is the angle between the center of the tibia to the femur as shown in Fig. 7(b). The choice of using a patellar analog that is not fixed to the joint keeps the patellar analog perpendicular to the radius arm for the full range of motion and also maintains the longest length for the radius arm.

4   Design of the Physical Prototype

Once the viability of the design was confirmed, construction of a physical prototype joint was used to physically test the design against the model that was created. Figure 8 shows the constructed prototype of the proposed bio-inspired knee joint (patella analog not shown) in several states of rotation.

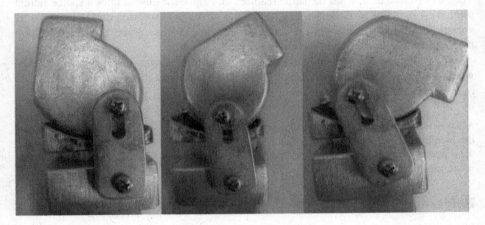

Fig. 8. Prototype knee progressing through the range of motion. The pneumatic actuator maintains contact and can apply a changing force between the tibial head and femoral head portions of the joint.

Like the human knee, we chose to mimic the LCL and MCL to provide the stability of the joint. Also shown in Fig. 8, the placement of the linkage changes as joint rotates adding stability through the range of motion of the joint.

Slip velocity has been linked to high wear; testing has shown that the transitions from rolling to sliding and high slip velocity lead to high wear. Conversely, low slip velocity lowers wear of the surface. By modulating the contact force of the joint, in the case of the prototype using a pneumatic cylinder, we can lower the transmission time between rolling and sliding as well as decrease the slip velocity of the surface, which will increase the life of the joint [14, 15]. The pneumatic cylinder fits inside the tibia bone and adds minimal weight to the joint.

The tibia and femur bones attach to the knee using a 44.5 mm^2 tube stock. The tibia bone and femur bones are 38 mm aluminum T-slot extrusion. The condylar joint was created from machined aluminum stock based on profiles created using the MRI scans.

426 A.G. Steele et al.

The contact surfaces were lined with strips of UHMW plastic film, which helps reduce the coefficient of friction and gives the joint a smooth wear resistant surface.

The design eliminates the cruciate ligaments, but retains the different curved profiles of the lateral and medial femoral condylar surfaces. The intercondylar tubercules (center of tibia) have been expanded upon and used as a rail to help direct the femur in the case of high out of plane forces acting on the joint. For the prototype joint, the patellar surface has been machined in order to guide the patellar analog similarly to a biological femur.

5 Conclusions

In this paper, we present a bio-inspired design alternative to the hinge joint. The design mimics the human condylar knee joint and has applications in mobile and agile robotics. The bio-inspired design has the same features that make the human knee a such efficient joint. This includes its compactness, high level of shock absorption due to its dynamic variation of pressure between the articular surfaces and its curved profiles, and high mechanical advantage due to its moving center of rotation.

The prototype condylar hinge joint has a significantly better mechanical advantage when compared to the typical hinge joint of the same size. This has the potential to reduce weight through a reduction in the joint actuator size.

Acknowledgments. The authors acknowledge the support of the Mechanical and Materials Engineering Department in the Maseeh College of Engineering at Portland State University and the support of the UK Engineering and Physical Sciences Research Council (EPSRC) under grant reference EP/P022588/1.

References

1. Orlowski, M.C.: DARPA Robotics Challenge (DRC), June 2015
2. Ling, Z.-K., Guo, H., Boersma, S.: Analytical study on the kinematic and dynamic behaviors of a knee joint. Med. Eng. Phys. **19**(1), 29–36 (1997)
3. Masouros, S.D., Bull, A.M.J., Amis, A.A.: Biomechanics of the knee joint. Orthop. TRAUMA **24**(2), 84–91 (2010)
4. Fekete, G.: Kinetics and Kinematics of the Human Knee Joint under Standard and Non-standard Squat Movement. Ghent University (2013)
5. Siegwart, R., Nourbakhsh, I.R., Scaramuzza, D.: Introduction to Autonomous Mobile Robots, 2nd edn. MIT Press, Massachusetts (2011)
6. Etoundi, A.C.: A bio-inspired condylar hinge joint for mobile robots. In: International Conference on Intelligent Robots Systems, September 2011
7. Etoundi, A., Burgess, S., Vaidyanathan, R.: A bio-inspired condylar hinge for robotic limbs. J. Mech. Rob. **5**, 8 (2013)
8. Burgess, S.C., Etoundi, A.C.: Performance maps for a bio-inspired robotic condylar hinge joint. J. Mech. Des. **136**(11), 115002-1–115002-7 (2014)
9. Khan, H., Featherstone, R., Caldwell, D.G., Semini, C.: Bio-inspired knee joint mechanism for a hydraulic quadruped robot. In: 2015 6th International Conference on Automation, Robotics and Applications (ICARA), pp. 325–331 (2015)
10. Esat, I.I., Ozada, N.: Articular human joint modelling. Robotica **28**, 321–339 (2009)

11. Carey, R.E., Zheng, L., Aiyangar, A.K., Harner, C.D., Zhang, X.: Subject-specific finite element modeling of the tibiofemoral joint based on CT, magnetic resonance imaging and dynamic stereo-radiography data in Vivo. J. Biomech. Eng. **136**, 041004 (2014)
12. Fekete, G., et al.: Sliding-rolling ratio during deep squat with regard to different knee prostheses. Acta Polytech. Hung. **9**(5), 20 (2012)
13. Sylvester, A.D., Mahfouz, M.R., Kramer, P.A.: The effective mechanical advantage of A.L. 129-1a for knee extension. Anat. Rec. **294**, 1486–1499 (2011)
14. Halonen, K.S., et al.: Importance of patella, quadriceps forces, and depthwise cartilage structure on knee joint motion and cartilage response during gait. J. Biomech. Eng. **138**, 071002 (2016)

Predator Evasion by a Robocrab

Theodoros Stouraitis[1], Evripidis Gkanias[1], Jan M. Hemmi[2],
and Barbara Webb[1(✉)]

[1] School of Informatics, Institute of Perception, Action and Behaviour,
University of Edinburgh, Edinburgh EH8 9AB, UK
bwebb@inf.ed.ac.uk
[2] School of Biological Sciences, University of Western Australia,
Crawley, WA 6009, Australia
jan.hemmi@uwa.edu.au

Abstract. We describe the first robot designed to emulate specific perceptual and motor capabilities of the fiddler crab. An omnidirectional robot platform uses onboard computation to process images from a 360° camera view, filtering it through a biological model of the crab's ommatidia layout, extracting potential 'predator' cues, and executing an evasion response that also depends on contextual information. We show that, as for real crabs, multiple cues are needed for effective escape in different predator-prey scenarios.

1   Introduction

Fiddler crabs are a semi-terrestrial marine species of crab which use visual information to make behavioural decisions such as escaping from predators, selecting mates, signalling conspecifics and detecting their burrows [5,6,16,21]. We explore the mechanisms of crab behaviour using a biorobotic approach (Fig. 1).

Courtship and burrow defence are regulated from the stimuli detected below the horizon, while the stimuli for initiating the evasion behaviour are detected above the horizon [6]. This paper focuses on the evasion behaviour of fiddler crabs, as its visuomotor characteristics, response stages and context dependence have been extensively studied [5].

An important characteristic of fiddler crabs is their ability to move in any direction while maintaining a fixed orientation or rotating around their centre of mass independently of translation. This allows them to escape rapidly from a potential threat independently of their pose. Another striking feature is their two panoramic compound eyes, located on eyestalks on the front part of the carapace, each with a 360° view. The eyes are composed from a large number of ommatidia in a particular layout, specialised for detection of behaviourally

---

T. Stouraitis and E. Gkanias contributed equally to this work and jointly assert first authorship.

This work was in part supported by an Australian Research Council (ARC) Discovery Grant (DP160102658).

© Springer International Publishing AG 2017
M. Mangan et al. (Eds.): Living Machines 2017, LNAI 10384, pp. 428–439, 2017.
DOI: 10.1007/978-3-319-63537-8_36

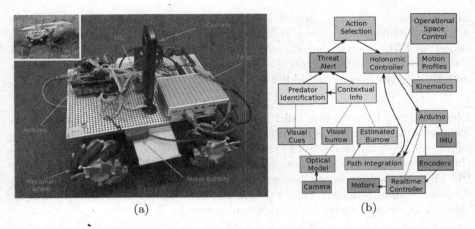

(a)   (b)

Fig. 1. (a) The Robocrab (inset: a real crab). (b) System schematic. Arrows: communication of data; dashed lines: hierarchical component relations.

relevant cues [15], despite the low resolution [5]. In addition, the fact that crabs live on mud flats simplifies the identification of the visual horizon and alignment of the eye equator to it [21].

The initiation of the evasion response depends not only on the location of the stimuli above the horizon but on additional cues and context [3]. The response itself can also differ. When a fiddler crab identifies a threatening stimulus but does not have a burrow, it first freezes, then runs in the opposite direction to the stimulus and rotates around its centre of mass to place the threatening stimulus on one of its sides [7]. On the other hand, burrow-holder fiddler crabs always run towards their burrow under the presence of any threatening stimulus [2], and will then pause again outside the burrow before making a descent.

In the following sections we describe the robot hardware choices, then the implementation of motion control, path integration, visual processing and behavioural intelligence. This is followed by the results of experiments on the evasion response of the robot, using the same paradigm as has been used for crabs. We discuss the conclusions that can be drawn and the plans for future work.

2   Methods

Robot Platform. Fiddler crabs have complex legged actuation, but for the purposes of mimicking escape, the crucial feature is their ability to execute omnidirectional motions on a 2D plane (the mud flat), described by a 3D motion vector (x, y, θ), where x and y define the position in the 2D space and θ the orientation. Thus, the robot platform used is omnidirectional, utilising Mecanum wheels, in which the sideways forces exerted by the rollers attached to the wheels can be used in combination to obtain any direction of motion (see Sect. 2). Each wheel is actuated by a DC motor. The actual size of the robot is x10 factor larger than the real fiddler crab.

Sensory Systems. Each motor is equipped with an optical encoder used for odometry, equivalent to proprioceptive input for leg motion in the crab. We added an IMU (inertia measurement unit) to replicate the statocyst organ of the fiddler crabs [14]. The robot is equipped with a Ricoh Theta S camera to imitate the eyestalks of the fiddler crabs, and provide a 360° field of view.

Electronics, Processing and Communications. The robot platform includes a custom Arduino board (Duemilanove) and the respective I/O expansion shield with interfaces to the sensors and the motors, which is used for real-time processing. The main processing unit is a fanless lightweight (250 grams) PC which is used for image processing on the fly, on-board motion control and decision making processes. The PC and Arduino communicate via serial port, through an asynchronous communication interface.

Motion Controller. Figure 2b summarises the method implemented to enable a pose command in the operational space of the robot x, y, θ to result in the corresponding movement of the robot base, following the methods in [13,18,19]. Every time a new pose command is received by the tracker, the motion profile module is called to provide a sinusoidal motion profile to implicitly control the accelerations of the system. The output is the interpolated substeps for each of the three dimensions x, y, θ. Next, the complete trajectory is tracked utilising the operational space controller which is running at 15 Hz with a fixed duty cycle for each substep. This uses three PD controllers one for each dimension x, y and θ to reach any target in the 3D space, plus a feedforward term to reduce the reaction time. This produces the desired robot velocities $\dot{x}, \dot{y}, \dot{\theta}$ which then need to be transformed to motor/wheel velocities $\omega 1, \omega 2, \omega 3, \omega 4$ through kinematic

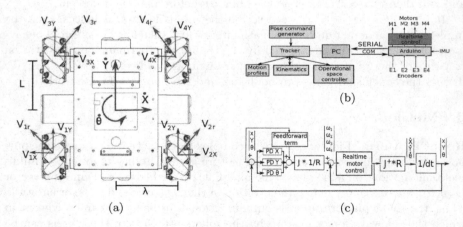

(a)   (b)   (c)

Fig. 2. (a) Robot frame, operation space velocities and linear velocities at the mounting points of the wheels, due to angular velocities of the wheels. (b) System software architecture. (c) Control structure including operational space and realtime controller.

modelling of the robot (Fig. 2a). These are sent to the Arduino, on which the realtime controller realises the specified motor angular velocities on the motors using 200 Hz PID control for each wheel.

Path Integration. Fiddler crabs use path integration to maintain an accurate estimate of their position relative to their burrow while foraging. In [8,9] Layne claims that crabs can accurately path integrate solely due to their leg proprioceptor sensors. However, crabs also possess an advanced statocyst organ that is capable of sensing accurately linear and angular accelerations [14]. Hence, we decided to utilise both the wheel encoders as well as an IMU to localise the Robocrab based on the work presented in [1,10,20].

Utilising motors' optical encoders and inverse kinematics, the Robocrab's velocities in the 3D space x, y, θ are estimated. Using acceleration and orientation information from the Inertia Measurement Unit (IMU), a second estimate of Robocrab's velocities is obtained. The two estimates are fused with a complementary filter to obtain a robust estimate of Robocrab's velocities, that is integrated to get the pose of the robot relative to its burrow.

Modelling the Compound Eye. The spatial distribution of ommatidia in the compound eye is described by the sampling resolution and every ommatidium has its own optical resolution (i.e. receptive field). We use the model from [15] (supplied to us by the authors) which describes the optical resolution using a Gaussian distribution for each ommatidium, while for the sampling resolution a combination of inverse sine functions is used. This model consists of 7,971 ommatidia positions in spherical coordinates (elevation, azimuth and radius). To create our twin-eye model we just drop the samples from the medial zone, duplicate the lateral zone, and invert the y axis to create the half-right eye. Finally, we merge the two half-eyes to create the twin-eye of Fig. 3a. This model has 9,740 ommatidia in total, 4,870 for each eye. To reduce the compution cost

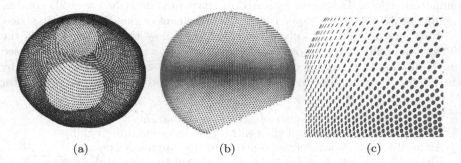

(a)   (b)   (c)

Fig. 3. Sampling and optical resolution of the twin-eye model. (a) represents the 3D model of the twin-eye sampling resolution. In (b) we represent the optical resolution of the left half-eye using Gaussian contours on the $Y - Z$ axes. This is more visible in (c), where we plot a closer view of it.

of applying a Gaussian filter for each ommaditium, we instead filter the image before sampling from it using two different filters: for the central pixels we use *Gaussian blurring* to incorporate information from all the adjacent pixels of every sample; and for the peripheral samples we use *rotational blurring* to imitate the Gaussian blurring on the fisheye-image. To visualise the result we use Voronoi diagrams to fill sets of pixels with the colour of their corresponding ommatidia (Fig. 4).

Fig. 4. A fully panoramic 360° crab's view image, created using Voronoi diagrams and the twin-eye sampling resolution model.

Visual Cue Extraction. Fiddler crabs typically live in areas with open, clear skies, hence a predator above the horizon, typically a bird, can be detected by its high contrast against the bright sky [21]. In the lab the segmentation of the artificial predator and burrow from the cluttered background is achieved by using a distinctive colour for these objects and utilising a simple colour filter from OpenCV as shown in Fig. 5. Such filtering provides a noisy segmentation signal, similar to a natural one. A Jacobian matrix V_j maps the pixel space to the ommatidia space. The set of activated ommatidia is described as a 2D contour within the 3D manifold shaped by the 3D placement of the ommatidia in space.

First, we assume that any detected object below the horizon level is the burrow and only objects above the horizon could be a potential predator. Based on the finding in [4,16,21] a set of visual cues that are assumed to be used by the fiddler crabs in order to identify whether the detected object is a predator or not are extracted:

– Ommatidia number: Number of activated ommatidia.
– Elevation: The elevation of the centroid of the ommatidia region.
– Azimuth: The azimuth of the centroid of the ommatidia region.
– Elevation velocity: Angular vertical velocity of the ommatidia region [16].
– Azimuth velocity: Angular horizontal velocity of the ommatidia region [16].
– Looming: The rate the activated ommatidia region grows [11].
– Flickering: The rate of activation and deactivation of the ommatidia region within a window of x frames [16].

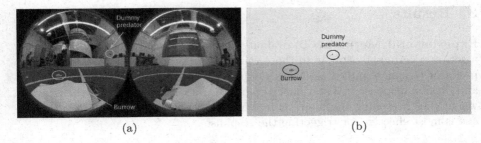

(a)   (b)

Fig. 5. (a) Captured fisheye images of a cluttered environment, with pink burrow and predator. (b) Stitched panoramic image after it has been colour filtered and passed through the optics of the eye model. (Color figure online)

Behavioural Intelligence. In Fig. 1b we provide a schematic representation of the overall system, with the following behavioural intelligence components:

Predator Identification: First, it is necessary to identify the predator given the visual cues. In [3,4,16,21] biologists present quantitative results that relate the triggering of the evasion with the respective set of visual cues. Consequently, for the robot, the predator identification block uses some of the visual cues and if one or more of these quantities exceed its respective upper threshold value, then the detected object is classified as predator.

Contextual Information: In [12,17] it has been demonstrated that fiddler crabs and other species of crabs are able to incorporate contextual information in their decision making scheme and alter their actions according to them. The contextual submodule for the Robocrab tracks whether it has a burrow or not, which will determine the direction of the evasive response; and whether the burrow location information should be taken from the path integration module or the visual detection of the burrow.

Threat Alert: In [2], fiddler crabs are shown to evaluate their risk level and compare it with the cost of an evasion action to decide whether they should perform an action or not. The threat alert module obtains the contextual information and predator characteristics from the two previously introduced submodules and updates the threshold limits according to the risk level in a inverse proportional fashion. In our current implementation the risk level is a linear function of the distance to the burrow. Additionally, the threat alert submodule declares whether a threat is present or not and initiates the evasion response.

Action Selection: The action sets are basically separated into two layers. The first distinction is between the burrow-holder and non-burrow holder, determined by contextual information. The second layer includes actions corresponding to the stages of the evasion response as in Sect. 1. Thus, depending on Robocrab's evasion stage, the respective action is selected. In a burrow-holder scenario, the action selection module coordinates the motion control with the localisation and the visual information of the burrow location to return to its burrow. For a non-burrow holder, it coordinates the motion control module with the visual based predator location to run away from the threat.

3   Results

Experimental Methods. To evaluate the performance of the robot we run a set of experiments similar to those reported for crabs in [3], where a dummy predator (a sphere pulled along a straight track, see Fig. 6) is used to stimulate and evasion response in a stationary crab. We track the evolution of the relevant visual cues preceding the moment in which the robot reacts to examine which cue is most effective for triggering the response.

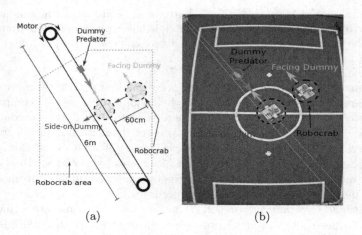

(a)   (b)

Fig. 6. Sketch (a) and the original top-view representations (b) of the experimental set-up, illustrating the four sets of experiments with the robot being placed under and off track and facing the dummy or side-on the dummy.

In the first two scenarios we look at the 'home-run' response, and in the second at the 'burrow-descent' response, which in the crab and our model involves different visual cues. Specifically, we set the thresholds for the home-run stage at 4 activated ommatidia, $0.020 \, \mathrm{rad \, s^{-1}}$ horizontal or $0.018 \, \mathrm{rad \, s^{-1}}$ vertical velocity; and for burrow-descent at 20 activated ommatidia, or a looming rate of 1 (doubling the size in one frame) with a minimum of 5 activated ommatidia. The thresholds were tuned based on preliminary experiments to obtain similar behaviour to fiddler crabs and they are indicated by the horizontal-dashed line in the figures.

For the home-run, we test the robot from two positions, either 60 cm offset from the track of the dummy with the dummy moving at a constant speed of $62 \, \mathrm{cm \, s^{-1}}$, or directly under the track of the dummy, with a speed of $47 \, \mathrm{cm \, s^{-1}}$, as in [3]. In each case we test with the robot either facing towards the dummy's approach direction or side-on, and at two different heights (38 cm and 51 cm) of the stimulus (notice that the height of centre of the camera lens mounted on the robot is at 26 cm distance from the ground). This produces a total of eight test conditions, and we run 5 trials for each condition. For the burrow-descent, we test with the robot under the track and the dummy at a lower height (32 cm), again facing or side-on to the predator, and with two different approach speeds, $62 \, \mathrm{cm \, s^{-1}}$ or $15 \, \mathrm{cm \, s^{-1}}$. This produces 4 further conditions for each of which we again run 5 trials.

Home-Run, Offset from the Predator Track. This simulates a scenario in which a predator flies along the horizon passing by the crab. When the robot is side-on to the dummy, for both heights of the dummy, we see that the triggering is caused by the horizontal velocity (see Fig. 7b). A higher horizontal velocity is obtained for the higher dummy, leading to a more reliable reaction when the dummy is more distant (see Fig. 10). This could be a result of the limited number of ommatidia available for the higher elevations of the eye model. The vertical velocity does not provide any useful information, and at the reaction frame, the number of activated ommatidia is only around 2 for the lower dummy, and 1 for the higher dummy, i.e., well below threshold. When the robot is facing the predator approach direction, we do not see much difference, with the horizontal velocity still providing the most information (Fig. 7a). The robot is able to track the dummy from a longer distance/time due to the higher sampling resolution in the frontal vs. lateral field. However, it reacts when the dummy is at a shorter distance (Fig. 10) and the higher dummy triggers a later reaction.

Home-Run, Under the Predator Track. This case resembles a predator flying directly above a crab. With the robot side-on to the predator, at both

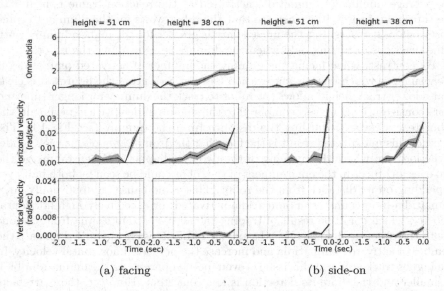

(a) facing   (b) side-on

Fig. 7. Visual cue evolution during the 2 sec preceding a response during off-track home-run experiments with two heights and directions of the dummy predator.

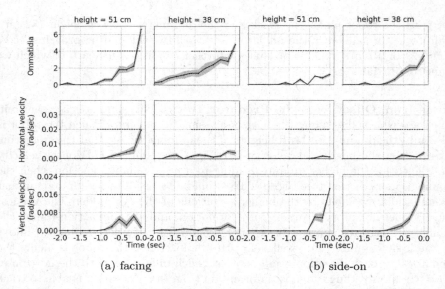

Fig. 8. Visual cue evolution during the 2 sec preceding a response during under-track home-run experiments with two heights and directions of the dummy predator.

dummy heights, the vertical velocity triggers the escape, while the horizontal velocity does not provide useful information. The ommatidia values are similar to the off-track experiments with a difference in the lower height scenario, where the average number of ommatidia activated at the reaction frame is nearer to the threshold. In Fig. 10 we can see that the robot waits for the dummy to come closer before it reacts, but the uncertainty is greater for the higher dummy, with some reactions occurring only when the dummy is right above the robot.

By contrast, when facing the dummy the number of activated ommatidia is the visual cue which triggers the evasion (see Fig. 8a). From the figures, it is clear that a lower track-height allows the robot to track the dummy for longer time and more consistently. However, compared to the side-on under-track experiment, the robot reacts earlier regarding the distance from the dummy (see Fig. 10). The cause of this unexpected result is the fact that the frontal area of our eye's model is the location where the merging of the two eyes happens. When we merge the two eyes an object in the environment of the robot could be seen by both half-eyes depending on its distance from the robot. This could cause an abnormally large ommatidia-blob (number of ommatidia activated) or excessively high horizontal velocity, which could be result of trading the activated ommatidia from the one eye to the other in almost each subsequent frame. This effect could duplicate the number of activated ommatidia and increase the perceived horizontal velocity. In the higher track scenario this issue is even more apparent, as the ommatidia-blob is smaller and therefore its detection is less consistent. However, these artefacts appear due to limitations of the twin-eye model, which were a trade-off to use only one 360° camera as described in Sect. 2.

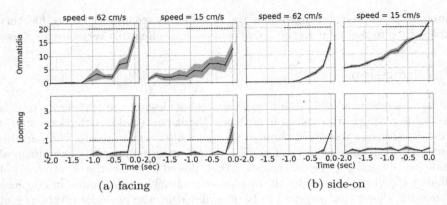

(a) facing   (b) side-on

Fig. 9. Visual cue evolution during the 2 sec preceding a burrow descent response, with two speeds and directions of the dummy predator.

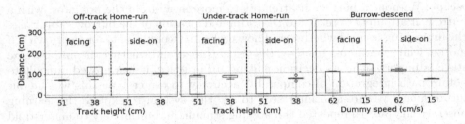

Fig. 10. The distance of the dummy from the robot at the reaction frame for every experimental scenario.

Burrow Descent. As described in the methods, we finally examined the evolution of visual cues relative to the 'burrow-descent' stage of the crab's behaviour. This assumes the crab has reached its burrow and now requires a stronger threat to make the costly move of entering it, and as described in the experimental methods, relies on different cues. When the robot is side-on to the predator, Fig. 9b illustrates that in a fast dummy attack the robot evasion is caused by the looming, while in a slow one the number of activated ommatidia will trigger the behaviour when the dummy is close enough to the robot. Following the experiments in [11], this could be a reasonable reaction scenario. In Fig. 10 we can see that the robot reacts in a greater distance from the predator in a fast attack rather than a slow one.

On the other hand, aligning the frontal axis to the track gives unexpected results. As for the under-track experiments, the looming and the activated ommatidia number are over-activated because of the twin-eye stitching. Figure 9a shows that slow dummies activate the looming, which is unrealistic, but can be explained if the robot sees the dummy with one eye in one time-step and with both eyes in the next. Additionally, Fig. 10 shows that there is great uncertainty about the reaction distance, while in the fast-dummy scenario the robot has very limited time to react due to the above confusion, leading to a very late response.

4   Discussion

We have demonstrated for the first time a robot design, the Robocrab, that is a suitable platform to explore hypotheses about the evasion behaviour of real crabs. It perceives the world through the same filter as a crab eye, can perform omnidirectional movement, and can be tested under comparable experimental conditions. In this paper we showed how this allowed us to test the effectiveness of different visual cues to initiate the evasion response. The behavioural intelligence framework allows this response to be modulated by the presence or absence of a burrow, and the estimated distance of the burrow, based on path integration. Qualitative results of the evasion behaviour are demonstrated with videos[1].

We plan to further explore the visual experience of the Robocrab during escape. Without a burrow, it attempts to move away from the predator; with a burrow, it attempts to return to the estimated location of the burrow. This raises the interesting issue of whether and how crabs may compensate for their own self motion while continuing to assess the visual threat, and whether the initial freeze stage is important in its decision-making. In practice, it has proved difficult to obtain path integration of comparable accuracy to the crab so this will be our future focus. We are also interested to use this system to explore the learning observed in crabs to repeated threatening stimuli, and ultimately to understand the underlying neural processes of this complex, coordinated behaviour.

References

1. Borenstein, J., Feng, L.: Gyrodometry: A new method for combining data from gyros and odometry in mobile robots. In: IEEE Robotics and Automation, vol. 1, pp. 423–428 (1996)
2. Hemmi, J.M.: Predator avoidance in fiddler crabs: 1. Escape decisions in relation to the risk of predation. Anim. Behav. **69**(3), 603–614 (2005)
3. Hemmi, J.M.: Predator avoidance in fiddler crabs: 2. The visual cues. Anim. Behav. **69**(3), 615–625 (2005)
4. Hemmi, J.M., Pfeil, A.: A multi-stage anti-predator response increases information on predation risk. J. Exp. Biol. **213**(9), 1484–1489 (2010)
5. Hemmi, J.M., Tomsic, D.: The neuroethology of escape in crabs: from sensory ecology to neurons and back. Curr. Opin. Neurobiol. **22**(2), 194–200 (2012)
6. Land, M., Layne, J.: The visual control of behaviour in fiddler crabs: I. Resolution, thresholds and the role of the horizon. J. Comp. Physiol. A. **177**(1), 81–90 (1995)
7. Land, M., Layne, J.: The visual control of behaviour in fiddler crabs: II. Tracking control systems in courtship and defence. J. Comp. Physiol. A **177**(1), 91–103 (1995)

---

[1] www.youtube.com/playlist?list=PLwwwvIsWO7M_SbmdCKl1siL9ajmSHzohw

8. Layne, J., Barnes, W.J.P., Duncan, L.M.J.: Mechanisms of homing in the fiddler crab *Uca* rapax 1. Spatial and temporal characteristics of a system of small-scale navigation. J. Exp. Biol. **206**(24), 4413–4423 (2003)
9. Layne, J., Barnes, W.J.P., Duncan, L.M.J.: Mechanisms of homing in the fiddler crab *Uca* rapax 2. Information sources and frame of reference for a path integration system. J. Exp. Biol. **206**(24), 4425–4442 (2003)
10. Nagatani, K., Tachibana, S., Sofne, M., Tanaka, Y.: Improvement of odometry for omnidirectional vehicle using optical flow information. In: Proceedings of the 2000 IEEE/RSJ International Conference on IROS 2000, vol. 1, pp. 468–473 (2000)
11. Oliva, D., Tomsic, D.: Visuo-motor transformations involved in the escape response to looming stimuli in the crab *Neohelice (= Chasmagnathus) granulata*. J. Exp. Biol. **215**(19), 3488–3500 (2012)
12. Raderschall, C.A., Magrath, R.D., Hemmi, J.M.: Habituation under natural conditions: model predators are distinguished by approach direction. J. Exp. Biol. **214**(24), 4209–4216 (2011)
13. Rojas, R., Förster, A.G.: Holonomic control of a robot with an omnidirectional drive. KI-Künstliche Intelligenz **20**(2), 12–17 (2006)
14. Sandeman, D.C.: Dynamic receptors in the statocysts of crabs. Fortschritte der Zoologie **23**(1), 185 (1975)
15. Smolka, J., Hemmi, J.M.: Topography of vision and behaviour. J. Exp. Biol. **212**(21), 3522–3532 (2009)
16. Smolka, J., Zeil, J., Hemmi, J.M.: Natural visual cues eliciting predator avoidance in fiddler crabs. Proc. Roy. Soc. Lond. B: Biol. Sci. **278**(1724), 3584–3592 (2011)
17. Tomsic, D., de Astrada, M., Sztarker, J., Maldonado, H.: Behavioral and neuronal attributes of short-and long-term habituation in the crab *Chasmagnathus*. Neurobiol. Learn. Memory **92**(2), 176–182 (2009)
18. Tsai, C., Tai, F., Lee, Y.: Motion controller design and embedded realization for mecanum wheeled omnidirectional robots. In: 2011 9th World Congress on Intelligent Control and Automation (WCICA), pp. 546–551. IEEE (2011)
19. Viboonchaicheep, P., Shimada, A., Kosaka, Y.: Position rectification control for mecanum wheeled omni-directional vehicles. In: The 29th Annual Conference of the IEEE Industrial Electronics Society, IECON 2003, vol. 1, pp. 854–859. IEEE (2003)
20. Yi, J., Zhang, J., Song, D., Jayasuriya, S.: IMU-based localization and slip estimation for skid-steered mobile robots. In: IEEE/RSJ International Conference on Intelligent Robots and Systems, IROS 2007, pp. 2845–2850. IEEE (2007)
21. Zeil, J., Hemmi, J.M.: The visual ecology of fiddler crabs. J. Comp. Physiol. A. **192**(1), 1–25 (2006)

MantisBot Changes Stepping Speed by Entraining CPGs to Positive Velocity Feedback

Nicholas S. Szczecinski[(✉)] and Roger D. Quinn

Case Western Reserve University, Cleveland, OH, USA
nss36@case.edu

Abstract. This paper demonstrates and analyzes how CPGs can entrain joints of a praying mantis robot (MantisBot) to positive velocity feedback resulting in a duration change of a leg's stance phase. We use a model of a single leg segment, as well as previously presented design techniques to understand how the gain of positive velocity feedback to the CPGs should be modulated to successfully implement the active reaction (AR) during walking. Our results suggest that the AR simplifies the descending control of walking speed, naturally producing the asymmetrical changes in stance and swing phase duration seen in walking animals. We implement the AR in neural circuits of a dynamic network that control leg joints of MantisBot, and experiments confirm that the robot modulates its walking speed as the simple model predicted. Aggregating the data from hundreds of steps in different walking directions show that the robot changes speed by altering the duration of stance phase while swing phase remains unaffected, as seen in walking animals.

Keywords: Synthetic Nervous System · Locomotion · Praying mantis · Central pattern generator · Active reaction · Positive feedback

1 Introduction

Insects control the velocity of their legs and joints when propelling their bodies [1, 2]. This makes sense for smoothly controlling the body's speed with respect to the ground (i.e. "body speed"). The duration of swing phase is generally constant and independent of body speed [3]. Thus, changing the body speed also changes the duty cycle of stepping, and different gaits emerge. At their fastest speeds, insects walk with an alternating tripod gait, in which three legs are on the ground at once. As speed decreases, the gait changes and more and more legs are in stance phase at the same time. This has the added benefit of distributing external forces among more legs, decreasing the support and propulsion force that each leg needs to provide [4].

How, then, is this asymmetrical phase duty accomplished in the animal? It is known that each joint of walking insects has its own central pattern generator (CPG), which contributes to ongoing rhythmic activity during walking [5, 6]. Related models of stick insect locomotor control networks accomplished this change by applying asymmetrical drive to the CPGs, changing the duration of retraction relative to protraction [7].

© Springer International Publishing AG 2017
M. Mangan et al. (Eds.): Living Machines 2017, LNAI 10384, pp. 440–452, 2017.
DOI: 10.1007/978-3-319-63537-8_37

Neurobiological studies with stick insects, however, suggest that descending signals that control walking speed target sensory pathways, rather than rhythm-generating signals [8]. More recent results from *Drosophila* with transfemoral amputation show that the CPGs in the cut leg do not change speed when the other legs do, but instead they oscillate at the maximum walking speed all the time [9]. This suggests that the CPGs in the leg entrain to leg-local sensory signals by prolonging stance phase.

One such leg-local sensory phenomenon that may contribute is the active reaction (AR). The AR is a positive-feedback reflex in which flexing the FTi joint causes excitation of the flexor muscle, followed by excitation of the extensor [10]. In reduced preparations, the AR is elicited by a combination of movement information from the femoral chordotonal organ (fCO), which monitors the FTi joint's rotation, and load information from the femoral campaniform sensilla (fCS) [11]. However, stimulating the fCO and fCS separately reduces the reliability of the reflex. Stimulating the fCO alone sporadically elicits an AR, while stimulating the fCS alone reliably elicits a delayed AR. The authors of [11] suggest that loading information may increase the gain of positive velocity feedback. This hypothesis led to the network design in this paper, which is tuned to reproduce results like those from the animal. We believe the AR is the CPG entraining to local positive feedback, and thus refer interchangeably to these phenomena throughout the text.

This paper uses previously presented design tools [12] to analyze how the AR affects the stability of walking at a continuum of body speeds, and how it should be

Fig. 1. MantisBot secured to a frame. The left middle leg is taking steps on a block. The green line on the block is 15 cm long, for scale. (Color figure online)

tuned. Section 2 presents our methods, including the neuron and synapse models, the control network connectivity, a simulation of a single leg segment from MantisBot, and how it enabled us to tune the AR. Section 3 presents our results, showing that the AR enables MantisBot (Fig. 1) to step at a continuum of speeds, while the duration of swing phase remains constant. Section 4 discusses the implications of these results for the study of the neural control of walking in both insects and robots.

2   Methods

2.1   Neuron and Synapse Models of the Synthetic Nervous System

MantisBot is controlled by a Synthetic Nervous System (SNS), a continuous-time dynamical neural simulation. SNS dynamics are computed with the neuromechanical simulator AnimatLab 2 [13]. The AnimatLab Robotics Toolkit enables the SNS to issue motor commands to a robot and receive sensory feedback over a serial connection. This system performs real-time neural control of MantisBot. The neural model is a simplified Hodgkin-Huxley model [7, 13, 14]. The details can be found in [12], but a summary is provided here.

Each neuron has three dynamical variables: V, the membrane voltage, m, the activation of a persistent sodium channel, and h, the deactivation of the same sodium channel. This model does not generate spikes. The dynamics of the model are as follows:

$$C_{mem}\frac{dV}{dt} = G_{mem} \cdot (E_{rest} - V) + g_{syn} \cdot (E_{syn} - V) + G_{NaP} \cdot m \cdot h \cdot (E_{NaP} - V)$$
$$\frac{dm}{dt} = \frac{m_\infty(V) - m}{\tau_m(V)}$$
$$\frac{dh}{dt} = \frac{h_\infty(V) - h}{\tau_h(V)}$$

where C is capacitance, G is a static conductivity, E is a static reference voltage (i.e. reversal potential), τ is a time constant, and subscript mem stands for membrane, syn stands for synaptic, NaP stands for persistent sodium, and ∞ stands for steady state. Both $m_\infty(V)$ and $h_\infty(V)$ are sigmoidal functions, but $m_\infty(V)$ increases monotonically with V, and $h_\infty(V)$ decreases monotonically with V. In addition, $\tau_m \ll \tau_h$, which means that initial rises in voltage are first positively reinforced by m, and then resisted as h decreases over time. Neurons communicate via synapses by changing their conductivity:

$$G_{syn} = \begin{cases} 0, & \text{if } V < E_{lo} \\ G_{max} \cdot \frac{V-E_{lo}}{E_{hi}-E_{lo}}, & \text{if } V \geq E_{lo} \text{ and } V \leq E_{hi} \\ G_{max}, & \text{if } V > E_{hi} \end{cases}$$

where V is the presynaptic neuron's voltage, and G_{max}, E_{hi}, and E_{lo} are properties of the synapse.

This model has several advantages for robotic control and animal modeling: (1) the persistent sodium channel causes bursting dynamics, in which the membrane voltage V rapidly increases, plateaus, and diminishes over time. This is necessary for the robust oscillations of central pattern generators [15]; (2) the leaky integrator dynamics enable the construction of dynamic networks such as differentiators and integrators [16]; (3) the simplicity of the dynamics enables direct tuning of parameter values [12].

2.2 Single Joint Model

We built a model of a single leg segment (i.e. the tibia) and its control network to examine how positive velocity feedback may affect the CPG's phase, and thus generate an active reaction. The network is shown in Fig. 2. It is mostly the same as our previously-presented network [17], so we will focus our explanation on the novel components. New descending pathways are shown in green, and new sensory pathways are shown in cyan.

Descending pathways affect motion by changing the local processing of sensory feedback [9, 18]. In our robot, the servomotor can control its speed by altering how negative velocity feedback is processed, so issuing a motor speed command via the "Comm. Speed" neuron effectively does the same thing. Comm. Speed receives excitatory synaptic input from the "Range of Motion" neuron, such that each joint's speed is scaled by its range of motion. This ensures that in swing phase, each servomotor returns to the AEP in the same amount of time. When an insect slows its body speed, it only slows the rotation of its joints when in stance phase [3, 4]. The "Leg Speed" neuron reduces the sensitivity of the Comm. Speed neuron via a modulatory disinhibitory pathway [12], which is only active if the leg is in stance phase (i.e. "Leg Strain" is active). This reduces the servomotor's commanded speed.

Tests with our model showed that if the servomotor is commanded to move more slowly but the CPG is not also slowed down, then the servomotor cannot achieve its full range of motion, resulting in incorrect stepping (See Fig. 4Bii). Therefore, the CPG must also receive feedback about the motion of the joint, so it can *entrain* with the motion of the leg. Experiments with stick insects showed that velocity feedback alone entrains the CPG about 50% of the time, but load feedback alone or simultaneous load and velocity feedback entrain the CPG every time [11].

The authors of [11] suggested that the load feedback may increase the gain of the velocity feedback. In order to cause an AR without load feedback, the velocity feedback must have some "baseline gain". This is provided by the "Baseline Gain" neuron in Fig. 2. When the leg is loaded, this neuron is further excited, increasing the amplification of the velocity signal, and evoking an AR. In order to cause an AR without velocity feedback, the "Ext. Velocity" and "Flx. Velocity" neurons must also receive some depolarizing input from the strain sensors. The "Strain Increasing" neuron, which primarily computes the rate of change of the leg strain, also supplies a small tonic current in steady state, which is sufficient to entrain the CPG. Because the duration of swing

Joint Controller Commands Servomotor Velocity and Uses Positive Velocity Feedback to Entrain CPG to Walking Speed

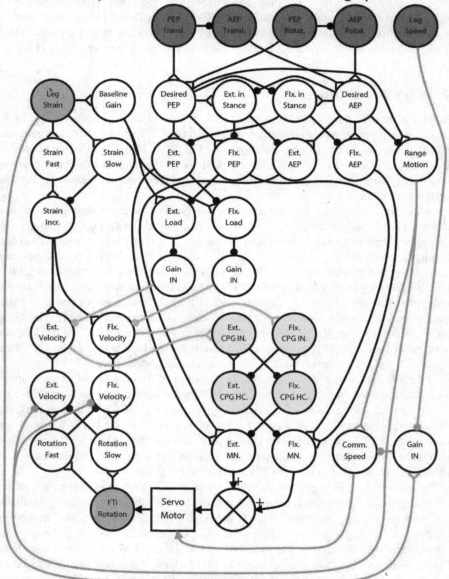

Fig. 2. Schematic of a control network for one joint of MantisBot. The overall function of the network is described in [17]. New descending pathways are drawn in green. New sensory feedback pathways are drawn in cyan. A full description is in the text. (Color figure online)

phase is the same no matter an insect's body speed [3], the swing phase velocity pathway is inhibited by the "Ext. PEP" or "Flx. PEP" neuron, depending on the direction of stepping, such that only the duration of stance phase is affected by velocity feedback.

2.3 δ and Tuning the Active Reaction

How can we tune the synaptic parameter values to obtain the insect-like CPG entrainment described in the previous section? The AR and other positive feedback mechanisms stabilize stance phase by ensuring that the leg does not enter swing until the leg is no longer propelling or supporting the body. Therefore, our sensory input must be strong enough to temporarily push the CPG into a stable stance phase (network shown in Fig. 3A). Simulations show that a weak stimulus will affect, but not halt oscillation, failing to produce a stable stance phase (Fig. 3Bi). However, a stronger stimulus will halt oscillation, and then cause an immediate transition once the stimulus disappears (Fig. 3Bii). This is precisely what is needed, but how do we quantify this effect?

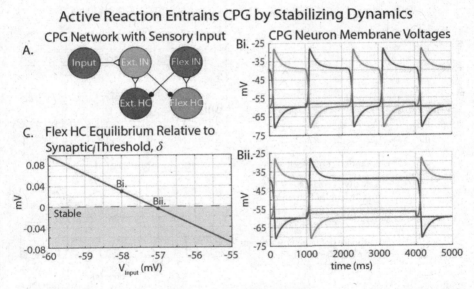

Fig. 3. Demonstration of how feedback may stabilize CPG dynamics. (A) A schematic of a four-neuron CPG model with one input neuron. (Bi) Weak sensory input will cause asymmetrical CPG bursting, but will not lock the phase. (Bii) Stronger sensory input will lock the CPG's phase. (C) Computing δ as a function of V_{input} reveals how strong the input must be to lock the CPG's phase.

CPG Entrainment Properties and Biological Data Determine the Strength of Velocity Feedback Necessary for Entrainment.

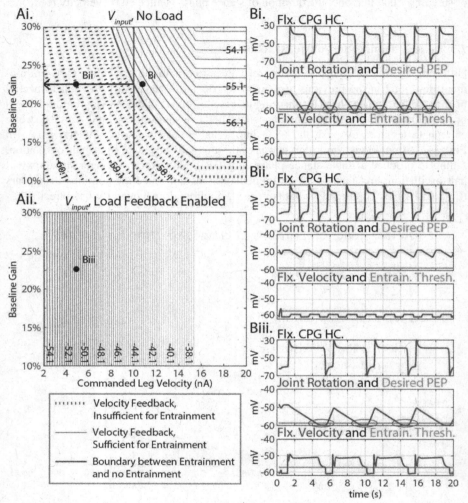

Fig. 4. Single-segment model informs the strength of positive velocity feedback to the CPGs. (Ai) To determine how strong velocity feedback should be to evoke an AR in 50% of cases, the magnitude of the velocity feedback, V_{input}, was calculated as the commanded leg velocity and baseline gain were changed. The contour plot reveals which parameter values produce stable entrainment. The baseline gain value that causes entrainment for 50% of commanded velocity values was chosen. (Aii) Another contour plot of the same system, but with load feedback during stance, verifies that the system always entrains to the sensory feedback. (Bi) High joint velocity and no load stably entrains the CPG. (Bii) Low joint velocity and no load feedback cannot stably entrain the CPG. (Biii) Low joint velocity and load feedback can entrain the CPG.

Previously we identified δ, a bifurcation parameter of our CPG that dictates its stability [12]. When $\delta > 0$, the CPG oscillates, and when $\delta < 0$, the CPG cannot oscillate. We can quickly calculate δ as a function of V_{input}, which dictates how large of a sensory input will stabilize the stance phase (Fig. 3C). For the network shown, $\delta < 0$ if $V_{input} > -57.1$ mV, meaning that we can use the positive feedback's amplitude to determine if it will stabilize stance phase.

Based on experiments with stick insects, the AR should occur 50% of the time without load feedback. We used our suite of design tools [19] to run a large batch of AnimatLab simulations of our single joint model to see how the leg velocity and baseline gain affected its ability to entrain to velocity feedback with and without load feedback. Figure 4Ai shows V_{input} plotted as a contour plot. If the CPG entrained with the sensory input, the contour is plotted as a solid line. If it failed to entrain, it is plotted as a dashed line. The boundary between these two regimes intersects with the command for 50% leg speed when the gain is 22.75%, meaning that the resting potential of the "Baseline Gain" neuron can be determined.

Finally, the tonic activity of the "Load Increasing" neuron must be determined such that the AR always occurs if load feedback is present, whether there is velocity feedback or not. Figure 4Aii shows another contour plot of the same simulation when load feedback is provided when the joint flexes. In this case, V_{input} no longer depends on the baseline gain, and is always large enough to stabilize stance phase. The plots in Fig. 4B illustrate this. In Fig. 4Bi, the servomotor is commanded to move fast in stance phase, and the velocity feedback alone is enough to ensure the joint rotation reaches the PEP. In Fig. 4Bii, the servomotor is commanded to move slowly in stance phase, and the velocity feedback does not surpass the entrainment threshold, and the range of motion is incomplete. In Fig. 4Biii, however, load feedback is enabled, and the same speed commanded in Fig. 4Bii entrains the CPG, and the joint rotation reaches the PEP. Note that "swing phase", or positive joint rotation, is the same no matter the conditions.

2.4   Robot and Experiments

The data in Figs. 5 and 6 were collected from experiments with our robot, MantisBot (Fig. 1). MantisBot's thorax was bolted to a stand that supported its weight. The left middle leg was then commanded to walk on a slippery block for 60-second trials. This is similar to the reduced preparation used to study insect locomotion [20]. In each trial, the leg was commanded to walk in a different direction ($-30°$, $-15°$, $0°$, $15°$, or $30°$ of body yaw per step) and at a different speed (15, 11.25, 7.5, 5.625, or 3.75 cm/s, which correspond to the stance phase foot speed that is 100%, 75%, 50%, 37.5%, and 25% of swing phase foot speed). Force feedback from a strain gage on the trochanter, and speed and position feedback from the motors coordinate the motion of the separate servomotors into directed locomotion [17].

CPG Entrainment Properties Enable MantisBot to Asymmetrically Change the Speed of Stance Phase

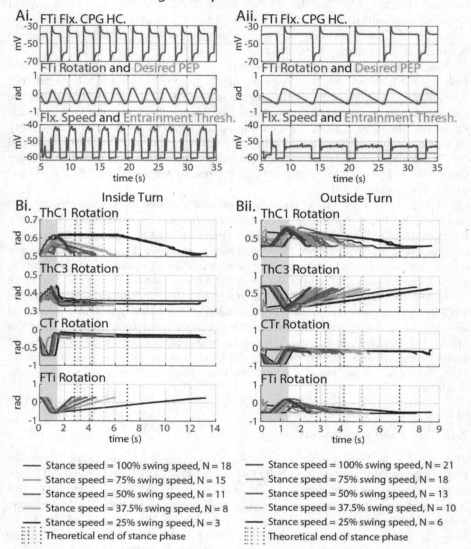

Fig. 5. The results from the single segment model apply directly to MantisBot, enabling it to change speed while the duration of swing phase remains constant. (A) Data from the robot, formatted as in Fig. 4B. The same entrainment occurs. (B) Data from all steps taken in specific directions. Each step is plotted, starting in swing phase at 0 s. The average swing phase duration is shaded in gray. Each walking speed is plotted in a different color. Dotted vertical lines indicate when lines of that color should end, based on the commanded walking speed. Note that each joint's range of motion does not depend on its speed in stance phase. (Color figure online)

MantisBot Changes Speed by Changing the Duty Cycle of Stepping

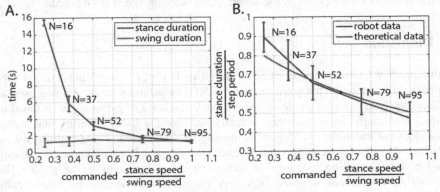

error bars are ± 1 standard deviation

Fig. 6. All 279 steps taken by the robot were aggregated to analyze gross trends in the data. (A) No matter the stance phase speed, the duration of swing phase is constant. At the robot's fastest speed, stance phase and swing phase have equal durations. As the robot walks more slowly, the duration of stance phase increases, as seen in animals. (B) The fraction of time spent in stance phase changes as intended with walking speed.

3   Results

Figure 5A shows data from single leg walking trials with MantisBot. They show the same data as in Fig. 4B. These plots show that the joints on MantisBot entrain to sensory input as in the single-joint model. No matter the commanded joint velocity, the CPG is entrained by sensory input to prolong stance phase and ensure the joint rotates all the way to its PEP. In contrast, swing phase's duration is constant.

Figure 5Bi and Bii each show plots of every step that MantisBot took in a particular direction. Each commanded speed is plotted in a different color. Each trace begins with the end of stance phase. Swing phase takes 1.4 s on average to complete, shaded in gray. Stance phase takes place from the end of swing phase to the dotted line of the matching color. For instance, the stance speed for traces in blue is 100% of the swing speed, meaning that stance phase should end 1.4 s after swing phase. One can see that the blue traces end near to the blue dotted line. Each speed's number of steps is listed in the key at the bottom of Fig. 5B. Note that each joint's range of motion is unaffected by the commanded speed, that is, that traces have the same amplitude, regardless of color.

Figure 6 aggregates all of the steps that MantisBot took (N = 279), bins them by commanded stance speed as a fraction of swing speed. Figure 6A shows that the duration of swing phase is constant no matter the body speed. This is critical for the emergence of the continuum of gaits seen in insects [4]. Figure 6B shows the duty cycle and the theoretical value, that is, if swing phase were exactly constant over all speeds, and the foot moved at exactly the commanded speed. The robot data is within one standard deviation for all stance speeds above 37.5% of the swing speed.

This suggests that the robot steps as intended in a gross sense, adjusting the proportion of time spent in stance phase as the commanded walking speed is changed.

4 Discussion

In this paper we used our suite of SNS design tools [19] to tune a positive velocity feedback reflex to model the active reaction (AR). The behavior of the AR was used as the intended behavior of a single-segment model, and sensory pathways were tuned to produce this result. These pathways actually stabilize the CPG, ensuring that stance phase does not end unless the leg stops moving or is unloaded. These pathways were added to MantisBot's stepping controller, enabling MantisBot to change its walking speed without changing its range of motion or the duration of swing phase. The resulting duty cycle of stepping was very close to what should be theoretically expected, showing that positive velocity feedback can indeed entrain the CPGs and stabilize stance phase over a continuum of body speeds.

The asymmetrical change in stance and swing duration as body speed changes is critical to producing the continuum of gaits seen in insects [4]. This continuum contributes to supporting external forces, since placing more legs on the ground at once distributes the forces over more legs. A robot must be able to react to external conditions, including external forces. For example, an agricultural robot's weight may increase as it harvests more crops, or its weight may decrease as it depletes fuel. Adapting to these changes is critical to real-world operation, and the work in this paper is a first step towards solving this problem via gait modification.

The network in this paper may also serve as a hypothetical structure underlying several neurobiological findings. It is known that the descending pathways that control an insect's walking speed and direction primarily affect sensory processing [8, 9, 18]. Our speed-control descending pathways limit the velocity of the servomotors; the servomotor, in turn, accomplishes this change by modulating negative velocity feedback in its controller. Ultimately, both the animal and MantisBot may change speed the same way, by increasing the strength of sensory feedback that resists motion in stance phase.

In this paper we called the AR a form of positive feedback. This phrasing is what is used in the neurobiological literature, because initiating stance phase motion, such as flexing the FTi joint or loading the leg, initiates and reinforces stance phase [10]. This is not the same as continuous positive velocity feedback acting on the motor neurons directly, which is also reported in the literature [21]. This mechanism, termed Local Positive Velocity Feedback (LPVF), can be used to minimize internal torques in an insect or robot with multiple legs on the ground at once [22]. In practice, this is implemented by adding the current position of the limb, passed through a high-pass filter, to the motor command. This is precisely what the "Flx. Velocity" and "Ext. Velocity" neurons calculate in our network (see Fig. 2). LPVF may be readily implemented in the future when MantisBot walks freely, with the legs mechanically coupled through the ground.

Our results highlight the importance of local sensory feedback for coordinating rhythmic motion. Recent findings in *Drosophila* show that the CPG that controls leg levation and depression will oscillate at its fastest possible speed (i.e. a 50% duty cycle) if the distal portion of the leg is amputated [9]. This suggest that in the animal, local sensory feedback is critical for prolonging stance phase when walking at slower speeds. Our results show how sensory feedback may bifurcate the CPG system into a stable state, ensuring that stance phase is the proper duration for the speed commanded by the brain. We believe that this work is more flexible and biologically accurate than providing asymmetrical drive to the different halves of the CPG [7], especially when considering that speed-controlling descending commands appear to target sensory feedback processing, not the CPG networks themselves [8].

References

1. Cruse, H.: Which parameters control the leg movement of a walking insect?: I. Velocity control during the stance phase. J. Exp. Biol. **116**, 343–355 (1985)
2. Gruhn, M., von Uckermann, G., Westmark, S., et al.: Control of stepping velocity in the stick insect Carausius morosus. J. Neurophysiol. **102**, 1180–1192 (2009). doi:10.1152/jn.00257.2009
3. Gabriel, J.P., Büschges, A.: Control of stepping velocity in a single insect leg during walking. Philos. Trans. A Math. Phys. Eng. Sci. **365**, 251–271 (2007). doi:10.1098/rsta.2006.1912
4. Foth, E., Graham, D.: Influence of loading parallel to the body axis on the walking coordination of an insect – I. Ipsilateral effects. Biol. Cybern. **47**, 17–23 (1983). doi:10.1007/BF00340065
5. Ryckebusch, S., Laurent, G.: Rhythmic patterns evoked in locust leg motor neurons by the muscarinic agonist pilocarpine. J. Neurophysiol. **69**, 1583–1595 (1993)
6. Büschges, A., Schmitz, J., Bässler, U.: Rhythmic patterns in the thoracic nerve cord of the stick insect induced by pilocarpine. J. Exp. Biol. **198**, 435–456 (1995)
7. Daun-Gruhn, S., Tóth, T.I.: An inter-segmental network model and its use in elucidating gait-switches in the stick insect. J. Comput. Neurosci. (2010). doi:10.1007/s10827-010-0300-1
8. Sauer, A.E., Büschges, A., Stein, W.: Role of presynaptic inputs to proprioceptive afferents in tuning sensorimotor pathways of an insect joint control network. J. Neurobiol. **32**, 359–376 (1997). doi:10.1002/(SICI)1097-4695(199704)32:4<359:AID-NEU1>3.0.CO;2-5
9. Berendes, V., Zill, S.N., Büschges, A., Bockemühl, T.: Speed-dependent interplay between local pattern-generating activity and sensory signals during walking in Drosophila. J. Exp. Biol. (2016). doi:10.1242/jeb.146720
10. Bässler, U.: Functional principles of pattern generation for walking movements of stick insect forelegs: the role of the femoral chordotonal organ afferences. J. Exp. Biol. **136**, 125–147 (1988)
11. Akay, T., Büschges, A.: Load signals assist the generation of movement-dependent reflex reversal in the femur-tibia joint of stick insects. J. Neurophysiol. **96**, 3532–3537 (2006)
12. Szczecinski, N.S., Hunt, A.J., Quinn, R.D.: Design process and tools for dynamic neuromechanical models and robot controllers. Biol. Cybern. (2017). doi:10.1007/s00422-017-0711-4
13. Cofer, D.W., Cymbalyuk, G., Reid, J., et al.: AnimatLab: a 3D graphics environment for neuromechanical simulations. J. Neurosci. Methods **187**, 280–288 (2010). doi:10.1016/j.jneumeth.2010.01.005

14. Hodgkin, A.L., Huxley, A.F., Katz, B.: Measurement of Current-voltage relations in the membrane of the giant axon of Loligo. J. Physiol. **116**, 442–448 (1952)
15. Selverston, A.I., Moulins, M.: Oscillatory neural networks. Annu. Rev. Physiol. **47**, 29–48 (1985)
16. Szczecinski, N.S., Hunt, A.J., Quinn, R.D.: Design methodology for synthetic nervous systems that control legged robot locomotion. Front. Neurorobot. (2017, in review)
17. Szczecinski, N.S., Getsy, A.P., Martin, J.P., et al.: MantisBot is a robotic model of visually guided motion in the praying mantis. Arthropod. Struct. Dev. (2017). doi:10.1016/j.asd.2017.03.001
18. Martin, J.P., Guo, P., Mu, L., et al.: Central-complex control of movement in the freely walking cockroach. Curr. Biol. **25**, 2795–2803 (2015). doi:10.1016/j.cub.2015.09.044
19. Szczecinski, N.S., Hunt, A.J., Quinn, R.D.: Design process and tools for dynamic neuromechanical models and robot controllers. Biol. Cybern. (2017). doi:10.1007/s00422-017-0711-4
20. Bässler, U.: The femur-tibia control system of stick insects–a model system for the study of the neural basis of joint control. Brain Res. Rev. **18**, 207–226 (1993)
21. Schmitz, J., Bartling, C., Brunn, D.E., et al.: Adaptive properties of hard-wired neuronal systems. Verh dt zool **88**(2), 95–105 (1995)
22. Schmitz, J., Schneider, A., Şchilling, M., Cruse, H.: No need for a body model: positive velocity feedback for the control of an 18-DOF robot walker. Appl. Bionics Biomech. **5**, 135–147 (2008). doi:10.1080/11762320802221074

EvoBot: Towards a Robot-Chemostat for Culturing and Maintaining Microbial Fuel Cells (MFCs)

Pavlina Theodosiou[1], Andres Faina[2], Farzad Nejatimoharrami[2], Kasper Stoy[2],
John Greenman[3], Chris Melhuish[4], and Ioannis Ieropoulos[1(✉)]

[1] Bristol BioEnergy Centre, Bristol Robotics Laboratory, Bristol, UK
`ioannis.ieropoulos@brl.ac.uk`
[2] Robot, Evolution, and Art Lab (REAL), IT University of Copenhagen, Copenhagen, Denmark
[3] Faculty of Health and Applied Sciences, University of the West of England, Bristol, UK
[4] Bristol Robotics Laboratory, Frenchay Campus, Coldharbour Lane, Stoke Gifford, Bristol, UK

Abstract. In this paper we present EvoBot, a RepRap open-source 3D-printer modified to operate like a robot for culturing and maintaining Microbial Fuel Cells (MFCs). EvoBot is a modular liquid handling robot that has been adapted to host MFCs in its experimental layer, gather data from the MFCs and react on the set thresholds based on a feedback loop. This type of robot-MFC interaction, based on the feedback loop mechanism, will enable us to study further the adaptability and stability of these systems. To date, EvoBot has automated the nurturing process of MFCs with the aim of controlling liquid delivery, which is akin to a chemostat. The chemostat is a well-known microbiology method for culturing bacterial cells under controlled conditions with continuous nutrient supply. EvoBot is perhaps the first pioneering attempt at functionalizing the 3D printing technology by combining it with the chemostat methods. In this paper, we will explore the experiments that EvoBot has carried out so far and how the platform has been optimised over the past two years.

Keywords: EvoBot · Microbial Fuel Cells · 3D-printer · Cathode · Rep-Rap

1 Introduction

The term robot has been defined by the International Organization for Standardization (ISO) as "*a programmable actuated mechanism with a degree of autonomy, which is able to move within its environment in order to perform intended tasks*". With this in mind a Rep-Rap 3D-printer was modified into a laboratory robot in order to inoculate and maintain Microbial Fuel Cells (MFCs), in a similar manner to the maintenance of bacterial cultures within a chemostat, and perform interactive experiments with minimum human supervision; this marked the start of EvoBot.

The first commercially available laboratory robot, the Robot Chemist, was marketed in 1959 with the aim to automate wet-chemical analytical procedures [1]. This was the first step towards laboratory automation practices. Twenty years later, lab robots were introduced into the pharmaceutical industry for drug analysis [2]. After a long period of adoption, robots are today playing a significant role in all aspects of laboratory

© Springer International Publishing AG 2017
M. Mangan et al. (Eds.): Living Machines 2017, LNAI 10384, pp. 453–464, 2017.
DOI: 10.1007/978-3-319-63537-8_38

procedures; from routine chemical analysis to drug development and DNA analysis. Autonomous laboratory robotics have generally advanced lab procedures since they offer: accuracy, speed, convenience and they are cost effective compared to labour cost. One research area that can directly benefit from such advances in laboratory robotics is Microbial Fuel Cells (MFCs).

Microbial Fuel Cells (MFCs) are bio-electrochemical devices that use microorganisms to generate electrical current through the oxidation of organic matter (i.e. anaerobic sludge, urine, acetate). For the last 17 years, MFCs have been closely associated with autonomous robots due to their capabilities of providing energy autonomy to biologically inspired robots such as Gastrobot [3], Ecobot I, II, III [4–6] and Row-bot [7]. Even though the above examples are proof-positive that energy autonomy is both plausible and feasible through the use of MFCs, research is still needed to reach full MFC potential and increase the capabilities of these artifacts, which are known as Symbots [8].

Due to the increasing demand for alternative ways of powering robots, MFC research has recently attracted a lot of interest that led to breakthrough discoveries [9, 10]; the research emphasis is now on the development of the next generation of improved and optimised MFCs for maximum power production. In this line of thought, EvoBot attempts to implement the 3D RepRap technology as a mechanism for printing organic and inorganic substrata as well as accurately dosing the biofilm (microbial community adhered to the anode electrode) of the fuel cell with organic matter in the same manner as a chemostat. The chemostat is a widely-used apparatus for culturing cells [11] that enables the experimental control of cell growth rate, in order to study the adaptive evolution of microbes and achieving dynamic steady-states [12]. The cell culture grows and evolves within the chemostat in the presence of a continuous flow of nutrients. The vessel retains a constant volume as an overflow system is in place (Fig. 1). The optimum aim of EvoBot is to enforce, monitor and interact with evolving systems and eventually produce optimally evolved/adapted MFCs which will have improved energy generation capabilities.

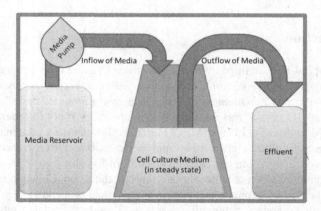

Fig. 1. Graphical representation of the chemostat method/biofilm reactor for culturing bacterial cells.

EvoBot has been developed to perform similarly to the chemostat and the work has been broken down into different phases. During Phase 1, EvoBot demonstrated interface and interconnection with an MFC. Shortly after, during Phase 2 a long term experiment with 9 MFCs was performed. EvoBot was maintained by using a feedback loop to control the nutrient supply rate to the fuel cells based on reduction in power below a threshold [13]. Now EvoBot is in Phase 3 where it has been optimised to host more MFCs on to the platform, using two direct-current (DC) pumps to supply media using and two syringes to distribute the media into the cells. The main aim of this work was to demonstrate the development of EvoBot as a modern-day robotic biofilm bioreactor continuous culture system (based on the "chemostat" approach) for studying MFCs and their optimisation for producing power.

2   EvoBot

EvoBot is a modular, open source, versatile, and affordable robotic platform, which has been developed to perform liquid handling experiments; in the context of the EVOBLISS EU project (FP7-ICT). Evobot has been used in seven different laboratories around the world for diverse applications, such as interaction with MFCs [13], moving droplets, and improving the quality of artificial life experiments and performing OCT scans [14].

EvoBot consists of an actuation layer on top, an experimental layer in the middle, and a sensing layer at the bottom as can be seen in Fig. 2. The actuation layer comprises the robot head and modules mounted on it. The modules are plugged into the head and are usually designed to perform an action on the experiment. However, they may also have sensor functionality e.g. OCT (optical coherence tomography) scanning, an imaging technique which allows for optical sectioning of the sample. Such a sample can be the electroactive anodic biofilm of an active MFC with transparent anode chamber.

Fig. 2.   Overview of the actuation, experimental, and sensing layer of the robot. EvoBot's actuation layer consists of a moving head on which various modules, such as syringe modules can be mounted. In this figure the experimental layer accommodates different vessels, and the camera at the bottom acts as the sensing layer by collecting experiment data.

The head which holds the modules can be moved in the horizontal plane. EvoBot's modularity allows for support of modules of different kinds for various applications. The experiment-dependent modules could entail syringe modules for liquid dispensing or aspirating, grippers to move the containers over the experimental layer or dispose dirty containers, an OCT scanner module to perform OCT scans, an extruder module to 3D print MFC parts, and other potential experiment-specific tools.

The experimental layer consists of a transparent Poly methyl methacrylate (PMMA) sheet on which reaction vessels (e.g. Petri dishes, well plates, beakers etc.) and/or MFCs are positioned. The actuation layer interacts with the experimental layer by filling or emptying a specific volume to/from a syringe, washing a syringe, or/and disposing dirty containers.

The robot frame is built from Aluminium profiles, and the experimental layer and actuation layer are mounted on it. The layers can be easily moved up or down on the robot frame with a cam lever mechanism.

Configuring EvoBot for different experiments is easy, as different types of modules can be easily removed or plugged at the appropriate position. The head is responsible for moving the modules in the x-y plane, while the modules have motors to move vertically. The robot head can accommodate syringe modules to aspirate or dispense liquid at 17 potential positions, and up to 11 syringes can be used simultaneously as the socket positions overlap with adjacent ones.

EvoBot's design is based on open-source 3D printers using an Arduino board and a RAMPS shield. This electronics design allows building on existing software for open-source 3D printers. A computer controls the robot by communicating with the Arduino through serial communication. The Arduino is connected to the RAMPS shield. The RAMPS shield controls the three stepper motors to move the robot head, the modules mounted on the robot head, as well as the two DC pumps.

In the experiment presented below EvoBot was primarily used for MFC inoculation and maintenance as well as a characterization tool akin to chemostat. This is where the novelty of the project lies as the interaction of biochemical systems with the 3D printing technology can lead to a Robot-Chemostat.

3   Materials and Methods

Hardware

The EvoBot base was elongated from 620 cm (EvoBot V 0.5) to 1000 cm (EvoBot V 1.0). The syringe modules were adapted to host 20 ml syringe rather than 5 ml syringe (EvoBot V 0.5), the pumps were placed on the external profile of the robot rather than on the robot head and a dispensing module was added on board. The camera was placed on the side of the robot, to record the feeding/maintenance of the MFCs.

Microbial Fuel Cells

Interface and Interconnection Experiment A standard analytical size two-chamber cubic MFC with anode and cathode liquid capacity of 25 mL and electrode total surface area of 270 cm^2 each, was placed on the EvoBot experimental arena. The anode and cathode electrode were connected with a 3 kΩ load (optimum load for that specific MFC).

Adaptation/Chemostat Mimicking Experiment A set of 18 Nanocure® printed MFCs were adapted for this experiment (Fig. 3). The anode of the MFCs was constructed as previously described [6]. A 3 mm custom made terracotta flat sheet, made following the same technique as previously reported, was used as the membrane [15]. For cathode electrode, a 5 ml of alginate based custom made conductive paste was deposited manually using a syringe directly onto the membrane and the mixture was solidified after 24 h exposure to air (Fig. 3). After the inoculation period, the MFCs were loaded with 1 kΩ.

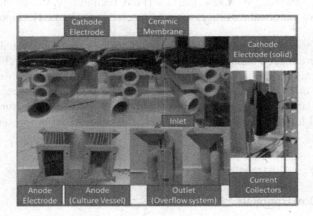

Fig. 3. Nanocure® printed Microbial Fuel Cells (MFCs). The anode compartment of the MFC acted like the culture vessel of a biofilm reactor/chemostat. Fresh medium entered through the inlet, whilst effluent was overflowing through the outlet.

Inoculation, Media and Delivery of Media

Interface and Interconnection Experiment The MFCs were inoculated with activated sludge supplied by Wessex Water Scientific Laboratory (Cam Valley, Saltford, UK) supplemented with 5 mM sodium acetate as the carbon energy (CE) source (Sigma-Aldrich, Dorset, UK). A DC pump was connected to the robot board and was used for pumping the enriched sludge into the anode chamber. The DC pump was calibrated and the activation time of the pump dictated the liquid volume which was dispensed. For this experiment, in order to fill the 50 ml beakers the pump was activated once a day for 148 s in total.

Adaptation/Chemostat Mimicking Experiment Initially the fuel cells were inoculated with activated sludge and then with effluent from active urine fed MFCs. After a total of two weeks inoculation, the cells left to starve for a week to ensure that no traces of urine were left in the anode. The experiment started with the introduction of sterile sodium acetate medium to the 9 MFCs and with sterile casein medium to the other 9

MFCs. The pH of the media was buffered to 7. The media were contained within two bottles; each bottle was connected to a DC pump and had an air filter (Fig. 6B). The tubing of the pump leads to the dispensing module of the robot (Fig. 6A, D). For each experimental cycle the pump was set to deposit and distribute the required amount of feedstock (overall volume 50 ml) to the specified beakers around the arena and then the syringe was collecting the liquid from the beaker and distributed it to the anode chambers of the fuel cells.

Data Acquisition

Interface and Interconnection Experiment Electrical output (voltage) was measured in real time using a PicoLog ADC-24 interface (Pico Technology, Cambridgeshire, UK). Both the EvoBot and the data-logger were connected in the same HP laptop computer.

Adaptation/Chemostat Mimicking Experiment The cells were individually connected and the data were logged using two multi-channels Agilent 34972A, LXI Data Acquisition/Switch Units (Farnell, UK). Both data loggers are connected to the computer however only one of them communicates with the Python interface, the other is recording the data using the Benchlink Datalogger 3 software to ensure continuous monitoring and logging of the MFC's performance even in the unforeseen event of a software crash.

3.1   Software

Development Environment

Interface and Interconnection Experiment Laboratory Virtual Instrument Engineering Workbench (LabVIEW) from National Instruments, Berkshire, UK was used to create the applications that interacted with the real-time data produced from MFCs. Using LabVIEW a multi-layer program was compiled in which a function was written that collected the data from the Picolog data logger and output that data as a string of values. LabVIEW was sampling the Picolog file every 1 min for the MFC voltage reading. The feeding scripts were written in Python language (Python 3.6.0). In the case where the MFC voltage dropped below the preset threshold limit that was set in LabView, the python script was activating the robotic head (actuation layer) to move above the MFC anode and deposit by pump 12.5 mL of inoculum. After the disposing of the liquid, the robot head homed itself.

Adaptation/Chemostat Mimicking Experiment Python script activates the pumps for a certain time frame in order to allow all the liquid to move from the end of the media tube to the tip of the dispensing module nozzle. Every 24 h the robot initiates the pumps to fill the pre-specified beakers with the media and when full the syringe module draws the liquid (5 ml) from the beakers and it starts dispensing it to the anode compartment of the MFCs. The coordinates of each item on the experimental layer is stored into Python dictionaries.

4   Results and Discussion

4.1   Interface and Interconnection of EvoBot with an MFC

In the first phase of the study, the work focused on investigating whether the voltage of an MFC can trigger the robotic arm of the EvoBot platform. To demonstrate this, a new abiotic MFC was set-up as described above. At the outset, the MFC was dry and contained no bacteria (therefore zero Voltage was recorded). The null voltage actuated the arm to move and deliver approximately 12.5 mL of activated sludge into the anode chamber (Fig. 4B). The 12.5 mL were deposited to the anode chamber of the MFC using a DC pump that was connected to a 50 ml bottle of activated sludge (Fig. 4C).

Fig. 4.   Experimental set-up of Phase 1 Experiment (Interface and Interconnection). [A] The robot and data-logger are connected with the laptop computer which acts as a real-time voltage display. **[B]** The photo shows the anode of the MFC, pressed against the membrane using a rectangular piece of inert styrene material and **[C]** the DC pump connected to the bottle containing the sludge.

After the robot had introduced 12.5 mL of activated sludge in the MFC (Fig. 5A), the voltage increased and stabilised at approximately 27 mV. Once the command "Feeding Threshold" was set to 20 mV the original experiment planned to let the voltage decrease as the microbes consumed the food in the sludge. However, since the time period for a freshly inoculated MFC, to fall under 20 mV, can take up to 24 h, 10 mL of sludge was manually removed to simulate food consumption (Fig. 5B); due to the disturbance when removing the anolyte a spike in voltage was observed. However, once the voltage decreased below the preset feeding threshold the robot arm was activated to move to the position above the MFC anode inlet and initiated the DC pump for 37 s, thus introducing 12.5 mL of new sludge medium into the MFC.

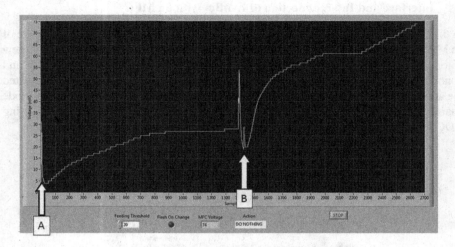

Fig. 5. Screen capture of the voltage increase. [A] Initially the MFC output was zero as the MFC was abiotic, this triggered the robotic arm to move to the fuel cell and activate the pump to feed the preset amount of inoculum (12.5 mL) to the anode. [B] At point B, 10 mL of the anolyte were manually removed to simulate food consumption and as a consequence, the voltage dropped. Since the MFC output fell below the preset threshold of 20 mV the robotic arm was activated and another 12.5 mL of inoculum were deposited to the anode; as a result, the voltage of the MFC continued to increase.

This particular experiment demonstrated the feedback loop between the MFC output (the voltage and power to maintain that voltage) produced by the microorganism living inside an MFC and the robotic controller, was able to activate a python script that initiated a set of given tasks to increase the voltage above the minimum threshold. This novel use of the EvoBot demonstrated that the robot was able to maintain the MFCs by reacting to the carbon energy source depletion within the anode which reflects on the power output. To sum up, this experiment successfully demonstrated for the first time that the electrical output recorded from an MFC in real-time can establish a feedback loop with the EvoBot.

4.2 Adaptation/Chemostat Mimicking Experiment

Following the successful demonstration of the interconnection between MFCs and EvoBot, the research moved onto Phase 2 using the platform to optimise the environmental parameters for faster growth of the cells on the electrode and maximum power transfer [13]. This work focused on optimising the substrate parameters that will be used to feed the MFCs and ultimately to improve their power performance. It is worth mentioning that the MFCs described here, are exactly the same as the ones used on EcoBot-III [6], which (as with all MFCs) are capable of relatively low, but continuous, levels of power. This is why for practical applications collectives of MFCs (i.e. stacks) are used and in the case of EcoBot-III the whole robot, including the central micro-controller, sensors and actuators, were powered by the 48 MFCs aboard (ca. 50uW/MFC). EvoBot platform was able to inoculate the otherwise empty/abiotic MFCs and note the power profiles developed over time with only EvoBot supporting the cells. This was the first long-term example of using the EvoBot platform for maintaining MFCs and operating interactive MFC experiments.

Having EvoBot driven experiments provides the fuel cells with automated feeding and hydration pulses, which are dictated by the voltage threshold, as well as continuous monitor and maintenance of the MFC, eliminating human intervention. This line of work provides a better understanding of what was needed in order to evolve the robot and increase its experimental capabilities. One of the needs was the enlargement of EvoBot's experimental arena, in order to hold more MFCs, allowing the parallel operation of more experiments as well as improvement of the syringe unit to avoid cross contamination. These issues were addressed and improved before performing the Phase 3 Adaptation/Chemostat mimicking experiment. The improved robot gave us the possibility to perform two different experiments using continuous feeding cycles (called the "chemostat" approach). The experimental set-up is presented below (Fig. 6).

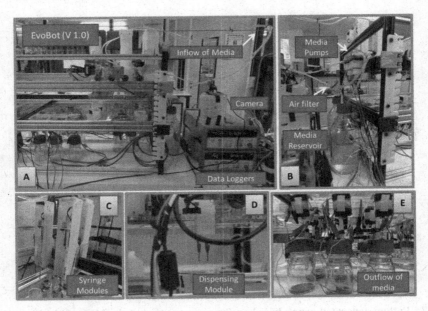

Fig. 6. EvoBot Version 1.0: Optimised based on the Chemostat approach for Phase 3 experiments. [A] EvoBot was able to host two experiments that count in total 18 MFCs. The data loggers were connected to the MFCs and the computer, and a camera was monitoring the experiment 24/7. **[B]** Similar to the chemostat the media reservoir was connected to the DC pumps and the tubes were connected to the **[D]** dispensing module. **[C]** The syringes were set to draw the liquid and dispense it to the culture vessel. **[E]** The waste perfusate was collected from the bottom into bottles (outflow stream) for further analysis.

In microbiology, the chemostat is the most common type of continuous culture device. It is an open system where its culture vessel maintains a constant volume to which fresh medium is added at a constant rate and an equal volume of spend culture is removed at the same rate and as a result, the growth rate is equal to the dilution rate ($\mu = D$) and the system reaches dynamic equilibrium. In the natural world, a plethora of biofilms are formed in continuous or periodic nutrient replenishing conditions and can be regarded as open systems as well [16]. Because the anodic biofilm electrodes are made from perfusible carbon veil the MFC falls within the general category of matrix perfusion systems. In the literature, it has been reported that such systems have similarities to a chemostat model [16]. Thus in this paper, we refer to the anode compartment of the fuel cell as the culture vessel (chemostat analogy).

Even though the system in the current study was a batch-fed culture system, i.e. not a continuous flow system like a conventional chemostat, the results nevertheless have shown that slow transitional repeat states can be maintained following further CE source supply following depletion (Fig. 7). These promising findings and this continuing line of work can provide useful insights into repeat batch fed microbial fuel cell systems, their behaviour as well as understanding of how to increase or optimize their power production capabilities.

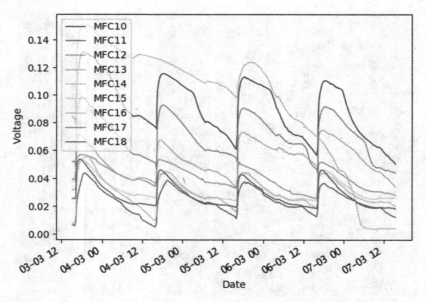

Fig. 7. Python generated graph based on the power output of the MFCs fed with casein. The graph shows the data produced within five days of EvoBot feeding the MFCs. The data show the reproducibility and stability of these cells when the fresh medium is fed to the biofilm daily. The non-periodic feeding of MFC 14, and consequent deterioration of performance was due to the liquid level of feedstock being below the lowest reach of the syringe needle; resulting in abnormal dispensing of volume (less frequent feeding). Undesired as this was, it demonstrates how depletion of the CE source affects bacterial metabolism, and therefore power output. In other words, it demonstrates the value-added of the automated maintenance, provided by EvoBot.

5   Conclusions

Robotic systems have many advantages for rapidly comparing MFC with different intrinsic factors (e.g. species of colonizing microbes and their ecological proportions; size/shape/design of MFC) as well as extrinsic physicochemical factors such as temperature, pH, pO_2, redox, osmotic pressure, type and concentration of nutrients. Biofilm reactors may offer many advantages compared to conventional chemostats so the two approaches (robot and biofilm bioreactor) can advance the biofilm and MFC research respectively. Thus, in order to continue developing the robot-chemostat, we need to more fully address the parameters of controlled physicochemical environments. At present, a pH module is being tested, which will be incorporated into the experiments. This will automatically test the inlet pH as well as the outlet pH of the anodic chamber in real time, to help understand the relationship between power output and anolyte pH. Also, it will give a new insight into the chemical and microbial transformations that take place within each chamber. Our aim is to continue using EvoBot to perform interactive experiments that will produce high quality reproducible data from multiple comparisons of conditions across a wide range of the physicochemical realm; data that would be difficult to achieve through conventional manual experimentation on MFCs and electroactive biofilms.

This work demonstrates the potential of using the EvoBot, an open-source modified RepRap 3D-printer, for the inoculation, maintenance and study of Microbial Fuel Cell (MFCs) systems. Furthermore, it presents the possibilities that this type of robot can be developed along the lines of robotically controlled environments integrated with MFC-based bioreactors with similarities as well as important differences to the well characterized planktonic chemostat. In addition, these experiments are the precursors to the development of a new class of living robots (Symbots), which can enforce, monitor and interact with evolving systems, such as MFCs. As a result, such a platform would not only benefit the MFC field but the robotics field as well since EvoBot can be used in the future as a maturing/optimising factory for MFCs that can ultimately be used as the power sources for small robots, including EvoBot itself that may be powered by the very same MFCs that it had built, inoculated and maintained; this would be a step in the direction of truly Living Machines!

Acknowledgments.   This work has been funded by the European Commission FP-7, grant # 611640.

References

1. Rosenfeld, L.: A golden age of clinical chemistry: 1948–1960. Clin. Chem. **46**, 1705–1714 (2000)
2. Bogue, R.: Robots in the laboratory: a review of applications. Ind. Robot Int. J. **39**, 113–119 (2011). doi:10.1108/01439911111106327
3. Wilkinson, S.: "Gastrobots''—benefits and challenges of microbial fuel cells in foodpowered robot applications. Auton. Robots **9**, 99–111 (2000). doi:10.1023/A:1008984516499

4. Ieropoulos, I., Greenman, J., Melhuish, C.: Imitating metabolism: energy autonomy in biologically inspired robots. In: Second International Symposium on Imitation of Animals and Artifacts, AISB 2003, Aberystwyth, Wales, pp. 191–194 (2003)
5. Ieropoulos, I., Melhuish, C., Greenman, J., Horsfield, I.: EcoBot-II: an artificial agent with a natural metabolism. Int. J. Adv. Robot Syst. **2**, 295–300 (2005). doi:10.5772/5777
6. Ieropoulos, I., Greenman, J., Melhuish, C., Horsfield, I.: EcoBot-III: a robot with guts. In: 12th International Conference on the Synthesis and Simulation of Living Systems, pp. 733–740 (2010)
7. Rossiter, J., Philamore, H., Stinchcombe, A., Ieropoulos, I.: Row-bot: an energetically autonomous artificial water boatman. In: Proceedings of the 2015 IEEE/RSJ International Conference on Intelligent Robots and Systems, pp. 3888–3893 (2015)
8. Melhuish, C., Ieropoulos, I., Greenman, J., Horsfield, I.: Energetically autonomous robots: food for thought. Auton. Robot. **21**, 187–198 (2006). doi:10.1007/s10514-006-6574-5
9. Ieropoulos, I., Ledezma, P., Stinchcombe, A., Papaharalabos, G., Melhuish, C., Greenman, J.: Waste to real energy: the first MFC powered mobile phone. Phys. Chem. Chem. Phys. **15**, 15312–15316 (2013). doi:10.1039/c3cp52889h
10. Rahimnejad, M., Adhami, A., Darvari, S., Zirepour, A., Oh, S.-E.: Microbial fuel cell as new technology for bioelectricity generation: a review. Alex. Eng. J. **54**, 745–756 (2015). doi:10.1016/j.aej.2015.03.031
11. La, M.J.: Technique de culture continue. Theorie et applications. Ann Inst Pasteur (Paris) **79**, 390–410 (1950)
12. Ziv, N., Brandt, N.J., Gresham, D.: The use of chemostats in microbial systems biology. J. Vis. Exp. 1–10 (2013). doi:10.3791/50168
13. Faíña, A., Nejatimoharrami, F., Stoy, K., Theodosiou, P., Taylor, B., Ieropoulos, I., EvoBot: An open-source, modular liquid handling robot for nurturing microbial fuel cells. In: Proceedings of the Artificial Life Conference 2016, pp. 626–633 (2016)
14. Nejatimoharrami, F., Faíña, A., Cejkova, J., Hanczyc, M.M., Stoy, K.: Robotic automation to augment quality of artificial chemical life experiments. In: Proceedings of the Artificial Life Conference 2016, vol. 1, pp. 634–635 (2016)
15. Ieropoulos, I., Theodosiou, P., Taylor, B., Greenman, J., Melhuish, C.: Gelatin as a promising printable feedstock for microbial fuel cells (MFC). Int. J. Hydrog. Energy **42**, 1783–1790 (2017). doi:10.1016/j.ijhydene.2016.11.083
16. Greenman, J., Ieropoulos, I., Melhuish, C.: Microbial fuel cells – scalability and their use in robotics. Appl. Electrochem. Nanotechnol. Biol. Med. I 239–290 (2011). doi:10.1007/978-1-4614-0347-0_3

Using Deep Autoencoders to Investigate Image Matching in Visual Navigation

Christopher Walker[1], Paul Graham[2], and Andrew Philippides[1($\boxtimes$)]

[1] Department of Informatics, Centre for Computational Neuroscience
and Robotics, University of Sussex, Brighton, UK
{chris.walker,andrewop}@sussex.ac.uk
[2] School of Life Sciences, Centre for Computational Neuroscience and Robotics,
University of Sussex, Brighton, UK
p.r.graham@sussex.ac.uk

Abstract. This paper discusses the use of deep auto encoder networks to find a compressed representation of an image, which can be used for visual navigation. Images reconstructed from the compressed representation are tested to see if they retain enough information to be used as a visual compass (in which an image is matched with another to recall a bearing/movement direction) as this ability is at the heart of a visual route navigation algorithm. We show that both reconstructed images and compressed representations from different layers of the auto encoder can be used in this way, suggesting that a compact image code is sufficient for visual navigation and that deep networks hold promise for finding optimal visual encodings for this task.

Keywords: Visual navigation · Insect-inspired robotics · Deep neural network · Autoencoder

1 Introduction

Navigation is an important ability for both natural and artificial agents [1]. When looking to nature for inspiration, engineers have turned to ants and bees as they use vision to navigate long distances through complex natural habitats despite limited neural and sensory resources [2–5]. They achieve this task by using retinotopic image-matching methods, which has inspired a range of bio-inspired algorithms (Ants: [6–8]; Bees: [3]; Review: [9]). We have previously shown that panoramic images can be used for navigation in desert ant-inspired algorithms even if images are low-resolution [10], processed through coarse visual filters modelled on parts of the drosophila visual system [11], or processed so that only the height of objects against the skyline is used [12]. This work not only demonstrates the robustness of using low-resolution images for navigation but also that they can be better than high-resolution images [10]. However, while we know that desert ants have low-resolution vision, we do not know how they encode images so that they are best-suited for navigation, nor do we know what visual encodings would be optimal for a navigating agent. In this paper, we investigate this question by examining the compressed visual encodings that arise from deep autoencoder networks trained on

© Springer International Publishing AG 2017
M. Mangan et al. (Eds.): Living Machines 2017, LNAI 10384, pp. 465–474, 2017.
DOI: 10.1007/978-3-319-63537-8_39

natural images. As autoencoders automatically derive low dimensional representations of the underlying data in an unsupervised manner, they are a relatively assumption-free methodology with which to investigate optimal visual encodings while also shedding light on insect visual systems.

The opportunity to use these methods arises because desert ant foragers are task specialists whose sole goal is to visually navigate between nest and food. We can therefore assume that their visual system has been honed by evolution for this task. Thus, we can use AI methodologies as statistical engines to investigate the optimal encodings for navigating with image matching. One such method is to use an autoencoder. Autoencoders are neural networks which are trained to reconstruct their input at the output and have a single hidden layer, usually much smaller in size than the input and output layers, which forces the network to learn a compressed representation of the input. Because these encodings represent statistical regularities from the visual world, they can be used to explore the visual computations that evolution might also have discovered allowing us to hypothesise about how an insect's visual pathway might process images.

Specifically, here we use deep autoencoder networks trained with images gathered by a robot equipped with a panoramic video camera navigating through a wooded environment. A deep autoencoder is an autoencoder with more than one hidden layer, which is particularly useful for this kind of task, as previous work has shown that different layers of the networks extract task-relevant features at different levels of abstraction [13]. As our task is navigation, we examine the encodings produced by different layers of an autoencoder network, to assess how well their output can be used to regain a bearing from a memorised image. This simple verification shows that the information needed for visual route navigation is retained in the encoding as this ability lies at the heart of our route navigation algorithms [7]. As a corollary, we can also examine if there is a compact encoding suited to robotic route navigation with our algorithm. While this work will thus aid robotic navigation, more importantly, it is a first step towards using deep learning to understand insect visual encodings.

2   Methods

2.1   Panoramic Images

The images used in this experiment are collected from a Unibot robot built by Creative Robotics Ltd (http://www.creative-robotics.com/?q=unibot) using a Kodak Pixpro SP360 panoramic camera fixed to the top of the robot. Video footage was recorded as the robot was driven along two different routes through wooded land on the University of Sussex campus at Falmer (sample video can be seen at: https://www.youtube.com/watch?v=f9fkPABQOhg). The video is unwrapped using the Pixpro SP360 desktop software so that the entire panorama is seen as a wide strip, where the forward direction of travel is always in the centre of the image and the far left and right edges are the view behind the robot. Individual frames are then extracted from the video footage which was recorded at 30 frames per second. The ffmpeg library is used with the default bicubic scaling algorithm to extract each frame and scale the output images to

A

B

Fig. 1. Unwrapped panoramic image collected from route on University of Sussex campus. First image is colour with dimensions of 360 × 90 pixels, as extracted from video. Second image is scaled monochrome with dimensions of 180 × 45 pixels, as used in all experiments.

180 × 45 pixels in monochrome, to reduce the complexity of the input to the network. Sample images can be seen in Fig. 1

2.2 Assessing Navigational Information via the Image Difference Function (IDF)

In our route navigation algorithms, we compare the current perceived view with remembered views with the objective of choosing the best direction to move next. Thus, to be useful for familiarity-based route navigation, images must be processed in such a way that we can reliably find the heading at which the current image best matches a stored view. A method which can assess whether processed images retain this property, and which is agnostic of the details of the route navigation algorithm, is the rotational image difference function (RIDF) [14]. The RIDF is based on the image difference function (IDF) a pixel-wise difference between two images X and Y defined as:

$$IDF(X, Y) = \frac{1}{P} \sum_i \sum_j |X(i,j) - Y(i,j)| \tag{1}$$

where $X(i,j)$ is the pixel in the i'th row and j'th column of image X and P is the number of pixels. The more similar the two images, the lower the IDF value will be. Here we are using the absolute pixel difference instead of the r.m.s. pixel difference originally used in [14].

In the above, X and Y are assumed to be aligned to a common heading. To get the RIDF, we rotate one of the images through 360° and find the minimum IDF value across all rotations. In this way we find the heading at which the current image best-matches the remembered view. If the images are sufficiently near each other and the encoding has persevered navigationally useful information, the best-matching heading of the current image will be similar to the heading of the remembered image and the RIDF has a characteristic V-shape around this minimum value [12, 14, 15]. To assess whether sufficient retinotopic information is retained in images encoded by deep networks, we thus compare images with rotated versions of themselves as the presence of the characteristic V is indicative of the presence of homing information.

2.3   Deep Autoencoder Networks

Autoencoder networks are a general class of network which try to produce as output a reconstruction of their input and are often used for dimensionality reduction by passing the input through a hidden layer which has a lower dimension than the input/output layers. Deep neural networks, which are defined as having more than one hidden layer, are suitable for use with the autoencoder technique as their multi-layered structure allows features of increasing complexity to be learned at different layers. Equally the autoencoder is suited to the deep network structure – which requires a great deal of training data – as the training data in this case is self-labelled as the autoencoder is trained to minimise the difference between the output and input.

Such *deep autoencoder networks* (Fig. 2) can thus be used to identify features of datasets with high-dimensionality, such that the data can be represented in a compressed form as low-dimensional codes [16, 17].

The autoencoder is formed of two distinct parts, an encoder which transforms the high-dimensional data into a low-dimensional code, and a decoder which reconstructs the original data from the code (Fig. 2). The code layer, which connects the encoder and decoder, has fewer features than the input creating a bottleneck. This forces the encoder to find a compact representation of the input data that the decoder can use to reconstruct the original input as accurately as possible. Here we use a fully connected autoencoder with an input image size of 180×45 pixels. The image is converted into a one-dimensional array, giving 8100 features at the input layer. The encoder has 5 hidden layers of decreasing size (4096, 2048, 512, 128, 64) with the decoder having the inverse. The layer sizes were chosen arbitrarily to reduce the network to 64 features at the narrowest layer. Networks were constructed and trained in the Tensor flow software (https://www.tensorflow.org/) using mini-batch gradient descent with an initial learning rate of 0.001 and the Adam optimiser [18] to speed up convergence and reduce overfitting.

The dataset contained 2707 images in a randomised order. The images were normalised and the first 100 were set aside for testing; the rest of the images formed the training data. The network was trained using mini-batches of 100 images for a total of 10 epochs, where an epoch is a full pass through all of the images in the training set. At

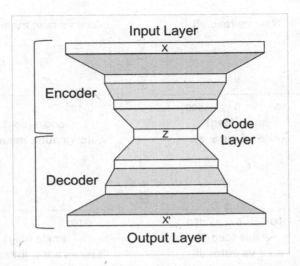

Fig. 2. Structure of a deep autoencoder network with fully connected hidden layers. The structure of the decoder is a mirror image of the encoder and each pair of equivalent layers share the same weights. An original image X is given as input and the network is trained to reconstruct X as X′ at the output layer. The narrow code layer at the centre forces the network to create a lower-dimensional encoding Z of X, from which it is able to reconstruct X′.

intervals during training the network is presented with the test set of images. For each test image the output from the network is saved as the reconstructed image. The network does not learn during the test phase, so the test images are new to the network each time and are not contained in the training set.

3   Results

In order to see if navigationally useful information is retained once an image is passed through the autoencoder, we use the RIDF metric to see if reconstructed images can be used to regain a heading. To do this, we take 5 new images, which are separate from the training and test images, and look at the RIDFs that are produced when these images are compared with rotated versions of themselves. This simulates rotation of the direction that a robot is facing, as if it were turning on the spot through 360°, to find the best-matching heading to its memory.

As a control condition, we first examine RIDFs from raw images. As can be seen in Fig. 3A, the pixel difference is lowest at the centre where there is no rotation, and increases rapidly as the current viewing direction is rotated away from its original orientation. This results in a characteristic V shape, indicating the image can be used to recover heading information from nearby positions. There is some variability in the widths of the V, which to some extent reflects the area over which the image can be used to regain a heading (broader V is indicative of a wider region). More importantly, the V is relatively smooth, especially towards the centre, which is good as the presence

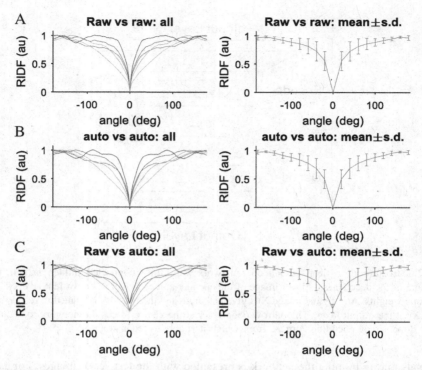

Fig. 3. Rotational image difference function (RIDF) plots showing A. original image vs original image, B. reconstructed image vs reconstructed image, C. original image vs reconstructed image. Data are shown for five test images, with images compared to the same image at all possible rotations. Left column shows individual RIDFs for each image. Right column shows mean with standard deviation error bars.

of multiple, deep, minima indicate there may be a problem with visual aliasing (that is, one location being confused with another).

We next compare the images that have been reconstructed by the autoencoder with themselves to see whether navigationally useful information has been retained (Fig. 3B). This is indeed the case and the RIDFs are very similar to those for the raw images and if anything, perhaps a little more consistent near the true heading. This bodes well for using such an autoencoder as the visual front-end of a robot. Further, if we compare the raw images with the reconstructed images, while as expected the minimum value is not zero, the correct heading is achieved, the RIDF shapes look very similar to the other two cases (Fig. 3C). This indicates that similar information is retained by the encoded network as is in the raw image. The implications are that an agent's perceptual system and memory system can work with different encodings, perhaps with perceptual input being minimally processed and memories being processed to a higher degree.

The reason for this can be seen if we compare the images and output after training (Fig. 4). Much of the high spatial frequency content is removed, leaving what looks like a low-resolution image. This is interesting as it was recently shown in simulation that navigation could be better with images whose resolution is on the order of degrees, similar to the ant's eye, than with high-resolution images [10]. It is thus perhaps no surprise that navigation is achievable with the reconstructed images, though the fact that one can use the raw image as a comparator further suggests that high spatial frequency information is somewhat redundant for this task.

A   B

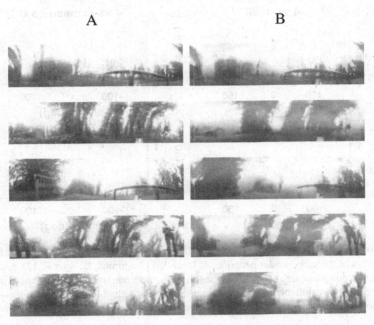

Fig. 4. Input and output images from the autoencoder. A. Original images supplied as test input to the network B. Reconstructed images from output of the deep autoencoder

While we have shown that the reconstructed images can be used for navigation, and thus that the compression in the central layers does not remove this information, the reconstructed images are of the same dimension as the originals. Using them instead of the originals does not therefore present a saving in terms of either memory (which scales with N, the number of pixels) or the computation needed to derive a heading from them (which scales roughly as N^2 depending on the method used). We next therefore assess the RIDFs which results from the three smallest layers of the encoder network, the 3rd 4th and 5th layers which have 512, 128 and 64 units respectively. These are show in Fig. 5A, B and C respectively.

Despite the over 10-fold reduction in dimension, the RIDFs remain and have smooth Vs around the correct heading (Fig. 5) even for the smallest layer (Fig. 5C). The RIDFs contain more spurious optima in headings away from the correct one than the reconstructed images (compare individual RIDFs, left column, in Fig. 5 at azimuths over 90 degrees from the centre with Fig. 3A, B, left column), suggesting visual

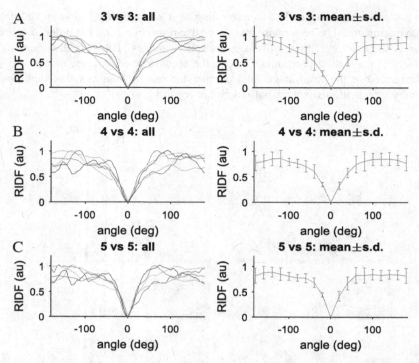

Fig. 5. Rotational image difference function (RIDF) plots derived from the outputs of the three smallest layers of the autoencoder network: A. layer 3, 512 neurons; B. layer 4, 128 neurons; C. layer 5, 64 neurons. Data are shown for five test images, with outputs compared to outputs for the same image at all possible rotations. Left column shows individual RIDFs for each image. Right column shows mean with standard deviation error bars.

aliasing could be a greater problem. However, the width of the Vs is if anything slightly greater than the reconstructed images. This is curious as the trend within the layers is for a slightly greater width for layer 3 (Fig. 5A) than 4 or 5 (Fig. 5B, C, respectively) perhaps suggesting there is an optimal size for image compression between raw and very low-resolution, again echoing [10].

4   Conclusion

In summary, we have shown that we can use a deep autoencoder to derive a compressed representation of natural images that preserves the information required to derive a heading. Further, the information persists at different levels of abstraction within the network and does not require the decoder part of the network. This work thus has implications for robotic navigation as well as understanding the insect visual pathway.

In terms of robotics, this work has implications for the visual processing that could be performed in our route navigation algorithms. our route navigation algorithm proceeds by training a neural network to encode a route [19] either by training a classifier to classify if an image is part of a route or not [20], or by learning the familiarity of the training images [7, 21]. Currently, aside from lowering the resolution and basic image normalization, we do not pre-process the images. This work suggests that if we pre-trained a deep autoencoder, we could use the initial layers to provide a compact representation of the images. We could then train these compact representations on-line to learn specific routes, which would result in much more efficient networks with encodings tuned to regularities in the natural habitat.

Turning to the implications for biology, at this stage, more analysis needs to be done to assess what features of the images are being encoded by the networks and whether these features could be plausibly extracted in the different layers of processing in the insect visual pathway. For instance, as the network is fully connected, the central layer encodings, which appear at least sufficient for homing, could be combining information from across the visual field. This hints at intriguing parallels with the wide-field integration which occurs after the initial two or three stages of visual processing in insect visual pathway, which will be borne out or disproved by analysis of the connectivity. In addition, due to the rather parallel nature of the first stage of visual pathways, we will additionally examine convolutional networks to assess what filters are learned as preliminary work produces similar results to the fully connected network presented here.

Acknowledgements. This work was funded by the Newton Agri-Tech Program RICE PADDY project (no. STDA00732). AP and PG were also funded by EPSRC grant EP/P006094/1.

References

1. Graham, P., Philippides, A.: Insect-inspired vision and visually guided behavior. In: Bhushan, B., Winbigler, H.D. (eds.) Encyclopedia of Nanotechnology. Springer, Netherland (2015)
2. Wehner, R., Räber, F.: Visual spatial memory in desert ants. Cataglyphis bicolor Experientia **35**, 1569–1571 (1979)
3. Cartwright, B.A., Collett, T.S.: Landmark learning in bees - experiments and models. J. Comp. Physiol. **151**, 521–543 (1979)
4. Wehner, R.: Desert ant navigation: how miniature brains solve complex tasks. Karl von Frisch lecture. J. Comp. Physiol. A **189**, 579–588 (2003)
5. Wehner, R.: The architecture of the desert ant's navigational toolkit (Hymenoptera: Formicidae). Myrmecol News **12**, 85–96 (2009)
6. Smith, L., Philippides, A., Graham, P., Baddeley, B., Husbands, P.: Linked local navigation for visual route guidance. Adapt. Behav. **15**, 257–271 (2007)
7. Baddeley, B., Graham, P., Husbands, P., Philippides, A.: A Model of Ant Route Navigation Driven by Scene Familiarity. PLoS Comput. Biol. **8**(1), e1002336 (2012)

8. Ardin, P., Peng, F., Mangan, M., Lagogiannis, K., Webb, B.: Using an insect mushroom body circuit to encode route memory in complex natural environments. PLoS Comput. Biol. **12**(2), e1004683 (2016)
9. Möller, R., Vardy, A.: Local visual homing by matched-filter descent in image distances. Biol. Cybern. **95**, 413–430 (2006)
10. Wystrach, A., Dewar, A., Philippides, A., Graham, P.: How do field of view and resolution affect the information content of panoramic scenes for visual navigation? A computational investigation. J. Comp. Physiol. A **202**, 87–95 (2016)
11. Dewar, A., Wystrach, A., Graham, P., Philippides, A.: Navigation-specific neural coding in the visual system of Drosophila. Biosystems **136**, 120–127 (2015)
12. Philippides, A., Baddeley, B., Cheng, K., Graham, P.: How might ants use panoramic views for route navigation? J. Exp. Biol. **214**, 445–451 (2011)
13. LeCun, Y., Bengio, Y., Hinton, G.: Deep learning. Nature **521**(7553), 436–444 (2015)
14. Zeil, J., Hofmann, M., Chahl, J.: Catchment areas of panoramic snapshots in outdoor scenes. J. Opt. Soc. Am. A **20**, 450–469 (2003)
15. Stürzl, W., Zeil, J.: Depth, contrast and view-based homing in outdoor scenes. Biol. Cybern. **96**, 519–531 (2007)
16. Hinton, G.E., Salakhutdinov, R.R.: Reducing the dimensionality of data with neural networks. Science **313**(5786), 504–507 (2006)
17. Bengio, Y., Courville, A.C., Vincent, P.: Unsupervised feature learning and deep learning: A review and new perspectives. CoRR, abs/1206.5538 (2012)
18. Kingma, D., Ba, J.: Adam: A method for stochastic optimization. arXiv:1412.6980 (2014)
19. Philippides, A., Graham, P., Baddeley, B., Husbands, P.: Using neural networks to understand the information that guides behavior: a case study in visual navigation. Artif. Neural Netw. **1260**, 227–244 (2015)
20. Baddeley, B., Graham, P., Philippides, A., Husbands, P.: Holistic visual encoding of ant-like routes: navigation without waypoints. Adapt. Behav. **19**, 3–15 (2011)
21. Baddeley, B., Graham, P., Philippides, A., Husbands, P.: Models of visually guided routes in ants: embodiment simplifies route acquisition. In: Jeschke, S., Liu, H., Schilberg, D. (eds.) ICIRA 2011. LNCS, vol. 7102, pp. 75–84. Springer, Heidelberg (2011). doi:10.1007/978-3-642-25489-5_8

3D-Printed Biohybrid Robots Powered by Neuromuscular Tissue Circuits from *Aplysia californica*

Victoria A. Webster[✉], Fletcher R. Young, Jill M. Patel,
Gabrielle N. Scariano, Ozan Akkus, Umut A. Gurkan, Hillel J. Chiel,
and Roger D. Quinn

Case Western Reserve University,
10900 Euclid Ave, Cleveland, OH 44106, USA
vaw4@case.edu
http://www.case.edu

Abstract. Biohybrid robotics offers the possibility of compliant, bio-compatible actuation and adaptive behavioral flexibility via the use of muscles as robotic actuators and neural circuits as controllers. In this study, neuromuscular tissue circuits from *Aplysia californica* have been characterized and implemented on 3D-printed inchworm-inspired bio-hybrid robots, creating the first locomotive biohybrid robots with both organic actuation and organic motor-pattern control. Stimulation via the organic motor-controller is shown to result in higher muscle tension and faster device speeds as compared to external electrical stimulation.

Keywords: Biohybrid devices · Organic actuators · Organic controllers · Biorobotics · *Aplysia californica*

1 Introduction

Organic materials are increasingly being combined with synthetic devices for the development of biohybrid robots. Some biohybrid and organic robots have been developed that utilize living muscle tissue or cells to crawl [1,2] or swim [3–5], whereas other biohybrid robots use living neurons to act as a controller for more traditional robotic platforms driven by motors [6,7]. Some devices even use microbial fuel cells to organically extract energy from the environment [8–10]. However, to date, a biohybrid robot with both organic actuation and organic control has not been developed.

Aplysia californica has previously been demonstrated as a possible source for material of biohybrid robots [11]. While the general robustness of tissue from *Aplysia* makes it a candidate material source, it also has an accessible neural system with large neurons and axons. This makes *Aplysia* a prime candidate for providing organic actuators and motor controllers. Additionally, while most biohybrid robots rely on external electrical stimulation, this allows us to investigate the effect of stimulation via the neural system. Previously, Lu et al. have

© Springer International Publishing AG 2017
M. Mangan et al. (Eds.): Living Machines 2017, LNAI 10384, pp. 475–486, 2017.
DOI: 10.1007/978-3-319-63537-8_40

demonstrated that modulatory effects on muscle force are observed when muscles are stimulated via the intact nervous system [12]. These effects may allow the muscle to exert higher forces than when the muscle is stimulated directly. In this study, neuromuscular tissue circuits have been isolated from *Aplysia californica* in which the actuating muscle remains innervated by the supporting ganglia. The effect of stimulating the muscle via an external electric field vs. stimulation via ganglia activity has been investigated. Additionally, 3D-printed inchworm-inspired biohybrid robots actuated by muscle and controlled by the ganglia have been developed (Fig. 1), along with a complete neuromuscular model to enable future device optimization.

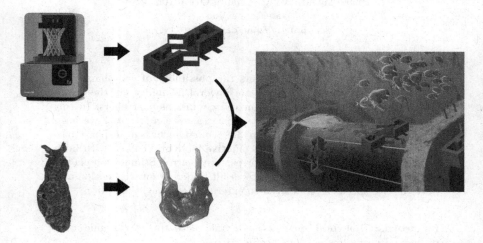

Fig. 1. 3D-printed biohybrid robots powered by neuromuscular tissue circuits from *Aplysia californica*. Top-left: *Aplysia californica* (photo of *Aplysia californica* by Jeff Gill), Top-right: SLA printer, Middle-left: The isolated I2 muscle for use as an actuator, Middle-right: Flexible inchworm-inspired 3D printed body. Such devices may one day provide bio-compatible environmental monitoring services (Bottom)

2   Materials and Methods

2.1   Animals

Aplysia californica were obtained from Marinus Scientific and South Coast Biomarine and maintained in artificial seawater (Instant Ocean, Aquarium Systems) in aerated aquariums at a temperature of 16C. 10 Animals weighing 200–450 g were sacrificed by injecting isotonic $MgCl_2$ (333 mM) at a volume in milliliters equal to 50% of the animals body weight in grams (i.e. 100–225 ml).

2.2   Neuromuscular Tissue Circuit Isolation

The neuromuscular tissue circuits (NTCs) were isolated from the animals in a fashion similar to that described previously by Webster et al. [11]. Two experimental preparations were used.

First, the I2 muscle was isolated independently. To access the muscle, a superficial (i.e. skin only) horizontal cut was made behind the rhinophores and perpendicular cut towards the animals jaw. Connective tissue limiting access to the buccal mass was cut away from the skin, and the esophagus exposed and cut above the cerebral ganglion. Carefully, connective tissue and nerves connected to the buccal mass were removed. The buccal mass was removed from the animal and placed in a separate dish in order to isolate the I2 muscle. In order to reveal the entire I2 muscle, the buccal ganglia were removed. The I2 muscle was isolated by making an initial superficial cut through the muscle then cutting along its insertion lines around the buccal mass. The muscle was then stored in *Aplysia* saline (0.46 M NaCl, 0.01 M KCl, 0.01 M MOPS, 0.01 M $C_6H_{12}O_6$, 0.02 M $MgCl_26H2O$, 0.033 M $MgSO_47H_2O$, 0.01 M $CaCl_22H_2O$, pH 7.5) for no more than 4 h prior to use.

In the second preparation, the buccal and cerebral ganglion were isolated along with the I2 muscle, maintaining neural innervation of the muscle. Following a similar protocol for the isolated I2 muscle dissection, surgical scissors and forceps were used to cut behind the rhinophores and down the body toward the jaw. The extraneous connective tissue was cut away to expose the buccal mass, esophagus, and cerebral ganglion. All nerves connected to the cerebral ganglion, except the cerebrobuccal connectives, were severed. The buccal mass was then cut out of the animal by two horizontal cuts at the base of the esophagus and mouth. The buccal mass and attached ganglia were then placed in a separate dish in order to dissect the innervated I2 muscle.

Six of the buccal nerves (bilateral BN1, BN2, BN3) not responsible for I2 innervation were severed. Additionally, all cerebral nerves except the two cerebrobuccal connectives were cut so as to remove all neuromuscular connections except the radular nerve attachment located underneath the I2 muscle. When it was possible to pull the I2 muscle away from the buccal mass, the left and right branches of the radular nerve were severed as far from the buccal ganglia as possible so as to avoid damaging the I2 nerves. Delicate cuts were made along the insertion line along the base of the I2, carefully avoiding the ganglia. Once the I2 muscle had been removed from the buccal mass, the entire neuromuscular circuit was guided around the remaining esophageal tissue. The neuromuscular circuit of I2 muscle, buccal ganglia, and cerebral ganglion was then stored briefly in *Aplysia* saline prior to experimentation.

2.3   Actuator Characterization

In order to determine the effect of external electrical stimulation vs. stimulation due to neural activity (induced by the application of carbamylcholine chloride (carbachol, ACROS Organics), the muscle tension during stimulation was assessed using a cantilever force measurement system [11].

After the I2 muscle was isolated from the buccal mass, the muscle was folded over itself to create a linear strand. One end of the muscular strand was threaded through and secured within the cantilever crossbar. The straightened muscle's other end was then attached to a metal cantilever with cyanoacrylate adhesive. Careful consideration was taken to ensure that there was minimal tension in the muscle during the attachment process so as not to strain the muscle or damage the attached ganglia. Once the I2 muscle was attached at both ends, the entire crossbar-cantilever structure was lowered into the cantilever well.

Electrical Stimulation. An external function generator was used to provide an electric field across the muscle via two platinum electrodes to produce rhythmic contractions (Fig. 2, left). The I2 muscle was initially stimulated at 5 V/cm until the I2 muscle fatigued and no longer responded to the electrical stimulation (1.5–2 h). Stimulation patterns were based on the stimulation activity seen in *Aplysia* during biting behavior. The muscle was stimulated at 20 Hz for a duration of 3 s followed by a 3 s rest [13]. The muscle force was then determined based on the deflection of the cantilever.

Carbachol Stimulation. Prior to I2 isolation, a 10 mM carbachol solution was prepared by dissolving 9 mg carbachol in 5 mL of *Aplysia* saline. This solution was used to stimulate rhythmic bite-like patterns in the cerebral ganglion [14].

Whereas carbachol stimulation to the cerebral ganglion initiates rhythmic motor patterns, application of the chemical to the buccal ganglia results in activity suppression, and application directly to the muscle results in sustained contraction. In order to apply chemical stimulation to the cerebral ganglion, but not the buccal ganglia or I2 muscle, a variation of the cantilever well used for electrical stimulation was fabricated, featuring an isolation well for the cerebral ganglion (Fig. 2, right). The cerebrobuccal connectives exit the isolation well via a small notch to provide signal transmission to the buccal ganglia. Before testing, a small amount of vacuum grease was applied to the notch to provide an

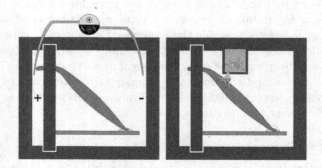

Fig. 2. Left: Schematic illustration of muscle characterization using external electrical stimulation. Right: Schematic illustration of muscle characterization using carbachol stimulation of the cerebral ganglion.

impermeable seal under the cerebrobuccal connective once the muscle-ganglia complex was in place.

Due to the delicate nature of the cerebrobuccal connectives, it was important to support the weight of the cerebral ganglion while transferring the cantilever crossbar into the testing well. As the cantilever crossbar was lowered into the cantilever well, the cerebral ganglion was lowered into the isolation well. The cerebrobuccal connectives were gently guided into the notch and lowered until they were in contact with the vacuum grease at the bottom of the notch. Once the structure was fully lowered into the well, vacuum grease was applied to the notch above the cerebrobuccal connective to create a complete barrier between the cantilever well and the isolation well.

The cantilever well was then filled with *Aplysia* saline while the isolation well was filled with the carbachol solution. Muscle contractions were recorded until contractions ceased at which time the ganglia isolation well was rinsed with *Aplysia* saline and following a 20 min rest, carbachol was reapplied. This was repeated for a total period of 2 h.

2.4   Inchworm-Inspired Biohybrid Robot Fabrication

Inchworm-inspired device bodies were printed using a Formlabs Form 1+ stereolithographic (SLA) printer with a flexible photocurable resin. Devices consisted of two bodies connected by a printed flexible spring. Each body was supported by a four angled, rigid legs, each of which had angled scales on the feet in order to provide anisotropic friction. As with the cantilever force measurement wells, the bodies were designed with an isolation well to allow for chemical stimulation of the cerebral ganglion. Neuromuscular tissue circuits were attached to the device by gluing each end of the muscle to one of the bodies. In preparations including the cerebral ganglion, a small amount of vacuum grease was placed at the base of the notch in the isolation well, the cerebrobuccal connectives were placed through the notch such that the cerebral ganglion was in the isolation well, and vacuum grease was added to seal the notch. Devices were stimulated either via external electrodes (6 s, 5 V/cm, 20 Hz, 50% duty cycle), or by application of 10 mM carbachol to the cerebral ganglion.

3   Simulation

A dynamical model of the inchworm-inspired biohybrid robot powered by NTCs has been developed in MATLAB in order to predict the performance of the robot when the ganglia is stimulated with carbachol. To approximate neuromuscular dynamics observed in the tissue circuit a stable heteroclinic channel (SHC) [15, 16] has been implemented as the neural controller.

3.1   Neuromuscular Model

A three node SHC model based on the work of Shaw et al. [15] has been implemented to model muscle activation and the neural activity of the ganglia in the tissue circuit (Fig. 3A).

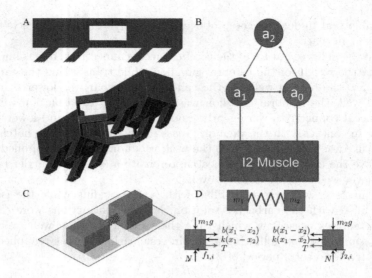

Fig. 3. A: 3D model of the inchworm model showing the side view (top) and isometric view showing the textured feet (bottom). B: Simplified neural model using a stable heteroclinic channel. C: Simplified schema of the inchworm model. D: Free Body Diagrams of both bodies in the inchworm model

The activity in each node is represented by the neural state variables a_i:

$$\frac{da_0}{dt} = \frac{1}{\tau_a} \left(a_0 \left(1 - a_0 - \gamma a_1 \right) + \mu \right) + \epsilon \left(x_r - S_0 \right) \sigma_0 \tag{1}$$

$$\frac{da_1}{dt} = \frac{1}{\tau_a} \left(a_1 \left(1 - a_1 - \gamma a_2 \right) + \mu \right) + \epsilon \left(x_r - S_1 \right) \sigma_1 \tag{2}$$

$$\frac{da_2}{dt} = \frac{1}{\tau_a} \left(a_2 \left(1 - a_2 - \gamma a_0 \right) + \mu \right) + \epsilon \left(x_r - S_2 \right) \sigma_2 \tag{3}$$

The neural state variables are then rigidly bounded such that $a_i \in (0, 1)$. Therefore, when a_i is at its maximum any additional excitatory input has no effect and when a_i is at its minimum any additional inhibitory input has no effect. All parameter definitions are detailed in Tables 1 and 2.

The activation of the muscle has been simulated based on the activity of two of the three neural nodes of the SHC (a_0 and a_1):

$$\frac{du_0}{dt} = \frac{1}{\tau_m} \left(\left(a_0 + a_1 \right) u_{max} - u_0 \right) \tag{4}$$

The normalized force of the muscle is calculated using a linear model:

$$T_m = k_0 \phi \left(\frac{x - c_0}{w_0} \right) u_0 \tag{5}$$

where k_0 is the strength and direction of the I2 muscle, x is the normalized length of the muscle, c_0 is the minimum length of the I2 muscle, and w_0 is the maximum length of the I2 muscle. ϕ represents the length-tension properties of the muscle which are calculated as:

$$\phi(x) = -\kappa x (x-1)(x+1) \tag{6}$$

where κ is the length-tension curve normalization constant. The final force applied by the muscle to the physical model is then calculated by multiplying the normalized force by the maximum possible force (T_0):

$$T = T_0 T_m \tag{7}$$

Physical parameters, such as the maximum muscle force [17], were set based on values previously reported in the literature, whereas neural parameters were initially set based on those used by Shaw et al. for simulation of the complete feeding apparatus and subsequently adjusted to approximate temporal effects observed in the isolated set up (Table 1).

Table 1. Neuromuscular model parameters

Parameter	Value	Description
ϵ	0.2	Sensory feedback strength
γ	2.4	Inhibition coupling constant
κ	$3\frac{\sqrt{3}}{2}$	Length-tension normalization constant
μ	10^{-9}	Neural pool intrinsic excitation
τ_a	0.15	Neural pool time constant
τ_m	0.5	Muscle activation time constant
u_{max}	1	Maximum muscle activation
w	0.5	Width of length-tension curve
c	0	Minimum muscle length parameter
S_0	0.495	Node 0 proprioceptive neutral position
S_1	0.495	Node 1 proprioceptive neutral position
S_2	0.495	Node 2 proprioceptive neutral position
σ_0	1	Node 0 proprioceptive input sign
σ_1	−1	Node 1 proprioceptive input sign
σ_2	−1	Node 2 proprioceptive input sign
T_0	120	Maximum possible muscle force (mN)

3.2 Physical Model

The model of the inchworm robot was approximated as two point masses connected by a flexible spring (Fig. 3B). In order to simulate the frictional anisotropy provided by the asymmetric feet and legs, asymmetric coefficients of friction were implemented based on the direction that each body was moving. The dynamic equations of motion were developed based on the free body diagrams of each body (Fig. 3C) such that:

$$m_j \ddot{x}_j + b(\dot{x}_1 - \dot{x}_2) + k(x_1 - x_2) + f_{j,i} = T \tag{8}$$

where m_j is the mass of each body, b is the damping constant of the spring, k is the spring constant, $f_{j,i}$ is the friction force on body j with i specifying the coefficient of friction to be used based on direction (Table 2), and T is the force applied by the muscle. In simulation, the frictional force is calculated at each time step such that if the sum of all other forces on the body is less than static friction ($T < \mu_s m_j g$), friction directly counters the other forces, whereas if it exceeds the static friction, the sliding friction force is applied to the body in the direction opposite the body's motion ($f_{j,i} = \mu_{j,i} m_j g$). The physical parameters used in simulation are presented in Table 2.

Table 2. Physical parameters used for device simulation

Parameter	Value	Description
m_1	0.003	Mass of body 1 (kg)
m_2	0.003	Mass of body 2 (kg)
k	0.5	Spring stiffness (N/mm)
b	0.001	Spring damping (Ns/m)
μ_s	0.5	Coefficient of static friction
$\mu_{j,1}$	0.15	Forward coefficient of friction
$\mu_{j,2}$	0.25	Backwards coefficient of friction

Integration of all differential equations was performed using a fourth order Runge-Kutta method.

3.3 Simulation Results

Simulation of the inchworm robot indicated that a minimum force of 88 mN was needed to produce a forward step. With a peak contraction force of 88 mN, the robot moved 0.25 cm in 1 min, whereas with a peak contraction force of 120 mN it moved 0.85 cm in the same time.

4 Results and Discussion

4.1 The Effect of Stimulation Type on Actuator Performance

The effect of stimulation type on I2 tension was investigated with 6 isolated I2 muscles (3/stimulation type). For each muscle only one stimulation type was used and the peak muscle tension was calculated for each contraction. For each muscle these peak values were averaged for comparison (Fig. 4). Carbachol stimulation via the cerebral ganglion resulted in significantly higher muscle tension as compared to external stimulation ($p < 0.05$), with carbachol stimulation resulting in an average peak tension of 0.1 N, whereas electrical stimulation resulted in an average peak tension of 0.028 N.

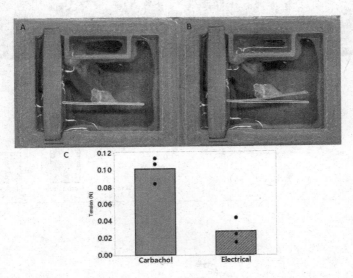

Fig. 4. A: Cantilever position while the muscle is relaxed. B: Cantilever position during contraction. C: A comparison of the average peak muscle tensions for carbachol or electrical stimulation. Dots represent the average peak muscle tension for each isolated NTC. Bars provide the mean peak muscle tension for each stimulation type.

4.2   Device Locomotion

Inchworm devices were tested with external electrical stimulation and chemical stimulation using carbachol applied directly to the cerebral ganglion innervating the muscle (Fig. 5A–C). Chemical stimulation resulted in faster locomotion than electrical stimulation with an average speed of 0.54 cm/min as compared to 0.21 cm/min (Fig. 5D). However, additional trials are needed to determine significance. The velocities of the carbachol stimulated devices are well predicted by the neuromuscular model of the carbachol stimulated system (Fig. 5E), falling within the range of simulated velocities.

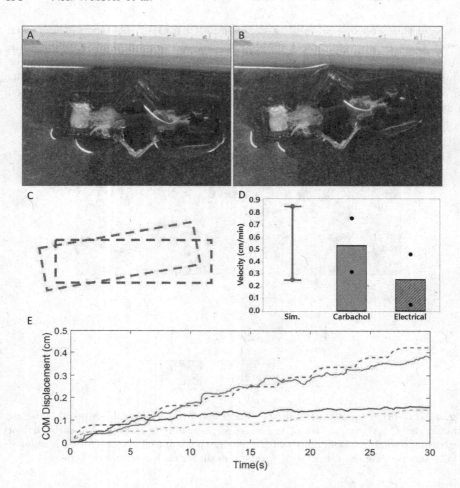

Fig. 5. A–C: Characteristic example of inchworm locomotion, A: starting position, B: final position, C: overlay of positions showing displacement. D: Comparison of device velocities using carbachol or electrical stimulation (N = 2). Black dots give the individual device velocities. Bars provide the mean velocity for each stimulation type. For comparison to the carbachol stimulated results the range between the minimum and maximum simulated values is provided (red interval). E: Traces of the position of the carbachol stimulated inchworm robots center of mass (COM) compared to the simulated results. Experimental results are shown by solid lines. Simulated results are shown by dashed lines for the maximum (blue) and minimum (orange) forces measured using the cantilever characterization system for carbachol stimulated NTCs. (Color figure online)

5   Conclusions

In this study, we have presented techniques for isolating intact neuromuscular tissue circuits from *Aplysia californica*. Additionally we have characterized the effect of stimulation type on muscle tension and demonstrated that neural stimulation results in better muscle performance than stimulation via an external electric field. This is significant because most existing biohybrid and organic devices rely on electrical stimulation to produce device locomotion. By doing so, such devices may not be achieving their full actuation potential. This result parallels existing literature on functional electrical stimulation showing that direct muscle stimulation results in lower force and higher fatigue than nerve stimulation [18]. It is important to note however, that the use of intact tissue for these devices presents a limitation in the ability to scale up production of such robots. To this end, techniques are needed to grow organized tissue circuits at the cellular level, allowing devices to be fabricated from cells harvested from *Aplysia* eggs sacks or biopsies.

Furthermore, we have demonstrated the use of neuromuscular tissue circuits to produce locomotion in biohybrid robots. These devices will serve as a foundation for the development of future organically controlled and actuated devices. Such devices may one day be used for searching for leaking pipelines, monitoring coral reef health, or even aquatic search and rescue operations. While the device presented utilizes an organic motor-controller, human intervention is still needed in the form of chemical stimulation to produce locomotion. Future work is needed to develop techniques for programming the neural circuitry to complete specific tasks in order to achieve autonomous locomotion.

Acknowledgments. The authors would like to thank Yanjun Zhang for assistance in actuator characterization. This material is based upon work supported by the National Science Foundation Graduate Research Fellowship under Grant No. DGE-0951783 and a GAANN Fellowship (Grant No. P200A150316). This study was also funded in part by grants from the National Science Foundation (Grant No. DMR-1306665), and the National Institute of Health (Grant No. R01 AR063701).

References

1. Cvetkovic, C., Raman, R., Chan, V., Williams, B.J., Tolish, M., Bajaj, P., Sakar, M.S., Asada, H.H., Saif, M.T.A., Bashir, R.: Three-dimensionally printed biological machines powered by skeletal muscle. PNAS **111**(28), 10125–10130 (2014)
2. Webster, V.A., Hawley, E.L., Akkus, O., Chiel, H.J., Quinn, R.D.: Effect of actuating cell source on locomotion of organic living machines with electrocompacted collagen skeleton. Bioinspiration Biomim. **11**(3), 036012 (2016)
3. Nawroth, J.C., Lee, H., Feinberg, A.W., Ripplinger, C.M., McCain, M.L., Grosberg, A., Dabiri, J.O., Parker, K.K.: A tissue-engineered jellyfish with biomimetic propulsion. Nat. Biotechnol. **30**(8), 792–797 (2012)
4. Park, S.-J., Gazzola, M., Park, K.S., Park, S., Di Santo, V., Blevins, E.L., Lind, J.U., Campbell, P.H., Dauth, S., Capulli, A.K., Pasqualini, F.S., Ahn, S., Cho, A., Yuan, H., Maoz, B.M., Vijaykumar, R., Choi, J.-W., Deisseroth, K., Lauder,

G.V., Mahadevan, L., Parker, K.K.: Phototactic guidance of a tissue-engineered soft-robotic ray. Science **353**(6295), 158–162 (2016)

5. Williams, B.J., Anand, S.V., Rajagopalan, J., Saif, M.T.A.: A self-propelled bio-hybrid swimmer at low Reynolds number. Nat. Commun. **5**(3081) (2014)

6. Ferrández, J.M., Lorente, V., DelaPaz, F., Cuadra, J.M., Álvarez-Sánchez, J.R., Fernández, E.: A biological neuroprocessor for robotic guidance using a center of area method. Neurocomputing **74**(8), 1229–1236 (2011)

7. De Santos, D., Lorente, V., De La Paz, F., Manuel Cuadra, J., Lvarez-Snchez, J.R., Fernández, E., Ferrández, J.M., Ferrández, J.M.: A client-server architecture for remotely controlling a robot using a closed-loop system with a biological neuroprocessor. Robot. Auton. Syst. **58**(12), 1223–1230 (2010)

8. Wilkinson, S.: 'Gastrobots' - benefits and challenges of microbial fuel cells in food powered robot applications. Auton. Robot. **9**(2), 99–111 (2000)

9. Philamore, H., Rossiter, J., Stinchcombe, A., Ieropoulos, I.: Row-bot: an energetically autonomous artificial water boatman. In: IEEE International Conference on Intelligent Robots and Systems, pp. 3888–3893 (2015)

10. Ieropoulos, I., Melhuish, C., Greenman, J., Horsfield, I.: EcoBot-II: an artificial agent with a natural metabolism. Int. J. Adv. Robot. Syst. **2**(4), 295–300 (2005)

11. Webster, V.A., Chapin, K.J., Hawley, E.L., Patel, J.M., Akkus, O., Chiel, H.J., Quinn, R.D.: *Aplysia californica* as a novel source of material for biohybrid robots and organic machines. In: Lepora, N.F.F., Mura, A., Mangan, M., Verschure, P., Desmulliez, M., Prescott, T.J.J. (eds.) Living Machines 2016. LNCS, vol. 9793, pp. 365–374. Springer, Cham (2016). doi:10.1007/978-3-319-42417-0_33

12. Lu, H., McManus, J.M., Cullins, M.J., Chiel, H.J.: Preparing the periphery for a subsequent behavior: motor neuronal activity during biting generates little force but prepares a retractor muscle to generate larger forces during swallowing in Aplysia. J. Neurosci. **35**(12), 5051–5066 (2015)

13. Hurwitz, I., Neustadter, D., Morton, D.W., Chiel, H.J., Susswein, A.J.: Activity patterns of the B31/B32 pattern initiators innervating the I2 muscle of the buccal mass during normal feeding movements in Aplysia californica. J. Neurophysiol. **75**(4), 1309–1326 (1996)

14. Susswein, A.J., Rosen, S.C., Gapon, S., Kupfermann, I.: Characterization of buccal motor programs elicited by a cholinergic agonist applied to the cerebral ganglion of Aplysia californica. J. Comp. Physiol. A Sens. Neural Behav. Physiol. **179**(4), 509–524 (1996)

15. Shaw, K.M., Lyttle, D.N., Gill, J.P., Cullins, M.J., Mcmanus, J.M., Lu, H., Thomas, P.J., Chiel, H.J.: The significance of dynamical architecture for adaptive responses to mechanical loads during rhythmic behavior. J. Comput. Neurosci. **38**, 25–51 (2015)

16. Horchler, A.D., Daltorio, K.A., Chiel, H.J., Quinn, R.D.: Designing responsive pattern generators: stable heteroclinic channel cycles for modeling and control. Bioinspiration Biomim. **10**(2), 26001 (2015)

17. Yu, S.N., Crago, P.E., Chiel, H.J.: Biomechanical properties and a kinetic simulation model of the smooth muscle I2 in the buccal mass of Aplysia. Biol. Cybern. **81**, 505–513 (1999)

18. Mortimer, J.T.: Motor Prostheses In: Comprehensive Physiology 2011, Supplement 2: Handbook of Physiology, The Nervous System, Motor Control, pp. 155–187 (1981)

Self-organising Thermoregulatory Huddling in a Model of Soft Deformable Littermates

Stuart P. Wilson[(✉)]

University of Sheffield, Sheffield, UK
s.p.wilson@sheffield.ac.uk

Abstract. Thermoregulatory huddling behaviours dominate the early experiences of developing rodents, and constrain the patterns of sensory and motor input that drive neural plasticity. Huddling is a complex emergent group behaviour, thought to provide an early template for the development of adult social systems, and to constrain natural selection on metabolic physiology. However, huddling behaviours are governed by simple rules of interaction between individuals, which can be described in terms of the thermodynamics of heat exchange, and can be easily controlled by manipulation of the environment temperature. Thermoregulatory huddling thus provides an opportunity to investigate the effects of early experience on brain development in a social, developmental, and evolutionary context, through controlled experimentation. This paper demonstrates that thermoregulatory huddling behaviours can self-organise in a simulation of rodent littermates modelled as soft-deformable bodies that exchange heat during contact. The paper presents a novel methodology, based on techniques in computer animation, for simulating the early sensory and motor experiences of the developing rodent.

Keywords: Self-organisation · Thermoregulation · Huddling · Shape-matching

1 Introduction

In cold environments, many species of endotherms keep warm by huddling together. The huddle is a self-organising system, where simple interactions between individuals collectively give rise to group-level emergent properties [1]. Emergent properties of huddling include (i) a capacity for thermoregulation amongst the group which exceeds that of the individual [2], (ii) a second-order critical phase transition from close aggregation at low environment temperatures to dispersion at high environment temperatures [3], and (iii) self-sustaining group dynamics at intermediate temperatures known as 'pup flow', whereby individuals continuously cycle between the warm core and cold periphery of the huddle [4].

Agent-based modelling has helped to establish that these group-level properties can emerge from simple rules of interaction between individuals [5,6]. Models of huddling have typically studied self-organisation from one of two directions. First, models using simulated or physical robots (even bean bags) have helped

© Springer International Publishing AG 2017
M. Mangan et al. (Eds.): Living Machines 2017, LNAI 10384, pp. 487–496, 2017.
DOI: 10.1007/978-3-319-63537-8_41

establish the importance of the animal morphology in determining how inter-actions between the body and the environment affect group-level aggregation patterns [7–9]. Second, models based on the thermodynamics of heat exchange have instead described individuals simply as gas particles bouncing around in a chamber [1,10]. These two levels of description yield complementary insights into self-organisation in the huddle. Additional insight may be gained by con-structing an agent-based model of huddling that is constrained by a combination of morphological and thermodynamic factors.

The aim of the current paper is to simulate the thermodynamics of rodent huddling interactions, in groups of agents whose body morphologies deform inside the huddle in a physically plausible way. To simulate soft-body deforma-tion, the model relies on an algorithm called shape-matching, which has found broad applicability in computer animation [11,12]. The simulation is shown here to recreate the phase transition in huddling as a function of the ambient temper-ature, as measured experimentally by [3]. The modelling approach could thus be used to generate complex naturalistic patterns of sensory and motor input for models of neural development [13–17], which are expected to vary systematically with the environment temperature as an underlying control parameter.

2   Methods

A model of self-organising thermoregulatory interactions between rodent litter-mates is presented. First, a description of the rodent as a three-dimensional soft deformable body is presented, based on the computer animation technique known as meshless deformation by shape-matching [11]. Second, a model of the interactions between individuals that give rise to huddling behaviours, formu-lated originally for simulations of two-dimensional rat pups [6], is extended to direct the movements of the three-dimensional pups.

2.1   Modelling Littermates as Soft Deformable Bodies

The arena is defined as a floor with a cylindrical boundary wall of radius $r_{arena} = 2.0$. Each of $N = 12$ 'pups' is represented as a cloud of points arranged on a sphere of radius $r_{pup} = 0.25$. Points are spaced (almost) equidistantly on the sphere by a re-mapping after first spacing the points equally on the face a cube with 7^2 points on each of six faces, yielding $n = 294$ points in total.

At each point is a sphere of radius $r_{point} = 0.05$ and associated with each point is a 3D position vector $\mathbf{p} = [p_x, p_y, p_z]$, a 3D velocity vector $\mathbf{v}$, and a mass m. The locations of the points can be displaced under external forces, e.g., gravity, and by contact of the sphere with a boundary wall or with a sphere belonging to another pup.

The pup is simulated as a soft-deformable body using the shape-matching algorithm of [11]. Essentially, the algorithm minimizes the discrepancy between the current point cloud (after displacement of the points caused by contacts and/or external forces) and the original point cloud (i.e., a spherical arrangement) by gradient descent. Thus, at each timestep the shape-matching algorithm works against the deformation of the body caused by contacts to restore the pup to its original shape.

For full details of the shape-matching algorithm the interested reader is referred to [11], and a brief overview is provided here. The idea is to specify on each iteration a goal position for each point in the point cloud, and to move the points in the direction towards these goal positions. Goal positions are chosen to minimize the discrepancy between the shape defined by the initial arrangement of the point cloud, $\mathbf{x}^0 = [\mathbf{p}_1, \mathbf{p}_2, \ldots, \mathbf{p}_n]$, and that defined by the current positions of the points $\mathbf{x}$. Specifying the goal positions involves finding the rotation matrix $\mathbf{R}$ which minimizes the following expression:

$$\sum_i m_i (\mathbf{R}(\mathbf{x}_i^0 - \mathbf{t}_0) + \mathbf{t} - \mathbf{x}_i)^2,$$

where $\mathbf{t}_0$ and $\mathbf{t}$ are translation vectors. The appropriate rotation matrix can be defined as,

$$\mathbf{R} = \mathbf{A}(\sqrt{\mathbf{A}^T \mathbf{A}})^{-1},$$

where $\mathbf{A} = \sum_i m_i(\mathbf{x}_i - \mathbf{x}_{cm})(\mathbf{x}_i^0 - \mathbf{x}_{cm}^0)^T$, and where $\mathbf{t}_0 = \mathbf{x}_{cm}^0$ and $\mathbf{t} = \mathbf{x}_{cm}$ are shown by [11] to be the centre of mass of the initial point cloud and the current point cloud, respectively. Then, goal positions are determined by

$$\mathbf{g}_i = \mathbf{R}(\mathbf{x}_i^0 - \mathbf{x}_{cm}^0) + \mathbf{x}_{cm}.$$

A vector of external forces $\mathbf{f}_{ext}$ can be applied to the point cloud on each iteration, and the effects of gravity can be simulated by applying a constant external force in the $-z$–direction; here $\mathbf{f}_{ext} = [f_x, f_y, f_z - 10]$.

After timestep h, the points are moved towards their goal locations, under the influence of the external force, by

$$\mathbf{x}_i(t + h) = \mathbf{x}_i(t) + h\left(\mathbf{v}_i(t) + \alpha\frac{\mathbf{g}_i(t) - \mathbf{x}_i(t)}{h} + h\frac{\mathbf{f}_{ext}(t)}{m_i}\right),$$

and velocities are updated as $\mathbf{v}_i(t + h) = h(\mathbf{x}_i(t + h) - \mathbf{x}_i(t))$.

The parameter $\alpha \in \{0, 1\}$ corresponds to the stiffness of the body. Setting $\alpha = 1$ defines a rigid body, and setting $\alpha < 1$ allows the body to deform. Linear deformations, i.e., shear and stretch, can be controlled by an additional parameter $\beta \in \{0, 1\}$, which determines the extent to which goal positions are defined in terms of the optimal rotation matrix $\mathbf{R}$, or the optimal linear transformation $\mathbf{B} = \mathbf{A}\sum_i m_i(\mathbf{x}_i^0 - \mathbf{x}_{cm}^0)(\mathbf{x}_i^0 - \mathbf{x}_{cm}^0)^T$, and as such the modified term $\beta\mathbf{B} + (1-\beta)\mathbf{R}$ replaces $\mathbf{R}$ in the computation of the goal locations, $\mathbf{g}_i$.

The algorithm can be further extended to allow for non-linear deformations of the body, i.e., twisting and bending, by redefinition of the matrices $\mathbf{B}$ and $\mathbf{R}$ to include quadratic terms (see [11]). The soft-deformable body of the pup was simulated using this quadratic extension, representing the assumption that pup bodies can bend and twist inside the huddle. A deformation coefficient of

$\beta = 0.5$, a rigidity constant of $\alpha = 0.7$, and a simulation timestep of $h = 0.01$ were used.

At the beginning of a simulation, the initial location of each pup at time $t = 0$ is set to be the center of the arena plus a random displacement in the xy–plane, and each pup, j, is dropped into the arena from a unique height, $\mathbf{x}_{cm}^0 = [r_x, r_y, j]$, where $r_x \in \{-0.25, +0.25\}$ and $r_y \in \{-0.25, +0.25\}$. This ensures that after a few timesteps, all pups will have dropped to the floor (under gravity) to form an initial cluster at the center of the arena.

When the sphere in the point cloud comes into contact with the arena floor $(p_z - r_{point} < 0)$ the corresponding velocity vector is set in the positive z direction and the point is displaced such that $p_z = r_{point}$. When the sphere comes into contact with the arena boundary $((|p_x| + r_{point})^2 + (|p_y| + r_{point})^2 > r_{arena}^2)$, the velocity is averaged with a vector pointing normal to the orientation of the boundary wall ($v_{wall} = [-\cos\phi, -\sin\phi, 0]$, for $\phi = \arctan2(p_y, p_x)$), and the corresponding position vector is displaced in this new direction.

If the sphere at a point belonging to pup j comes into contact with that at a point belonging to pup $k \neq j$, i.e., $\|\mathbf{p}_{j,i} - \mathbf{p}_{k,i}\| < 2r_{point}$, and the point of pup j is above that of k, the velocity of the point belonging to pup j is set to $\mathbf{v}_{j,i} = \mathbf{p}_{k,i} - \mathbf{p}_{j,i}$ and its position is set to $\mathbf{p}_{j,i} = \mathbf{p}_{j,i} + (\mathbf{p}_{j,i} - \mathbf{p}_{k,i})(2r_{point} - \|\mathbf{p}_{j,i} - \mathbf{p}_{k,i}\|)$. This assymmetry in collision detection/resolution reduces an anomaly that continuously maintained contacts can otherwise cause a pair of pups to 'float' as if unaffected by gravity.

Once any collisions with the arena boundaries and collisions between pups have been resolved, and the affected points have been displaced accordingly, the dynamics of the shape-matching algorithm are applied.

2.2   Thermoregulatory Huddling Behaviours

At this point we have described a method for simulating how a pile of soft bodies can be dropped into a simple arena, how they will deform upon contact with one another, and how they will subsequently come to rest either by stacking a top one another or by rolling across the arena floor. However, to actively form and maintain a huddle, movements of the bodies must be self-directed. Here we first describe the dynamics of thermal exchange with the environment and between bodies that are in contact, before showing how 'homeothermotaxic' [6] behaviours of the individuals can direct pups that are in contact towards or away from one another, giving rise to self-organising thermoregulatory huddling.

The ambient temperature of the arena is T_a, which is constant for the duration of the simulation. Each pup maintains a dynamic body temperature T_b, and has a preferred temperature $T_p = 0.6$, which is used to represent a desired body temperature, e.g., 37 °C. The sphere located at each point in the cloud used to define the pup is now considered to be a thermal-tactile 'afferent'. By default, each afferent measures the ambient temperature T_a. However, if the afferent is in contact with the afferent of another pup, T_c is set to the current value of T_b of the other pup.

$T_a = 0.2$

$T_a = 0.5$

$T_a = 0.8$

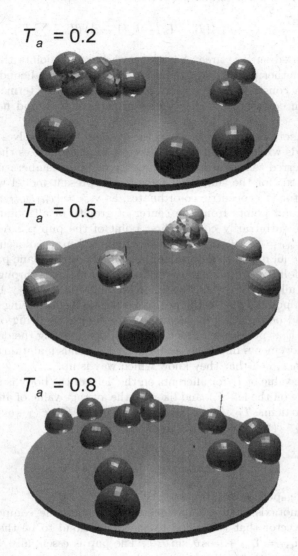

Fig. 1. Typical aggregation patterns self-organised at three ambient temperatures. Pups, simulated as soft-deformable bodies, huddle together in large stable groups in cold environments $T_a = 0.2$ (a stable cluster of 7 pups in the top left corner was maintained in this simulation), cycle between smaller subgroups at $T_a = 0.5$ (transient clusters of size 5 and 2 are visible), and disperse at higher temperatures $T_a = 0.8$. The body temperature of each pup T_b is indicated by the colour. Note that the preferred ambient temperature is $T_p = 0.6$. (Color figure online)

According to the huddling model of [6], the body temperature is updated by

$$\frac{dT_b}{dt} = G - k_1 A(T_b - T_a) - k_2(1 - A)\left(T_b - \sum T_c\right),$$

where the exposed surface area A is the proportion of points that are not in contact with another pup, $k_1 = 0.9$ and $k_2 = 1.0$ are thermal conductances and $G = 0.055$ is a constant rate of thermogenesis. The three terms on the right of this equation correspond to heat generation, heat loss, and heat exchange, respectively.

Homeothermotaxis, as described in the model of [6], involves moving continually forwards while turning in the direction that minimizes the discrepancy between a preferred temperature and the current body temperature. To define homeothermotaxis for the simulated pups in the present model we first define a reference vector in egocentric co-ordinates, as $\mathbf{v}_{\mathrm{ref}} = [x_{\mathrm{ref}}, y_{\mathrm{ref}}, z_{\mathrm{ref}}]$, which is of unit length and points from the center of gravity of all points $\mathbf{x}_{\mathrm{cm}}$ in the direction of the (arbitrarily chosen) first point of the pup $\mathbf{p}_0$. A target direction vector $\mathbf{v}_{\mathrm{target}} = [\cos\theta, \sin\theta, 0]$ is then computed in an allocentric reference frame, using the [6] model to determine the angle θ (in the plane parrallel to the arena floor) as follows. All points are rotated around an axis through $\mathbf{x}_{\mathrm{cm}}$ that is aligned to the (allocentric) z–axis by angle $\arctan2(-y_{\mathrm{ref}}, -x_{\mathrm{ref}})$. Following this transformation, points lying in the positive x are defined as being on the 'right' of the body, and points lying in negative x are defined as being on the 'left' of the body. Note that performing this calculation before using the derived quantities to direct movements implicitly assumes that animals maintain an egocentric frame of reference, i.e., that they know which way is up.

The average value of T_c for afferents on the 'left' of the body is used to define the temperature on the left T_L, and likewise the average value of afferents on the 'right' is used to define T_R. According to [6], we define $S_L = (1+\exp(-\sigma T_L(T_p - T_b)))^{-1}$ and $S_R = (1 + \exp(-\sigma T_R(T_p - T_b)))^{-1}$, and

$$\Delta\theta = \arctan\left(v\frac{S_L - S_R}{S_L + S_R}\right),$$

where v sets the speed of rotation of θ.

Homeothermotaxis is then implemented by setting the components of the external force vector that is applied to the point cloud to be the components of $\mathbf{v}_{\mathrm{target}}$, i.e., $\mathbf{f}_{\mathrm{ext}} = \mathbf{f}_{\mathrm{ext}} + [\cos\theta, \sin\theta, 0]$. The pup is essentially 'blown' in the direction of contacts that will bring its body temperature closer to the preferred temperature. Note that the only information exchange between pups occurs when they are in contact, and that homeothermotaxis involves no distal sensing.

3   Results

The model was implemented in c++ using standard libraries. OpenGL was used for visualisation. The armadillo linear algebra library was used, mainly for calculation of the (pseudo-) matrix inverse operations required for computing $\mathbf{R}$

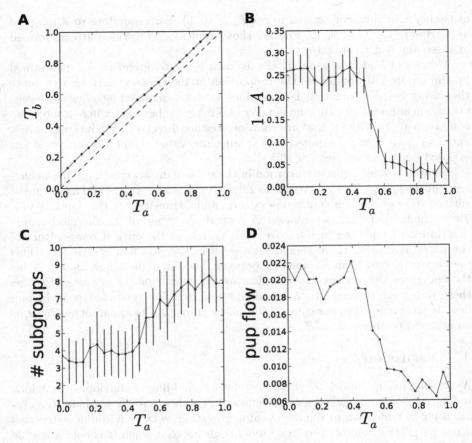

Fig. 2. Metrics of self-organising huddling. **A.** The average body temperature T_b is maintained above the ambient temperature T_a, when the ambient temperature is low. **B.** The average exposed surface area A gives a metric of huddling $1 - A$, which reveals a phase transition from huddling to non-huddling as the ambient temperature is raised. **C.** The average number of independent groups of clustered individuals increases as the ambient temperature is raised, showing that individuals group together in the cold and separate when it is warm. **D.** A metric of pup-flow, defined in terms of the variability in the exposed surface area, reveals a peak at $T_a \approx 0.4$, as predicted by earlier models.

in the shape-matching algorithm (arma.sourceforge.net). Python was used for analysis. The code is available by request.

Each simulation consisted of iterating the dynamics of the model through 1500 timesteps, and the first 500 timesteps (as pups fell into an initial pile) were discarded from analysis. At a given ambient temperature, ten simulations were run, each with the pups initialised at different random initial locations and different random initial orientations, and means and standard deviations were computed for each metric of huddling. Simulations were conducted at each of twenty four ambient temperatures $T_a \in \{0, 1\}$, corresponding to a range of approximately 0 °C to 50 °C. Figure 1 shows typical aggregation patterns formed at three ambient temperatures.

Metrics of huddling were (i) the average body temperature T_b maintained by the group, (ii) one minus the proportion of the exposed surface area (note that using $1 - A$ means that larger values indicate increased aggregation levels, i.e., stronger huddling), (iii) the number of different subgroups, where a subgroup comprises all individuals that are connected either directly or via intermediaries, and (iv) 'pup flow', computed as the absolute value of the derivative of the exposed body surface (see [10,16]).

Figure 2 reveals a phase transition in the self-organisation of huddling behaviours in the group. This is revealed as raised body temperatures and low exposed surface areas at cool ambient temperatures, and a transition to the opposite profile as the temperature is increased. A central prediction of the thermodynamic description of huddling in related models is that at the critical temperature of the phase transition there should be a peak in the pup flow metric [16]. This prediction derives from an analogy between huddling (measured as $1 - A$) as the energy of the system and pup flow as the heat capacity of the system, as these terms are defined in statistical physics [10]. The predicted peak in pup flow at the critical temperature of the phase transition is apparent in Fig. 2 at an ambient temperature of $T_a \approx 0.4$.

4 Discussion

A self-organising model of thermoregulatory huddling behaviours, in which rodent pups were modelled as soft-deformable bodies, was shown here to recreate a phase transition in the degree of aggregation as the environment temperature was varied. Simply put, the model confirms that when it is cold a huddle will form and when it is warm the pups will disperse; at the critical temperature between huddling and dispersion, pup flow peaks, as predicted by earlier related models. This result confirms the behaviour of the earlier models [6,10], which simulate the pups as two-dimensional rigid bodies, and suggests that a temperature-mediated phase transition is a robust feature of the self-organising interactions assumed by these more abstract descriptions of huddling.

The present result adds to a program of work in which thermoregulatory huddling has provided a lens on a range of questions about self-organisation. In what might be thought of as a *zero*-dimensional model, the litter were described using a single equation, allowing the effects of huddling on the evolution of a population of litters to be investigated in simulation [18]. In what might be thought of as a *one*-dimensional model, the litter were described as a system of magnetic spins fixed in position on a lattice, allowing the thermodynamics of huddling interactions to be understood with reference to the theory of Ising spin-glass models and statistical physics [10]. In what might be thought of as a *two*-dimensional model, the litter were described as a Vicseck system [19], as particles moving around in a 2D arena, and this model provides an important existence-proof that realistic aggregation patterns can emerge under basic assumptions about the physical interactions between animals [6]. The present model can be thought of as a further extension to *three*-dimensions, which may be used to explore how morphological factors constrain self-organisation in real systems of interacting animals.

The patterns of movement generated by animals during early postnatal development constrain the nature of the sensory signals that drive brain development, in particular in highly plastic structures such as the neocortex. For rats and mice, thermoregulatory huddling behaviours dominate the behavioural repertoire during the first three postnatal weeks, when the thermal physiology is maturing and the neocortex is most plastic. Movement of the animal introduces correlations between sensory receptors *within* a sensory modality, e.g., whisker deflection patterns are correlated by the movement of the face relative to tactile stimuli. Self-organising map models, based on local competitive interactions between neurons and Hebbian learning, explain how such correlations shape cortical circuitry, e.g., explaining somatotopic alignment between representations of whisker movement direction in the developing primary somatosensory (barrel) cortex [14]. As well as constraining the statistical structure within a modality, huddling also introduces patterns of correlation *between* modalities that might be captured by the same mechanisms of cortical map self-organisation in multimodal areas. For example, an orienting movement that allows a cold animal to contact a warm littermate on its left introduces correlations between touch (i.e., subsequent contacts made on the left side of the body), vision (i.e., a left-to-right optic flow pattern), audition (i.e., the sound of a contact, or the subsequent vocalisation of the littermate on the left), and correspondingly distinct vestibular and olfactory patterns. Importantly, during huddling these multimodal correlations are experienced in the context of thermo-tactile reward, hence via huddling cortical map self-organisation may be interrogated in a situation where the *function* of the maps may be clearly defined.

The rodent huddle has been suggested to be an ideal 'biohybrid system', in which synthetic littermates might interact with real animals in order to collectively thermoregulate [20]. To this end, ongoing work with the three-dimensional huddling model is investigating the relative contribution and interplay between a number of factors thought to influence the emergence of the huddle, including differences in body shape, relative body mass, and differences in the rates of thermogenesis, e.g., between males and females. Systemmatic manipulation of these factors in controlled simulations may reveal how the patterns of sensory and motor experience in the huddle constrain the inputs to the developing rodent brain. Reproducing similar mechanisms of brain development in the articifical neural networks of a synthetic littermate might enable it to learn to control and manipulate the self-organisation of a biohybrid huddle.

References

1. Schank, J.C., Alberts, J.R.: The developmental emergence of coupled activity as cooperative aggregation in rat pups. Proc. Biol. Sci. **267**(1459), 2307–2315 (2000)
2. Gilbert, C., McCafferty, D., Le Maho, Y., Martrette, J.M., Giroud, S., Blanc, S., Ancel, A.: One for all and all for one: the energetic benefits of huddling in endotherms. Biol. Rev. **85**(3), 545–569 (2010). Cambridge Philosophical Society
3. Canals, M., Bozinovic, F.: Huddling behavior as critical phase transition triggered by low temperatures. Complexity **17**(1), 35–43 (2011)
4. Alberts, J.R.: Huddling by rat pups: ontogeny of individual and group behavior. Dev. Psychobiol. **49**(1), 22–32 (2007)
5. Schank, J.C., Alberts, J.R.: Self-organized huddles of rat pups modeled by simple rules of individual behavior. J. Theor. Biol. **189**(1), 11–25 (1997)
6. Glancy, J., Gross, R., Stone, J., Wilson, S.P.: A self-organising model of thermoregulatory huddling. PLoS Comput. Biol. **11**(9), e1004283 (2015)
7. Canals, M.R., Rosenmann, M., Bozinovic, F.: Geometrical aspects of the energetic effectiveness of huddling in small mammals. Acta Theriol. **42**(3), 321–328 (1997)
8. May, C.J., Schank, J.C., Joshi, S., Tran, J., Taylor, R.J., Scott, I.E.: Rat pups and random robots generate similar self-organized and intentional behavior. Complexity **12**(1), 53–66 (2006)
9. Schank, J.C.: The development of locomotor kinematics in neonatal rats: an agent-based modeling analysis in group and individual contexts. J. Theor. Biol. **254**(4), 826–842 (2008)
10. Wilson, S.P.: Self-organised criticality in the evolution of a thermodynamic model of rodent thermoregulatory huddling. PLoS Comput. Biol. **13**(1), e1005378 (2017)
11. Müller, M., Heidelberger, B., Teschner, M., Gross, M.: Meshless deformations based on shape matching. ACM Trans. Graph. **24**(3), 471–478 (2005)
12. Urashima, H., Wilson, S.P.: A self-organising animat body map. In: Duff, A., Lepora, N.F., Mura, A., Prescott, T.J., Verschure, P.F.M.J. (eds.) Living Machines 2014. LNCS, vol. 8608, pp. 439–441. Springer, Cham (2014). doi:10.1007/978-3-319-09435-9_55
13. Stafford, T., Wilson, S.P.: Self-organisation can generate the discontinuities in the somatosensory map. Neurocomputing **70**, 1932–1937 (2007)
14. Wilson, S.P., Law, J.S., Mitchinson, B., Prescott, T.J., Bednar, J.A.: Modeling the emergence of whisker direction maps in rat barrel cortex. PLoS ONE **5**(1), e8778 (2010)
15. McGregor, S., Polani, D., Dautenhahn, K.: Generation of tactile maps for artificial skin. PLoS ONE **6**(11), e26561 (2011)
16. Wilson, S.P., Bednar, J.A.: What, if anything, are topological maps for? Dev. Neurobiol. **75**(6), 667–681 (2015)
17. Bednar, J.A., Wilson, S.P.: Cortical maps. Neuroscientist **22**(6), 604–617 (2016)
18. Glancy, J., Stone, J.V., Wilson, S.P.: How self-organization can guide evolution. Roy. Soc. Open Sci. **3**(11) (2016)
19. Vicsek, T., Czirók, A., Ben-Jacob, E., Cohen, I., Shochet, O.: Novel type of phase transition in a system of self-driven particles. Phys. Rev. Lett. **75**, 1226–1229 (1995)
20. Wilson, S.P.: The synthetic littermate. In: Lepora, N.F., Mura, A., Krapp, H.G., Verschure, P.F.M.J., Prescott, T.J. (eds.) Living Machines 2013. LNCS, vol. 8064, pp. 450–453. Springer, Heidelberg (2013). doi:10.1007/978-3-642-39802-5_63

Bio-inspired Tensegrity Soft Modular Robots

D. Zappetti[(⊠)], S. Mintchev, J. Shintake, and D. Floreano

Laboratory of Intelligent Systems, Ecole Polytechnique Fédérale
de Lausanne (EPFL), 1015 Lausanne, Switzerland
davide.zappetti@epfl.ch

Abstract. In this paper, we introduce a design principle to develop novel soft modular robots based on tensegrity structures and inspired by the cytoskeleton of living cells. We describe a novel strategy to realize tensegrity structures using planar manufacturing techniques, such as 3D printing. We use this strategy to develop icosahedron tensegrity structures with programmable variable stiffness that can deform in a three-dimensional space. We also describe a tendon-driven contraction mechanism to actively control the deformation of the tensegrity modules. Finally, we validate the approach in a modular locomotory worm as a proof of concept.

Keywords: Modular robots · Soft robotics · Tensegrity structures

1 Introduction

The quest for reconfigurable modular robots dates back to von Neumann's early speculations that one would need a dozen types of simple modules fabricated in millions and floating in an agitated medium [1] to Fukuda's cellular robots composed of self-contained mobile units that could assemble to carry out diverse tasks [2]. Over the past 30 years, this challenge has been tackled by the growing fields of modular robotics [3] and swarm robotics. While there is not enough space here to review those two fields, it is important to point out that modular and swarm systems fall in two categories: systems made of complex units that can individually move or perform some tasks and systems made of simple units that can move or perform tasks only when they are assembled with other units. In this paper we are concerned with the latter type of systems where morphofunctional properties emerge from the assembly of multiple units, as in biological multi-cellular organisms. Since most modular artificial systems are made of rigid components, they display some scalability challenges [4]: they may require precise motion for docking of terminal units, the morphological space may be limited to simple geometries, and the resulting structure may be relatively rigid for robust and safe interaction with the environment.

On the other hand, biological multi-cellular systems are made of cells that display different degrees of stiffness: very soft, such as fat cells, very stiff, such as bone cells, to directional stiffness, such as hair cells, and variable stiffness, such as muscle cells, for example. A soft modular robot made of programmable-stiffness modules could display physical compliance providing safer and better interaction with the environment and

M. Mangan et al. (Eds.): Living Machines 2017, LNAI 10384, pp. 497–508, 2017.
DOI: 10.1007/978-3-319-63537-8_42

span a larger morpho-functional space, including curved bodies, soft bodies, such as worms, and complex bodies, such as vertebrates.

Recently, some authors have described examples of soft modular robots and most of them make use of pneumatic actuation and inflatable modules [5–7]. J.Y. Lee et al. [8] proposed a Soft Robotic Blocks kit (SoBL) composed of three basic types of pneumatically actuated soft modules. Each type of module implements a single type of motion (i.e. translation, bending or twisting) and can be assembled in branch-like structures to achieve more complex motion patterns. However, pneumatic actuation requires several pumps and complex network of pipes that make the system complex and bulky for a high number of independent controlled modules. Non-pneumatic approaches have also been used. Yim et al. [9] described a chain of soft cubes that could return to a previously designed 3-D shape after being stretched. Wang et al. [10] describe a variable stiffness actuated hinge that can be used to form modules capable of being assembled in various different deployable structures. B. Jenett et al. [11] describe the use of two-dimensional compliant modules assembled in three-dimensional structures. The stiffness and density of the assembly can be programmed locally based on the number and the orientation of the modules. Some of us have recently described a soft multi-cellular robot [12] where each cell is a ring made of PDMS at desired stiffness value with embedded magnets. The cells could connect in chains and, by automatically selecting the appropriate cell stiffness and the position of the magnetic contact points, they could approximate a large variety of desired curved morphologies by magnetic repulsion or attraction of the magnets in adjacent cells. The cells have also been functionalized by adding an internal network of nitinol fibers connecting opposite points of the ring that could contract the cell in desired directions, akin to muscle cells [13]. However, the system was constrained to two dimensions.

Here we propose a design principle for three-dimensional soft modules based on tensegrity structures inspired by the cytoskeleton of biological cells. Tensegrity structures are lightweight, can undergo large deformations generated by internal or external forces, and can resist large compressive forces. We describe a method for designing and assembling three-dimensional modules with planar manufacturing technologies. Moreover, the stiffness of the cells can be programmed by changing some parameters of the manufacturing process. We also describe a method to add contraction movement by means of tendon-driven actuation and validate the approach in a proof-of-concept crawling, multi-modular worm.

The paper is organized as follows: in Sect. 2, the tensegrity approach is introduced. In Sect. 3 the main tensegrity structure module is selected, and materials, manufacturing tools and processes are described. The tendon-driven contractive actuation that can be added to the module is also presented. In Sect. 4 a proof of concept soft modular crawling robot is presented. Finally, in Sect. 5 we discuss future work.

2   Tensegrity in Biological and Robotic Systems

In order to develop soft modules with programmable stiffness for our modular robot, we took inspiration by the mechanical structure of multicellular organisms cells. These organisms are heterogeneous systems made of cells with very diverse functions and mechanical properties. However, every cell has a cytoskeleton, which is an extremely complex network made of two types of interconnected fibers [14]: rigid microtubules and bendable actin filaments (Fig. 1a). This architecture, which is responsible for the shape, stiffness, and strength of the cell can be formally described as a tensegrity structure [15].

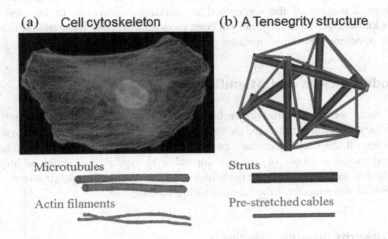

(a)   Cell cytoskeleton   **(b) A Tensegrity structure**

Microtubules   Struts

Actin filaments   Pre-stretched cables

Fig. 1. (a) Image of a cell obtained with fluorescence microscopy. In blue is the nucleus of the cell and at the bottom a sketch of the two main constituents of the cytoskeleton: microtubules and actin filaments [14]. (b) Example of a tensegrity structures: the icosahedron tensegrity structure composed of six struts and 24 cables. (Color figure online)

The term "tensegrity" has been coined by architect R. Buckminster Fuller to describe a structure that maintains its mechanical integrity throughout the pre-stretching of some elements constantly in tension (called "cables" in red in Fig. 1b) connected in a network with other elements constantly under compression (called "struts" in green in Fig. 1b) [15]. In the cytoskeleton, the microtubules and the actin filaments function as struts cables, respectively. The cytoskeleton has different values of stiffness (e.g. resistance to external loads) according to the level of pre-stretch of its actin filaments (i.e. the higher the pre-stretch, the higher the stiffness of the cytoskeleton) [15]. Similarly, the stiffness of a tensegrity structure with a given network configuration depends on the pre-stretch of its cables. If the pre-stretch is properly tuned, the tensegrity structure has low stiffness in all directions (e.g., as in elastomers and living matter), and display a behavior akin to soft matter [15].

In addition to describing the cytoskeleton of the cells, the tensegrity model has been used to describe several structures at many scales of the life [16]. An example at the macroscopic level is the skeletal system of the human body, where bones, tendons and muscles are elements continuously under tension or compression to keep the body in its mechanical integrity [16, 17]; an example at the nanoscale is some proteins and basic molecules that maintain integrity with tensegrity structure [17]. The scalability of the tensegrity model is therefore an important asset for modular robot design because it could be employed at different levels, although manufacturing methods may vary according to the chosen scale.

The use of tensegrity structures to develop biologically inspired robots has already been suggested by Haller et al. [18], while Rieffel et al. suggested the use of tensegrities to achieve "mechanical intelligence" [19]. Yet another example of a tensegrity robot is SUPERball developed at NASA [20], which is able to roll, even on rough terrains, with an optimized control of the six embedded actuators. However, to the best of the author's knowledge, there have not yet been proposals of using tensegrity structures as a design principle for simple modules of a modular, functional robot.

3   Module Design and Manufacturing

In this paper we aim at studying robotic modules in the scale of few centimeters because they can be manufactured using affordable and simple manufacturing methods, incorporate off-the-shelf electronic components, and eventually be assembled into functional modular robots of a size comparable to typical household or inspection robots. In this section we describe the selection of the main tensegrity structure module, the choice of materials, and the manufacturing method.

3.1   Tensegrity Structure Selection

Different tensegrity structures can be obtained according to the number of struts and cables, the network configuration (e.g. struts and cables positions in the three-dimensional space), the cables' stiffness and how much they are pre-stretched [21]. In this study, for sake of simplicity, we assume that all the cables of the tensegrity structure have the same pre-stretch value.

Three main criteria have been applied to select the main tensegrity structure for a modular tensegrity robot. The first criterion is that the main soft module should be able to deform (e.g. stretch or contract) in all directions to augment the morphological diversity of the assembly and comply with objects and surfaces in a three-dimensional space. The second criterion is that the tensegrity structure should involve the smallest possible number of struts and cables to ease the manufacturing and assembling. The third criterion is to favor network configurations whose inner volume is not crossed by any cable or strut in order to place and protect a useful payload (e.g. actuator, microcontroller, energy storage, sensor, depending on the cell type).

To assess the first criterion, we had to consider that tensegrity structures can display different values of stiffness along different directions. However, the more the structure is symmetric, the more it will exhibit similar mechanical properties along different directions [21]. The tensegrity structures that use the most symmetric networks are those with an almost spherical shape [22]. Among these, to assess the second criterion, we selected the tensegrity structure that has the smaller number of struts and cables: the icosahedron tensegrity (Fig. 2) [20, 22]. It is composed of 6 identical struts and 24 cables of equal length. The cables are organized in 8 equilateral triangles interconnected by 12 vertices and distributed in 4 parallel opposite pairs (in Fig. 2b the pairs of

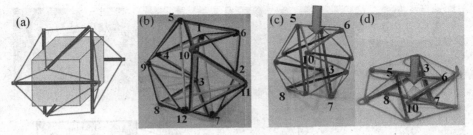

Fig. 2. (a) Icosahedron tensegrity structure with a grey cube at the geometrical center to better display the inner cubic volume not crossed by any cable or strut. (b) Icosahedron tensegrity prototype with the 4 couples of parallel equilateral triangular faces marked with 4 different colors. (c) An external load is applied along the orthogonal direction to the two parallel triangular faces (a "collapsibility direction"). (d) The collapsed structure. (Color figure online)

triangles are marked with four different colors). Furthermore, the icosahedron tensegrity structure has an inner cubic volume that is not crossed by any cable or strut (Fig. 2a), satisfying the third criterion.

When the icosahedron structure is compressed along a direction orthogonal to any of the four triangle pairs (Fig. 2c), it displays maximum deformability and can be collapsed to a flat configuration (Fig. 2d). These four directions, which we name "collapsibility directions", allow the structure to deform in three-dimensional space.

3.2   Design and Manufacturing

Instead of manufacturing every single component separately, here we propose to manufacture all cables as a single flat network that can be folded into a three-dimensional structure and subsequently filled with struts. The cable network can be rapidly manufactured using inexpensive 3D printers.

Two different materials are used for cables and struts. The cables require an elastic printable material that can withstand a wide range of pre-stretch and deformations of the module without losing its elastic properties; the struts instead require a stiffer material that can withstand compressive forces without buckling [21]. For cables, we use NinjaFlex, an elastic material compatible with commercially available fused deposition modeling 3D printers that can withstand 65% of elongation at yield. For struts, we use pultruded carbon rods (that are commercially available in different diameters and length) with a longitudinal Young's modulus of approximately 90 GPa, which is sufficient to withstand compression and buckling.

To design the flat cable's network, a 3D CAD model of the tensegrity module is realized (Fig. 3a). The model is composed of the 8 equilateral triangles of the icosahedron tensegrity assembled with joints at the 12 vertices. The unfolded flat network is obtained by eliminating the 6 struts and disconnecting the cables at two vertices (see Fig. 3b), and then rotating the triangles around the joints until obtaining a flat network's configuration (see Fig. 3c and d). The nodes of the network are marked according to the corresponding vertex of the 3D tensegrity model. The two vertices that

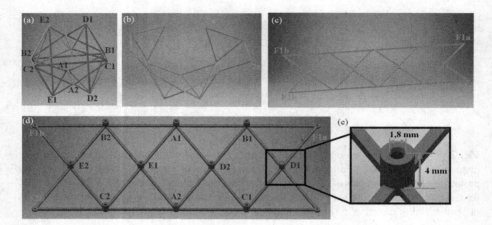

Fig. 3. (a) 3D CAD model of a tensegrity icosahedron. In yellow are marked the two vertices disconnected to unfold the cable's network. (b-c) The unfolding sequence obtained rotating the triangles around the joints in the vertices, in yellow are marked the 4 nodes that will be overlapped to generate vertices F1 and F2. (d) Flat cable's network with housings in the vertices to insert the carbon rods. (e) Detailed view of one of the housings. (Color figure online)

were disconnected in software to unfold the network will be overlapped to close it back during the physical assembly.

To assemble the rigid struts in the elastic network, cylindrical housings are designed at all the nodes (see Fig. 3d and e). The housings are 4 mm high and have an inner diameter equal to the one of the carbon rods in order to ease the insertion. Thanks to this type of design, no additional connection elements or adhesives are required to connect the struts, which will be kept in place and secured by the pre-stretch of the cables after the physical assembly.

The cable network is 3D printed with a Desktop 3D printer LulzBot TAZ 5. After 3D printing, the carbon rods are assembled following the illustration in Fig. 3d: the two vertices of a rod are inserted in the two housings with the same letter starting from the housing "A" (e.g., the ends of the first strut will be inserted in A1 and A2 housings) and following the alphabetical order until the sixth and final strut "F" which has to be inserted in four housings (one side of the carbon rod is inserted in the two F1a-b housings and the other in the two F2a-b).

The icosahedron tensegrity prototype in Fig. 2c is made of an elastic network of thickness and width of 1 mm and with cables long 4.75 cm each, has a height of 7.8 cm and can be approximated to a sphere of the same diameter, therefore occupying a volume of about 248 cm^3. The printing process requires approximately 30 min for the icosahedron elastic network with an infill of 100% and 0.25 mm of vertical resolution. When collapsed along any of the four different directions, the structure reduces its volume by 84% to about 40 cm^3 (Fig. 2d).

The stiffness of the tensegrity modules mainly depends on the stiffness of the cable networks that deforms when the module is compressed. Hence, by modifying the cable's stiffness, it is possible to tune the elastic behavior of the entire module. This can be achieved by changing some design parameters, such as the thickness, width, material of the cables or their pre-stretch, which is defined by the ratio between struts and cables lengths. All cables have the same length and, if shortened, they become more stretched during the assembly (when the struts length is kept constant) and result into a stiffer module. Different prototypes with different pre-stretches and thicknesses of the cables have been manufactured and tested. The pre-stretch ranges from less than 1% up to 30% (after which the manual assembly becomes difficult) and the thickness from 1 mm to 3 mm. Every module has been compressed 50% of its height and for every module a load versus compression curve has been obtained. The Fig. 4a shows increasing stiffness of the modules with increasing values of pre-stretch at a constant thickness of 1 mm, although there is no sensitive variation over 15%. Figure 4b shows higher stiffness with increasing thickness.

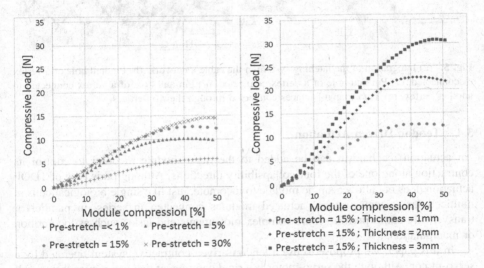

Fig. 4. (a) Compressive load versus module compression of manufactured modules with different pre-stretch and fixed 1 mm thickness. The compression tests have been conducted with an Instron universal testing machine. (b) Modules with 15% pre-stretch and different thicknesses.

3.3 Connectivity

A connection mechanism is required to physically connect modules into a modular robot that can display some functionality. Connection points may also serve as a medium to transmit information and energy throughout the robot. Furthermore, the number and type of connection points in a module may affect the morphology and functionality of the robot. However, at this stage we use only a simple mechanical latching that allows us to assemble cells into a proof-of-concept functional robot.

The mechanical latching system connects vertices and faces of different modules. The system is integrated into the flat cable's network configuration and therefore manufactured with all the other cables during the 3D printing process (see Fig. 5a).

The system consists of a pin and a hole at every vertex. The pin has a slightly larger diameter than the hole and when forced into it, it remains thanks to the friction between the lateral walls of pin and hole. An opposite pressure is required to detach them. The pin and hole can be inserted inside one another when the vertex is not connected to another vertex (see Fig. 5b and c) or can be latched to another vertex with each pin inserted in the hole of the other vertex (see Fig. 5d). Two triangular faces of two modules can be connected latching their 6 vertices in 3 pairs (see Fig. 5e).

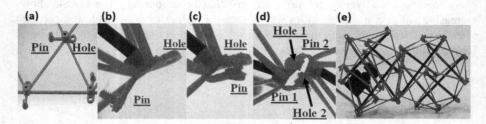

Fig. 5. (a) Integration of the latching system in the cable's network. (b) Pin and hole of a vertex disconnected. (c) Pin and hole of a vertex connected. (d) Pin and hole of a vertex connected to another vertex. (e) Two triangular faces connected through their 6 vertices.

3.4 Tendon-Driven Actuation

An actuation mechanism can be added to the main module in order to control its contraction along one of the four collapsibility directions. Although this type of 1DOF actuation does not allow a single module to locomote, just like single biological cells in multicellular systems, several actuated modules connected in series, in parallel or transversally could perform more complex movements and tasks, such as locomotion or manipulation.

In this paper, we propose to use a tendon-driven contractive system operated by a servo-motor. Although the servo-motor is a rigid component that can reduce the overall softness of the system [23] and the volume reduction in fully collapsed state, the chosen motor has a small volume (5 cm^3) compared to that of the module (248 cm^3). The servo is strategically placed in a modified strut with rectangular housing. This design avoids rigid connections between two struts and the rigid servomotor and preserve the tensegrity nature of the structure and its deformability [21].

The tendon-driven actuation contracts the icosahedron tensegrity by simultaneously pulling two opposite triangular faces towards the geometrical center of the structure along the collapsibility direction (see Fig. 6a). The contractive mechanism comprises six tendons that connect each vertex of the selected triangular faces to a pulley that is activated by a servo-motor (see Fig. 6b). The pulley is placed at the geometrical center of the icosahedron structure (see Fig. 6b). When the servo is activated, the pulley starts to rotate wrapping the tendons, which in turn produce a contraction of the icosahedron (see Fig. 6c).

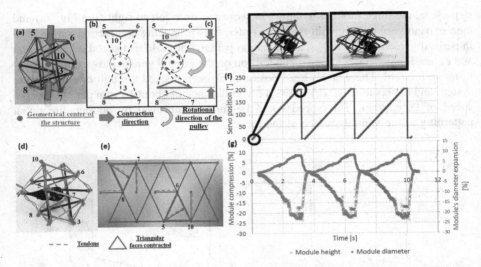

Fig. 6. (a) The arrows represent the contractive directions of the two triangular faces marked in yellow. While the red dot represents the geometrical center of the tensegrity structure toward which the two faces are contracted. (b) The six dashed lines represent the six tendons connecting the two triangular faces to the pulley placed at the geometrical center of the structure. (c) When the pulley is rotated, the six tendons are all pulled at the same time contracting the two triangular faces. (d) The assembled module with marked in dashed yellow lines the six tendons connecting the 6 vertices of the two opposite triangular faces (marked in blue) to the pulley in the geometrical center of the structure. (e) The unfolded network with the tendons. (f) Graph of the servo position driving signal generated by an Arduino Uno. (g) Corresponding graph of the module compression and lateral expansion recorder through motion capture system (i.e. Optitrack system recording at 240 Hz). (Color figure online)

The six-tendon design can be directly included in the design process of the module's elastic network. The six tendons have one end attached to the vertices of the triangular faces and the other end free (see Fig. 6d and e).

The kinematic tests of the actuated module show that a compression of about 25% (the negative pick at about 3.5 s in Fig. 6g) and a lateral diameter expansion of 9% of the module height can be achieved. Improved design or different actuation technologies could further increase the contraction of the module, if required.

4 A Crawling Modular Robot

We exploited the specific kinematic of the actuated module (i.e. lateral expansion when compressed) to develop a simple crawling modular robot using peristaltic locomotion as a proof-of-concept. The robot is composed of 3 actuated modules connected along the actuated axes with mechanical latching. The three modules are controlled with an external Arduino Uno board using three different signals (Fig. 7a). A driving signal controls the contraction and expansion of a module along the actuated axis. At each step cycle, the three modules contract in sequence from left to right (see Fig. 7a) and then expand very rapidly with the same order, thus producing contraction waves used in peristaltic locomotion [24]. This actuation pattern with different speed in contraction and expansion generates a directional friction on the ground which allows the assembly to move forward. The worm's head position has been tracked through motion capture system and the result (Fig. 7b) shows a travel distance of about 15 mm per cycle and a speed of 90 cm/min. This speed could be further improved by adding directional patterning to the ventral surface of the robot.

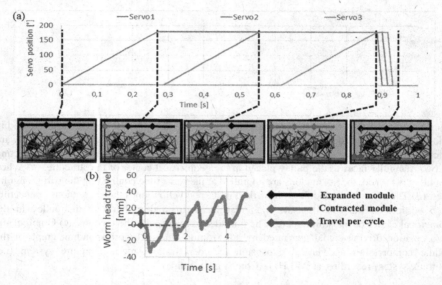

Fig. 7. (a) Contraction snapshots sequence of a peristaltic movement cycle compared with the servo position driving signal given by the Arduino board. (b) Graph of the worm's head movement in the longitudinal direction. Every cycle the worm moves forward of about 6.5 mm.

5   Conclusions and Future Work

In this paper, we presented a design principle for novel, bio-inspired tensegrity soft modular robots. Although the proposed design displays many desirable features that could lead to the assembly of a variety of more complex robots with a diverse set of behaviors, other tensegrity structure modules could be considered for specialized functions within a heterogeneous modular robot. However, several challenges remain to be addressed. In addition to a connection system that provides more functionalities than simple mechanical latching, as shown in this paper, the system could benefit from a better integrated actuation system to replace the conventional servo-motor used here as proof of concept: a possible solution could be a contraction system made of shape memory alloys that fit the reticular structure of the modules.

Furthermore, autonomous modular robots require at least a sensing system, an internal signaling system, a programmable control unit, and an energy system.

For sensing, we are currently considering replacing the elastic material of the cable network with conductive elastic materials that change conductivity when stretched, thus, enabling a simple form of proprioception. More conventional sensors, such as infrared and vision could be hosted within the cell body to perceive the environment with relatively little interference from the cable and struts.

For internal communication, the tensegrity approach could exploit a phenomenon known as mechanotransduction in biology, which is used by cells to activate bio-chemical processes or gene expression. This form of communication could enhance or replace digital electrical communication by propagating mechanical disturbances that alter the function and behavior of the cells. Finally, a microcontroller and energy storage could be placed as payloads in the empty volume of the cells (as shown in Fig. 2a).

Tensegrity simulation tools, such as the NASA Tensegrity Robotics toolkit [25], could be adapted to design and even evolve [20] modular tensegrity robots.

The scalability of tensegrity structures, from stadium domes to biological cells, opens the possibility of conceiving multi-cellular robots at a smaller scale by means of inkjet printing [26], MEMS fabrication [27], or 3D pop-up micro-structures [28].

We believe that, despite the many challenges that remain to address, the biological inspiration from multi-cellular organisms and the recruitment of novel soft robotic technologies and materials could lead to the generation of diverse, resilient, and scalable modular robots based on tensegrity structures.

References

1. Burks, A.: Essays on Cellular Automata. University of Illinois Press, Urbana (1970)
2. Fukuda, T., Nakagawa, S., Kawauchi, Y., Buss M.: Self organizing robots based on cell structures – CEBOT. In: IEEE International Conference on Robotics and Automation, pp. 145–150, 31 October–2 November 1989
3. Moubarak, P., Ben-Tzvi, P.: Modular and reconfigurable mobile robotics. Robot. Auton. Syst. 60(12), 1648–1663 (2012)
4. Yim, B.M., Shen, W.-M., Salemi, B., Rus, D., Moll, M., Lipson, H., Klavins, E., Chirikjian, G.S.: Modular self-reconfigurable robot systems: challenges and opportunities for the future. IEEE Robot. Autom. Mag. 14(1), 43–52 (2007)
5. Morin, S.A., et al.: Elastomeric tiles for the fabrication of inflatable structures. Adv. Funct. Mater. 24(35), 5541–5549 (2014)
6. Vergara, A., Lau, Y., Mendoza-Garcia, R.-F., Zagal, J.C.: Soft modular robotic cubes: toward replicating morphogenetic movements of the embryo. PLoS ONE 12(1), e0169179 (2017)
7. Onal, C.D., Rus, D.: A modular approach to soft robotics. In: 4th IEEE RAS/EMBS International Conference on Biomedical Robotics and Biomechatronics (BioRob), Rome, Italy, pp. 1038–1045, 24–27 June 2012
8. Lee, J.Y., Kim, W.B., Choi, W.Y., Cho, K.J.: Soft robotic blocks: introducing SoBL, a fast-build modularized design block. IEEE Robot. Autom. Mag. 23(3), 30–41 (2016)

9. Yim, S., Sitti, M.: SoftCubes: toward a soft modular matter. In: IEEE International Conference on Robotics and Automation (ICRA), Karlsruhe, Germany, pp. 530–536, 6–10 May 2013
10. Wang, W., Rodrigue, H., Ahn, S.-H.: Deployable soft composite structures. Sci. Rep. **6** (2016)
11. Jenett, B., et al.: Digital morphing wing: active wing shaping concept using composite lattice-based cellular structures. Soft Robot. **4**, 33–48 (2016)
12. Germann, J., Maesani, A., Pericet, Camara R., Floreano, D.: Soft cell for programmable self-assembly of robotic modules. Soft Robot. **1**(4), 239–245 (2014)
13. Germann, J.: Soft cells for modular robots. Ph.D. dissertation, nr. 6217, EPFL, CH (2014)
14. Alberts, B., Bray, D., Lewis, J., Raff, M., Roberts, K., Watson, J.D.: Molecular Biology of the Cell. Garland Publishing, New York (1983)
15. Ingber, D.E., Wang, N., Stamenovic, D.: Tensegrity, cellular biophysics, and the mechanics of living systems. Rep. Prog. Phys. (2014)
16. Ingber, D.E.: The architecture of life. Sci. Am. **278**(1), 48–57 (1998)
17. Krause, F., Wilke, J., Vogt, L., Banzer, W.: Intermuscular force transmission along myofascial chains: a systematic review. J. Anat. **77**(12) (2006)
18. Yu, C., Haller, K., Ingber, D., Nagpal, R.: Morpho: a Self-deformable modular robot inspired by cellular structure. In: IEEE/RSJ International Conference on Intelligent Robot and Systems, Nice, France, 22–26 September 2008
19. Rieffel, J., Trimmer, B., Lipson, H.: Mechanism as mind - what tensegrities and caterpillars can teach us about soft robotics. In: Artificial Life: Proceedings of the Eleventh International Conference on the Simulation and Synthesis of Living Systems, Cambridge, MA, pp. 506–512, 5–8 August 2008
20. Bruce, J., Caluwaerts, K., Iscen, A., Sabelhaus, A.P., SunSpiral, V.: Design and evolution of a modular tensegrity platform. In: IEEE International Conference on Robotics and Automation, (ICRA), Hong Kong, China, 31 May– 7 June 2014
21. de Oliveira, M.C., Skelton, R.E.: Tensegrity Systems, Chap. 1. Springer, New York (2009). doi:10.1007/978-0-387-74242-7
22. Agogino, A., SunSpiral V., Atkinson, D.: SuperBall bot – structure for planetary landing and exploration. Final Report for the NASA Innovative Advanced Concepts, NASA Ames Reasearch Center, Intelligent Systems Division, July 2013
23. Umedachi, T., Vikas, V., Trimmer, B.A., et al.: Softworms: the design and control of non-pneumatic, 3D-printed, deformable robots. Bionspir. Biomim. **11**(2) (2016)
24. Kanu, E.N., Daltorio, K.A., Quinn, R.D., Chiel, H.J.: Correlating kinetics and kinematics of earthworm peristaltic locomotion. In: 4th International Conference of Living Machines, Barcelona, Spain, pp. 92–96, 28–31 July 2015
25. Mirletz, B.T., Park, I., Quinn, R.D., Sunspiral, V.: Towards bridging the reality gap between tensegrity simulation and robotic hardware. In: IEEE/RSJ International Conference on Intelligent Systems (IROS), Hamburg, Germany, 28 September–2 October 2015
26. Singh, M., Haverinen, H.M., Dhagat, P., Jabbour, G.E.: Inkjet printing-process and its applications. Adv. Mater. **22**(6), 673–685 (2010)
27. Gad-el-Hak, M.: The MEMS Handbook. CRC Press, Boca Raton (2010)
28. Whitney, J.P., Sreetharan, P.S., Ma, K.Y., Wood, R.J.: Pop-up book MEMS. J. Micromech. Microeng. **21**(11) (2011)

Consciousness as an Evolutionary Game-Theoretic Strategy

Xerxes D. Arsiwalla[1(✉)], Ivan Herreros[1], Clement Moulin-Frier[1], and Paul Verschure[1,2]

[1] Synthetic Perceptive Emotive and Cognitive Systems (SPECS) Lab,
Center of Autonomous Systems and Neurorobotics,
Universitat Pompeu Fabra, Barcelona, Spain
x.d.arsiwalla@gmail.com

[2] Institució Catalana de Recerca i Estudis Avançats (ICREA), Barcelona, Spain

Abstract. The aim of this article is to highlight the role of consciousness as a survival strategy in a complex multi-agent social environment. Clinical approaches to investigating consciousness usually center around cognitive awareness and arousal. An evolutionary approach to the problem offers a complimentary perspective demonstrating how social games trigger a cognitive arms-race among interacting goal-oriented agents possibly leading to consciousness. We begin our discussion declaring the functions that consciousness serves for goal-oriented agents. From a functional standpoint, consciousness can be interpreted as an evolutionary game-theoretic strategy. To illustrate this, we formalize the Lotka-Volterra population dynamics to a multi-agent system with cooperation and competition. We argue that for small population sizes, supervised learning strategies using behavioral feedback enable individuals to increase their fitness. In larger populations, learning using adaptive schemes are more efficient. However, when the network of social interactions becomes sufficiently complex, including the prevalence of hidden states of other agents that cannot be accessed, then all aforementioned optimization schemes are rendered computationally infeasible. We propose that that is when the mechanisms of consciousness become relevant as an alternative strategy to make predictions about the world by decoding psychological states of other agents. We suggest one specific realization of this strategy: projecting self onto others.

Keywords: Conscious agents · Evolutionary games · Complex systems · Social interactions

1 Introduction

Understanding the nature of consciousness has been an outstanding scientific puzzle at the crossroads of neuroscience and artificial intelligence. In clinical practice, assessments of consciousness (and its disorders, in patients) are based on calibrations of cognitive awareness and arousal [7], leading to a two dimensional operational definition of clinical consciousness. Both awareness and arousal

© Springer International Publishing AG 2017
M. Mangan et al. (Eds.): Living Machines 2017, LNAI 10384, pp. 509–514, 2017.
DOI: 10.1007/978-3-319-63537-8_43

are required in order to generate action and perception in any multi-agent environment. In turn, these action-perception loops are used for maintaining and regulating the agent's homeostatic needs and drives. This leads to goal-oriented behaviors that optimize objective functions or drives based on the agent's predictions of states of the physical world as well as predictions of behavioral states of other agents [2,10]. The former comprises the agent's world model, while the latter, the agent's social or exteroceptive intentional model. Both of these are required in order to engage in social cooperation and competition required for survival.

It has been proposed that evolutionary pressures on social dynamics of interacting agents leads to the emergence of consciousness, which is a process for predicting hidden intentional states of other agents (and self) in order to generate social cooperative and competitive behaviors necessary to optimize an agent's survival drives in a world with limited resources [2,8,10]. In fact, for all known organic life forms, biochemical arousal is a necessary precursor supporting the hardware necessary for cognition. In turn, evolution has shaped cognition in such a way so as to support the organism's basic survival as well as higher-order drives achievable via cooperation and competition in a multi-agent world. Awareness and arousal thus work together in conscious agents.

In this work, we extend the above socio-evolutionary perspective to interpret consciousness as an evolutionary game-theoretic strategy for decoding hidden states of other agents in order to optimize one's own fitness while cooperating or competing with other agents. An evolutionary approach to the problem offers a complimentary perspective demonstrating how social games trigger a cognitive arms-race among interacting goal-oriented agents possibly leading to consciousness. To formalize these concepts, we extend the Lotka-Volterra population dynamics to a multi-agent system including cooperation and competition and qualitatively compare different fitness strategies as the complexity of the multi-agent system scales up.

2   The Function of Consciousness in a Social World

Based on phenomenological considerations, our discussion above suggests at least two generic categories of complexity metrics (see [1,4,5] for a discussion on complexity measures) to label states of consciousness, those associated to computational capabilities (related to mechanisms of awareness) and those referring to autonomous action or survival drives (related to mechanisms to arousal) [3]. In this section we take a functional approach to consciousness, interpreting it as a game-theoretic strategy. We argue that in addition to the two above-mentioned complexity classes, there is a third one: social complexity; resulting in a three dimensional state space for calibrating consciousness.

Following [10] and [2] let us begin with a discussion on the function of consciousness in biological agents and how it could have become necessary in the context of evolution. From this perspective, consciousness is a solution to the problem of autonomous goal-oriented *action* with intentionality priors in a multi-agent social environment. In order to act in the physical world an agent needs to determine a behavioral procedure to achieve a goal state; that is, it has to answer the HOW of action. In turn this requires the agent to solve the following five problems: (1) Define the motivation for action in terms of its needs, drives and goals; that is, the WHY of action. (2) Determine knowledge of objects it needs to act upon and their affordances in the world, pertaining to the above goals; that is, the WHAT of action. (3) Determine the location of these objects, the spatial configuration of the task domain and the location of the self; that is, the WHERE of action. (4) Determine the sequencing and timing of action relative to dynamics of the world and self; that is, the WHEN of action. (5) Estimate hidden mental states of other agents when action requires cooperation or competition; that is, the WHO of action. In [10] this was proposed as the H5W problem that biological agents need to solve in order to act in the world and that consciousness consists of the mechanisms that enable this.

Note that while the first four of the above questions suffices for generating simple goal-oriented behaviors, the last of the Ws (the WHO) is of particular significance as it involves intentionality, in the sense of estimating the future course of action of other agents based on their social behaviors and psychological states. However, because mental states of other agents that are predictive of their actions are hidden, they can at best be inferred from incomplete sensory data such as location, posture, vocalizations, social salience, etc. As a result the agent faces the challenge to univocally assess, in a deluge of sensory data those exteroceptive and interoceptive states that are relevant to ongoing and future action and therefore has to deal with the ensuing credit assignment problem in order to optimize its own actions. Furthermore, this results in a reciprocity of behavioral dynamics, where the agent is now acting on a social and dynamical world that is in turn acting upon itself. Hence, it was suggested in [10] that consciousness is associated to the ability to maintain a transient and autonomous memory of the virtualized agent-environment interaction, that captures the hidden states of the external world, in particular, the intentional states of other agents and the norms that they implicitly convey through their actions.

3 Consciousness as an Evolutionary Game-Theoretic Strategy

From a game-theoretic perspective, a conscious agent is trying to solve the H5W problem while being engaged in social cooperation-competition games with other agents, who are trying to solve their own H5W problem in a world with limited resources. In a scenario with only a small number of other agents, a given agent might use statistical learning approaches to learn and classify behaviors of the few others agents in that game. For example, multiple robots interacting to learn naming conventions of perceptual aspects of the world [9]. In this case, the multi-agent interaction has to be embodied so that one agent can interpret which specific perceptual aspect the other agent is referring to (by pointing at objects). Another example is the emergence of signaling languages in sender-receiver games based on replicator dynamics described by David Lewis in 1969 in his seminal book, Convention. However, in both these examples, strategies that evolutionarily succeed when only few players are involved, are no longer optimal in the event of an explosion in the number of players [6].

As in the above examples, when considering a social environment comprising a large number of agents trying to solve the H5W problem, machine learning strategies for reward and punishment valuations may soon become computationally unfeasible for an agent's processing capacities and memory storage. Therefore, for a large population to sustain itself in an evolutionary game involving complex forms of cooperation and competition would require strategies other than solely behavior-based machine learning. One such strategy involves modeling and inferring intentional states of itself and that of other agents. Emotion-driven flight or fight responses depend on such intentional inferences and so do higher-order psychological drives. The mechanisms of consciousness enable such strategies.

It might be an interesting question to investigate in a game-theoretic setting whether this strategy also brings a multi-agent ecosystem to a Nash equilibrium because in a game where say two species attain consciousness, the population pay-offs in cooperation and competition games between them are likely to reach one of possible equilibria due to the recursive nature of intentional inferences, where an agent attempts to infer the inferences of other agents about its own intentions. In summary, interpreting consciousness as a game-theoretic strategy highlights the role of complex social behaviors inevitable for survival in a multi-agent world. From an evolutionary standpoint, social behaviors result from generations of cooperation-competition games, with natural selection filtering unfavorable strategies. Presumably, winning strategies eventually got encoded into anatomical mechanisms, such as emotional responses. The complexity of these behaviors depends on the ability of an agent to make complex social inferences. This suggests social complexity as a measure of an agent's social intelligence.

4   Formalizing Social Dynamics

As a first attempt to study how social games shape evolutionary strategies we formalize a dynamical model of a multi-agent population that can cooperate or compete for hunting prey. This is an extension of Lotka-Volterra dynamics, where the prey population H is governed by the equation

$$\frac{dH}{dt} = rH\left(1 - \frac{H}{k}\right) - \frac{aH}{c+H}\sum_{i=1}^{n} L_i \tag{1}$$

For the moment, let each L_i denote a population of predators. The constants, r, denotes the birth rate of the prey population, k, denotes the maximum prey population in the absence of predation, a, denotes the strength of predation, and c is a saturating factor for very large prey populations.

We write the predator equations for each population L_i as follows:

$$\frac{dL_i}{dt} = \frac{baH}{c+H}L_i - dL_i + C_{ij}^{op}L_j - C_{ij}^{om}L_j + v_i(t) \qquad (2)$$

where the constant b takes into account the number of prey required for an incremental increase in predator population, and d denotes the death rate of predators. C_{ij}^{op} denotes weights of cooperation among predators, that help increase their survival rates, while C_{ij}^{om} denote weights of competition, that decrease their respective survival rates. We will assume that population i cannot cooperate and compete with population j at the same time. The functions $v_i(t)$ denote time-dependent inputs to the system of equations, which can be used to formulate learning and optimization strategies within the system of equations.

While the variables L_i in the above equations denote n different populations, in order to work with variables that refer to n individual agents, we define $p_i = L_i / \sum_{j=1}^{n} L_j$ as the fraction of the ith population. With respect to a multi-agent system of n distinct predators, the p_i denote relative survival probabilities of individual agents with respect to other agents (since $\sum_{i=i}^{n} p_i = 1$ by definition). To optimize its own fitness, each agent has to increase its relative survival probability by tuning weights (the components of C_{ij}^{op} and C_{ij}^{om}) in its social network of cooperation and competition with other agents. For small population sizes this can be achieved using behavioral feedback of actions of other agents, akin to supervised learning. More sophisticated learning algorithms such using adaptive models of the world may be implemented to improve results. However, like many machine learning techniques, these methods carry the curse of dimensionality. The performance of these learning methods drastically deteriorates when the number of agents n scales to very large proportions (due to heavy computational costs involved). In such cases, it is hypothesized that consciousness serves as an alternative strategy for learning which agents to cooperate with or compete with based on predictions of their (hidden) intentional states.

5   Discussion

Building on earlier work, in this paper we have taken the perspective that evolutionary pressures on social dynamics of interacting agents leads to the emergence of consciousness. Consciousness itself is seen as an evolutionary strategy for learning which agents to cooperate with or compete with based on predictions of their (hidden) intentional states. The missing ingredient in almost all contemporary artificial intelligence approaches to multi-agent interactions is the ability to predict intentional states of self and that of other agents, which are crucial for engaging in complex social behaviors. The role of consciousness is to provide a means to predict these intentional states. Even though the precise biophysical mechanisms by which consciousness solves this problem are still unknown, it might be worth investigating functional strategies that collectively constitute conscious processes. We propose one possible candidate for such strategies, namely, projecting self states (homeostatic levels, goal values, etc.) onto states of other 'similar' agents. This allows an agent to build a first approximation of intended behavioral predictions of other agents, which are then refined using actual behavioral feedback. In the multi-agent social dynamics introduced in this paper, we argued for the need for cooperation and competition strategies that are efficient to learn when social complexity scales up. For future work, it would be interesting to quantify the above-mentioned self onto others strategy and study its quantitative effects on multi-agent social models.

Acknowledgments. This work has been supported by the European Research Council's CDAC project: "The Role of Consciousness in Adaptive Behavior: A Combined Empirical, Computational and Robot based Approach" (ERC-2013- ADG 341196).

References

1. Arsiwalla, X.D., Verschure, P.F.M.J.: Integrated information for large complex networks. In: The 2013 International Joint Conference on Neural Networks (IJCNN), pp. 1–7 (2013)
2. Arsiwalla, X.D., Herreros, I., Moulin-Frier, C., Sanchez, M., Verschure, P.F.: Is Consciousness a Control Process? pp. 233–238. IOS Press, Amsterdam (2016). http://dx.doi.org/10.3233/978-1-61499-696-5-233
3. Arsiwalla, X.D., Herreros, I., Verschure, P.: On Three Categories of Conscious Machines, pp. 389–392. Springer International Publishing, Cham, Switzerland (2016). http://dx.doi.org/10.1007/978-3-319-42417-0_35
4. Arsiwalla, X.D., Verschure, P.: Computing information integration in brain networks. In: Wierzbicki, A., Brandes, U., Schweitzer, F., Pedreschi, D. (eds.) NetSci-X 2016. LNCS, vol. 9564, pp. 136–146. Springer, Cham (2016). doi:10.1007/978-3-319-28361-6_11
5. Arsiwalla, X.D., Verschure, P.F.: The global dynamical complexity of the human brain network. Appl. Netw. Sci. 1(1), 16 (2016)
6. Hofbauer, J., Huttegger, S.M.: Feasibility of communication in binary signaling games. J. Theor. Biol. 254(4), 843–849 (2008)
7. Laureys, S., Owen, A.M., Schiff, N.D.: Brain function in coma, vegetative state, and related disorders. Lancet Neurolog. 3(9), 537–546 (2004)
8. Moulin-Frier, C., Puigbò, J.Y., Arsiwalla, X.D., Sanchez-Fibla, M., Verschure, P.F.: Embodied artificial intelligence through distributed adaptive control: An integrated framework. arXiv preprint (2017). arXiv:1704.01407
9. Steels, L.: Evolving grounded communication for robots. Trends cogn. Sci. 7(7), 308–312 (2003)
10. Verschure, P.F.: Synthetic consciousness: the distributed adaptive control perspective. Phil. Trans. R. Soc. B 371(1701), 20150448 (2016)

Using the Robot Operating System for Biomimetic Research

Alexander Billington, Gabriel Walton, Joseph Whitbread,
and Michael Mangan[✉]

Lincoln Centre for Autonomous Systems, University of Lincoln, Lincoln, UK
mmangan@lincoln.ac.uk
http://staff.lincoln.ac.uk/mmangan

Abstract. Biomimetics seeks to reveal the methods by which natural systems solve complex tasks and abstract principles for development of novel technological solutions. If these outcomes are to either explain behaviour, or be applied in commercial settings, they must be verified on robot platforms in natural environments. Yet development and testing of hypothesis in real robots remains sufficiently challenging for many in this highly cross-disciplinary research field that it is often omitted from biomimetic studies. Here we evaluate whether the Robot Operating Systems (ROS) can address this issue taking desert ant navigation as a case study. We demonstrate and discuss both the strengths and weaknesses of the current ROS implementation with regard to the specific needs of biomimetic researchers of varying technical experience, and describe the establishment of our central code repository with user guides to aid novice users.

Keywords: Navigation · Insect · Robot · Robot Operating System · Path integration · Visual compass

1 Introduction

Biomimetics aims to develop novel technological solutions to engineering goals, such as autonomous navigation, by abstracting fundamental principles found in biological systems. The gold-standard test of solutions is their verification in the same natural environments in which animal data was recorded allowing direct comparison between animal and robot behaviour. Verification typically follows a two-stage process: initial model implementation and testing in simulated environments including time-consuming operations such as parameter tuning followed by final evaluation on a robot in the real world.

Despite adopting a common methodology the biomimetics research community has yet to adopt of a common set of technical tools, standards, and coding practices. For example, recent models of insect visual navigation (the case study used throughout this paper) have been developed and tested in MATLAB based 3D environments [1,2]; in C++ based simulations and image database studies

© Springer International Publishing AG 2017
M. Mangan et al. (Eds.): Living Machines 2017, LNAI 10384, pp. 515–521, 2017.
DOI: 10.1007/978-3-319-63537-8_44

[3]; and on android-powered mobile robots [4] making benchmarking problematic without code reimplementation. Furthermore, tools such as MATLAB lack features of dedicated simulation environments such as realistic lighting and physics. In addition, this fragmentation of tools often leaves the process of porting computational models from simulation environments to robot platforms cumbersome due to the level of code refactoring and optimisation required.

Facing similar challenges, the applied robotics research community developed a robot software framework known as the Robotic Operating System (ROS). ROS aims to provide a robot software development pipeline allowing seamless transfer of code, developed and tested in modern simulation packages, to real world robot platforms. This is supported by an active development community in which code sharing is encouraged, allowing iterative development and easy benchmarking of models. The standardised communications protocols defined by ROS has the potential to lower the barrier to entry for new users as they can access all the features of ROS with knowledge of a single programming language.

Here we evaluate the utility of ROS for biomimetic studies of navigation using desert ant navigation as a case study. Rather than assessing performance of the described navigation models per se, we instead report on the ease, efficacy, and suitability of the ROS framework for use in this highly cross-disciplinary research field. We intend for this paper to act as a guide for researchers of all technical proficiencies who may consider adopting ROS and code sharing practices in their biomimetic studies.

2   Methods

2.1   ROS Overview

At a high-level, ROS can be considered as a middleware with a clearly defined set of protocols allowing communication between sensors, actuators and control scripts independent of the specific hardware. It provides low-level device control, implementation of commonly used functionality, message-passing between processes, and package management shielding developers from these details. Control scripts (known as nodes in ROS) interact with other nodes along with sensors and actuators using a well-defined publisher/subscriber scheme. For example, a control node may subscribe to sensory streams such as camera data and publish motors command outputs. This control code can be tested in simulation or on a real robot with little to no amendments. ROS supports powerful vision libraries such as OpenCV and is tightly integrated with simulation environments such as Gazebo. The C++, Python and LISP programming languages are well supported with some support for Java via ROSJava. Detailed information about ROS can be found at http://www.ros.org.

In this work, we evaluate the ROS Indigo package using a standard desktop computer (3.5 GHz Intel i5, 16 GB RAM with dedicated GPU) running Ubuntu 14.04, OpenCV 2.4.8, and Gazebo 2.2.3. Code was written in Python 2.7 using the Spyder IDE and all scripts are available with a wiki guide to usage at https://github.com/LCAS/A.N.T.

2.2   Generating Custom Simulated Environments

We generated two simple 3D environments: a featureless space mimicking the desert salt-pans where ants are known to navigate by path integration (PI) [5] and a single landmark in an otherwise featureless space in which Judd and Collett [6] reported visual homing behaviours (Fig. 1D). The Gazebo simulation tool (http://gazebosim.org) which is closely integrated with ROS was used which provides physics, lighting and visualization of the robot model within a 3D space.

2.3   Generating Custom Robots

We constructed a robot model allowing sensing, movement and interaction of a simulated robot in the simulation environment. ROS uses the Universal Robotic Description Format (URDF) and the XML Marco Format (Xacro) to model all components and connections between components comprising the robot. For example, the URDF would describe the chassis and how it is connected to the wheels. Many common robot systems (e.g. PR2, Baxter, Turtlebots) are made available for download from the ROS website however for many biometric studies a custom robot is required.

A custom robot-base was created to emulate the sensory experience of a desert ant (Fig. 1B insert). Specifically, a box-shaped chassis was augmented with an internal odometer (a standard sensory module within ROS), three 120° field of view cameras providing panoramic vision (grey protrusion from chassis), and four wheels mounted on the side of the chassis to generate movement. As ROS does not provide inherent localisation for the robot, a laser scanner was added so that the gmapping package (http://wiki.ros.org/gmapping) could be used to track the robot position for subsequent plotting (note that this data was never provided to the navigation models described below).

2.4   Generating Navigation Behaviours

Path Integration. Desert ants are famed for their PI abilities (Fig. 1A) whereby they keep track of their position relative to the origin by continuously monitoring their speed and orientation changes (for a review see [5]). We implemented a simple trigonometry based PI controller in ROS using the standard odometer module providing realistic distance and orientation sensing (including noise) from wheel encoder sensors.

Visual Compass. Ants are also experts at relocating previously visited locations using surrounding visual cues: a behaviour termed visual homing (VH) [7]. Here we replicate the visual approach to a target observed by Judd and Collett [6], using the same visual compass (VC) method described by Wystrach et al. [8]. Specifically, we supply the robot with a set of visual snapshots oriented towards to the nest at random points (Fig. 1D upper). The homeward direction is retrieved by rotating on the spot until the best match is found between the current view and the stored views using the pixel-wise sum-squared error, before moving in that direction. By iteratively finding the best match and stepping in that direction the target location will be recovered.

2.5 Robot Platform

Finally, we conducted robot trials to assess the ease with which ROS allows code, created for a custom robot model in simulation, could be applied to a standard robot platform in an office environments. Experiments were performed using a Turtlebot robot with a wireless camera with fisheye lens providing panoramic vision (Fig. 2A)

3 Results

Figure 1B and D show the successful replication of ant-like PI and VH behaviours in simulation and Fig. 2B, C, and D demonstrates the same code controlling a robot in an office environment.

4 Discussion

We present a proof of concept study demonstrating that the Robot Operating System provides all the necessary components for biomimetic studies of navigation. This ranges from code development and testing using modern simulation tools complete with physics and lighting engines, to the verification of the same software on real world robot platforms without the need for refactoring or optimisation of the codebase. The primary benefit of wide adoption of this paradigm by the biomimetic community would be the streamlining of work-flow allowing rapid of hypothesis deployment onto robots. Secondary benefits such as model benchmarking and distributed development can be achieved through the use of modern code-sharing practices such as open code repositories, shared data-sets, and agreed benchmarking protocols. Such benefits are already being realised in the applied robotics field.

One possible barrier to the wide uptake of ROS may be a steep learning curve perceived by biomimeticists with limited programming experience. For example, new users may be faced with learning to use a new operating system, a new programming language, and the ROS communications protocols concurrently. This may be compounded for novices that may have difficulty accessing support that is largely provided by the user community through wikis and forums, in comparison to dedicated support of more commercial software. Yet, we would argue that as with increased adoption, the community will increasingly support itself with code examples, user guides, and dedicated user groups making uptake easier.

It should be noted that we identified some limitations in the current ROS tool-chain that affect its utility to studies of navigation specifically. For example,

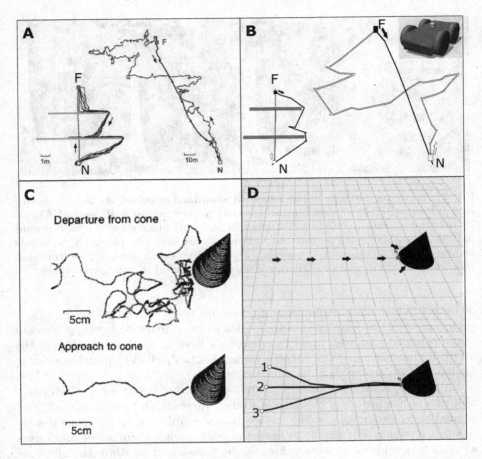

Fig. 1. Simulating ant behaviours in ROS *A.* PI behaviours observed in desert ants (adapted from [9]). The right path shows a long range path, and the left the update of PI after obstacles are placed in the direct path home. (N and F represent Nest and Feeder locations throughout) *B.* Replication of PI behaviours using custom ant-sized robot model (shown in insert) in simulation. Outward paths were under direct control of a user using the tethered PC, and obstacle avoidance triggered by laser scanner *C.* Visual approach behaviours observed in ants (adapted from [6]). (Top) An example learning walk with nestward facing phases when memories are thought to be stored, and (Bottom) a subsequent homing path *D.* Replication of visual approach behaviour. (Top) Stored memories marked by black arrows, (Bottom) homing paths from three locations using the visual compass method

there exists no built-in functionality to record the path of the robot independent of a map - posing problems if testing models in open environments such a featureless salt pans. We circumvented this issue by adding a laser scanner to our ant model and using the gmapping tool to record ground-truth but this may be impractical for many studies. Moreover, while Gazebo allows for easy creation

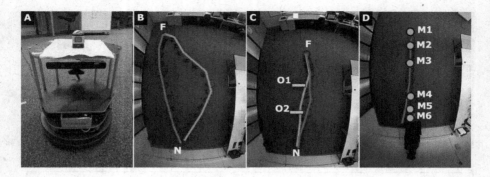

Fig. 2. Applying antlike behaviours on standard a robot *A*. Turtlebot with Kodak PixPro SP360 wireless camera providing panoramic vision *B. and C.* Classic PI behaviours demonstrated on Turtlebot in standard office environment (outward path in green, homeward path in red. O1 and O2 are obstacles placed in the robots homeward path at the start of homing). *D.* Visual approach using VC on Turtlebot (blue dots are the memory locations, red line is robot path) (Color figure online)

of simple indoor 3D environments, it does not easily extend to large outdoor environments with complex structures such as trees, dynamic lighting. We note that projects to address these specific needs have been commenced (e.g. the MORSE simulator for ROS, https://youtu.be/k2wVj0BbtVk) promising better alignment with the needs of biomimeticists.

In this work, we have focused on the navigation of desert ants, which has a well-established history of biomimetic studies. However, the described work flow, simulation environments, robot models, and control schema are generalisable to other systems. For example, bees and ants navigate similar environments so models developed for ants can easily be transferred to study the alternate species with the creation of a new robot model. Comparison across a wider range of species (e.g. mammals) should be similarly easy to implement offering great benefits for the field of comparative cognition. In tandem, identified solutions to engineering goals can be validated on various robot platforms with ease, speeding the time to application.

In conclusion, we are convinced that ROS represents the best currently available platform for hypothesis development, testing and validation of biomimetic studies of navigation. This benefit is already being realised in the applied robotics field with many algorithms now freely available for implementation, adaption and benchmarking in ROS. To this end, and to alleviate some of the start-up issues highlighted, we have established a central software repository and user guide to support uptake of ROS for biomimetic studies of insect navigation (https://github.com/LCAS/A.N.T).

References

1. Baddeley, B., Graham, P., Philippides, A., Husbands, P.: Holistic visual encoding of ant-like routes: navigation without waypoints. Adapti. Behav. **19**(1), 3–15 (2011)
2. Ardin, P., Mangan, M., Wystrach, A., Webb, B.: How variation in head pitch could affect image matching algorithms for ant navigation. J. Comp. Physiol. A **201**(6), 585–597 (2015)
3. Möller, R.: A model of ant navigation based on visual prediction. J. Theor. Biol. **305**, 118–130 (2012)
4. Kodzhabashev, A., Mangan, M.: Route following without scanning. In: Wilson, S.P., Verschure, P.F.M.J., Mura, A., Prescott, T.J. (eds.) Living Machines 2015. LNCS, vol. 9222, pp. 199–210. Springer, Cham (2015). doi:10.1007/978-3-319-22979-9_20
5. Collett, M., Collett, T.S.: How do insects use path integration for their navigation? Biol. Cybern. **83**(3), 245–259 (2000)
6. Judd, S.P.D., Collett, T.S.: Multiple stored views and landmark guidance in ants. Nature **392**(6677), 710–714 (1998)
7. Zeil, J.: Visual homing: an insect perspective. Curr. Opin. Neurobiol. **30.22**(2), 285–293 (2012)
8. Wystrach, A., Mangan, M., Philippides, A., Graham, P.: Snapshots in ants? New interpretations of paradigmatic experiments. J. Exp. Biol. **216**(10), 1766–1770 (2013)
9. Wehner, R.: Desert ant navigation: how miniature brains solve complex tasks. J. Comp. Physiol. A **189**(8), 579–588 (2003)

Modeling of Bioinspired Apical Extension in a Soft Robot

Laura H. Blumenschein[1]([⊠]) [iD], Allison M. Okamura[1] [iD],
and Elliot W. Hawkes[1,2] [iD]

[1] Stanford University, Stanford, CA 94305, USA
{lblumens, aokamura}@stanford.edu,
ehawkes@engineering.ucsb.edu
[2] University of California Santa Barbara, Santa Barbara, CA 93106, USA

Abstract. Artificial apical extension in a soft robot, inspired by biological systems from plant cells to neurons, offers an interesting alternative to movement forms found traditionally in robots. Apically extending systems can move effectively in some environments that impede traditional locomotion. Artificial apical extension has been realized using a continuous stream of surface material, thin-walled, flexible plastic, which is everted at the tip by internal pressure. Understanding artificial apical extension as a form of movement requires a model to describe and predict the capabilities of the system. Unlike many other forms of movement, the model includes components that are dependent on the previous path in addition to path-independent terms associated with actuation. The model draws inspiration from biological models of apical extension and mechanical models of compliant Bowden cable actuation, and is verified though a series of tests on physical systems that isolate each term of the model.

Keywords: Soft robotics · Bioinspiration

1 Introduction

Apical extension is a method of movement found in nature, and is not typically realized in an artificial form. Movement by apical extension is defined by lengthening of a body from a base with volume added almost exclusively at the most distal point. This behavior can be seen in nature in individual cells, such as pollen tubes [1] and neurons [2], as well as at larger scales, in trailing vines. We developed a system that achieves artificial apical extension by passing circumferentially compacted material through the core of the existing body in a continuous stream to the tip, where it is everted and expanded. This is similar to the actuation used in everting toroidal robots, but leads to a semi-permanent body deformation instead of a continuous reconfiguration [3]. The movement of material is propelled by internal pressure pushing at the point of eversion (Fig. 1). Material eversion has also been used in plant root inspired soil-penetrating robots, but without significant length change of the system [4].

Artificial apical extension offers an interesting alternative to the various forms of locomotion already found in bio inspired robotic systems [5, 6]. By achieving movement through a semi-permanent extension of the body from the tip, apical extension

© Springer International Publishing AG 2017
M. Mangan et al. (Eds.): Living Machines 2017, LNAI 10384, pp. 522–531, 2017.
DOI: 10.1007/978-3-319-63537-8_45

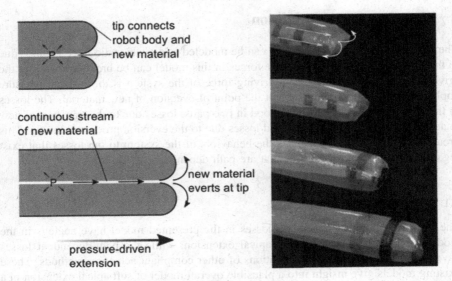

Fig. 1. Implementation of artificial apical extension in a soft robot. A continuous stream of new material is passed through the interior of the soft robot body and then is everted at the tip to extend the existing soft robot body. The material is a thin walled plastic. Extension is driven by the internal pressure.

allows movement of the most distal point without relative movement of the body with respect to the environment. This allows movement in environments that would be insufficient to support more traditional methods of locomotion (e.g. across a gap) or would be detrimental to locomotion (e.g. across a very sticky substrate). The extension of the body along the path creates the additional benefit of a conduit that can transfer material or information.

While locomotion can continue as long as there is an energy source available [7], apical extension, and other forms of movement by volume addition, are limited by the material that needs to be added to increase the surface area. Both the availability of the raw materials and the ability to transport those materials limits the extension of the system. This tradeoff means that while movement by extension can access a wide variety of environments, the range of movement it can achieve will be limited by the path taken. Thus, to understand the capabilities of a pressure-driven extending soft robot, we developed a model that relates the driving force of the soft robot to its dynamic extension. The inherent compliance of the soft extending robot creates a challenge for developing a simple model that describes the behavior sufficiently enough in order to be useful for understanding the system.

2 Modeling Apical Extension

The soft extending robot movement can be modeled as a quasi-static force balance due to the low inertia of the system. The forces in this model can be broken down into the driving force and the losses. The driving force of the system is the internal pressure applied to the cross-sectional area at the point of eversion of new material. The losses in the system can be further understood in two parts: losses due to the transport of new material from the base to the tip, and losses due to the eversion process at the tip. This breakdown fundamentally isolates the behaviors of the system to the losses that exist regardless of path and the losses that are path dependent.

2.1 General Model Form

The form of the path-independent losses in the presented model have analogs in the models of natural pressure-driven apical extension, while the path-dependent losses have analogs in the governing equations of other compliant actuation methods. These existing models give insight into a plausible overall model of soft apical extension in a soft robot.

In plant cell expansion, the equation describing the expansion based on the driving force, turgor pressure, is the Lockhart-Ortega equation, which relates the rate of volume change to the difference between the turgor pressure and yield pressure [8]. A modification of this viscoplastic model of plant cell extension described by Green relates pressure to a linear extension rate:

$$r = \varphi(P - Y)^n \tag{1}$$

where r is the linear extension rate, Y is the yield pressure below which no extension occurs, P is the internal pressure, φ is the extensibility, and n is a power term close to unity [9]. The model describes a monotonically increasing relationship between pressure and extension rate with an offset. For our model, we solve for internal pressure and write in terms of forces, giving

$$PA = YA + \left(\frac{1}{\varphi}v\right)^{\frac{1}{n}} A \tag{2}$$

where A is the cross-sectional area and v is the tip velocity, replacing the normalized rate, r. This equation provides two terms in the quasi-static force balance, a yield pressure and a velocity-dependent term.

This model of pressure-driven natural extension does not include path dependence. It is suspected that path dependent losses will exist and the material transport in the soft robot most closely resembles compliant Bowden cable actuation, where a cable is pulled inside a flexible conduit. We therefore turn to this model for the path dependent part of the model. In Bowden cable actuation, the path of the conduit determines the tension loss based on a modified Capstan equation:

$$T_{out} = T_{in}e^{\mu\frac{L}{R}}, \tag{3}$$

where T_{out} is the output tension, μ is the friction coefficient, L is the length of the path, R is the radius of curvature of the path, and T_{in} is the input tension [10]. This equation describes an exponential relationship between tension and length. Increasing the curvature, or decreasing the radius of curvature, causes the exponent to increase. In the soft extending robot, tension is replaced by the force due to internal pressure. There are some key differences between the soft extending robot and Bowden cable actuators that make a direct application of the Capstan equation difficult. In our apical extension soft robot, material being transported, although compacted, presses against the inside wall of the conduit, so there will be additional friction forces due to keeping this material compacted. In addition, the axial stiffness of the soft extending robot is much lower than even a compliant Bowden cable, so under compressive-axial loading, buckling may occur in the system causing an absolute change in curvature that is not accounted for by the Capstan equation. We therefore expect an exponential fit versus curved length with an additional term accounting for the length-dependent normal force applied by the internal material. The modified, Capstan-based model of the path dependence is:

$$PA = \mu_s wL + Ce^{\frac{\mu_c L}{R}} \tag{4}$$

where P is the pressure needed to extend, A is the cross-sectional area at the point of eversion, C is a coefficient based on the exponential fit, μ_c is the coefficient of friction due to curvature, R is the radius of curvature, L is the length of the soft robot path, μ_s is the length-dependent friction coefficient, and w is the normal force exerted per unit length. This path-dependent model accounts for both the curvature factor (first term) and the length-dependent normal force factor (second term). This model accounts for a single constant curvature path. However, in the full model, each curvature in the path requires its own term, leading to a sum over the curved lengths.

Combining the path-dependent terms (Eq. 2) and independent terms (Eq. 4), the full hypothesized model of the quasi-static force balance becomes

$$PA = [Path\,Independent] + [Path\,Dependent]$$

$$PA = \left[YA + \left(\frac{1}{\varphi}v\right)^{\frac{1}{n}}A\right] + \left[\mu_s wL + \sum_i Ce^{\frac{\mu_c L_i}{R_i}}\right], \tag{5}$$

which describes an input force due to pressure as the sum of a yield term, a velocity-dependent term, a path-length-dependent term, and a curvature-dependent term for each curvature in the path.

2.2   Model Components and Experimental Parameter Identification

The form for this model was verified through a series of experiments to isolate each of the components and find the relationship to internal pressure. Each component was mitigated by lessening the contributing behavior of the soft robot, to eliminate that component's effect on the force balance and leave only the effect of the component being tested. However, terms associated with eversion regardless of speed were present in all results involving extension of the soft robot. In each test the soft extending robot was constructed of thin-walled polyethylene tubing (50–80 μm wall thickness, Elkay Plastics) pressurized with air.

Path Independent Losses. Hypothesized path-independent losses (Eq. 5) include static and velocity dependent terms as derived from a modified Lockhart-Ortega equation.

Static Extension. The static extension term was tested independently by slow extension over short straight sections, which mitigates velocity, length, and curvature dependencies. To determine whether the static term is a yield force or yield pressure, and under what conditions the term changes, the soft extending system was extended into a slanted gap. The yield force is related to the yield pressure through the cross-sectional area,

$$F_y = YA. \tag{6}$$

Thus, by modifying the cross-sectional area, the yield force and pressure were identified independently. During the test, the pressure was increased incrementally until extension began. The soft robot extended until the pressure (P) was no longer above the pressure needed to overcome the static term, due to the decreasing gap area. A measurement was made of the achieved gap height (h) and the pressure was incremented to begin extension again. This process was continued until 80% of bursting pressure was achieved. For a gap between two flat surfaces, we estimate the cross-sectional area as a shape with the height of the gap while maintaining the circumference of the free inflated tube. This results in:

$$A = \frac{\pi}{4}h^2 + \pi\left(r - \frac{h}{2}\right)h \tag{7}$$

where A is the cross-sectional area of the soft robot body within the gap, r is the free inflated radius of the soft robot body and h is the gap height.

The results of the experiment for soft extending robots with a wall thickness of 80 μm and three different values of r, 1.3 cm, 2.4 cm, and 4.0 cm, are shown in Fig. 2. The data shows a linearly proportional relationship between inverse cross-sectional area and pressure, indicating that there is a constant yield force related only to the eversion of new material at the tip – yield force is independent of the cross-sectional area that the driving pressure acts on. The value of the yield force, the minimum applied force needed to extend, for our system was calculated to be 5.9 N.

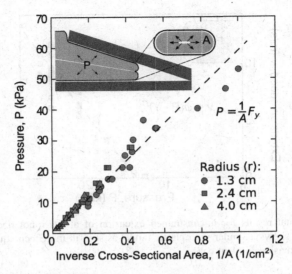

Fig. 2. Experimental results for extension of a soft robot body into a slanted gap. Colors indicate data from robots with different free inflated radii, r. The linear relationship between inverse cross-sectional area and pressure indicates a constant yield force is present.

Velocity-Dependent Extension. The velocity dependency was measured in a test over a short straight section. As already discussed, we do not expect this test to separate the rate dependency from the static yield term, but the velocity dependency should be isolated from the path-dependent terms. A soft robot body, with a radius of 1.3 cm, wall thickness of 80 μm, and length of approximately 1 m fully extended, was extended without external resistance. A large pressure vessel was used to maintain a constant pressure over the length of the extension, without restricting the flow. The pressure within the vessel was increased to the test point pressure while the soft robot body was prevented from extending by applying a force at the tip to stop material everting. The soft robot body was then released, and the movement was recorded by a high-speed overhead camera for a set of pressures between yield and burst pressure. The average velocity of the movement was calculated from the video. The data, shown in Fig. 3, follows the behavior of the modified Green equation (Eq. 2) with a yield pressure and a monotonically increasing relationship between pressure and extension velocity. The coefficients found for the model were a yield pressure, Y, of 13.3 kPa, an extensibility, φ, of 0.75 $\frac{m}{s}$ kPa^{-n} and a power term, n, of 1.1.

It is hypothesized that this viscoplastic relationship is due to losses arising from material deformation during eversion at the tip, not losses due to fluid flow within the body or flow limit at the input. Increasing the diameter of the valve providing flow to the soft robot body had no effect on the rate of extension. Calculations of skin drag due to turbulent fluid flow indicate that at the Reynolds numbers of the tests lengths of over 30 m and flow rates of 8 m/s would be necessary to see any significant pressure drop over the length of the body. Conversely, when the thickness of the material was

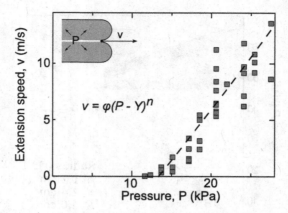

Fig. 3. Experimental results for unconstrained extension of a soft robot body at a constant pressure. The data shows a similar viscoplastic model as seen in the Green equation for rate of pressure driven extension in plant cells.

modified, the yield and extensibility both changed, suggesting that material properties determine the rate dependency.

Path-Dependent Losses. Hypothesized path dependent losses (Eq. 5) include path length and path-curvature-dependent terms as inspired by the Capstan equation as used in compliant actuation systems.

Length Dependency. A test was performed to find the relationship between pressure required to extend and length. To isolate this phenomenon, a rigid tube with inner diameter equal to the free inflated diameter was lined with the material of the soft extending robot, and a DC motor pulled additional material of the same inflated diameter and thickness through the interior at a constant speed. A normal force is produced between the inner and outer material by the compacting of the inner material to fit in the smaller diameter transport conduit. The resulting shear force on the rigid tube due to friction was measured by a six-axis force sensor (ATI Nano17) mounted between the rigid tube and stationary surface. The relationship between length of material in the tube and force is linear as seen in Fig. 4. The data shown is for soft extending robot material with an inflated radius of 2.4 cm and wall thickness of 80 μm. The value found for $\mu_s w$ was 0.8446 N/m. This value indicates that the length dependency is small, as for each additional meter of length, the pressure required to extend only increases by 1.9 kPa, which is 2.7% of the bursting pressure.

Curvature Dependency. To determine the effect of curved paths on the losses, a soft robot was extended between two curved surfaces at a slow rate. A soft extending robot with inflated diameter of 2.4 cm and wall thickness of 50 μm was used. The radius of the test was set at 8.7 cm, 10.4 cm, and 14.9 cm, in order to give a range of radii relative to the soft robot inflated diameter. Similar to the test to determine the yield term, the pressure was increased until extension began and then held at that pressure. When the soft robot could no longer extend at the current pressure, a measurement of

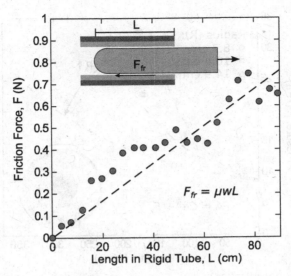

Fig. 4. Experimental results measuring the friction due to straight tube material transport. The linear relationship matches with the predicted normal force proportional to length.

the total length was recorded and the pressure was incremented again. The experiment continued until either the max desired pressure or max desired length was achieved.

The results of this experiment (Fig. 5) show the expected exponential relationship between curved length and pressure to extend, with an offset due to the yield force term. The length-based losses are insignificant at the length of 1 m, so that term was not included in the model fit to the data. Over the three curvatures tested, the losses decreased as the radius of curvature increased, as expected. The value of the exponential fit coefficient, C, was fairly constant across the different curvatures. However, the friction coefficients, μ_c, varied between 0.141 and 0.088 over the curvatures tested. In the unmodified Capstan equation, the friction coefficient is expected to be constant across curvatures and the exponent would vary only based on the radius of curvature. This result indicates that while a general exponential model fits the data well, more tests will need to be done to fully understand the relationship between the path curvature and the coefficients of the exponential term.

2.3 Model Implications

This preliminary model of pressure-driven artificial apical extension has important implications for apical extension as a model for robot movement. The pressure required for extension depends on the path taken as a combination of both length and curvature. Apical extension is thus limited in the workspace that it can reach compared to other methods for robotic movement. However, losses due to length are relatively small. This result was tested with a maximum straight length extension, conducted at a slow speed and with turns minimized. The soft robot reached a length of 72 m, only stopped by the

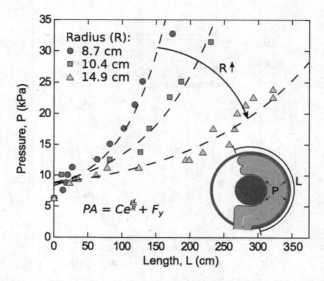

Fig. 5. Experimental results for constant curvature extension over three curvatures showing the expected exponential fit between length and pressure. As the radius of curvature is increased, the losses are decreased for the same length extension.

size of the pressure vessel available to contain the new material at the base. Further, the curvature loss results indicate that extension paths should keep the curvature of turns as low as possible.

3   Summary

We presented a preliminary model of a bioinspired apically extending soft robot. The quasi-static force balance model leads to understanding of the movement capabilities and limitations of the apically extending robot as compared to traditional forms of robot movement. The proposed model incorporates rate-based extension models from biological apically extending cells to describe the path-independent losses and a modified Capstan equation from compliant Bowden cable actuation to explain the path-dependent losses. Through separation of the model components, the full model was verified for both path-dependent and path-independent terms. The model can be further used to explain and predict the response of the soft extending robot as it moves along different paths.

Acknowledgements. This work was supported in part by National Science Foundation grant 1637446 and the National Science Foundation Graduate Fellowship Program.

References

1. Palanivelu, R., Preuss, D.: Pollen tube targeting and axon guidance: parallels in tip growth mechanisms. Trends Cell Biol. **10**, 517–524 (2000)
2. Dent, E.W., Gertler, F.B.: Cytoskeletal dynamics and transport in growth cone motility and axon guidance. Neuron **40**, 209–227 (2003)
3. Orekhov, V., Hong, D.W., Yim, M.: Actuation mechanisms for biologically inspired everting toroidal robots. In: IEEE/RSJ International Conference on Intelligent Robots and Systems (2010)
4. Sadeghi, A., Tonazzini, A., Popova, L., Mazzolai, B.: Robotic mechanism for soil penetration inspired by plant root. In: IEEE International Conf. on Robotics and Automation (ICRA), pp. 3457–3462 (2013)
5. Ma, K.Y., Chirarattananon, P., Fuller, S.B., Wood, R.J.: Controlled flight of a biologically inspired, insect-scale robot. Science **340**, 603–607 (2013)
6. Seok, S.: Design principles for energy-efficient legged locomotion and implementation on the MIT cheetah robot. IEEE/ASME Trans. Mechatron. **20**, 1117–1129 (2015)
7. Alexander, R.M.: Principles of Animal Locomotion. Princeton University Press, New York (2003)
8. Lockhart, J.A.: An analysis of irreversible plant cell elongation. J. Theor. Biol. **8**, 264–275 (1965)
9. Green, P.B., Erickson, R.O., Buggy, J.: Metabolic and physical control of cell elongation rate in vivo studies in nitella. Plant Physiol. **47**, 423–430 (1971)
10. Kaneko, M., Yamashita, T., Tanie, K.: Basic considerations on transmission characteristics for tendon drive robots. In: 5th International Conference on Advanced Robotics, pp. 827–832 (1991)

A Biomechanical Characterization of Plant Root Tissues by Dynamic Nanoindentation Technique for Biomimetic Technologies

Benedetta Calusi[1,2], Francesca Tramacere[2], Carlo Filippeschi[2], Nicola M. Pugno[1,3,4(✉)], and Barbara Mazzolai[2(✉)]

[1] Laboratory of Bio-Inspired and Graphene Nanomechanics, Department of Civil, Environmental and Mechanical Engineering, University of Trento, 38123 Trento, Italy
nicola.pugno@unitn.it
[2] Center for Micro-BioRobotics, Istituto Italiano di Tecnologia, Viale Rinaldo Piaggio 34, 56025 Pontedera, Italy
barbara.mazzolai@iit.it
[3] Ket-Lab, Italian Space Agency, Via del Politecnico snc, 00133 Rome, Italy
[4] School of Engineering and Materials Science, Queen Mary University of London, Mile End Road, London E1 4NS, UK

Abstract. In this work we present a study on mechanical properties of *Zea mays* primary roots. In order to have an accurate overview of the root structure, three different regions have been analyzed: the outer wall, the inner part, and the root cap. We used a dynamic nanoindentation technique to measure the elasticity modulus of root tissues in correspondence of different distances from the root tip. A sample holder was built to test the tip and a method conceived to separate the outer wall of the root from the inner part. As determined by dynamic nanoindentation, we measured the storage modulus of plant roots over 1–200 Hz range. We found that the values of the storage modulus along the outer wall are higher with respect to the central core. Moreover, the inner core and root cap seem to be similar in terms of elasticity modulus. This study aims to shed light on the mechanical properties of roots that significantly affect root movements and penetration capabilities. The gathered data on mechanical response and adaptive behaviours of natural roots to mechanical stresses will be used as benchmarks for the design of new soft robots that can efficiently move in soil for exploration and rescue tasks.

Keywords: Plant root tissues · Mechanical properties · Nanomechanics · Root-inspired robots

1 Introduction

It is well known that plant roots are an excellent example of efficiency for soil penetration and exploration with optimal adaptation strategies. Specifically, roots can adapt their growth direction through their soft tissues. Therefore, the penetration capabilities in conjunction with the flexibility of root structure can inspire new design and fabrication of soft robots [1]. In this work we present a preliminary study on the mechanical

© Springer International Publishing AG 2017
M. Mangan et al. (Eds.): Living Machines 2017, LNAI 10384, pp. 532–536, 2017.
DOI: 10.1007/978-3-319-63537-8_46

properties of *Zea mays* primary root, at level of its outer wall, inner part, and cap. For this aim, we used a dynamic nanoindentation technique.

2 Materials and Methods

2.1 Planting and Sample Preparation

We tested 3/4-day old *Z. mays L.* seeds. The seedlings were grown on filter paper with tap water and kept into a growth chamber at 25 °C. We measured the mechanical properties in correspondence of three different regions of the root at different frequencies. The nanoindentation experiments were carried out at room temperature employing an iNano indentation system (Nanomechanics, Inc.) and using the 'Dynamic Flat Punch Complex Modulus' test with a routine method for biomaterials [2–4]. We exploited indenter tips having 109 and 198.5 μm diameter. All the root regions were investigated in 1–200 Hz frequency range. To avoid root dehydration, the nanoindentation measurements were carried out in distilled water. The measurements were made on roots (~1 cm in length) at different distances from the tip (2, 3, 4, 5, 6, and 7 mm) in correspondence of the wall, the inner core, and the root cap (Figs. 1 and 2). To avoid any movements of the root during the tests, we fixed the samples on the bottom of the holder by means of attack (Loctite). Due to the complex geometry of root tip, an *ad hoc* sample holder with two inclinations was built to perform tests in the root tip area (Fig. 1b). In order to extract the root core, we made a circumferential incision of the root at the base of the seed, thus the outer wall can be easily separated from the inner core (Fig. 1d).

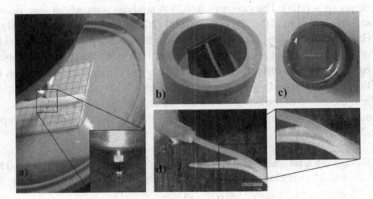

Fig. 1. View of a nanoindentation experiment. (a) A nanoindentation test. The zoom shows the tip of the indentation system; (b) and (c) a sample holder used to test mechanical properties of root tissues near to the cap, and in outer and inner areas, respectively; (d) separation procedure of the outer wall from the inner core.

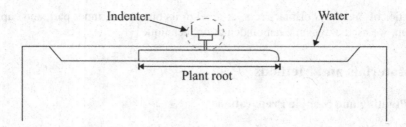

Fig. 2. Schematic of the setup used for testing along the root at different distances from the tip.

2.2 Statistical Analysis

We performed one-way ANOVA to quantify the storage modulus (E) changes with respect to the distance from the tip and at different tissue level (i.e. outer and inner tissues, inner tissue and near the cap) for each frequency. Moreover, we exploited two-way ANOVA (without interaction) to test the effect of the distances and tissues for each frequency.

The statistical analysis is at the 95% confidence level.

3 Results

3.1 Nanoindentation Measurements: Results of Statistical Analysis

We obtained significant E difference between inner and outer tissue measurements at all distances from the tip for each test frequency, except for the inner core measurements at 1 Hz frequency ($p > 0.05$). The one-way ANOVA analysis pointed out significant differences ($p < 0.05$) at 2 mm ($10 \div 200$ Hz), 3 mm (200 Hz), 4 mm ($35 \div 200$ Hz), 5 mm ($1 \div 200$ Hz), and 7 mm ($1 \div 200$ Hz) distance. On the contrary, the comparison between the inner core measurements at each distance from the tip and the measurements near the cap showed no significant differences ($p > 0.05$) for each frequency. Moreover, the two-way ANOVA results showed significant E differences ($p < 0.05$) within the group at all the distances and the group of the tissue levels (i.e. inner core and outer wall).

The results of the nanoindentation tests along the three root regions are reported in Tables 1, 2 and 3.

Table 1. Storage modulus (mean ± SD) of the root cap for all frequencies (1, 3, 10, 15, 35, 85, 200 Hz). A total of 7 indentations on 4 roots were performed.

Freq. (Hz)	E (MPa)
200	4.23 ± 1.53
85	3.9 ± 1.29
35	3.65 ± 1.22
15	3.39 ± 1.15
10	3.27 ± 1.13
3	2.81 ± 1.01
1	2.49 ± 0.93

Table 2. Storage modulus (mean ± SD) along the inner core at different distances from the tip (2, 3, 4, 5, 6, 7 mm) for all frequencies (1, 3, 10, 15, 35, 85, 200 Hz). The measurements number for each distance is reported.

Freq. (Hz)	E (MPa)					
	2 mm (no. 9)	3 mm (no. 12)	4 mm (no. 14)	5 mm (no. 12)	6 mm (no. 7)	7 mm (no. 5)
200	4.54 ± 1.74	4.63 ± 1.53	4.21 ± 1.15	3.33 ± 1.15	3.23 ± 1.19	2.77 ± 0.83
85	4.29 ± 1.69	4.39 ± 1.45	3.97 ± 1.12	3.17 ± 1.08	3.04 ± 1.15	2.61 ± 0.85
35	4.13 ± 1.67	4.23 ± 1.41	3.83 ± 1.1	3.04 ± 1.06	2.93 ± 1.14	2.49 ± 0.84
15	3.98 ± 1.64	4.08 ± 1.37	3.69 ± 1.09	2.91 ± 1.06	2.83 ± 1.12	2.38 ± 0.84
10	3.9 ± 1.63	4 ± 1.34	3.63 ± 1.09	2.85 ± 1.06	2.78 ± 1.11	2.33 ± 0.85
3	3.54 ± 1.54	3.65 ± 1.27	3.3 ± 1.05	2.58 ± 1	2.54 ± 1.05	2.1 ± 0.84
1	3.25 ± 1.5	3.38 ± 1.25	3.09 ± 1.07	2.4 ± 0.98	2.38 ± 1.04	1.9 ± 0.85

Table 3. Storage modulus (mean ± SD) along the outer wall at different distances from the tip (2, 3, 4, 5, 6, 7 mm) for all frequencies (1, 3, 10, 15, 35, 85, 200 Hz). The measurements number for each distance is reported.

Freq. (Hz)	E (MPa)					
	2 mm (no. 5)	3 mm (no. 13)	4 mm (no. 8)	5 mm (no. 22)	6 mm (no. 14)	7 mm (no. 15)
200	7.32 ± 2.22	5.99 ± 1.7	5.65 ± 1.92	5.49 ± 1.3	4.36 ± 1.26	4.82 ± 1.14
85	7.01 ± 1.99	5.64 ± 1.69	5.43 ± 1.92	5.27 ± 1.26	4.16 ± 1.2	4.6 ± 1.1
35	6.63 ± 1.89	5.31 ± 1.61	5.18 ± 1.87	5.01 ± 1.22	3.95 ± 1.15	4.39 ± 1.1
15	6.23 ± 1.75	4.99 ± 1.51	4.94 ± 1.83	4.75 ± 1.19	3.74 ± 1.1	4.17 ± 1.11
10	6.03 ± 1.65	4.83 ± 1.46	4.82 ± 1.81	4.62 ± 1.18	3.64 ± 1.07	4.07 ± 1.11
3	5.23 ± 1.39	4.15 ± 1.22	4.32 ± 1.66	4.05 ± 1.1	3.21 ± 0.96	3.62 ± 1.06
1	4.66 ± 1.27	3.63 ± 1.08	3.93 ± 1.51	3.65 ± 1	2.94 ± 0.86	3.35 ± 0.99

4 Discussion

The E smaller values at the inner tissue with respect to the outer tissue suggest that plant roots could be made by a stiff coating around a soft core, which works as a 'soft skeleton'. This would represent a cushioning layer that contributes to protect the root cap from external abrasion and friction, together with mucilage secretion from the same area. On the other hand, plant roots penetrate soil by growing at the apical region, thus a stiff tissue near the tip could enhance the movements into a medium. Moreover, the E higher value at distances close to the tip and the tissue softness just behind the tip could explain the adaptive behaviour during penetration. Since soils could have several barriers, e.g. rocks, roots may activate a response to circumvent them, when the penetration is not the optimal strategy at low energy cost. Yet, roots buckle to avoid obstacles, e.g. a tip-to-barrier angle, was observed in [5].

5 Conclusions

In this work, we used the dynamic nanoindentation technique to study the mechanical properties of root tissues. Since new cells are continuously created in the apical region, the cellular differentiation is at an early stage close to the root tip. Therefore, the results of the mechanical tests could reveal root penetration strategies during growing from the tip. Our results suggest that roots could consist of an internal 'soft skeleton' and a stiffer outer wall. This combination of soft and stiff materials may enhance plant roots to simultaneously penetrate and adapt to soil constraints. Future developments of these preliminary results will be the study of the whole plant tissues in various plant species to better understand the role of mechanical tissue properties in penetrating soil at different impedance.

Acknowledgments. This study was partially founded by the PLANTOID project (EU-FP7-FETOpen grant n. 29343). N.M.P. is supported by the European Research Council PoC 2015 "Silkene" No. 693670, by the European Commission H2020 under the Graphene Flagship Core 1 No. 696656 (WP14 "Polymer Nanocomposites") and under the FET Proactive "Neurofibres" No. 732344.

References

1. Sadeghi, A., Mondini, A., Del Dottore, E., Mattoli, V., Beccai, L., Taccola, S., Lucarotti, C., Totaro, M., Mazzolai, B.: A plant-inspired robot with soft differential bending capabilities. Bioinspir. Biomim. **12**(1), 015001 (2016). doi:10.1088/1748-3190/12/1/015001
2. Herbert, E.G., Oliver, W.C., Pharr, G.M.: Nanoindentation and the dynamic characterization of viscoelastic solids. J. Phys. D: App. Phys. **41**(7) (2008). doi:10.1088/0022-3727/41/7/074021
3. Pharr, G.M., Oliver, W.C., Brotzen, F.R.: On the generality of the relationship among contact stiffness, contact area, and elastic modulus during indentation. J. Mater. Res. **7**(3), 613–617 (1992). doi:10.1557/JMR.1992.0613
4. Herbert, E.G., Oliver, W.C., Lumsdaine, A.: Measuring the constitutive behavior of viscoelastic solids in the time and frequency domain using flat punch nano indentation. J. Mater. Res. **24**(3), 626–637 (2009). doi:10.1557/jmr.2009.0089
5. Popova, L., Tonazzini, A., Di Michele, F., Russino, A., Sadeghi, A., Sinibaldi, E., Mazzolai, B.: Unveiling the kinematics of the avoidance response in maize (*Zea mays*) primary roots. Biologia **71**(2), 161–168 (2016). doi:10.1515/biolog-2016-0022

Biomimetic Creatures Teach Mechanical Systems Design

Matthew A. Estrada, John C. Kegelman, J. Christian Gerdes,
and Mark R. Cutkosky[✉]

Stanford University, Stanford, CA 94305, USA
cutkosky@stanford.edu
http://me112.stanford.edu

Abstract. For over a decade, mechanical creatures have formed the basis for the final project in a large engineering class on mechanical systems design. Each year a different real or fictitious animal provides the inspiration and design requirements for synthesizing, fabricating, and analyzing mechanisms and power transmission systems to achieve a plausible biomimetic motion. We explore why biomimetic creatures are particularly suited to learning about mechanisms, and discuss implementation details and pedagogical insights based on our experience.

1 Introduction

There is something about mechanical creatures that captures the imagination. The concept has arisen in multiple cultures with famous examples that include Vaucanson's duck [2], Leonardo da Vinci's roaring lion [9], automatons from China and Korea, and many recent examples ranging from Disney Animatronics to El Pulpo Mecanico[1] and the Strandbeests of T. Jansen.[2] As entertaining as they are to watch, they are even more fun to work on, which may explain why the lion was one of few designs that da Vinci actually completed and why so many inventors have been willing to spend countless hours over the centuries designing, building and perfecting their machines.

For similar reasons, biomimetic creatures have become a staple of Mechanical Systems Design (ME112) at Stanford University, an undergraduate course that has grown over the last 15 years from an enrollment of 60 to around 150. ME112 covers standard mechanical engineering topics including transmissions, motors, efficiency, power and linkages. The final project challenges teams of 3–4 students to design, build, test, and analyze battery-powered systems propelled by a small electric motor coupled to a transmission and linkages.

Animals provide an ideal motivation for the project. Most animals have appendages that move periodically with some repeated and nontrivial motion through space. Without robotic controllers, the preferred way to animate a mechanical animal is to design a linkage, which can be powered through gears,

---

[1] http://www.elpulpomecanico.com/.
[2] http://www.strandbeest.com/.

© Springer International Publishing AG 2017
M. Mangan et al. (Eds.): Living Machines 2017, LNAI 10384, pp. 537–543, 2017.
DOI: 10.1007/978-3-319-63537-8_47

chains, cams, etc. For this reason, all of the examples mentioned above involve mechanisms design. Disney has also developed software to facilitate the design of animatronic mechanisms [3]. In nature too, reconciling the demands of environmental loads with actuator constraints is accomplished through mechanisms. A typical example is the limb morphology that allows frog muscle to operate near the velocity for maximum power during swimming [8].

In the following sections we discuss the pedagogical motivations for bioinspired mechanical creatures and present implementation details associated with delivering the course. We conclude with insights regarding what is most effective in this approach.

Fig. 1. Recent creatures from ME112: A kangaroo reproduces a pentapedal gait using its tail [11], a duck swims in the Stanford pool [5], a sloth traverses an overhead canopy wire [10], and an emu exhibits bipedal locomotion [6].

2   Pedagogical Motivation

Focus on Class Content. It can be difficult to develop design problems that allow open-ended solutions while encouraging students to make use of a body of content. If the solution space is too large, students may focus on approaches that have nothing to do with the material the instructors wish to emphasize. Imposing artificial constraints, on the other hand, can lead to disengagement if the students feel they would approach the task differently in the "real world." For example, realistic examples from manufacturing machinery are often hard to formulate in a way that does not seem "dry" and that does not lead students to wonder if unrelated solutions (e.g. using stepper motors) might work better.

In contrast, bioinspired mechanical creatures are already somewhat whimsical and lend themselves to approaches appropriate to a class involving motors and mechanisms. Past examples have included artificial gibbons that retrieve berries for medicinal purposes, ferrets that flush out snakes on a space station, or giant stag beetles that can retrieve a lost artifact. Students generally enjoy and accept these premises, putting the focus on application of course material instead of "lawyer solutions" that satisfy the rules but not the intent of the project. Figure 1 shows a few other examples from recent years.

Practice with Abstraction. Motion in biology is much more complex than the motions created with typical planar mechanisms. Mimicking the motion of an animal requires students to think about what aspects of that motion warrant their attention. (A similar "bioabstraction" process applies to most bioinspired robotics research.) Specifically, they must examine the problem at different levels of abstraction that span the space between the actual motion and the limited motions of different mechanisms. This process also creates an interplay between qualitative analysis of the motion through observation, selection of an appropriate type of mechanism, and the detailed design of a chosen mechanism. While there are many analytical tools available for detailed design, type selection is inherently qualitative, exposing students to the art and science of engineering that exists in real world problems. Frequently students will go back and forth between videos of animals and articles from the zoology literature and coupler curves and velocity plots that they generate with their own models and simulations.

Since abstraction is inherently qualitative, only informal guidelines are used to judge biomimesis. First, anything that walks successfully is more biomimetic that something which doesn't. Students also receive credit for their efforts to understand the motions of biological creatures and produce comparable motions with mechanisms. No requirement is set on the biological justification nor on mechanism specifics such as the number of DoF.

Opportunities for Creativity and Collaboration. We pick a new animal and problem statement each year, following a long-standing philosophy from the Stanford Design Group that projects should be (i) novel so that the "best solution" is not known a priori to either the students or the teaching staff and (ii) framed in terms of design requirements and not as a competition, so that there is no disincentive to share ideas and materials [4]. Animals created in ME112 include turtles, lemurs, ferrets, hippogriffs, dinosaurs, sea elephants, sloths, dragons, alligators, insects, rabbits, kangaroos and in 2017, walking birds inspired by a paper at Living Machines 2016 [1].

Projects are always time consuming, and students are invariably more motivated by projects where they can display creative touches that differentiate their solution from others in the class. Projects involving animal motion support differentiation not only in the design of the overall gait and leg mechanisms but also in complementary features of the animal. Past examples include a hippogriff with origami-like wings that folded and unfolded with alternating steps, a rabbit that used a cam mechanism to leave droppings every few steps and a kangaroo fabricated entirely with 3D printed parts. In these cases, students expressed gratitude for the opportunity to bring their own interests or ideas into a class project.

Analysis and Diagnostics. Adding a physical design project to the mechanical systems class is intended to give students practice connecting the theory with the physical world. The projects offer ample opportunities to bring simple analysis to bear on all manner of diagnostic problems from motors that draw too much current to feet that have gone out of phase due to excessive backlash where a crank is connected to a shaft. Regardless of project outcome, students engage in the analysis process to compare theoretical predictions against real-world performance, deduce reasons for discrepancies, and brainstorm areas for improvement. Facilitating this process requires coaches with considerable experience in mapping analytical concepts to observed physical behavior and availability during times students are struggling with their designs (generally after business hours).

3   Project Implementation

The course begins each year with a few focused laboratory exercises including dissecting and analyzing a manual automobile transmission and characterizing a motor to understand how performance is affected by operating conditions. There is also a short mid-term project that typically involves the design of a wheeled vehicle and transmission for some light-hearted challenge. The course then switches focus to mechanisms, guided by a four-week final project.

Systems Integration and Power Flow. Often we contextualize the role of analysis by tracing power flow through the system, as depicted in Fig. 2. By

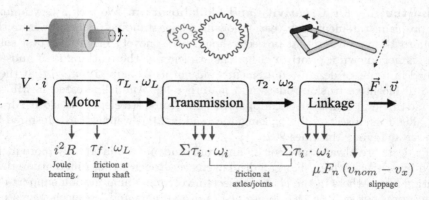

Fig. 2. Power flow through a biomimetic creature. Inefficiencies are present at each stage. "Useful work" at the output is typically $< 10\%$ of the input energy.

emphasizing the purpose of each mechanical stage at a higher level, we hope to instill in students the flexibility and confidence to adapt their particular choice of actuators or transmissions to the specifics of future engineering challenges.

The motors are small, brush-type DC motors and typically no more than 60% efficient at best. Students conduct tests to determine the internal resistance, R, electromagnetic constant, k, and an estimate of the friction torque, T_f. With these constants they can determine motor speeds for maximum power or maximum efficiency, assuming steady-state operation ($V - iR - k\omega = 0$).

The second source of power loss is in the transmission and linkages. The low-cost prototypes have plain sleeve bearings so losses are chiefly due to Coulomb friction and theoretically independent of speed. Hence it is worthwhile to select an overall transmission ratio that allows the motor to run at an efficient speed, subject to constraints on providing adequate torque and allowing the mechanical creature to move at a desired pace.

Reconciling Theory with Practice. While the analysis of motors, transmissions and linkages is straightforward, it is always a surprise to students to discover how difficult it can be to correlate theory and simulation with empirical results for a self-propelled machine. Building a mechanism also provides intuition about problems arising from transmission angle, manufacturing tolerances, singularities and elastic deflections in theoretically rigid links.

In addition, although planar mechanisms have been studied for at least 200 years, the synthesis of mechanisms can be challenging, especially when there are many other constraints on packaging, weight, structural integrity, etc. In contrast to animatronics [3], mobile creatures need to have all their mechanisms contained within their bodies and not hidden underneath a stage. Hence many of the initial solutions from an atlas of coupler curves, three-point synthesis

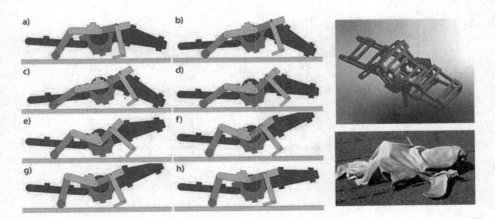

Fig. 3. Left: Working Model2D^{TM} simulation of the motion of an elephant seal on land (termed galumphing in the literature [7]). Right: CAD model of structure and final prototype *en costume*.

methods, etc. are not practical. As a result, students use a mix of kinematic and dynamic simulations and prototypes fabricated from increasingly robust materials. A common solution is to use laser-cut Masonite, acrylic or aircraft plywood in combination with plastic or metal screw posts for joints. Figure 3 shows this typical progression from simulation to prototype. Ultimately, the animals need to be assembled and tested to see whether they move as expected and whether the actual forces and velocities match those predicted.

Diagnostics. As roboticists know, it is much harder to get a dynamic simulation to produce accurate, or even realistic ground interaction forces than to produce a plausible motion (because $f = m \cdot dv/dt$, many different force profiles can produce the same velocity). In addition, ground forces depend critically on the friction and elasticity of the feet and the ground. Slippage is often significant, which leads students to experiment with different shoe materials and foot shapes. To get an estimate of the ground forces, we have built an inexpensive force plate that measures normal and tangential forces at 300 Hz with an accuracy of approximately 5 gf (Fig. 4). Details of the open-source design are posted online.[3] Other common difficulties include overall balance in three dimensions and unexpected dynamic effects due to the feet not working perfectly in phase as a result of backlash, etc. All of these issues provide ample opportunity for discussion and design coaching.

4   Closing Observations

Students in a large undergraduate course come with diverse scholastic backgrounds and varying levels of experience and affinity for rigorous analysis and

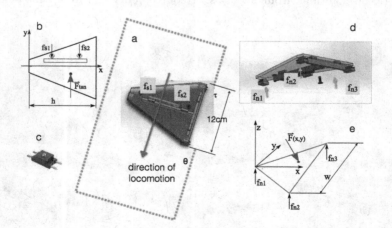

Fig. 4. Force plate for small biomimetic creatures: (a) plan view of plate on bench top, (b) tangential forces, (c) sensor: Honeywell FSS005WNGB (5 total), (d) view from below showing normal force sensors at corners of plate (e) force diagram.

---

[3] http://bdml.stanford.edu/SimpleForcePlate.

creative projects. The biomimetic creatures project encourages them to expand beyond their comfort zones into open-ended engineering problems that have no right answer and yet are rooted in fundamental analytical practice. Many students remark that this is the first course they encounter that requires integration of engineering tools and creative design. Mimicking biology with compound machines provides a relatable goal that facilitates the bridging of theory and practice. A biomimetic design project presents dual emphases on getting something to work and understanding *why* it works. The groups develop an attachment to their creations such that by the time of the final demonstration, students and instructors alike are rooting for their success. Regardless of project outcome, students engage in analysis to compare theoretical predictions against real-world performance, deduce reasons for discrepancies, and brainstorm areas for improvement. Nascent designers and engineers are encouraged when numbers derived from engineering approximations reasonably match physical outcomes.

References

1. Abourachid, A., Hugel, V.: The natural bipeds, birds and humans: an inspiration for bipedal robots. In: Lepora, N.F.F., Mura, A., Mangan, M., Verschure, P.F.M.J.F.M.J., Desmulliez, M., Prescott, T.J.J. (eds.) Living Machines 2016. LNCS, vol. 9793, pp. 3–15. Springer, Cham (2016). doi:10.1007/978-3-319-42417-0_1
2. Bedini, S.A.: The role of automata in the history of technology. Technol. Cult. **5**(1), 24–42 (1964)
3. Coros, S., Thomaszewski, B., Noris, G., Sueda, S., Forberg, M., Sumner, R.W., Matusik, W., Bickel, B.: Computational design of mechanical characters. ACM Trans. Graph. (TOG) **32**(4), 83 (2013)
4. Faste, R., Roth, B.: The design of projects and contests-the rules of the game. J. Rob. Mech. **10**, 7–13 (1998)
5. Handa, K., Lopez, B., McMordie, J., Solis, M.: If it swims like a duck. Technical report, Stanford University, Mech. Eng. Dept. (2014). https://undergrad.stanford.edu/programs/pwr/publications-prizes-and-awards/hoefer-prize-winners
6. Porter, C., Fearon, B., Devon, M., Stevens, M.: Bipedal robot design report. Technical report, Stanford University, Mech. Eng. Dept. (2017). https://undergrad.stanford.edu/programs/pwr/publications-prizes-and-awards/hoefer-prize-winners
7. Renouf, D., Lawson, J.: Play in harbour seals (phoca vitulina). J. Zool. **208**(1), 73–82 (1986)
8. Richards, C.T., Clemente, C.J.: Built for rowing: frog muscle is tuned to limb morphology to power swimming. J. R. Soc. Interface **10**(84), 20130236 (2013)
9. Rosheim, M.: Leonardo's Lost Robots. Springer Science & Business Media, New York (2006)
10. Soong, A., Cooper, E., Cooper, A., Lin, K.: Slothy. Technical report, Stanford University, Mech. Eng. Dept. (2014)
11. Young, J., Le Roux, A., Sarsona, J., Oro, A.: Respect the pouch. Tech. rep., Stanford University, Mech. Eng. Dept. (2016). https://undergrad.stanford.edu/programs/pwr/publications-prizes-and-awards/hoefer-prize-winners

Soft Fingers with Controllable Compliance to Enable Realization of Low Cost Grippers

Keng-Yu Lin and Satyandra K. Gupta[✉]

Department of Aerospace and Mechanical Engineering,
University of Southern California, Los Angeles, CA 90089, USA
{kengyuli,skgupta}@usc.edu

1 Introduction

Grippers are needed to manipulate objects using robotic arms. In many applications, grippers need to meet the following three criteria:

- Grippers need to conform to shape of objects being manipulated to handle irregularly shaped objects.
- Grippers need to apply force on the object or resist the gravity force (e.g., need to be able to lift heavy objects).
- Grippers need to be low cost to enable new applications.

Traditional multi-finger grippers with built-in sensors are able to the meet the first two requirements. However, they tend to be expensive. Robotic grippers based on the jamming of granular material are able to meet the first and third requirements. However, they cannot apply large compressive force on the object. Recently there has been significant interest in grippers with soft fingers [1–9].

We believe that grippers with soft fingers can meet all three requirements listed above. This paper reports redesign of a conventional soft finger by integrating a structural member that can be used to control its compliance. Our focus is on a design that is easier to manufacture and low cost. We have built a three-fingered pneumatically actuated gripper that only requires simple pneumatic actuation. Our gripper is attached to a UR5 robot arm. The robot arm is able to lift heavy objects using the new gripper.

2 Approach

We are interested in developing a design that is pneumatically actuated to ensure that it is compatible with the existing soft finger designs. To design the structural member with controllable compliance, we decided to only pursue design concepts that can be actuated by air either with pressure or vacuum.

© Springer International Publishing AG 2017
M. Mangan et al. (Eds.): Living Machines 2017, LNAI 10384, pp. 544–550, 2017.
DOI: 10.1007/978-3-319-63537-8_48

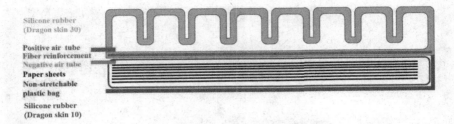

Silicone rubber
(Dragon skin 30)

Positive air tube
Fiber reinforcement
Negative air tube
Paper sheets
Non-stretchable
plastic bag
Silicone rubber
(Dragon skin 10)

Fig. 1. Sectional view of compliant finger.

The main idea behind our soft robotic fingers lies in embedded sealed chamber which is made up of a non-stretchable polymer sheet. The chamber is filled with floating compliant sheets. When air is pumped out to create vacuum inside the sealed chamber, floating sheets inside the sealed chamber adhere to each other, increasing the friction due to an increase in the contact area. When the structural member is set in low stiffness mode, floating sheets can easily slide with respect to each other due to the sheet separation created by positive air pressure. If air is vacuumed out of the chamber, the sheets are stacked together. This prevents sheets from sliding over each other. This leads to the structural member appearing stiff under bending loads. By controlling the vacuum inside the chamber, we can control the stiffness of the structural member and the finger.

Figure 1 shows the design concept. The design has been realized by using in-mold assembly process. Molds were printed using FDM process. Figure 2 shows the traditional soft finger. Figure 3(a) shows the soft finger and structural element with controllable stiffness. Soft finger is actuated using positive air pressure. Structural element is actuated using negative air pressure. Figure 3(b) shows the final assembly of the finger. Our design allows easy replacement of sheet media to control stiffness.

Fig. 2. Traditional soft finger which has only positive air pressure actuator.

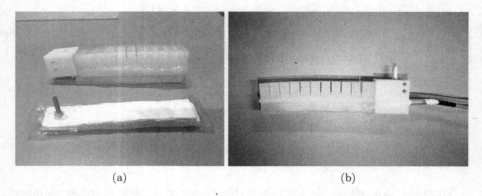

(a)   (b)

Fig. 3. Compliant soft finger which has both positive air pressure actuator, and negative air pressure actuator. (a) The finger has 3D printed base, (b) the finger is covered with Silicone rubber.

3   Results

Figure 4 shows how the soft finger behaves under low stiffness mode. Figure 5 shows that the finger is unable to apply much force on a test cantilever beam under low stiffness mode. Figure 6 shows that the finger is able to apply significant force under the high stiffness mode. Figure 7 shows the side by side comparison of low stiffness and high stiffness modes. We estimate that under the high stiffness mode the finger is able to apply 4.6 times more force than the low stiffness mode.

Fig. 4. Bending test of a single compliant soft finger kept in low stiffness mode under positive air pressure.

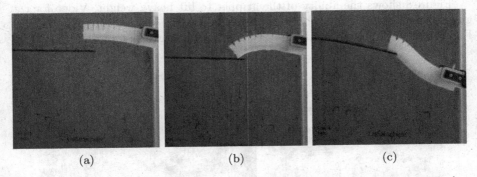

(a)   (b)   (c)

Fig. 5. (a) Initial position of a soft compliant finger with respect to a cantilever beam. (b) The finger was kept in low stiffness mode after actuation, and moved to touch the cantilever beam. (c) The finger was moved down by the robotic arm along the vertical axis.

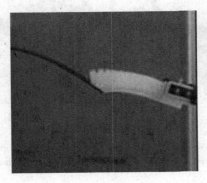

Fig. 6. The soft compliant finger was kept in high stiffness mode after actuation, and it was moved down by robotic arm along the vertical axis.

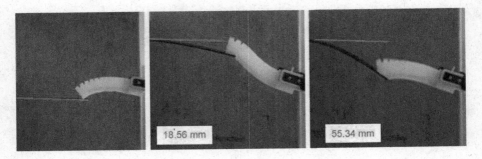

Fig. 7. Force exerted by finger in low stiffness mode is 0.72 N, and the force exerted by finger in high stiffness mode is 3.34 N.

Figure 8 shows the ability of the gripper to lift heavy weight. Video for the gripper mounted on UR-5 robot can be seen at https://youtu.be/YutKQs6W95g. Figure 9 shows different kinds of objects that can be grasped by this gripper.

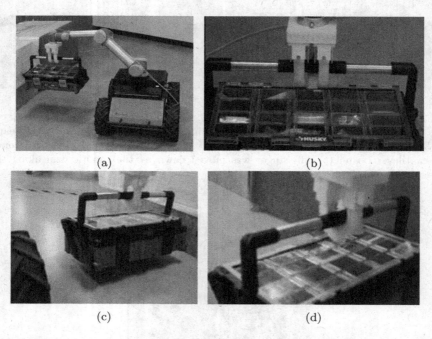

(a)

(b)

(c)

(d)

Fig. 8. Heavy payload grasped by compliant finger. The toolbox is weight 4.5 Kg. The compliant fingers are kept in high stiffness mode after actuation and can pick the toolbox.

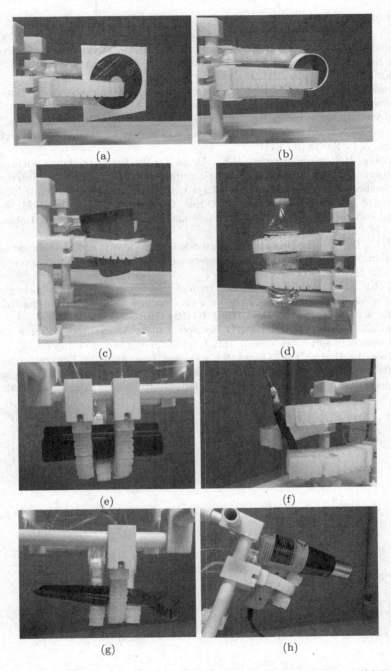

Fig. 9. Objects grasped by the compliant fingers: (a) CD, (b) Teflon tape, (c) Porcelain cup, (d) Plastic bottle, (e) Laptop battery, (f) Screwdriver, (g) Wrench, and (h) Heat gun.

References

1. Amend, J.R., Brown, E., Rodenberg, N., Jaeger, H.M., Lipson, H.: A positive pressure universal gripper based on the jamming of granular material. IEEE Trans. Robot. **28**(2), 341–350 (2012)
2. Deimel, R., Brock, O.: A novel type of compliant and underactuated robotic hand for dexterous grasping. Int. J. Robot. Res. **35**(1–3), 161–185 (2016)
3. Galloway, K.C., Becker, K.P., Phillips, B., Kirby, J., Licht, S., Tchernov, D., Wood, R.J., Gruber, D.F.: Soft robotic grippers for biological sampling on deep reefs. Soft Robot. **3**(1), 23–33 (2016)
4. Homberg, B.S., Katzschmann, R.K., Dogar, M.R., Rus, D.: Haptic identification of objects using a modular soft robotic gripper. In: 2015 IEEE/RSJ International Conference on Intelligent Robots and Systems (IROS), pp. 1698–1705. IEEE (2015)
5. Manti, M., Hassan, T., Passetti, G., D'Elia, N., Laschi, C., Cianchetti, M.: A bioinspired soft robotic gripper for adaptable and effective grasping. Soft Robot. **2**(3), 107–116 (2015)
6. Wall, V., Deimel, R., Brock, O.: Selective stiffening of soft actuators based on jamming. In: 2015 IEEE International Conference on Robotics and Automation (ICRA), pp. 252–257. IEEE (2015)
7. Yap, H.K., Lim, J.H., Goh, J.C.H., Yeow, C.H.: Design of a soft robotic glove for hand rehabilitation of stroke patients with clenched fist deformity using inflatable plastic actuators. J. Med. Dev. **10**(4), 044504 (2016)
8. Zhou, X., Majidi, C., O'Reilly, O.M.: Soft hands: An analysis of some gripping mechanisms in soft robot design. Int. J. Solids Struct. **64**, 155–165 (2015)
9. Johnson, L., Bruck, H.A., Gupta, S.K.: Design, fabrication, and characterization of a soft multi-fingered hand. In: ASME 2016 International Mechanical Engineering Congress and Exposition, American Society of Mechanical Engineers (2016)

Self-organisation of Spatial Behaviour in a Kilobot Swarm

Calum Imrie[(⊠)] and J. Michael Herrmann

School of Informatics, Institute for Perception, Action and Behaviour,
University of Edinburgh, 10 Crichton St, Edinburgh EH9 8AB, UK
c.c.imrie@sms.ed.ac.uk, mherrmann@staffmail.ed.ac.uk

Abstract. Applications of robotic swarms often face limitations in sensing and motor capabilities. We aim at providing evidence that the modest equipment of the individual robots can be compensated by the interaction within the swarm. If the robots, such as the well-known Kilobots, have no sense of place or directionality, their collective behaviour can still result in meaningful spatial organisation. We show that a variety of patterns can be formed based on a reaction-diffusion system and that these patters can be used by the robots to solve spatial tasks. In this contribution, we present first results for applications of this approach based on 'physically realistic' Kilobot simulations.

1 Introduction

The design and control of robot swarms is often inspired by biological systems. However, simple, inexpensive robots may lack the sensory or motor capabilities of their intended biological counterparts such that following patterns that realise behavioural goals would not achievable by individual robots. It is an interesting option to use the interaction among the robots as a source of information such that the swarm dynamics compensates the limitations of the individuals. In this way it may also be possible to reduce the effects of obstruction and interference amongst the robots. Related phenomena have been studied in social insects and even single living cells, where, as a paradigm, simple local rules lead to complex behavioural patterns in the swarm, which can enable decision making and improve efficiency [8].

Here we consider the Kilobot [9], a popular robotic swarm platform, which is a three-pronged robot with the two back prongs mechanically connected to vibrational motors. The Kilobot can sense ambient light by a single sensor, and it can send and receive messages by means of IR signals to other robots in the immediate neighbourhood. Ref. [3] uses diffusive information to navigate Kilobots for random walks and shows that this approach allows control to improve the exploration behaviour by tuning the parameters to optimise either in an open or a closed environment.

We present here first results for a robot swarm controlled by a reaction-diffusion (RD) system which is studied for a collective of Kilobots in a few

© Springer International Publishing AG 2017
M. Mangan et al. (Eds.): Living Machines 2017, LNAI 10384, pp. 551–561, 2017.
DOI: 10.1007/978-3-319-63537-8_49

simple decision-making tasks. The Kilobot's lack of directionality suggests the use of this approach, because the RD equations (3) do not involve gradients with respect to spatial variables, and the diffusion terms do not require a directional comparison of potentials carried by neighbouring robots. The next section describes the pattern formation algorithm, which will be followed by the experimental setup, results and finally a discussion of applications and future work within the approach.

Fig. 1. A Kilobot simulated in ARGoS.

2   Turing Patterns

Reaction-diffusion (RD) systems produce stripe-like or honeycomb-like Turing patterns [10] if two substances are spreading with different diffusion constants such that the inhibitor is vanishing faster than the activator that in turn has caused its production. The RD equations

$$\frac{\partial u}{\partial t} = D_u \triangle u + f(u, v)$$
$$\frac{\partial v}{\partial t} = D_v \triangle v + g(u, v) \tag{1}$$

describe the spatiotemporal dynamics of an activator u and an inhibitor v. The Laplace operator, $\triangle = \frac{\partial^2}{\partial x^2} + \frac{\partial^2}{\partial y^2}$, represents the diffusion of u and v, typically with diffusion constants $D_u < D_v$. The dynamics can be realised e.g. by substances in solution, but a large variety of natural systems that follow this dynamics is known [1]. The functions $f(u, v)$ and $g(u, v)$ are the reaction models for the respective potentials and may vary for the different applications. Often, the activation function is non-linear and the inhibition is linear. We will use the FitzHugh-Nagumo model, i.e. choose the reaction terms, $f(u, v)$ and $g(u, v)$, as

$$f(u, v) = \lambda u - \alpha u^2 - (1 - \alpha)u^3 - \sigma v$$
$$g(u, v) = u - v \tag{2}$$

Whether the system produces spot-like or stripe-like patterns depends, resp., on the presence or absence of a quadratic term in the reaction models, which is governed in Eq. 2 by the parameter α. Moreover, dynamical patterns such as moving spirals are known to exist in RD systems.

The application of RD systems in robotics is just in its beginning. Morphogenetic robotics [6] uses Gene Regulatory Networks (GRN) in order to control multi-robot systems [5]. Also for the Kilobot problem attempts have been made to realise the potential of this approach. In principle, two options can be used: The robots can either realise the RD dynamics by estimating the robot concentration from the mutual distances and slow down, or speed up in order to change the pattern. We will follow here instead the simpler option to simulate the RD dynamics by exchanging messages with neighbouring robots to account for the diffusion in Eq. 1, while the behavioural consequences become apparent only after a pattern has formed. The clustering behaviour (Sect. 4.2, see also Refs. [2,7] for earlier attempts) can be used as a precondition for the simulation, because it can gather all robots to enable a message exchange among all robots.

3   Experimental Setup

The Kilobot swarm is simulated using ARGoS (see Fig. 1) using the Kilobot plug-in which includes their limited messaging capabilities. The Kilobot broadcast at each time step their u and v values and update their own values depending on the received messages. Due to limited messaging, each robot will receive a varying number of messages per time step. This is taken into account in the redefinition of the Laplace operator (1) for a discrete set of robots:

$$\triangle u = -ru_0 + \sum_{i=1}^{r} u_i \tag{3}$$

The sum runs over all neighbours of the robot and compares their potential with its own, u_0. The operator (3) is stochastic, because the robot configuration that sends messages within a time step will typically deviate from a regular grid. It is possible to scale the diffusion constants in order to reduce the effect of the variable distances between neighbours, but we prefer Eq. 3 as it tends to produce a pattern as if the robots were in a regular formation, rather than relative to the embedding space. In addition to the messaging limitation, the messages themselves use limited information, namely 8 bit per time step, to implement the diffusive interaction. This can become problematic if small values of the potentials are to be broadcasted.

In addition, the RD dynamics needs to be tolerant to discretisation effects. In our ARGoS simulation, 225 Kilobots are run, usually in a 15×15 grid, with an initial distance between the robots such that a communication with eight immediate neighbours is possible. A border is present so that when motion is introduced to the Kilobots they do not disperse completely, i.e. given enough time they will encounter another Kilobot. The initial values for the potentials were randomly chosen from $[0, 0.1]$. Uniform noise with a range $[-0.0005, 0.0005]$ was added to u to escape from chimeric states. Larger noise in combination with the discretisation noise may result in unstructured patterns, or transitions across multiple ground states which may impede subsequent decision making.

4   Results

4.1   Pattern Formation in Stationary Swarm

Figure 2 shows an example run of the set of simulations ran for a stationary swarm, and that a spotted pattern can emerge. Figure 3 displays the location of the Kilobots as well as their activation values. The frequency of the spots in the pattern can be controlled through D_u, which is demonstrated in Fig. 4. The spots will also remain stable over a long period of time, see Fig. 5.

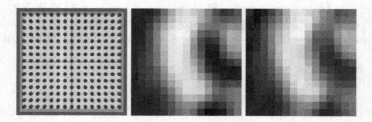

Fig. 2. The simulation is of the stationary setup described in Sect. 3. The first image is of the setup itself. The second and third images are the Kilobots' u and v values, resp., at the end of the simulation. These values were normalised to clearly see the differences between highly and weakly 'activated' Kilobots. In this case the largest values for u and v were, resp., 0.078 and 0.068. It can be seen quite clearly that there is a single spot, as to be expected since there is the inclusion of a quadratic term which will make the system favour spotted patterns. Parameters: $\alpha = 1$, $\sigma = 1$, $D_u = 0.5$, $D_v = 1.0$ and $\Delta t = 0.2$.

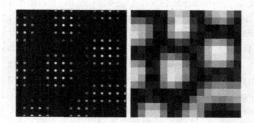

Fig. 3. Example of a robot configuration in the physical world space. The images on the right are showing each Kilobot's activation and inhibition, resp., and each circle represents the actual Kilobot's place in the boxed environment, which is shown in Fig. 2.

The type of pattern can also be modified, and it is possible to create stripe-like patterns within these simulations. Under the same setup as before but through modifying α and D_u the system can prefer striped patterns, which is shown in Fig. 6. It does, however, take a significantly longer time for stable stripes to form in comparison to spot pattern formation. Once formed though, the pattern will remain stable, see Fig. 7.

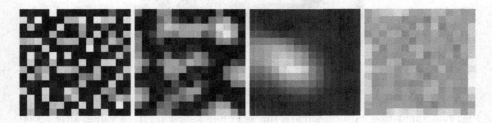

Fig. 4. The simulations demonstrates the effect of increasing D_u. After the simulations the values were normalised, and the images left of the graph shows the Kilobots' u values. As D_u increases, getting closer to the value of D_v, the spotted pattern becomes less spatially frequent, and also each spot becomes larger in size. Parameters: Similar to that in Fig. 2, except $D_u = 0.1, 0.3, 0.7$ and 0.9 (from left to right).

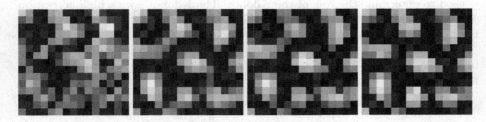

Fig. 5. Stationary simulation where each image displays the normalised u values. The images are after 100, 400, 700 and 1000 time steps. The evolution of the spotted formations began as a larger spot which over time broke apart to form smaller stable spots. Parameters: Similar to that in Fig. 2 except $D_u = 0.3$.

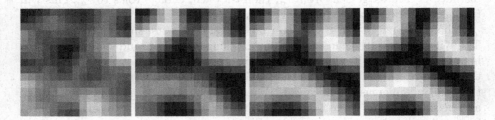

Fig. 6. The results here show a simulation where $D_u = 0.55$, $D_v = 1.0$, $\Delta t = 0.02$, $\alpha = 0$ and $B = 1$. All other parameters are set as that in Fig. 2. The images displays the activation, u, of the simulated Kilobots. From left to right, images show time steps 500, 2000, 3000 and 4000. A pattern containing curved stripes is produced around time step 1000 and is then solidified by time step 4000.

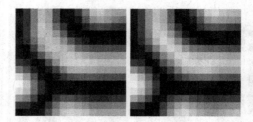

Fig. 7. A simulation with the same setup as in Fig. 6 showing the u values after 50000 and 600000 time steps. It can be seen that the pattern, once formed, remains fixed.

4.2   Clustering

A nearly regular configuration of the swarm can be achieved if the robots follow a form of preferential attachment where detachment is also allowed. The system is setup so that it favours spots, and after stable spots are formed a threshold value is used to determine the role of the Kilobot. The Kilobot's with high activation will be the centre of these clusters, and the weakly activated will converge towards the highly activated. Let's denote these roles as r_a and r_m, resp.

The adapt their speed by multiplication dependent upon the number of neighbours they hear from, and their respective roles. The power sent to the motors for r_a and r_m, denoted as p_a and p_m, are updated as follows

$$p_a = p_a a_1^{n_a} a_2^{n_m},$$
$$p_m = p_m m_1^{n_a} m_2^{n_m}, \tag{4}$$

where n_a and n_m are the number of messages received by a Kilobot with the role r_a and r_m respectively, and a_1, a_2, m_1, and m_2 are constants. $a_2 > 1 > a_1 > 0$ and $m_2 > 1 > m_1 > 0$. Random motion is generated for both r_a and r_m. The power can never be below a minimum value, p_{min}. Kilobots will speed up if they do not hear from a r_a, and slow down if they do. If r_m is within a certain distance of a message sent from r_a, then they will change roles to r_a. This will further encourage stable clustering.

The constants in Eq. 4 will determine several aspects, such as compactness. The Kilobots have the capability to push a single Kilobot, and so if m_1 is set to be a gradual decline it will promote compactness. The Kilobots will converge on those that are highly active, as they will be surrounded by r_as themselves, and thus will not increase their speed. When Kilobots of role r_m join the clusters and become r_a, then there is more potential for the Kilobots to receive messages from r_a, and remain as a cluster, see Fig. 8.

4.3   Stripe Segmentation

The Turing pattern constructed by the robots can be controlled to some extent. The wavelength of the striped pattern can be modified through the parameter D_u. This would lead to the swarm of having the ability to distinctly separate

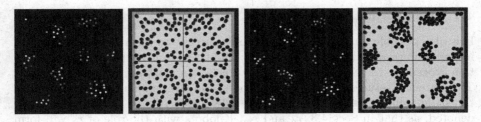

Fig. 8. The Kilobots were randomly spatially distributed and only sent and received messages for the first 2500 time steps. If a Kilobot's u exceeds a threshold θ, then they are r_a and their LED turns red. Else they are r_m. Afterwards, the Kilobots will speed up if they do not receive messages from r_a, and slow down otherwise. The top two images displays the simulation after 2500 time steps, and the bottom two from time step 4000. Parameters: $D_u = 0.6$, $D_v = 1.0$, $\lambda = 0.97$, $\alpha = 1$, $\theta = 0.04$, $p_{min} = 0.01$, $a_1 = 0.1$, $a_2 = 1.1$, $m_1 = 0.9$, $m_2 = 1.1$, and $\Delta t = 0.2$.

itself into subswarms. In this experiment, D_u was set to form at most two full parallel stripes through the swarm. Kilobots that have activations u below a certain threshold, would begin to randomly move until they arrive near other robots, then they slow down.

Once the stripes are formed the Kilobots will cluster into distinct groups. Figure 9 displays an example simulation of this. In this particular case the a stripe was curving, meaning that the other stripe would barely form as it was parallel to this stripe. The initial layout of the Kilobots have them all facing forward. Though the stripes can form in different directions, which will result in different cluster formations, the stripes will always remain still, thus the other Kilobots will tend to stop around them as they will be consistently slowing down within the stripes presence.

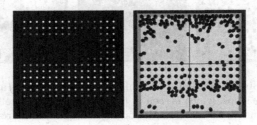

Fig. 9. The stripes here were formed within 15000 time steps (left). After the formation of the pattern, the Kilobots with low u values randomly moved and would slow down when in contact with other Kilobots, which can be seen after 370 time steps of stripes formation (right). Parameters: $D_u = 0.63$, $\alpha = 0$ and $\Delta t = 0.5$.

4.4   Ring Formation

It is also possible for the Kilobots, to some success, to form rings based on the information provided by the RD system. Similar to that in the stripe segmentation procedure, the Kilobots initially act solely as a platform for the RD system. The system is setup so that it favours spots, and after stable spots are formed a threshold value is used to determine the role of the Kilobot. These roles will be denoted, as that in Sect. 4.3, r_a and r_m. Kilobots with the role of r_a will form the shape of the ring, and r_m will fill in the gaps.

Random motion is generated for r_m while for r_a there is a probability of 0.2 to perform a random movement, while otherwise the movement depends on the messages received. If r_a receives a message from another active robot r_a' then it shall turn left, otherwise it will move forward. Due to the lossy communication, it is unlikely that both of the two active Kilobots will receive a message from each other on the same time step, and thus will result in a repulsion force. They follow the same power updates as that in Eq. 4, but here $a_1 > 1 > a_2 > 0$.

The setup described above takes advantage of the spotted pattern created since the spots, i.e. r_a, will immediately repel each other to create a loose ring. The other Kilobots will then, through random movement, converge onto these loose structures. The size of the ring is controlled through a_1 and a_2, as this determines the initial repelling force within each of the spots produced.

The majority of spots will lead to ring formations, though difficulty arises with spots that are present at the borders. These will mostly transform into semi-rings and more cluster like groupings. Figure 10 displays an example simulation on a randomly spatially distributed swarm.

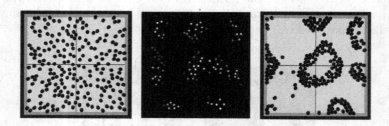

Fig. 10. The Kilobots were initially randomly spatially distributed and were stationary for the first 10^4 time steps, which is seen in the left image. The spots formed after this period can be seen in the middle. The Kilobots were then given a role dependent upon u. If $u > \theta$ then they are r_a, else r_m. r_m would move randomly and will slow down if they hear from a r_a, and speed up otherwise. r_a will speed up and turn left if they hear from a r_a, and slow down and move forward otherwise. r_a also has a small chance of moving randomly. Parameters: $D_u = 1.1$, $D_v = 2.0$, $\alpha = 0$, $\lambda = 0.97$, $\theta = 0.135$, $a_1 = 1.8$, $a_2 = 0.8$, $m_1 = 0.2$, $m_2 = 1.05$, and $\Delta t = 0.05$.

5   Discussion

This work has demonstrated not only are Turing Patterns possible in a Kilobot swarm, even with their communication difficulties, pattern formation can be decided upon (spots or stripes) through parameter choice, and how these patterns form, such as the number of spots. There had been work conducted into the sensitivity of the parameters, in particular how the parameters affect the pattern. One of the main parameters is the diffusion ratio between D_u and D_v, and this can be used as a control parameter for the pattern (spots, stripes; a third type, inverted spots, was not considered here).

This ability to produce stable patterns can then lead to various applications. The spotted pattern, see Fig. 5, can be controlled with respect to the number of spots, and this produces distinct teams within the swarm. This would be ideal for team-based tasks, including swarm exploration as this would allow groups of robots to detach from the swarm and conduct their exploration within their own group.

The stripe segmentation used a very basic rule for velocity. It was to show how the stripes can separate a swarm and that the wavelength can be controlled to determine the number of splits. Once the stripes have been formed then it is relatively simple to have control of the swarm by allocating the movement to those below/above a threshold that can be found analytically. The Kilobots can also communicate the maximum value of the potential during formation of a pattern, and use this information to determine the threshold (e.g. as half the maximum). This can also split them into types which again can be communicated to each other, and thus have more specific segmentation behaviours. For example, all non active Kilobots should avoid active Kilobots and form their own cluster.

The other formation demonstrated was the ring formation. The wall borders hinder ring formation, but as stated in Sect. 3, the borders contain the Kilobots' Brownian motion and eliminate the chance of swarm dispersement. One solution to this would be the use of lighting, which the Kilobots could detect and move away from. This would provide a non-obstacle repellent that would keep the Kilobots contained and potentially increase the success of ring formation. Another approach would be to change the Kilobots' motion so that it can home in early on either Kilobots that have the role r_a or have seen r_a. While this is difficult given the lack of directional information, it is possible to gain this information using past data. Interestingly this is where the RD system can provide additional information such as diffusivity, which will allow the Kilobots to move with the diffusion, thus a smoother spatial transition.

In order to realise the experiments with real Kilobots, finite battery life needs may be problematic. Obtaining straight stripes (Fig. 9), would be around two hours real time, which considerably exceeds the typical battery life of 20 min, although in most of our experiments the robots move only for part of the time. Therefore other robotic platforms may be preferable. A more capable robotic swarm would also have improved communication abilities, and thus reduce the time of pattern formation due to lossy communication.

Critical behaviour within a swarm can lead to improved results within a swarm, and work has been conducted into parameter selection specifically for particle swarm optimisation [4]. Turing patterns themselves require the correct parameters to be chosen to be able to form, thus they too will have a critical point w.r.t. the particular pattern formed. The critical behaviour in this swarm would allow the swarm to easily transition between one pattern to the other, thus increasing the range of information and behavioural capabilities of the entire swarm.

6   Conclusion

This investigation has demonstrated through preliminary analysis that the Kilobots are capable of constructing Turing patterns via a RD system through message passing. Furthermore, they have the capability of creating both spotted and striped patterns with control of how the patterns will be formed. This could lead to applications such as swarm separation and team formation. Through the information from the final pattern that was formed, the Kilobots have the capability to cluster and to form teams and boundaries. Although the Turing patterns offer only a few formation types, further work will show that a larger manifold of behaviours will be achievable by scheduling or adapting the parameter values or in combination with other techniques.

References

1. Adam, J.A.: Mathematics in Nature: Modeling Patterns in the Natural World. Princeton University Press, Princeton (2006)
2. Correll, N., Martinoli, A.: Modeling and designing self-organized aggregation in a swarm of miniature robots. Int. J. Robot. Res. **30**(5), 615–626 (2011)
3. Dimidov, C., Oriolo, G., Trianni, V.: Random walks in swarm robotics: an experiment with kilobots. In: Dorigo, M., Birattari, M., Li, X., López-Ibáñez, M., Ohkura, K., Pinciroli, C., Stützle, T. (eds.) ANTS 2016. LNCS, vol. 9882, pp. 185–196. Springer, Cham (2016). doi:10.1007/978-3-319-44427-7_16
4. Erskine, A., Joyce, T., Herrmann, J.M.: Parameter selection in particle swarm optimisation from stochastic stability analysis. In: Dorigo, M., Birattari, M., Li, X., López-Ibáñez, M., Ohkura, K., Pinciroli, C., Stützle, T. (eds.) ANTS 2016. LNCS, vol. 9882, pp. 161–172. Springer, Cham (2016). doi:10.1007/978-3-319-44427-7_14
5. Guo, H., Meng, Y., Jin, Y.: A cellular mechanism for multi-robot construction via evolutionary multi-objective optimization of a gene regulatory network. BioSystems **98**(3), 193–203 (2009)
6. Jin, Y., Meng, Y.: Morphogenetic robotics: an emerging new field in developmental robotics. IEEE Trans. Syst. Man Cybern. C **41**(2), 145–160 (2011)
7. Li, W., Gauci, M., Groß, R.: Turing learning: a metric-free approach to inferring behavior and its application to swarms. Swarm Intell. **10**(3), 211–243 (2016)
8. Oh, H., Shirazi, A.R., Sun, C., Jin, Y.: Bio-inspired self-organising multi-robot pattern formation: a review. Robot. Auton. Syst. **91**, 83–100 (2017)

9. Rubenstein, M., Ahler, C., Nagpal, R.: Kilobot: a low cost scalable robot system for collective behaviors. In: IEEE International Conference on Robotics and Automation (ICRA), pp. 3293–3298 (2012)
10. Turing, A.M.: The chemical basis of morphogenesis. Philos. Trans. R. Soc. Lond. B **237**(641), 37–72 (1952)

Bio-inspired Design of a Double-Sided Crawling Robot

Jong-Eun Lee[1], Gwang-Pil Jung[2], and Kyu-Jin Cho[1]($\boxtimes$)

[1] Seoul National University, Seoul 08826, Republic of Korea
{yhjelee,kjcho}@snu.ac.kr
[2] Seoul National University of Science and Technology, Seoul 01811, Republic of Korea
gwangpiljung@gmail.com

Abstract. A CardBot is a crawler with a thin card-sized structure, which has a limit in crawling when turned upside down. A double-sided CardBot presented in this paper is a robot that can crawl even when it is turned upside down because it can crawl on both sides. By adding one more robot body on a single-sided CardBot and sharing a motor to drive both slider cranks, a low height double-sided robot can be made. This 19 mm high, 26.39 g robot can crawl at a speed of 0.25 m/s. Thanks to the low body height, the robot can explore narrow gaps. Experiments were conducted to compare the running performance between the single-sided CardBot and the double-sided CardBot. Compared with a single-sided CardBot, only little degradation of the performance occurs due to implementation of a double-sided driving. The design has been adapted to reduce the friction, but the weakness of the shared joint due to the interaction of both cranks remains an unsolved problem. The structure of the robot will be modified to provide better performance in the future.

Keywords: Legged locomotion · Double-sided · Hexapod

1 Introduction

Insects in the nature explore diverse and complex terrains. Many robots have been designed for the purpose of exploring regions that are hard to access otherwise by mimicking insects. HAMR[3] [1], DASH [2], DynaRoACH [3], iSprawl [4], Mini-Whegs [5], RHex [6] are the representing crawlers inspired by nature.

When exploring rough terrain, an insect will often find itself turned upside down, and therefore would need the ability to turn itself back to the right position. Usually insects self-right by flapping wings or twisting a body to move the center of mass to a favorable position. Like some insects, robots that have the ability to self-right have recently been developed [7, 8].

Another way to crawl from an inverted position is constructing a structure that can run on both sides. Inspired by creatures that can always land on the right position after

---

The original version of this chapter was revised. The acknowledgement was added. The erratum to this chapter is available at 10.1007/978-3-319-63537-8_62

M. Mangan et al. (Eds.): Living Machines 2017, LNAI 10384, pp. 562–566, 2017.
DOI: 10.1007/978-3-319-63537-8_50

a fall, a robot has been developed that can run regardless of the direction of landing. A CardBot with a thin card-sized body has been introduced to make a double-sided crawling robot. The robot crawls based on a single-sided slider-crank mechanism. As seen in Fig. 1(b), it consists of a thin body and motors lie on it. The height of a basic robot with one transmission and body is 15 mm. Since having a low height, this robot can pass through low placed narrow gaps.

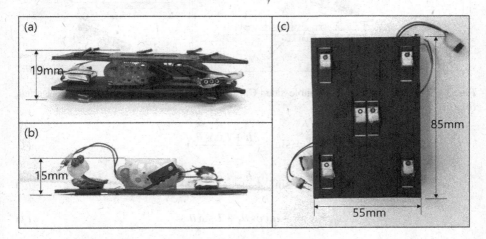

Fig. 1. Double-sided CardBot: A 26.39 g, 85 mm × 55 mm × 19 mm crawling robot. Side view (a), Single-sided CardBot (b) and top view (c).

To allow the robot to operate on either of its sides, robot bodies are attached to the top and bottom surfaces of the motor part. Double-sided traveling robots contain a motor that represent a large portion of their total height. Since a flat robot body is added, it is possible to capitalize on the advantages of a flat body shape, particularly occupied space, meaning double-sided traveling functionality only adds a height of 4 mm. Final height of the double-sided crawling robot is 19 mm.

In the following sections, design of this robot, crawling performance test, conclusion, and future work will be described.

2   Design

The robot consists of a lower body, a motor part, and an upper body. The lower body and the upper body have the same structure, which contain slider, crank, and leg links as shown in Fig. 2(a). The upper and lower cranks are connected with a single rotational joint for the purpose of simultaneous operation. As the rotating motor pushes the upper and lower slider crank to the negative x-direction, leg links are protruded. When pushed in positive x-direction, slider cranks let the leg links intrude. As a result of the protrusion and intrusion of leg links, the robot can crawl in the positive x-direction. Since the leg links are protruding in the same direction, the robot moves in the same direction from both sides.

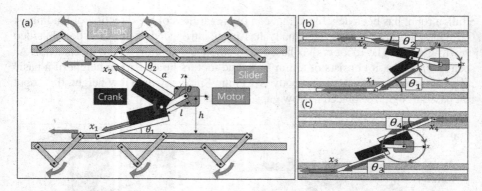

Fig. 2. Schematic view of the Double-sided CardBot (a) and friction forces of the slider crank (b), (c).

$$\theta_1 = \theta_3 = \sin^{-1}\left(\frac{h + l\cos\theta}{a}\right) \tag{1}$$

$$\theta_2 = \theta_4 = \sin^{-1}\left(\frac{h - l\cos\theta}{a}\right) \tag{2}$$

$$x_1 = x_3 = -a\cos\theta_1 - l\sin\theta \tag{3}$$

$$x_2 = -a\cos\theta_1 - l\sin\theta \tag{4}$$

$$x_4 = a\cos\theta_1 - l\sin\theta \tag{5}$$

Due to the Limited torque of the motor, the friction of slider crank has to be minimized to achieve better driving performance. Figure 3 indicates the location of the slider for the designs indicated in Fig. 2(b) and (c). Phase lag between upper and lower body is shown in both graphs. Assuming the same amount of force transmitted to the crank which is indicated as red arrow, vertical component of the force determines the magnitude of the frictional force between the slider and robot body which is indicated as blue arrow. Friction of the slider is important in the phase when leg links protrude because large torque is transmitted to the motor in this phase. Vertical force in the protraction phase in Fig. 2(b) is larger than in Fig. 2(c) when the motor rotates in clockwise. In addition, the force component in the vertical direction will temporarily deform the structure and is unable to deliver sufficient torque to the floor.

As a result, the final design was determined as Fig. 2(c). However, in this design, fatigue occurs frequently because the cranks interfere with each other more than in Fig. 2(b). For this reason, the experiment in the next chapter has been conducted with the design of Fig. 2(b).

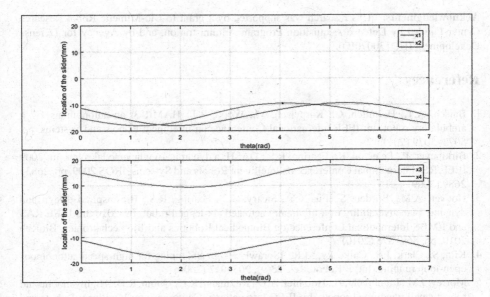

Fig. 3. Location of the upper slider and lower slider in x direction

3   Experimental Result

Running sequence of the robot was recorded using a high-speed camera(Miro eX4) at 1000 fps. Measure time for the robot traveled 0 cm, 5 cm, 10 cm, and 20 cm is 0 s, 0.20 s, 0.43 s, and 0.79 s respectively. As a result, the speed of the Double-sided CardBot is measured as 0.25 m/s. Speed of the Single-sided CardBot is also measured by high speed camera as 0.27 m/s.

4   Conclusion and Future Work

This paper presents a robot, with a height of 19 mm, and a length of 85 mm, that can travel on both sides.

This design performs at a slower speed when compared to the speed of a robot that the body attached only at one side, 0.25 m/s and 0.27 m/s respectively. As a result, this both-sided traveling robot can operate on both sides at the cost of a slight speed reduction. As mentioned previously in design section, speed reduction was caused by increased friction due to torque and increased weight of about 4 g.

In the future, this design will further change, so that the upper and lower plates may rotate with of less friction, further improving the running performance. In addition, an appropriate payload will be applied to minimize vertical movement and improve driving performance.

Acknowledgments. This research was supported by a grant to Bio-Mimetic Robot Research Center Funded by Defense Acquisition Program Administration, and by Agency for Defense Development (UD130070ID).

References

1. Baisch, A.T., Heimlich, C., Karpelson, M., Wood, R.J.: HAMR 3: an autonomous 1.7 g ambulatory robot. In: IEEE International Conference on Intelligent Robots and Systems, pp. 5073–5079 (2011)
2. Birkmeyer, P., Peterson, K., Fearing, R.S.: DASH: a dynamic 16 g hexapedal robot. In: 2009 IEEE/RSJ International Conference on Intelligent Robots and Systems, IROS 2009, pp. 2683–2689 (2009)
3. Hoover, A.M., Burden, S., Fu, X.Y., Sastry, S.S., Fearing, R.S.: Bio-inspired design and dynamic maneuverability of a minimally actuated six-legged robot. In: 2010 3rd IEEE RAS and EMBS International Conference on Biomedical Robotics and Biomechatronics, BioRob 2010, pp. 869–876 (2010)
4. Kim, S., Clark, J.E., Cutkosky, M.R.: ISprawl: design and tuning for high-speed autonomous open-loop running. Int. J. Robot. Res. **25**(9), 903–912 (2006)
5. Morrey, J.M., Lambrecht, B., Horchler, A.D., Ritzmann, R.E., Quinn, R.D.: Highly mobile and robust small quadruped robots. In: IEEE International Conference on Intelligent Robots and Systems, pp. 82–87 (2003)
6. Saranli, U., Buehler, M., Koditschek, D.E.: RHex: a simple and highly mobile hexapod robot. Int. J. Robot. Res. **20**(7), 616–631 (2001)
7. Li, C., Kessens, C.C., Young, A., Fearing, R.S., Full, R.J.: Cockroach-inspired winged robot reveals principles of ground-based dynamic self-righting. In: IEEE International Conference on Intelligent Robots and Systems, pp. 2128–2134 (2016)
8. Jung, G.P., Casarez, C.S., Jung, S.P., Fearing, R.S., Cho, K.J.: An integrated jumping-crawling robot using height-adjustable jumping module. In: Proceedings - IEEE International Conference on Robotics and Automation, pp. 4680–4685 (2016)

A Closed Loop Shape Control
for Bio-inspired Soft Arms

Dario Lunni[1,2]([✉]), Matteo Cianchetti[1], Egidio Falotico[1], Cecilia Laschi[1],
and Barbara Mazzolai[2]

[1] The BioRobotics Institute, Scuola Superiore Sant Anna, Polo SantAnna Valdera,
Via Rinaldo Piaggio 34, Pontedera, 56025 Pisa, Italy
{d.lunni,m.cianchetti,e.falotico,c.laschi}@santannapisa.it
[2] Center for Micro-BioRobotics, Italian Institute of Technology,
Polo SantAnna Valdera, Via Rinaldo Piaggio 34, Pontedera, 56025 Pisa, Italy
{barbara.mazzolai,dario.lunni}@iit.it

Abstract. We present a model-based approach for the control of the
shape of a tendon-driven soft arm. The soft robotic structure, which is
inspired by an octopus arm, has variable section that allows to obtain
variable curvature when actuated. The main goal of our control system
is to obtain a target curvature at a desired section of the arm. The controller combines input shaping and feedback integral control in order to
overcome modeling errors and constant disturbances. Simulations show
the coupling between the control loop and a dynamic model of the arm.

1 Introduction

Industrial robotics scenarios are characterized by the use of classic rigid robots
able to achieve optimal solutions in known environment. Yet, in recent years the
need of robots able to adapt in partially known or unstructured environments
has been growing. For this reason, flexibility and adaptability skills are more
and more required, leading to the development of soft robots having uncommon
properties for rigid systems. Soft robots are usually built using unconventional
materials, sensors and actuators that allow them to obtain a theoretically infinite
number of degrees of freedom. This high deformability gives them abilities to
grasp and manipulate objects of unknown shapes and sizes. Another important
aspect for soft robots lies in the reduction of risk of damage of the environment
in contact with the robot. Nature can be taken as inspiration for new ideas of
design for soft robots having better performances [1].

One of the most challenging aspects in soft robotics is control. The classical
model-based approaches are usually not followed because of difficulties in modeling high deformable structures, characterized by strong non-linearities. Usually
the models used refer to constant curvature continuum manipulators such as in
[2] or [3]. Some bio-inspired robots with non-constant curvature are present in
literature (e.g. [4]), but usually the control is based on machine learning techniques [5]. Attempts to design model-based controls are present in literature

© Springer International Publishing AG 2017
M. Mangan et al. (Eds.): Living Machines 2017, LNAI 10384, pp. 567–573, 2017.
DOI: 10.1007/978-3-319-63537-8_51

(e.g. [6]), but usually these approaches are not focused on the control of the shape of the arm.

In our case, the system under study is a tendon-driven soft manipulator inspired by the octopus arm. Because of the structure of the robot, the shape of the arm is characterized by an increasing curvature from the base when actuated. We propose a closed loop control system to obtain a desired curvature on a specific section, with the aim to control the shape of our robot in the planar case. The main idea of a generic manipulation task is represented in Fig. 1. From a target section the curvature of the backbone is higher than a desired one, so from that section to the tip the contact is guaranteed and this can enable manipulation tasks.

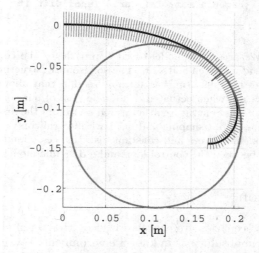

Fig. 1. Example of arm-object interaction. The blue circle can represent an object of estimated curvature and the arm is actuated in order to touch the object using the backbone from the highlighted section to the tip. (Color figure online)

The control system integrates a feed-forward technique with a closed control loop for disturbance rejection. Section 2 presents a steady-state model used to compute the feed-forward component, while Sect. 3 describes the control architecture. We discuss the obtained results in Sect. 4 through MATLAB simulations. Finally, Sect. 5 reports conclusive remarks and future perspectives of this work.

2   Model

The model used to simulate the system behavior and to study the control design are developed exploiting a Cosserat approach following [4,7] for the planar case. In particular we here describe the steady-state model, used to compute the feed-forward component of our control system.

The arm is actuated by two cables connected to the tip of the arm and passing through the arm to reach the base of the system. The actuators are coplanar and at the same distance respect to the midline. Pulling the tendon a bending movement of the arm is generated. The tension of the cable generates a point load on the arm where the cable is anchored and a distributed load along the arm. The concentrated load is equal to the cable tension and tangent to it $T_i \mathbf{t}_{ci}(T_i > 0)$, where T_i is the tension of the i-th cable and $\mathbf{t}_{ci}$ is a unit vector tangential to the backbone. The distributed load is proportional to the curvature of the cable $\mathbf{w}_i(s) = T_i d\mathbf{t}_{ci}/dS_c$ where $i \in \{1, 2\}$ identifies one of the two cables and S_c represent the arc length of the cable. The parameter s is used to identify the s-th section, considering the arm as composed by an infinite number of rigid bodies.

From the equilibrium and constitutive equations of this system [8] it is possible to derive the relations between the cable tension and the curvature $K(s)$ and the axial deformation $q(s)$ of the backbone at section s:

$$\begin{cases} K'(s) = -K(s)\dfrac{E\pi R^3(s)R' + 2\alpha y_c(s)y_c'(s)}{EJ(s) + y_c^2\alpha} + \dfrac{y_c'\beta}{EJ(s) + y_c^2\alpha} \\ K(L) = \dfrac{b\beta}{EJ(s)} \end{cases} \quad (1)$$

$$q(s) = \frac{K(s)y_c\beta - \alpha}{EA(s)} \quad (2)$$

where $R(s)$ is the radius of the section s of the arm, $R' = dR/ds$ is the derivative of the radius of the section, $y_c' = dy_c/ds$ is the derivative of the position of the cable, $\alpha = T_1 + T_2$ and $\beta = T_1 - T_2$, T_1 and T_2 are the tensions of the tendons. $J(s)$ is the moment of inertia of the section s, E is the Young Modulus of the material and $A(s)$ is the area of the section.

3   Control Architecture

The main architecture of the control system used for the shape control is composed by two parts: a filtered feed-forward component and a feed-back integral control. The feed-forward component is based on the inversion of the steady-state model of the system that is then filtered through an "input shaping", then the loop is closed using an integral control.

3.1   Filtered Feed-Forward Control

As already developed in [8] we propose a feed-forward loop using inversion of the steady-state model of the system [4]. Starting from Eq. 1 it is possible to rewrite the system as (Fig. 2):

$$\begin{cases} K'(s) = K(s)B(s, \alpha, \beta) + C(s, \alpha, \beta) \\ K(L) = K_l(\beta) \end{cases} \quad (3)$$

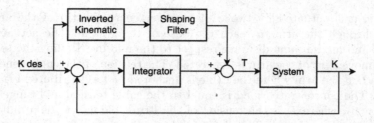

Fig. 2. Control system scheme

that admits a closed analytical solution as:

$$K(s) = e^{\int_s^L -B(\sigma)d\sigma} \left(K_l - \int_s^L C(\xi)e^{\int_\xi^L B(\sigma)d\sigma}d\xi \right) \tag{4}$$

We integrated this equation in s, and calculated the tension value to obtain the desired curvature at the desired section s_{des} imposing $K(s_{des}) = K_{des}$. To obtain a reduction of the vibrations of the system, the feed-forward signal is then filtered using an "input shaping" technique. Firstly, we studied the dynamic response of the model. This response has been approximated to the one of a second order linear system. Starting from these approximated values of natural frequency and damping an input filter has been designed [9] (Fig. 3).

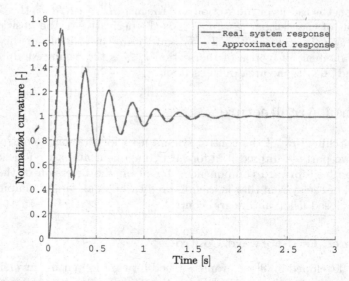

Fig. 3. Normalized response of the arm to a step signal of tension for a target section. The approximated response has been calculated minimizing the quadratic error between the responses.

3.2 Integral Control

To overcome constant errors, we designed a feed-back integral control. The closed loop control was tuned in order to work in a quasi-static condition. Approximating the robot with a second order system it is possible to obtain the resonance frequency of the system and then tune the integral control in order to not influence the system dynamics. In this way the system is able to overcome constant model errors without taking the system to the instability.

4 Results and Discussion

We tested the control system through MATLAB simulation. The model developed in [7] was used to simulate the behavior of the real system, while we used the steady-state model [4] to obtain the model inversion. The arm used was characterized by a length of 100 mm, 15 mm of maximum radius at the base and 5 mm of minimum tip radius. The material used to built the arm is ECOFLEX™ silicone (00-30 type), characterized by a density of $1070 \, kg/m^3$ and a Young Modulus of 110 kPa. Figure 4 shows the results of simulations comparing the controller containing just the inverted model and the one with the whole controller. An error in the model inversion was considered to demonstrate the effectiveness of the integral action. In particular we considered an underestimation of 10% of the Young Modulus in the model inversion. In the last simulation, to test the robustness of the control system to measurement disturbances, some random noise at 100 Hz and unitary amplitude has been added to the measured curvature.

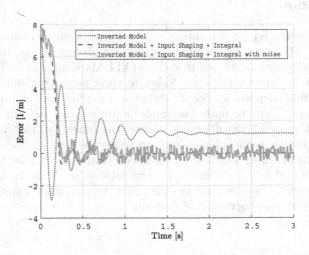

Fig. 4. Error dynamics comparison in presence of model error. The dotted line represents the response using the simple inverted model. The dashed line shows the response adding the integral component and the input shaping filter. The continuous line shows the behaviour adding also noise on the measure.

As expected, the feed-forward component is not able to take the system to the desired curvature in presence of error on the model. On the other hand the dashed line shows the response of the arm in presence of the whole controller and the dotted line shows the response of the same controller adding the random noise on the measured curvature. Firstly, it is visible the rejection of the error introduced in the model used for the inversion. Secondly, we can see a reduction of the response vibrations thanks to the input shaping filter. Finally, we can appreciate also the good response of the system in presence of random noise on the measure that does not affect the stability of the system. It is necessary to say that the curvature of the arm on the sections different from the chosen one is not controlled, this is because we are trying to control a system with an infinite number of degree of freedom with a single control variable. In general, the results confirm the effectiveness of the control design approach. In particular an increase of the robustness of the control was proved. This aspect is fundamental in the soft robotics field because of the difficulties of a correct model characterization. For this reason in case of a model-based control the open loop can lead to bad results. It is clear that one of the most important aspect to be taken into account to obtain results comparable with the model-free approach is the robustness of the controller. The second aspect that should be underlined is the target of our controller. The shape control becomes fundamental for soft robots because of their high deformability and their peculiarity to be easily used in contact with the environment. This new perspective, coupled with the intrinsic adaptability of soft structures can have important developments in soft robots applications.

5   Conclusion

This paper presents an implementation of a closed loop to control the shape of a soft arm. The architecture of the system is presented and the behaviour of the complete system is shown through MATLAB simulations. These results encourages us to continue on this way following these next steps: improvements in simulations and 3D expansion, deeper theoretical studies and experimental tests on the real robot. First, to make the model more realistic we will consider the presence of a low level control loop to control the tension of the cable. Second, the obtained results in the planar open to paves the way to the study of the expansion of the 3D case. Third, from the theoretical point of view the next step is to perform a stability analysis of the control system. Finally, the idea is to implement the control system on the real robot coupled with a suitable sensing system to compute the curvature of the arm.

References

1. Lee, C., et al.: Soft robot review. Int. J. Control Autom. Syst. **15**(1), 3–15 (2017)
2. Camarillo, D.B., et al.: Mechanics modeling of tendon-driven continuum manipulators. IEEE Trans. Rob. **24**(6), 1262–1273 (2008)
3. Trivedi, D., Lotfi, A., Rahn, C.D.: Geometrically exact models for soft robotic manipulators. IEEE Trans. Rob. **24**(4), 773–780 (2008)
4. Renda, F., et al.: A 3D steady-state model of a tendon-driven continuum soft manipulator inspired by the octopus arm. Bioinspiration Biomim. **7**(2), 025006 (2012)
5. Braganza, D., et al.: A neural network controller for continuum robots. IEEE Trans. Rob. **23**(6), 1270–1277 (2007)
6. Penning, R.S., et al.: Towards closed loop control of a continuum robotic manipulator for medical applications. In: 2011 IEEE International Conference on Robotics and Automation (ICRA). IEEE (2011)
7. Renda, F., et al.: Dynamic model of a multibending soft robot arm driven by cables. IEEE Trans. Rob. **30**(5), 1109–1122 (2014)
8. Giorelli, M., et al.: A two dimensional inverse kinetics model of a cable driven manipulator inspired by the octopus arm. In: 2012 IEEE International Conference on Robotics and Automation (ICRA). IEEE (2012)
9. Singh, T., William, S.: Input shaping/time delay control of maneuvering flexible structures. In: Proceedings of the 2002 American Control Conference, vol. 3. IEEE (2002)

Learning Modular Sequences in the Striatum

Giovanni Maffei[1], Jordi-Ysard Puigbò[1(✉)], and Paul F. M. J. Verschure[1,2]

[1] Laboratory of Synthetic, Perceptive, Emotive and Cognitive Science (SPECS),
DTIC, Universitat Pompeu Fabra (UPF), Barcelona, Spain
jordiysard@gmail.com
[2] Catalan Research Institute and Advanced Studies (ICREA), Barcelona, Spain

Abstract. The execution of habitual actions is thought to rely on
the exploitation of procedural motor memories. These memories encode
motor commands as organized in functional sequences with well defined
boundaries in the Striatum. Here, we present a biophysical model of the
striatal network composed by inhibitory medium spiny neurons (MSNs)
governed by anti-hebbian STDP. We show that these two features allow
for learning an arbitrary sequence through multiple exposures to cortical inputs and reproducing it under a single, non-specific excitatory
drive. Our results shed light on the computational properties of biologically plausible inhibitory networks and suggest a simple, yet effective
mechanism of behavioral control through striatal circuits.

1 Introduction

The execution of habitual action sequences is thought to rely on the exploitation of procedural motor memories. These memories encode motor commands as
organized in functional modules with well defined boundaries [3]. Medium Spiny
Neurons (MSNs) within the Dorso-Lateral Striatum (DLS) could play a major
role from the perspective of motor learning where transient activation seem to
encode the representation of an overtrained motor sequence [4]. In addition,
anti-hebbian Spike-Time Dependent Plasticity (STDP) observed in the striatal
network could contribute in shaping connectivity patterns favoring sequencing
and chunking [5]. We investigate this hypothesis by using a spiking model of the
striatal microcircuit, composed by MSNs, to learn sequences through an anti-
hebbian STDP rule. We show that after learning, the induced plastic changes
lead to an asymmetric connectivity pattern. This asymmetry favors transient
sequential suppression and activation MSNs, that encode elementary bits of the
sequence. The acquired sequence can be recalled by globally exciting the network with a non specific input which induces connectivity-driven state-switching
around hetero-clinic orbits, a necessary property of this type of dynamical systems [1]. Our preliminary results suggest that the interaction of laterally connected MSNs and a simple anti-hebbian STDP rule are sufficient conditions for
the network to acquire and reproduce a sequence of neural activity, in line with
physiology [5] and anatomy [6] of the striatal microcircuit.

© Springer International Publishing AG 2017
M. Mangan et al. (Eds.): Living Machines 2017, LNAI 10384, pp. 574–578, 2017.
DOI: 10.1007/978-3-319-63537-8_52

2   Methods

We test the ability of a biophysically plausible spiking inhibitory network composed by 300 neurons to learn to reproduce a temporal sequence of inputs by mimicking the transient activity found in the striatal microcircuit [4]. To do this we implement a spiking neural model [7], whose dynamics are governed by the following equations:

$$C\frac{dV_i}{dt} = I_i - g_L(V_i - E_L) - g_{Na}m_{\inf}(V_i - E_{Na}) - g_k n_i(V_i - E_k) \qquad (1)$$

$$\frac{dn_i}{dt} = \frac{(n_{\inf}(V_i) - n_i)}{\tau} \qquad (2)$$

We select the parameters so to emulate the behavior of striatal MSNs (Fig. 1A) [8]. Values for each parameter are detailed in Table 1. For a complete description of each parameter see [7,8].

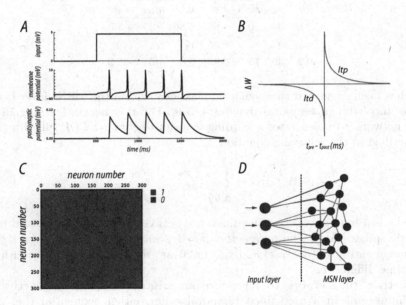

Fig. 1. Methods. A. Behavior of a single MSN under 5 mV stimulation. **B.** Anti-hebbian STDP curve **C.** Connectivity matrix for naive network (63% random connectivity). **D.** Schematic representation of the network including input layer, MSN layer and readout layer.

We choose a connectivity pattern between MSNs units (Fig. 1C) defined by random initial weights and a connection probability of 63%, so to mimic the lateral inhibitory connectivity found in the striatal microcircuit [6]. As in previous literature [8] the synaptic dynamics follow the Rall model, described by Eqs. (3) and (4):

$$\tau_g \frac{dV_j}{dt} = \theta(V_i(t) - V_{th}) - V_j(t) \tag{3}$$

$$I_i(t) = I_i^c + \sum j(-K_{synij}V_j(t)(V_i(t) - V_{syn}) \tag{4}$$

For a complete description of each parameter see [8].

We define the input (I^c) to the network in order to emulate the excitatory drive of cortical neurons (Fig. 1D). In particular we implement 10 excitatory input units characterized by constant activity between 3.6 and 8.1 mV (depending on the experimental condition) and a fixed gaussian noise with $\sigma = 2$. The MSNs network is connected to the input layer so that each input unit excites an equal number of MSNs (=30) in a topologically organized way.

Table 1. Network parameters

I^c	g_k	g_{Na}	g_L	E_{Na}	E_k	E_L	$m_{\inf}$	$n_{\inf}$	tau	C
5	10	20	8	60	−90	−80	−20	−25	1	1
	a_+	a_-	τ_+	τ_-	τ_g	v_{th}	v_{syn}			
	0.2	30	75	6600	0.8	-40	-65.9			

It has been suggested that synaptic plasticity within the MSNs layer is subject to anti-hebbian temporal dynamics [5,9]. For this reason learning in the MSNs network is implemented according to an anti-hebbian STDP rule (Fig. 1B), as described in the following equation:

$$\frac{dw_i}{dt} = W_i(\Delta t_s) = \begin{cases} \eta p e^{-\frac{\Delta t_s}{\tau_{post}}}, & for\, \Delta t_s > 0, \\ \eta q e^{\frac{\Delta t_s}{\tau_{pre}}}, & for\, \Delta t_s < 0 \end{cases} \tag{5}$$

where w_i is a function of the time difference between the pre-synaptic and post-synaptic spikes (Δt_s). The parameters $p(= 1.)$ and $q(= -1.)$ define the sign of the change and τ_{post} (=20 ms) and τ_{pre} (=20 ms) the time dependency with the spike-time difference.

We train the network to reproduce an arbitrary sequence by activating each input unit in a predefined temporal order, either sequential (i.e. from unit 1 to unit 10) or random, for 20 repetitions. After learning we test the acquired sequence by driving homogeneously the network with a constant input (=3.6 mV). We also test the generalization of the sequence to a new speed by providing a higher input (=8.1 mV).

3   Results

We train the network until convergence is reached (i.e. 20 sequence repetitions) by providing a cycling 10-element sequence as input, both for sequential and

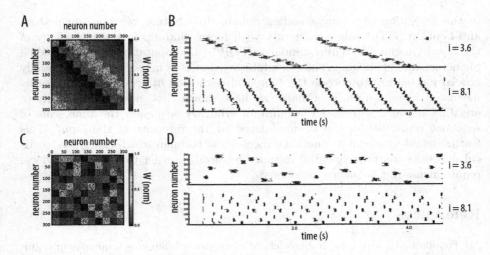

Fig. 2. Results. A. Weight matrix after learning a linear sequence of inputs. **B.** Raster plot of the linear sequence generalized to different time scale depending on the input magnitude. **C.** Same as A. for a random sequence of inputs. **D.** Same as B. for a random sequence of inputs.

random permutations (see Methods). Figure 2A and C show two examples of a trained connectivity matrix. Moreover, we test the ability of the network to reproduce the trained sequence by providing constant input at two different intensities. Figure 2B and D show how a sufficient excitatory input to the network triggers the reproduction of the trained sequence (top), while increasing input intensity reproduces the sequence at higher rates (bottom).

4   Discussion

We have proposed a novel computational model that could contribute to shed light on the neural mechanisms underlying sequential memories in the brain. Learning of arbitrary sequences has been subject of extensive study in excitatory populations (i.e. in the hippocampus) [10], and it has been previously shown in rate based models of inhibitory units [1]. However, little attention has been devoted to the computational features of spiking inhibitory networks (see [8] for an exception), whereas their understanding is crucial for unveiling the mechanisms governing learning in sub-cortical areas (i.e. the basal ganglia). For this reason we selected neuron parameters that allowed us to reproduce the biophysical properties of striatal MSNs, characterized by slow depolarization time constants, slow firing rate and absence of sub-threshold oscillations [8]. We used a connectivity matrix that reflects the local striatal connectivity pattern, where 63% of one MSNs are recurrently connected via inhibitory synapses [6]. Note that the existence of lateral inhibition in the Striatum is object of controversy, thus providing a demonstration of its functional implications could contribute

578   G. Maffei et al.

to the resolution of a long standing debate. In addition, we implemented an
anti-hebbian STDP rule consistently with recent findings in the physiology of
the basal ganglia [5]. While standard STDP is a computationally well studied
phenomenon, anti-hebbian rules in inhibitory networks may represent an equally
crucial mechanism to encode the temporal resolution of acquired memories by
means of dis-inhibition [9]. Finally, we have shown that, despite the ability of
an MSN network to robustly acquire an arbitrary sequence, the time scale of
sequence reproduction can be modulated by the intensity of the input. This
feature could represent a functional mechanism through which excitatory cortical networks control sequential memory retrieval in order to match behavioral
requirements (i.e. sequence of actions).

References

1. Fonollosa, J., Neftci, E., Rabinovich, M.: Learning of chunking sequences in cognition and behavior. PLoS Comput. Biol. **11**(11), e1004592 (2015)
2. Gage, G.J., Stoetzner, C.R., Wiltschko, A.B., Berke, J.D.: Selective activation of
   striatal fast-spiking interneurons during choice execution. Neuron **67**(3), 466–479
   (2010)
3. Jin, X., Costa, R.M.: Shaping action sequences in basal ganglia circuits. Curr.
   Opin. Neurobiol. **33**, 188–196 (2015)
4. Rueda-Orozco, P.E., Robbe, D.: The striatum multiplexes contextual and kinematic information to constrain motor habits execution. Nat. Neurosci. **18**(3), 453–
   460 (2015)
5. Fino, E., Venance, L.: Spike-timing dependent plasticity in the striatum. Froniters
   Synaptic Neurosci. **2**, 6 (2010)
6. Chuhma, N., Tanaka, K.F., Hen, R., Rayport, S.: Functional connectome of the
   striatal medium spiny neuron. J. Neurosci. **31**(4), 1183–1192 (2011)
7. Izhikevich, E.M.: Dynamical Systems in Neuroscience. MIT press, Cambridge
   (2007)
8. Ponzi, A., Wickens, J.: Sequentially switching cell assemblies in random inhibitory
   networks of spiking neurons in the striatum. J. Neurosci. **30**(17), 5894–5911 (2010)
9. Kleberg, F.I., Fukai, T., Gilson, M.: Excitatory and inhibitory STDP jointly tune
   feedforward neural circuits to selectively propagate correlated spiking activity.
   Frontiers Comput. Neurosci. **8**, 53 (2014)
10. Melamed, O., Gerstner, W., Maass, W., Tsodyks, M., Markram, H.: Coding and
    learning of behavioral sequences. Trends Neurosci. **27**(1), 11–14 (2004)

Spermbots: Concept and Applications

Mariana Medina-Sánchez[1(✉)], Veronika Magdanz[1], Lukas Schwarz[1],
Haifeng Xu[1], and Oliver G. Schmidt[1,2]

[1] Leibniz Institute for Solid State and Materials Research, IFW Dresden, Institute for Integrative
Nanosciences, Helmholtzstrasse 20, 01069 Dresden, Germany
m.medina.sanchez@ifw-dresden.de
[2] Chemnitz University of Technology, Reichenhainer Str. 70, 09107 Chemnitz, Germany

Abstract. Biohybrid systems are promising solutions in micro- and nanobio-technology due to the possibility to combine exciting biological properties of living microorganisms/cells (e.g. sensing and taxis mechanisms), and the controllability of man-made microstructures. Here we present the development of tubular and helical spermbots, a concept that refers to a sperm-based micro-robot. The recent achievements include the capture, guidance and release of motile and immotile sperm cells by artificial magnetic microstructures (micro-tubes, microhelices or four-armed microtubes). These approaches are interesting for potential applications in *in vivo* assisted fertilization and targeted drug delivery. The characteristics, challenges and possibilities are discussed in detail throughout this work.

Keywords: Spermbot · Assisted fertilization · Drug delivery

1 Introduction

The development of highly functional autonomous microdevices is an emerging research field that mainly aims at biomedical applications by the use of remote-controlled microsensors and actuators. They can be driven by chemical fuels [1], surface tension gradients [2], ultrasound [3], magnetic [4, 5] or electric fields [6]. The use of such microdevices in *in situ* sensing, drug delivery, theranostics or micro-surgery has led to interesting progress such as the drilling into tissue and cells [7, 8], the isolation of cancer cells, DNA and proteins [9], operation of an artificial micromotor inside the stomach of a mouse [10], and the motion of rod shaped nanomotors inside living cells by ultrasound [3]. However, the efficient operation in physiological environments and the avoidance of toxic fuels remain the main challenges. To overcome these problems, artificial bacterial flagella [4] and flexible magnetic swimmers [11] have been created to mimic biological flagella in shape and dimension as they can tolerate a large range of conditions and propel efficiently in low-Reynolds-number regimes. They are rigid helical [12] or straight flexible [13] elements that propel by non-reciprocal motion. Flexible magnetic microswimmers require an oscillating field in order to be actuated. Alternatively, a further approach makes use of biological motors designed by nature, implementing whole cells as actuators. Bacteria and contractile muscle cells have been

© Springer International Publishing AG 2017
M. Mangan et al. (Eds.): Living Machines 2017, LNAI 10384, pp. 579–588, 2017.
DOI: 10.1007/978-3-319-63537-8_53

harnessed as propulsion sources in various approaches for the development of such bio-hybrid systems that may perform robotic tasks on the microscale in the future [14]. Muscle-powered microdevices were first developed based on an assembly of cardio-myocytes, i.e. beating heart cells, and fabricated microstructures that served as the scaffold to which the contracting cells were attached [15]. In a similar fashion, cardio-myocytes were cultured along a polydimethylsiloxane filament to create a slender tail that performed a bending wave by the synchronous contraction of the cells [14]. Motile bacteria have also been explored for their potential in developing hybrid microbiorobots. They have been utilized in creating disk-shaped motors that turned due to their geomet-rical design in which bacteria got stuck and pushed the corners of the disk leading to a rotational motion similar to a ratchet [16]. Furthermore, bacteria's ability to act as bacterial carpets that move fluid or push microparticles forward was demonstrated [17]. Such microorganisms are especially attractive as components in microrobots because of their sensing and taxis abilities that offer various directional control mechanisms [18]. For instance, magnetotactic bacteria have been explored for their potential as micro-carriers in blood vessels and for cancer therapy [19, 20]. Likewise, chemotaxis [22], pH-taxis [23], phototaxis [41], and electrokinetic control [24] have been applied as direc-tional control mechanisms of bacterial microrobots. Spermatozoa have also been employed for developing hybrid systems due to their swimming, sensing, and cargo/delivery capabilities that are appealing mainly for applications such as assisted fertili-zation and drug delivery [25], which will be describe along this article as well as their challenges and opportunities. Although most of the above mentioned hybrid microrobots do not need an external power source for their propulsion, their activity range is restricted to physiological conditions (temperature and pH tolerance of the respective microor-ganisms/cells) and they may require an optimal design and material selection that allows them to navigate through a biological environment without their functionality being affected.

2   Spermbots and Their Applications

Here, we present hybrid micromotors based on the integration of sperm cells and man-made microstructures, also named spermbots [26, 27]. Such microstructures are fabri-cated in sizes in the nano- and micrometer range to help sperm cells carry out their function in a physiological microenvironment, when there are few sperms (oligo-spermia) [26] or when sperms are not strong enough to reach the fertilization site (asthe-nospermia) [27]. These two sperm deficiencies are the main causes of male infertility, diagnosed in about 40–50% of the cases [28].

In our first spermbot approach (to counter oligospermia), motile bovine sperm cells enter magnetic microtubes and propel them forward, while the direction is controlled by an external magnetic field (Fig. 1A). First, the velocity of the sperm-driven micro-tubes was investigated in dependence of the microtube radius, the penetration of the cell inside the microtube [26] and the length of the microtubes [29]. Several control mech-anisms were developed for remote-controlled motion of these tubular spermbots. For example, temperature is a natural factor that can be used to control the velocity of the spermbots. Within the temperature tolerance of sperm cells, which is in the range of 5–40 °C, the spermbots can be sped up or slowed down repeatedly [26]. It was observed that the confinement of the sperm flagella plays a crucial role in the reduction of the velocity of the spermbots in comparison to freely swimming cells. Therefore, shorter microtubes with larger radius are beneficial for the performance of these microbiorobots (Fig. 1B). Directional control mechanisms are offered by external permanent magnets to steer the spermbots to certain locations. This was demonstrated to be useful for on-chip separation of the sperm-driven microbiorobots from a mixture of sperm cells and microtubes [26]. A setup of electromagnetic coils was used to demonstrate the accurate closed loop control of the sperm-driven micromotors in two dimensions [30]. Regarding the release of the single sperm cells from the microtubes, a remote-controlled method was developed involving thermoresponsive materials. The temperature-sensitive polymer poly-N-isopropylacrylamide (PNIPAM) was modified to become photosensitive and thus it could be patterned by photolithographic processes. Furthermore, its swelling temperature was adjusted to 38 °C, which is the optimum temperature for bovine sperm cells and oocytes. Thin films of this polymer were fabricated by spin-coating and photolithography, and coated with nanometer-thin layers of titanium and iron for magnetic control. These multilayered thin film structures were able to fold into microtubes and capture single sperm cells below 38 °C. Upon a small temperature increase to about 38 °C, the polymeric microtubes unfolded and released the sperm cells to the surroundings (Fig. 1C, D) [31].

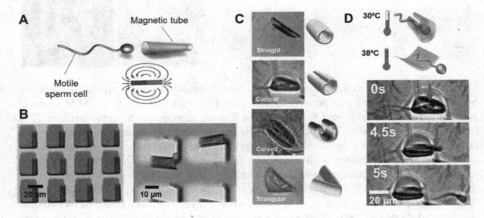

Fig. 1. Tubular spermbots: (A) Schematic of magnetic microtubes for the capture and guidance of single motile sperm cells, reprinted with permission from [26]. Copyright 2013 Wiley. (B) Array of 20 μm microtubes fabricated by roll-up nanotechnology (left) and optical image of bovine sperm cell captured inside a magnetic microtube (right), reprinted with permission from [29]. Copyright 2015 Wiley. (C) Schematic and optical images of thermoresponsive rolled-up polymeric microtubes with a sperm inside, reprinted with permission from [31]. Copyright 2016 Wiley. (D) Image series of bovine sperm cell upon release from thermoresponsive microtube. The red arrow points at the sperm head.

For the second case of male infertility (asthenospermia), when the sperms are immotile, we proposed artificial micropropellers to transport cells with motion deficiencies to help them succeed in fertilization (Fig. 2A). We demonstrated that metal-coated polymeric microhelices, fabricated by 3D laser lithography (Fig. 2B) are suitable to transport sperm cells without compromising their viability (indicated by a hypoosmotically swelled tail after transporting the sperm towards the oocyte, Fig. 2C). We managed to capture, transport, and release single immotile living sperm cells in fluidic channels that allow mimicking physiological conditions (Fig. 2D). For that, we employed suitable selection methods for capturing an immotile but functional sperm cell by using the well-established hypoosmotic swelling test (HOS) [32]. This method evaluates the functional integrity of the sperm's cell membrane by determining the ability of the membrane to maintain the equilibrium of osmotic pressure inside the sperm compared to its environment by swelling, which results in a curling of its tail (Fig. 2E).

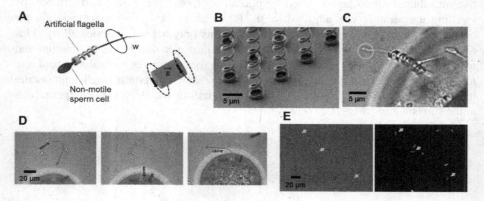

Fig. 2. Helical spermbots: (A) Schematic of a captured immotile sperm by a remotely controlled magnetic helix, reprinted with permission from [27]. Copyright 2016 ACS. (B) SEM image of polymer microhelices fabricated by Direct Laser Writing, reprinted with permission from [27]. Copyright 2016 ACS. (C) Transport of viable sperm by a helical magnetic micromotor to the oocyte (retained viability indicated by the hypoosmotically swelled tail of the successfully transported sperm), reprinted with permission from [27]. Copyright 2016 ACS. (D) Image sequence showing the sperm transport, approach to the oocyte and release (red arrows), reprinted with permission from [25]. Copyright 2017 Wiley. (E) Sperm cell viability after HOS test: bright field micrograph of sperm cells in HOS medium (left), (b) fluorescence image of its live/dead staining to demonstrate the compliance of sperm swelling and viability (right). Yellow arrows indicate the swelled and live sperm cells, reprinted with permission from [27]. Copyright 2016 ACS. (Color figure online)

Sperm cells are also very attractive for other applications such as drug delivery. We have demonstrated a sperm-hybrid micromotor for cargo-delivery to potentially treat diseases of the female reproductive system (Fig. 3A). Here we used a laser-printed microstructure coated with an asymmetric layer of iron to provide the sperm with magnetic guidance (Fig. 3B, C). The structure also serves as mechanical release trigger, based on a flexure mechanism that opens a cavity for the sperm to be released once it

Fig. 3. Sperm-tetrapod for drug delivery: (A) Schematic of magnetic microstructure (tetrapod) for the capture, guidance and release of drug-loaded sperm cells, (B) SEM image of polymeric tetrapods fabricated by Direct Laser Writing, (C) Motile sperm coupled to a tetrapod, (D) Image sequence of the sperm release after hitting a HeLa cell cluster, and (E) Overlaid z-stack images of sperm fusion with HeLa cells, showing local drug release.

hits the wall of a tumor spheroid. We found that sperm cells are perfect candidates to carry and deliver drugs, specifically to cancers that are in the cervix, ovaries or uterus, because sperms are naturally adapted to swim in this environment. Sperms also have the ability to fuse with somatic cells, which allows them to release the drug inside the cells, avoiding the dilution of the drug in body fluids (Fig. 3D, E). They are also apt microswimmers due to their particular tail beating that is ideal for tissue infiltration. Also, we found that sperms can load a great amount of drug (15 pg per sperm) and transport it for a long time without loss and without adverse effects on sperm motility. Sperms, as biocompatible cellular carriers, neither express pathogenic proteins, nor proliferate to form undesirable colonies, in contrast to other cells or microorganisms, making this biohybrid system a promising biocompatible drug delivery platform [33].

Such sperm-hybrid systems can also be used to treat other diseases like endometriosis, pelvic inflammatory diseases and ectopic pregnancies. They can also be engineered to carry genes, mRNA, or imaging contrast agents, individually or in a swarm, performing cooperative work to solve complex tasks in the future.

3   Towards *in vivo* Application

To achieve *in vivo* fertilization, several aspects should be considered, for example the different biological barriers that spermbots have to face on their journey to the oocyte, the interaction of their synthetic component with the surrounding environment, proper sperm selection, the necessary amount of sperms to ensure the pass through the cumulus cells and zona pellucida that surround the oocyte and consequent oocyte membrane fusion. Other important considerations are the *in vivo* tracking of those micromotors in order to precisely control their trajectory to deliver the sperm in the area of interest

(Fig. 4). New designs of magnetic guidance and propulsion vehicles are also indispensable, first to improve the transport/guidance efficiency, and second to increase the sperm coupling probability, which is crucial to transport multiple sperms nearby the oocyte and guarantee the fertilization by one of them. These micromotors can be flexible, rigid or dynamic, but need to be designed to move efficiently in the physiological environment (Fig. 4).

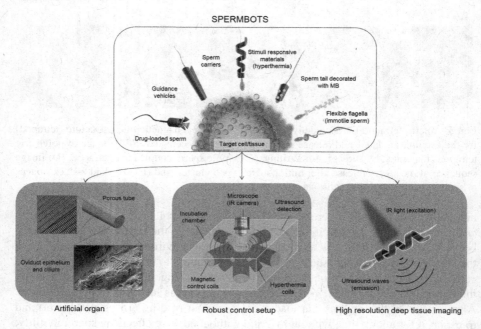

Fig. 4. Schematic including new designs for sperm-micromotors, including thermo-responsive polymer tubes, helical carriers, flexible artificial flagella, magnetic microparticles, among others are explored to carry and/or guide single or multiple sperm cells to accomplish fertilization or drug delivery, reprinted with permission from [25]. Copyright 2017 Wiley. Model of a 3D artificial organ (left); prospective setup including microscope, magnetic controller, incubation chamber and hyperthermia setup (center), and a suggested technique for high resolution deep tissue *in vivo* imaging (i.e. optoacoustic imaging) (right).

Another important aspect is how the spermbots approach the oocyte. As it is well known, the oocyte is surrounded by a thick layer of cumulus cells that are important for oocyte maturation. These cells are tightly packed by hyaluronic acid links that become softer when the oocyte matures. The sperm cell membrane contains a series of enzymes that digest this extracellular matrix. This, plus their swimming capabilities, allows the sperm cells to pass through this natural barrier. Thus, the spermbot assembly should be able to accomplish the same task in an efficient way. To achieve that, for example, the synthetic part of the spermbot can be loaded with enzymes that can be delivered during its journey to the oocyte or once the sperm reaches it to facilitate their passage.

Further experiments mimicking the geometry and physiology of the oviduct are also necessary in order to understand the behavior of spermbots in more realistic conditions. This can elucidate information such as sperm-binding to the oviduct for sperm selection and maturation. Physiological parameters such as temperature gradients, backflows or chemical gradients generated by the oocyte are of great interest to develop highly functional spermbots that can potentially be applied *in vivo* in the near future.

In order to apply such biohybrids *in vivo*, two important obstacles must be overcome, namely the tracking of spermbots, involving advanced 3D control and visualization *in vivo*, and the degradation of the synthetic components after accomplishing their task [40]. There have been numerous efforts to control magnetic microstructures in a 3D environment by using a set of electromagnets that generate rotating magnetic fields that allow the magnetic control of microobjects in three dimensions [34]. Moreover, engineers are developing robust control systems to improve the micromotors' maneuvering. They include for example image recognition algorithms to define specific trajectories that micromotors should follow. However, there is still a long way to realize a complicated process like automatic cell capture and release. There are also some requirements that must be fulfilled to control/track the coupled sperm cells in *in vivo* environments, such as real-time, non-invasive, deep-tissue and high resolution imaging.

Researchers in the bioimaging field are working into improving sensitivity and selectivity with more specific contrast agents for methods like infrared emission, X-ray analysis, ultrasound and magnetic resonance imaging, among others. For example, photoacoustic imaging of biological entities is one possible solution. It is a very new area that couples light excitation to acoustic detection via the photoacoustic effect to yield images of optical absorbance. This effect was first described in 1994 by R.A. Kruger [35]. The sample is illuminated by pulsed light and then acoustic waves generated from illuminated absorbing regions are detected by acoustic sensors. Indeed, current photoacoustic imaging tools are able to image samples with 150 μm resolution at the full cross section of a mouse [36]. Nonetheless, much effort is currently being undertaken to augment the image resolution. In order to enhance the absorbance properties of the sample, researchers are making use of nanomaterials such as metal-based nanoparticles. These nanomaterials show specific plasmon resonances which enable selective imaging of the analyte of interest as their respective absorbance spectra are unique and differ from the biological matter [37].

Finally, regarding the biocompatibility and biodegradability of micromotors, there have been some reported works where biopolymers or other biodegradable materials such as Mg/ZnO, Mg/Si, and Zn/Fe have been successfully employed in synthetic micromotors that disintegrate after a short period of time. However more exhaustive analysis of materials biocompatibility and biodegradability are required according to the application [38, 39].

4 Conclusions

In this paper, we present novel types of 'living machines' which are composed of spermatozoa as the biological component and a micro-sized artificial component. The already achieved capture, assisted transport and release of motile and immotile sperm cells by tubular and helical microstructures demands improvement of efficiency and reproducibility of the sperm coupling step, as well as spermbot swimming performance. This will be achieved by further optimization of the respective microstructure geometry and material composition, the magnetic actuation system, and robust remote control by tailored software. The next step will be to accurately mimic physiological conditions in order to create an environment suitable for micromotor-assisted *in vitro* fertilization or drug delivery. This will be tackled by the design and fabrication of an artificial organ, including epithelial cells under cell culture conditions and by mimicking the oviductal fluid properties. This setup will serve as a platform for fertilization and drug delivery experiments, but also to investigate mechanisms of sperm capacitation, hyperactivation and sperm selection in the presence of an oocyte, as well as extensive risk evaluations. In a final step, the artificial organs will serve to explore and validate methods of real-time deep tissue high resolution imaging and spermbot tracking as the final step to prepare for further *in vivo* application. Once these challenges are met, *in vivo* experiments with spermbots to achieve fertilization or targeted drug delivery inside living organisms can be pursued, maintaining as far as possible sperm's initial abilities.

References

1. Smith, E.J., Schulze, S., Kiravittaya, S., Mei, Y., Sanchez, S., Schmidt, O.G.: Lab-in-a-tube: detection of individual mouse cells for analysis in flexible split-wall microtube resonator sensors. Nano Lett. **11**, 4037–4042 (2011). doi:10.1021/nl1036148
2. Zhao, G., Stuart, E.J.E., Pumera, M.: Enhanced diffusion of pollutants by self-propulsion. Phys. Chem. Chem. Phys. **13**, 12755–12757 (2011). doi:10.1039/c1cp21237k
3. Wang, W., Li, S., Mair, L., Ahmed, S., Huang, T.J., Mallouk, T.E.: Acoustic propulsion of nanorod motors inside living cells. Angew. Chemie Int. Ed. **53**, 3201–3204 (2014). doi:10.1002/anie.201309629
4. Ghost, A., Fischer, P.: Controlled propulsion of artificial magnetic nanostructured propellers. Nano Lett. **9**, 2243–2245 (2009). doi:10.1021/nl900186w
5. Tottori, S., Zhang, L., Qiu, F., Krawczyk, K.K., Franco-Obregõn, A., Nelson, B.J.: Magnetic helical micromachines: fabrication, controlled swimming, and cargo transport. Adv. Mater. **24**, 811–816 (2012). doi:10.1002/adma.201103818
6. Loget, G., Kuhn, A.: Electric field-induced chemical locomotion of conducting objects. Nat. Commun. **2**, 1–6 (2011). doi:10.1038/ncomms1550
7. Solovev, A.A., Xi, W., Gracias, D.H., Harazim, S.M., Deneke, C., Sanchez, S., Schmidt, O.G.: Self-propelled nanotools. ACS Nano **6**, 1751–1756 (2012). doi:10.1021/nn204762w
8. Srivastava, S.K., Medina-Sánchez, M., Koch, B., Schmidt, O.G.: Medibots: dual-action biogenic microdaggers for single-cell surgery and drug release. Adv. Mater. **28**, 832–837 (2016). doi:10.1002/adma.201504327

9. Balasubramanian, S., Kagan, D., Jack Hu, C.M., Campuzano, S., Lobo-Castañon, M.J., Lim, N., Kang, D.Y., Zimmerman, M., Zhang, L., Wang, J.: Micromachine-enabled capture and isolation of cancer cells in complex media. Angew. Chemie - Int. Ed. **50**, 4161–4164 (2011). doi:10.1002/anie.201100115

10. Gao, W., Dong, R., Thamphiwatana, S., Li, J., Gao, W., Zhang, L., Wang, J.: Artificial micromotors in the mouse's stomach: a step toward in vivo use of synthetic motors. ACS Nano **9**, 117–123 (2015). doi:10.1021/nn507097k

11. Khalil, I.S.M., Dijkslag, H.C., Abelmann, L., Misra, S.: MagnetoSperm: a microrobot that navigates using weak magnetic fields. Appl. Phys. Lett. **104**, 223701 (2014). doi: 10.1063/1.4880035

12. Gao, W., Feng, X., Pei, A., Kane, C.R., Tam, R., Hennessy, C., Wang, J.: Bioinspired helical microswimmers based on vascular plants. Nano Lett. **14**, 305–310 (2014). doi:10.1021/nl404044d

13. Singleton, J., Diller, E., Andersen, T., Regnier, S., Sitti, M.: Micro-scale propulsion using multiple flexible artificial flagella. In: IEEE International Conference on Intelligent Robots and Systems, pp. 1687–1692 (2011). doi:10.1109/IROS.2011.6048742

14. Williams, B.J., Anand, S.V., Rajagopalan, J., Saif, M.T.A.: A self-propelled biohybrid swimmer at low Reynolds number. Nat. Commun. **5**, 1–8 (2014). doi:10.1038/ncomms4081

15. Xi, J., Schmidt, J.J., Montemagno, C.D.: Self-assembled microdevices driven by muscle. Nat. Mater. **4**, 180–184 (2005). doi:10.1038/nmat1308

16. Di Leonardo, R., Angelani, L., Dell'Arciprete, D., Ruocco, G., Iebba, V., Schippa, S., Conte, M.P., Mecarini, F., De Angelis, F., Di Fabrizio, E.: Bacterial ratchet motors. Proc. Natl. Acad. Sci. **107**, 9541–9545 (2010). doi:10.1073/pnas.0910426107

17. Darnton, N., Turner, L., Breuer, K., Berg, H.C.: Moving fluid with bacterial carpets. Biophys. J. **86**, 1863–1870 (2004). doi:10.1016/S0006-3495(04)74253-8

18. Martel, S.: Bacterial microsystems and microrobots. Biomed. Microdevices **14**, 1033–1045 (2012). doi:10.1007/s10544-012-9696-x

19. Martel, S., Mohammadi, M., Felfoul, O., Lu, Z., Pouponneau, P.: Flagellated Magnetotactic bacteria as controlled MRI-trackable propulsion and steering systems for medical nanorobots operating in the human microvasculature. Int. J. Robot. Res. **28**, 571–582 (2009). doi: 10.1177/0278364908100924

20. Pouponneau, P., Leroux, J.C., Soulez, G., Gaboury, L., Martel, S.: Co-encapsulation of magnetic nanoparticles and doxorubicin into biodegradable microcarriers for deep tissue targeting by vascular MRI navigation. Biomaterials **32**, 3481–3486 (2011). doi:10.1016/j.biomaterials.2010.12.059

21. Martel, S., Tremblay, C.C., Ngakeng, S., Langlois, G.: Controlled manipulation and actuation of micro-objects with magnetotactic bacteria. Appl. Phys. Lett. **89**, 233904 (2006). doi: 10.1063/1.2402221

22. Kim, D., Liu, A., Diller, E., Sitti, M.: Chemotactic steering of bacteria propelled microbeads. Biomed. Microdevices **14**, 1009–1017 (2012). doi:10.1007/s10544-012-9701-4

23. Zhuang, J., Wright Carlsen, R., Sitti, M.: pH-taxis of biohybrid microsystems. Sci. Rep. **5**, 11403 (2015). doi:10.1038/srep11403

24. Steager, E.B., Sakar, M.S., Kim, D.H., Kumar, V., Pappas, G.J., Kim, M.J.: Electrokinetic and optical control of bacterial microrobots. J. Micromech. Microeng. **21**, 35001 (2011). doi: 10.1088/0960-1317/21/3/035001

25. Magdanz, V., Medina-Sánchez, M., Schwarz, L., Xu, H., Elgeti, J., Schmidt, O.G.: Spermatozoa as functional components of robotic microswimmers. Adv. Mater. 1–18 (2017). doi:10.1002/ADMA.201606301

26. Magdanz, V., Sanchez, S., Schmidt, O.G.: Development of a sperm-flagella driven micro-bio-robot. Adv. Mater. **25**, 6581–6588 (2013). doi:10.1002/adma.201302544
27. Medina-Sánchez, M., Schwarz, L., Meyer, A.K., Hebenstreit, F., Schmidt, O.G.: Cellular cargo delivery: toward assisted fertilization by sperm-carrying micromotors. Nano Lett. **16**, 555–561 (2015). doi:10.1021/acs.nanolett.5b04221
28. Sato, Y., Tajima, A., Tsunematsu, K., Nozawa, S., Yoshiike, M., Koh, E., Kanaya, J., Namiki, M., Matsumiya, K., Tsujimura, A., Komatsu, K., Itoh, N., Eguchi, J., Imoto, I., Yamauchi, A., Iwamoto, T.: An association study of four candidate loci for human male fertility traits with male infertility. Hum. Reprod. **30**, 1510–1514 (2015). doi:10.1093/humrep/dev088
29. Magdanz, V., Medina-Sánchez, M., Chen, Y., Guix, M., Schmidt, O.G.: How to Improve Spermbot Performance. Adv. Funct. Mater. **25**, 2763–2770 (2015). doi:10.1002/adfm. 201500015
30. Khalil, I.S.M., Magdanz, V., Sanchez, S., Schmidt, O.G., Misra, S.: Biocompatible, accurate, and fully autonomous: a sperm-driven micro-bio-robot. J. Micro-Bio Robot. **9**(3-4), 79–86 (2014)
31. Magdanz, V., Guix, M., Hebenstreit, F., Schmidt, O.G.: Dynamic polymeric microtubes for the remote-controlled capture, guidance, and release of sperm cells. Adv. Mater. **28**, 4048–4089 (2016). doi:10.1002/adma.201505487
32. Jeyendran, R.S., Van der Ven, H.H., Perez-Pelaez, M., Crabo, B.G., Zaneveld, L.J.: Development of an assay to assess the functional integrity of the human sperm membrane and its relationship to other semen characteristics. J. Reprod. Fertil. **70**, 219–228 (1984). doi: 10.1530/jrf.0.0700219
33. Xu, H., Medina-Sánchez, M., Magdanz, V., Schwarz, L., Hebenstreit, F., Schmidt, O.G.: Sperm-hybrid micromotor for drug delivery in the female reproductive tract. arXiv: 1703.08510 (2017)
34. Khalil, I.S.M., Magdanz, V., Sanchez, S., Schmidt, O.G., Misra, S.: Three-dimensional closed-loop control of self-propelled microjets. Appl. Phys. Lett. **103**, 172404/1-5 (2013).
35. Kruger, R.A., Kuzmiak, C.M., Lam, R.B., Reinecke, D.R., Del Rio, S.P., Steed, D.: Dedicated 3D photoacoustic breast imaging. Med. Phys. **40**, 113301 (2013). doi:10.1118/1.4824317
36. Neuschmelting, V., Lockau, H., Ntziachristos, V., Grimm, J., Kircher, M.F.: Lymph node micrometastases and in-transit metastases from melanoma: in vivo detection with multispectral optoacoustic imaging in a mouse model. Radiology **280**, 137–150 (2016)
37. Comenge, J., Fragueiro, O., Sharkey, J., Taylor, A., Held, M., Burton, N.C., Park, B.K., Wilm, B., Murray, P., Brust, M., Lévy, R.: Preventing plasmon coupling between gold nanorods improves the sensitivity of photoacoustic detection of labeled stem cells in vivo. ACS Nano **10**, 7106–7116 (2016). doi:10.1021/acsnano.6b03246
38. Tu, Y., Peng, F., André, A.A., Men, Y., Srinivas, M., Wilson, D.A.: Biodegradable hybrid stomatocyte nanomotors for drug delivery, ACS Nano. (2017). doi:10.1021/acsnano.6b08079
39. Chen, C., Karshalev, E., Li, J., Soto, F., Castillo, R., Campos, I., Mou, F., Guan, J., Wang, J.: Transient micromotors that disappear when no longer needed. ACS Nano **10**, 10389–10396 (2016). doi:10.1021/acsnano.6b06256
40. Medina-Sánchez, M., Schmidt, O.G.: Medical microbots need better imaging and control. Nature **545**, 406–408 (2017)
41. Vizsnyiczai, G., Frangipane, G., Maggi, C., Saglimbeni, F., Bianchi, S., Di Leonardo, R.: Light controlled 3D micromotors powered by bacteria. Nat. Commun. **8**, 15974 (2017). doi: 10.1038/ncomms15974

An Insect-Scale Bioinspired Flapping-Wing-Mechanism for Micro Aerial Vehicle Development

Kenneth C. Moses[1]([✉]), Nathaniel I. Michaels[1], Joel Hauerwas[1],
Mark Willis[2], and Roger D. Quinn[1]

[1] Biologically-Inspired Robotics Lab, Case Western Reserve University, Cleveland, OH, USA
{kcm7,nim9,jah301,rdq}@case.edu
[2] Biology Department, Case Western Reserve University, Cleveland, OH, USA
mark.willis@case.edu

Abstract. Steps have been taken to reduce the size and mass of a flapping-wing-mechanism previously designed for testing artificial *Manduca sexta* forewings. A modified scotch-yoke mechanism is implemented to convert continuous rotary motion into flapping-wing oscillatory motion with a desired inter-wing angle and stroke amplitude. The new device measures $33.0 \times 41.9 \times 33.4$ mm (maximum dimensions excluding the forewings) with a total mass of 15.5 g. Approximately 9.3 g of this total mass is attributed to the DC motor alone, which is significantly overpowered, indicating room for improvement. This reduction in size allows for more accurate forewing testing as it achieves proportions closer to the model organism (*Manduca sexta*). Furthermore, the inclusion of precise micro load cells in the testing apparatus is possible, as the total mass no longer exceeds the maximum loads permissible for these instruments. The design lends itself to employing a compliant yoke that stores and returns energy during stroke reversal thus improving the flapping mechanism's efficiency. As we make progress in mimicking the *Manduca sexta* thorax, we become closer in our goal of developing a flapping-wing micro aerial vehicle (FWMAV) with flight capabilities similar to the hawkmoth.

Keywords: Bioinspired · Thorax · FWMAV · Wing · Manduca sexta · Hawkmoth · Biomimicry

1 Introduction

With commercial, military, and recreational use of small-unmanned aircraft on the rise, it is increasingly necessary to minimize weight and power requirements. While fixed-wing micro air vehicles (MAV) have been popular for decades, this decade is seeing a drastic increase in the popularity of rotorcraft MAVs. Fixed-wing MAVs prove more energy efficient and have a longer range of use than rotorcraft MAVs, while rotorcraft MAVs outshine fixed-wing MAVs in maneuverability. However, flapping wing vehicles rival fixed-wing aircraft in both their power requirements and range, as well as approach the agility of rotorcraft [1].

© Springer International Publishing AG 2017
M. Mangan et al. (Eds.): Living Machines 2017, LNAI 10384, pp. 589–594, 2017.
DOI: 10.1007/978-3-319-63537-8_54

As we observe biology, we recognize that insect flight has many unique attributes, several of which coincide with the desirable traits of flapping-wing micro aerial vehicles (FWMAVs). Birds may be suitable for outdoor flight, but insects tend to outperform them in confined spaces. Furthermore, insects in general have a less complex anatomy with far fewer joints and muscles. A particular insect (the hawkmoth *Manduca sexta*) is known to have remarkable flight abilities and is easily reared within laboratories. Thus, it is commonly selected as a model organism for biomimicry [2]. A systematic approach to mimicking this animal is taken. The hawkmoth is partitioned into its various major components (forewings, hindwings, thorax, abdomen, and head) and each component is developed for future use on an FWMAV. This work is predicated on the idea that it is most important to accurately mimic the motion, structure, geometry and mass of the wings of the animal in order for the FWMAV to have similar flight characteristics [3]. The forewings were then selected to be the first component attempted to be replicated. Prior work has been done in this area by our research group as well as others [2, 4, 5].

The next logical component of interest is the mechanism that drives the wings. This is considered analogous to the hawkmoth's thorax. Numerous types of flapping-wing-mechanisms have been designed including compliant mechanisms that store and release energy during stroke reversal [6]. Many of these mechanisms use a crank-rocker type of drivetrain to achieve wing motions similar to the respective model organism. This preliminary work builds upon a test stand developed for measuring the performance of artificial forewings and presents an alternative to the common crank-rocker mechanism.

2   *Manduca sexta* Forewing Lift Assessment

Previous work attempted to replicate the *Manduca sexta* forewing using carbon fiber and Icarex™ as materials for the venation structure and membrane, respectively [7]. Additionally, a flapping test stand was designed and constructed in order to assess the performance of the engineered forewings. Data captured from stereoscopic high-speed videography provided the inter-wing angle (the angle between the left and right forewings), stroke amplitude, and forewing angle of attack of the hawkmoth. These parameters were employed in the design of the test stand. The test stand examines *Manduca sexta* forewing lift performance by utilizing a five-bar crank-rocker mechanism. The drivetrain and surrounding support structure is comprised of PLA 3D printed parts, DuPont™ Delrin® acetal machined components, and a 3000 RPM 10:1 Pololu Micro Metal Gearmotor. It weighs approximately 120 g and is $7.6 \times 7.6 \times 6.4$ cm in size, depending upon its configuration. The lift generated by the wings is measured in two different configurations (Fig. 1). First, by placing the test stand on top of a digital weigh scale and positioning the device such that the forewings extend beyond the scale's platform thus ensuring that downwash from the wings does not impart a force on the scale. Second, by attaching a 100-gram micro load cell to the base of the stand. These yielded similar results, however, the micro load cell allowed for higher frequency measurements than the digital weigh scale (80 samples per second versus 5 samples per second, respectively).

Fig. 1. The test stand in its two configurations for measuring lift. The stand on top of a Fairbanks digital weigh scale (Left). The stand affixed to a Phidgets 100-gram micro load cell.

While this flapping mechanism proved useful in comparing augmented artificial forewings, the design potentially hindered wing-wing interaction due to its inaccurate proportions. Wing-wing interaction is an important phenomenon that occurs in small flapping animals and transpires during the maximum and minimum inter-wing angles where the left and right wings are closest to one another and in some cases touch each another. Studies have shown that this phenomenon may aid in lift production and contribute to the inherit flight stability of the animals [8]. Additionally, the base of the structure had large supporting members that prevented any potential fluid interaction to occur between left and right forewings. In order to advance the study of the performance of these forewings a new device has been developed that remedies these issues.

3   Flapping-Wing-Mechanism Design and Fabrication

The goal of this redesign was to reduce the size of the flapping mechanism such that the locations of the forewings with respect to each other more closely match the *Manduca sexta* thorax and minimize the structure around the wings. A size reduction required an increase in the tolerance of the parts of the mechanism. As a result, many of the components that were previously 3D printed, now had to be CNC machined. Furthermore, several components were made out of aluminum (AL 6061) in order to maintain strength while decreasing major dimensions of the mechanism. A scotch-yoke mechanism was conceived to provide the necessary conversion from continuous rotary motion of the Pololu DC motor to flapping-wing oscillatory motion and maintain the observed hawk-moth inter-wing angles (Fig. 2). The distance from the two axes that the forewings rotate about decreased from 38.10 mm to 10.16 mm. The overall dimensions of the mechanism are 33.0 × 41.9 × 33.4 mm. This is approximately 1/8th the volume of the initial test stand. Additionally, the new mechanism weighs nearly 1/8th that of the previous test stand (15.5 g as compared to the 120.4 g).

Fig. 2. The new flapping-wing-mechanism based on a scotch-yoke device (quarter included for scale).

4   Summary and Future Work

Flapping-wing micro aerial vehicles (FWMAVs) inspired by insects have the potential to advance the capabilities of drone flight in a number of areas (robustness, maneuverability, efficiency, range). This project is intended to progress our understanding of insect flight with the goal of producing an MAV with similar features to the *Manduca sexta* hawkmoth by mimicking it. The work presented here provides necessary improvements for studying the performance of artificial forewings based on *M. sexta*. The flapping mechanism used to study these forewings was successfully condensed such that its dimensions are similar to the *M. sexta* thorax (Fig. 3), thus paving the way for a number of avenues to be explored in the future.

Following this preliminary work, modest refinements to a few of the components of the mechanism will be made. While all of the parts of the device have been fabricated and assembled, subtle tolerance adjustments are necessary to produce smooth motion during flapping. Once this is achieved, tests will be performed utilizing the previously configured 100-gram micro load cell for lift measurements and the results compared to those from the original test stand. Subsequently, a more advanced apparatus will be designed for taking measurements of additional degrees of freedom such as drag and lateral loads. This can be accomplished by incorporating multiple highly sensitive micro load cells in series. Developing an apparatus of this nature is possible due to the substantial reduction in mass of the flapping mechanism. Experiments with this device will be conducted within a controlled environment (such as a wind tunnel) yielding valuable information about the stability of the mechanism in certain conditions.

Advancements made regarding replicating other parts of the hawkmoth will be incorporated into the model. With the yoke design, we foresee achieving reductions in energy requirements and realizing the potential for alternatives to the DC motor as the

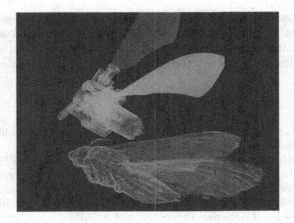

Fig. 3. Comparison of a 3D scan of a Manduca sexta hawkmoth to the new flapping-wing-mechanism.

power source. The potential for energy storage during stroke reversal is based off simulating the animal's tergal plate, which flexes and stores energy like a spring-mass system. The yoke design lends itself to implementing flexible compliant members for energy storage. In regards to replicating the animal's aerodynamic properties, progress has already been made in developing a shell (Fig. 4) for the flapping-wing-mechanism. Currently the shell is 3D printed from a scan of a *Manduca sexta*. Future iterations will be made of carbon fiber by producing a positive mold for forming the desired shape. In addition, work has begun in replicating the *M. sexta* hindwing which we believe will prove vital in generating the necessary aerodynamic forces for performing advanced flight maneuvers and aid in overall vehicle stability.

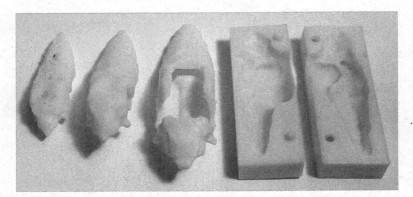

Fig. 4. Initial attempts at 3D printed hawkmoth shells made with an assortment of different types of 3D printers. From left to right: 1:1 Scale Shell made with Fortus 250mc, 1:1.25 scale shell made with Polyjet Objet350 Connex3, 1:1.5 scale shell made with Fortus 400mc, left half of negative mold made with Fortus 400mc, right half of negative mold made with Fortus 400mc.

The steps taken in this work will aid in observing the implications of wing-wing interaction during flapping. We hope to gather new information about the inherent stabilities within *Manduca sexta*. As a result, we are closer to reaching our goal of a biologically inspired FWMAV that can demonstrate the benefits of insect flight.

References

1. Ansari, S.A., Zbikowski, R., Knowles, K.: Aerodynamic modelling of insect-like flapping flight for micro air vehicles. Prog. Aerosp. Sci. **42**(2), 129–172 (2006)
2. DeLeón, N.E.: Manufacturing and Evaluation of a Biologically Inspired Engineered MAV Wing Compared to the Manduca sexta Wing Under Simulated Flapping Conditions. Air Force Institute of Technology (2011)
3. Combes, S.A.: Materials, Structure, and Dynamics of Insect Wings as Bioinspiration for MAVs. Encyclopedia of Aerospace Engineering. vol. 3 (2010)
4. Michaels, S.C., Moses, K.C., Bachmann, R.J., Hamilton, R., Pena-Francesch, A., Lanba, A., Demirel, M.C., Quinn, R.D.: Biomimicry of the *Manduca sexta* forewing using SRT protein complex for FWMAV development. In: Wilson, S.P., Verschure, P.F.M.J., Mura, A., Prescott, T.J. (eds.) Living Machines 2015. LNCS, vol. 9222, pp. 86–91. Springer, Cham (2015). doi: 10.1007/978-3-319-22979-9_8
5. Michaels, S.C.: Development and assessment of artificial Manduca sexta forewings: how wing structure affects performance. Case Western Reserve University (2015)
6. Zhang, C., Rossi, C.: A review of compliant transmission mechanisms for bio-inspired flapping-wing micro air vehicles. Bioinspir. Biomim. **12**(2) (2017). 025005
7. Moses, K.C., Michaels, S.C., Willis, M., Quinn, R.D.: Artificial Manduca sexta forewings for flapping-wing micro aerial vehicles: how wing structure affects performance. Bioinspir. Biomim. (2017)
8. Lehmann, F.-O., Pick, S.: The aerodynamic benefit of wing-wing interaction depends on stroke trajectory in flapping insect wings. J. Exp. Biol. **210**(Pt 8), 1362–1377 (2007)

Geometric Mechanics Applied to Tetrapod Locomotion on Granular Media

Yasemin Ozkan Aydin[1]([✉]), Baxi Chong[4], Chaohui Gong[4], Jennifer M. Rieser[1], Jeffery W. Rankin[2], Krijn Michel[2], Alfredo G. Nicieza[3], John Hutchinson[2], Howie Choset[4], and Daniel I. Goldman[1]

[1] School of Physics, Georgia Institute of Technology, Atlanta, GA, USA
{yasemin.ozkanaydin,jennifer.rieser,daniel.goldman}@physics.gatech.edu
[2] Royal Veterinary College, London, UK
jwrankin@gmail.com, {kmichel,JHutchinson}@rvc.ac.uk
[3] Universidad Oviedo, Asturias, Spain
agnicz@gmail.com
[4] Carnegie Mellon University, Pittsburgh, PA, USA
baxichong8@gmail.com, chaohuigong@gmail.com, choset@cs.cmu.edu

Abstract. This study probes the underlying locomotion principles of earliest organisms that could both swim and walk. We hypothesize that properly coordinated leg and body movements could have provided a substantial benefit toward locomotion on complex media, such as early crawling on sand. In this extended abstract, we summarize some of our recent advances in integrating biology, physics and robotics to gain insight into tetrapod locomotor coordination and control principles. Here, we observe crawling salamanders as a biological model for studying tetrapod locomotion on sloped granular substrates. Further, geometric mechanics tools are used to provide a theoretical framework predicting efficacious body motions on yielding terrain. Finally, we employ these coordination strategies on a robophysical salamander model traversing a sandy slope. This analysis of salamander-like robotic motion in granular media can be seen as a first application of how tools from geometric mechanics can provide insight into the character and principles of legged locomotion.

1 Introduction

Many studies of terrestrial locomotion have focused on interaction with rigid, flat laboratory trackways. However, organisms often live on or within complex terrain that is flowable, loose, and wet (e.g. sand, dirt, or mud). In addition, organisms with different limb morphologies and kinematics likely evolved terrestrial locomotion strategies as they moved from oceans and lakes onto beaches of these flowable materials. When investigating principles of biological locomotion, it has proved useful to build robots with analogous capabilities. These robophysical models allow for tests of function and suggest hypotheses for mechanical and nervous system control.

© Springer International Publishing AG 2017
M. Mangan et al. (Eds.): Living Machines 2017, LNAI 10384, pp. 595–603, 2017.
DOI: 10.1007/978-3-319-63537-8_55

To understand how vertebrates made the transition from water to land, we first analyzed the motion of mudskippers, modern analogs of early tetrapods, and showed that their tail is helpful for land locomotion, especially on sandy slopes [11]. Now we have begun to work on salamanders, another extant animal model for tetrapod evolution. Previous work focused on gait transition between swimming and walking by changing the frequency and amplitude of drive applied to the spinal cord of salamander robot [8]. As far as we know there has been little discussion on robustness and efficiency of tetrapod locomotion on deformable/flowable substrates. In this extended abstract, we aim to show experimentally and theoretically that coordinated body bending may improve the robustness of locomotion on challenging substrates. Recently, our group has developed a model using geometric mechanics (an area of research combining classical mechanics with differential geometry) to understand how shape changes relate to translations and rotations of the body during limbless locomotion on and in granular media [2,6,7,11]. In this extended abstract, we summarize our application of the model to legged locomotion providing further evidence that geometric mechanics can be used as a general framework and language for locomotion.

2   Experimental Setups and Robot Platform

2.1   Animal Experiments

Fire salamanders (*Salamandra salamandra*) (about 40 g, 15−20 cm long) (see Fig. 1-D) were placed at the end of long, narrow trackway (~10 cm wide, ~50 cm long) filled with a 3-cm-deep layer of ~0.2 mm diameter glass particles, see Fig. 1-B. A dark space (which the animal prefers) was provided at the other end of the track to encourage the salamanders to move. Five trials were collected for each of ten salamanders on level ground. The order in which individuals were used in experiments was randomized and animals were given at least a two minutes to rest after each trial. Videos of each experiment were recorded by three synchronized GoPro cameras (see Fig. 1-B). We observed that in the sand, animals adopted a gait in which at least three legs were on the ground at all times (see step diagrams and average step frequencies with standard deviation that shows the portion of a stride during which each foot is on the ground during one cycle in Fig. 3-B).

2.2   Robot Experiments

Using a *robophysical* approach [1], we built a salamander robot and have tested its performance on granular media. This open-loop, servo-driven, 3D-printed robot (450 g, ~40 cm long) has four limbs, an actuated trunk, and an active tail (Fig. 1-C, E). Each limb has two servo motors to control the vertical position and the step size of the limb. A joint in the middle of the body controls horizontal bending. The design and materials allow us to easily change morphological parameters of the body and legs.

Fig. 1. Experimental setups, robophysical model and salamander. (A) The Systematic Creation of Arbitrary Terrain and Testing of Exploratory Robots (SCATTER) [12] system allows for automated three axis manipulation of robot traversing a granular substrate with specified surface incline. (1) Optitrack vision capture system (2) 3-axis stage (3) Gripper (4) Fluidized bed trackway filled with granular media (5) Tilting actuator (B) Animal experimental setup (C) Robot side view and servo angles (D) Top view of the fire salamander (E) Top view of the robophysical model

We performed experiments to systematically test the importance of body morphology, limb coordination, tail use and body flexibility on yielding substrates using a fully-automated setup [12], and compared the successes and failures of the robot with that of the animals. Our setup consists of four vacuums to prepare a uniform loosely packed state [10] in the trackway filled with ∼1 mm diameter poppy seeds before each experiment, tilting actuators to create inclined granular environments, a 3-axis motor system (Copley) to control the gripper position and four cameras (Naturalpoint, Flex13, 120 FPS) to capture the position and orientation of the robot (Fig. 1-A).

3   Geometric Mechanics Model

Geometric mechanics is a mathematical tool that relates internal shape changes (i.e. the joint angles) to net changes in translation and rotation of the body in world frame for a given interaction model with environment [9]. Previously, it was used to analyze a wide range of locomotor behaviors such as limbless locomotion [4]. Here we describe how it is also applicable to legged locomotion under several choices of modeling assumptions. Our legged-locomotor (Fig. 1-E) uses a lateral sequence, diagonal-couplet walking gait that is similar to animal salamander gait (three legs are on ground at each time step Fig. 2-A). The leg angle, β, is defined as the angle between the distal end of the limb and a line that is perpendicular to body transverse axis (see Fig. 2-B). We used a sinusoidally-varying leg angle (Fig. 2-D) and assumed that four legs followed the same gait but differed by phase,

$$\beta_i = \beta(t + \Delta\phi_i) \quad \text{and} \quad A_i(t) = A(t + \Delta\phi_i) \quad \text{for} \quad i = 1, 2, 3, 4, \tag{1}$$

where β and A describe leg angle and activation and are given by

$$\beta = \begin{cases} sin(\frac{\pi}{2\tau_1}\tau) \\ cos(\frac{\pi(\tau-\tau_1)}{\tau_2-\tau_1}) \\ -cos(\frac{\pi(\tau-\tau_1)}{2(2\pi-\tau_2)}) \end{cases} \quad A = \begin{cases} air & if \ 0 < \tau < \tau_1 \\ ground & if \ \tau_1 < \tau < \tau_2 \\ air & if \ \tau_2 < \tau < 2\pi \end{cases} \tag{2}$$

To ensure the three legs are on the ground at the same time, the air phase of each leg must be coordinated in sequence. There are six combinations of leg phases that satisfy the tripod condition. Since the gait is periodic, we take FR as the first leg to move. The six combinations are listed in Fig. 3-A. We used the optimal body frame to represent the position and orientation of the robot, and the optimal frame is determined using the method described in [6]. This frame roughly sits at the average position of the two body links, and is aligned with the average orientation of the two body links. In the model, the position and orientation of the salamander head in world frame are defined as $\mathbf{g} = [x, y, \theta]$. We used shape space variables $\mathbf{r} = [\alpha \ \tau]$, where τ is leg phase and α is body angle, to fully characterize the leg gait. For a mechanical system, the body velocity can be expressed by $\mathbf{g}' = A(\mathbf{r})\dot{\mathbf{r}}$ where $A(\mathbf{r})$ maps the shape variables to the body velocity.

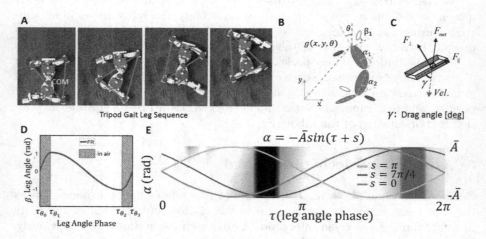

Fig. 2. Geometric mechanics applied to tetrapod locomotion on granular media. (A) Dorsal view of the sequence of support patterns during one stride of a lateral sequence, diagonal-couplet walk of robophysical model moving on granular media (B) Snapshot from the geometric mechanics model. The model has two body segments, a tail, and four limbs. The tail is inactive and always in contact with the ground. Leg colors define the activation, where red: legs are on ground and white: in air (C) The thrust and drag components of reaction force (Eq. 3) (D) An example nonlinear sinusoidal pattern of leg angle (E) Height function of the sinusoidal gait as a function of body angle α and leg angle phase τ. (Color figure online)

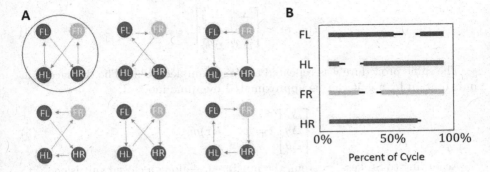

Fig. 3. Fire salamander's gait pattern. (A) Possible leg patterns of quadrupedal locomotion. In our studies the salamanders typically used the circled pattern. (B) Salamander step diagram H: hind, F: front, R: right, L: left, error bars indicated the standard
deviations; data from three animals walking on level sandy ground, 13 steps in total,
the average cycle duration is ∼1.5 s.

Granular resistive force theory (RFT) is used to model the movement of animals and robots in granular media [5,11]. Prior work has shown simulations using
granular RFT agreed well with actual robots [7,11]. We therefore used granular RFT to numerically derive local connections. In granular RFT, the resistive
force experienced on an infinitesimally small segment of a moving intruder is
decomposed into thrust and drag components. The reaction force applied on the
entire system is computed as,

$$\mathbf{F} = \int d\mathbf{F}_\perp + d\mathbf{F}_\parallel \tag{3}$$

where $\mathbf{F}_\perp$ and $\mathbf{F}_\parallel$ respectively denote forces perpendicular and parallel to a
segment and are a function of attack angle, γ (Fig. 2-C) and are independent
from the magnitude of speed [11]. The attack angle of a moving segment can be
computed from its body velocity. We assumed the motion of the salamander-like
robot in granular material is quasi-static, which means the total net force, $\mathbf{F}$
applied on the system is zero. The only unknown in Eq. 3 is the velocity of the
body frame. Given a shape velocity $\mathbf{r}$ the body velocity of the average body
frame can be computed numerically. To do that we uniformly sampled the two
dimensional shape space on a grid where $\tau \in (0, 2\pi)$ and $\alpha \in (-\pi/6, \pi/6)$.
At every point $(\alpha\ \tau)$ on the grid, we first locally perturbed the value of first
shape variable $\alpha = \alpha + \epsilon$, while holding the other constant. In practice, the
magnitude of perturbation was $\epsilon = 0.01$. This small change in shape produces a
small displacement $[\Delta x_1^b, \Delta y_1^b, \Delta \theta_1^b]$, measured in the body frame, which can be
computed using the RFT calculation. Next, we kept the value of the first shape
variable constant and perturbed second shape variable $\tau = \tau + \epsilon$. The resultant
displacement and rotation measured in the body frame was $[\Delta x_2^b, \Delta y_2^b, \Delta \theta_2^b]$.
These numerically calculated displacements served an approximation to the local
connection at shape (α, τ),

$$J(\alpha, \tau) \approx \begin{bmatrix} \Delta x_1^b & \Delta x_2^b \\ \Delta y_1^b & \Delta y_2^b \\ \Delta \theta_1^b & \Delta \theta_2^b \end{bmatrix} / \epsilon \tag{4}$$

The same procedure was repeated at every sampled point. The displacement and rotation by a gait can be approximated by a line integral:

$$\begin{bmatrix} \Delta x \\ \Delta y \\ \Delta \theta \end{bmatrix} := \oint_{gait} J(\mathbf{r}) d\mathbf{r} \tag{5}$$

According to Stokes Theorem, the line integral along a closed gait is equal to the area integral of the curl of $J(\mathbf{r})$ over the surface area enclosed by the gait:

$$\oint_{gait} J(\mathbf{r}) d\mathbf{r} = \iint_{gait} \nabla \times J(\mathbf{r}) d\alpha d\tau \tag{6}$$

The curl of connection vector field $\nabla \times J(\mathbf{r})$ is referred to as the height function. Figure 2-E shows the height function of the sinusoidal gait used in this study. The largest displacement per cycle can be determined by finding the zero set which has the largest area integral. Therefore, the optimal body-leg coordination is obtained with the largest surface integral in the height function.

4   Results

In this section, we demonstrate how the parameterized gait model can be used to model salamander motion on level granular media. We first tested the six leg-lifting sequences (Fig. 3-A) according to a static stability criteria such that the CoM should remain inside the support polygon formed by the leg contacts at each time step. Theoretical leg sequences shown in bottom row of Fig. 3-A contain moments in which the CoM falls outside of the supporting polygon. We compared the other three sequences (given in top row of Fig. 3-A) and we see both experimentally and theoretically that FR-HL-FL-HR, which is also used by animals leads to a longer step length (Fig. 4-A). We took the FR-HL-FL-HR as a leg-lifting sequence and computed the step length (half of the distance between two successive placements of the same foot) experimentally and theoretically associated with the gait pattern. The height function introduced in Sect. 3 was calculated assuming that the body angle is sinusoidal, i.e. $\alpha = -\bar{A} \sin(\tau + s)$ where τ is the leg phase and s is the body phase-shift. According to theory, the optimal gait would enclose most surface integral in the height function (see Fig. 2-E). Apparently, the optimal body phase-shift would be $7\pi/4$, which encloses the most red and least black regions in the height function. To verify our theory, we tested eight different body phase-shifts given in Fig. 4-C both in simulation and experiment and the results indicate that the optimal body phase-shift is close the $7\pi/4$. In addition to varying the phase-shift, we varied amplitude of body bending, α from 0 to 50°. We see also that body angle increases step length (see Fig. 4-B). The experimental results agree well with

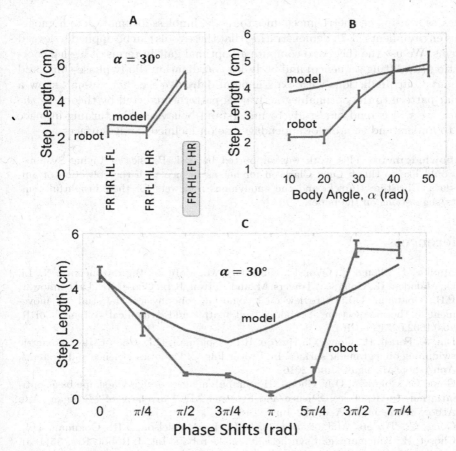

Fig. 4. Comparison of the model with robophysical experiments on level ground granular media. (A) Step lengths of the robot/model with four different leg sequences (FR: front right, FL: front left, HR: hind right, HL: hind left leg) (B) Step length vs. amplitude of body angle, α (C) Step length changes as a function of phase shift between body and leg angles for $\alpha = 30°$. Error bars denote standard deviation over 10 trials. (Color figure online)

the model for 40 and 50°. We posit that for the low amplitude region, the accumulation of the sand (especially on the posterior portion of the robot) decreases the step length, which is not considered in the model.

5   Conclusion

In this extended abstract, we summarized the application of techniques from geometric mechanics to model the locomotion of fire salamanders and a tetrapod robophysical model in a granular medium. Our study builds on our previous successes in using geometric mechanics to model limbless locomotion in homogeneous environments [3, 7], demonstrating that these tools can be applied to legged systems. We use the theory to compute an optimal gait in terms of leg-leg coordination (leg-lifting sequence) and body-leg coordination (body phase-shift) and compared our predictions with experimental data. We see that animals show a similar pattern to the optimal leg activation pattern predicted by theory. In the future, we will extend our study to understand behaviors like turning-in-place and to understand the role of appendage use on inclined sandy surfaces.

Acknowledgment. This work was supported by NSF Physics of Living Systems. We would like to thank Lucy Clarkson for her assistance with the collection of animal data, Christian Hubicki and the anonymous reviewers for their insightful comments/suggestions on the paper.

References

1. Aguilar, J., Zhang, T., Qian, F., Kingsbury, M., McInroe, B., Mazouchova, N., Li, C., Maladen, R., Gong, C., Travers, M. and Hatton, R.L., Choset, H., Umbanhowar, P.B., Goldman, D.I.: A review on locomotion robophysics: the study of movement at the intersection of robotics, soft matter and dynamical systems. CoRR, abs/1602.04712 (2016)
2. Dai, J., Faraji, H., Gong, C., Hatton, R.L., Goldman, D.I., Choset, H.: Geometric swimming on a granular surface. In: Proceedings of Robotics: Science and Systems, AnnArbor, Michigan, June 2016
3. Gong, C., Goldman, D.I., Choset, H.: Simplifying gait design via shape basis optimization. In: Robotics: Science and Systems XII, University of Michigan, Ann Arbor, Michigan, USA, 18–22 June 2016
4. Gong, C., Travers, M.J., Astley, H.C., Li, L., Mendelson, J.R., Goldman, D.I., Choset, H.: Kinematic gait synthesis for snake robots. Int. J. Robot. Res. **35**(1–3), 100–113 (2016)
5. Gray, J., Hancock, G.J.: The propulsion of sea-urchin spermatozoa. J. Exp. Biol. **32**(4), 802–814 (1955)
6. Hatton, R.L., Choset, H.: Geometric motion planning: the local connection, stokes theorem, and the importance of coordinate choice. Int. J. Robot. Res. **30**(8), 988–1014 (2011)
7. Hatton, R.L., Ding, Y., Choset, H., Goldman, D.I.: Geometric visualization of self-propulsion in a complex medium. Phys. Rev. Lett. **110**, 078101 (2013)
8. Ijspeert, A.J., Crespi, A., Ryczko, D., Cabelguen, J.-M.: From swimming to walking with a salamander robot driven by a spinal cord model. Science **315**(5817), 1416–1420 (2007)
9. Kelly, S.D., Murray, R.M.: Geometric phases and robotic locomotion. J. Robot. Syst. **12**(6), 417–431 (1995)
10. Li, C., Umbanhowar, P.B., Komsuoglu, H., Koditschek, D.E., Goldman, D.I.: Sensitive dependence of the motion of a legged robot on granular media. Proc. Natl. Acad. Sci. **106**(9), 3029–3034 (2009)

11. McInroe, B., Astley, H.C., Gong, C., Kawano, S.M., Schiebel, P.E., Rieser, J.M., Choset, H., Blob, R.W., Goldman, D.I.: Tail use improves performance on soft substrates in models of early vertebrate land locomotors. Science **353**(6295), 154–158 (2016)
12. Qian, F., Zhang, T., Daffon, K., Goldman, D.I.: An automated system for systematic testing of locomotion on heterogeneous granular media. In: International Conference on Climbing and Walking Robots (CLAWAR) (2013)

Bioinspired Grippers for Natural Curved Surface Perching

William R.T. Roderick[(✉)], Hao Jiang, Shiquan Wang, David Lentink, and Mark R. Cutkosky

Stanford University, Stanford, CA 94305, USA
wrtr@stanford.edu
http://bdml.stanford.edu, http://lentinklab.stanford.edu

Abstract. Perching and climbing as animals do is useful to aerial robots for extending mission life and for interacting with the physical world because flight is energetically costly. This paper presents the design and modeling of a claw or spine based gripper for perching on rough, curved surfaces. Drawing inspiration from the opposed grip techniques found in animals, we focus on the design considerations associated with surface geometry and preload. A model elucidates the relationship between these variables, and a mechanism demonstrates the effectiveness of the opposed grip technique.

1 Introduction and Bioinspiration

Drawing inspiration from animals, mechanical designers can construct aerial robots with the ability to perch and climb to save power, extend mission life, and interact with the physical world. Most work involving perching mechanisms has focused on performance on regular, predictable surfaces relevant in urban environments, such as glass and building exteriors [12]. However, opportunities to unlock new ways to study the environment motivate the study of landing on and interacting with irregular, natural surfaces, such as trees. Such capabilities could, for example, enable tracking animals and collecting environmental data in remote areas. The variation in irregular tree surfaces, both in terms of the varying surface geometry and texture, makes versatility and adaptability a keystone for perching on such surfaces.

Despite the complexity of tree surfaces, many animals land on and move around them with ease [12]. This reliability and effectiveness stems in part from their opposed grip method of attachment whereby the animal pulls its claws or spines together in opposite directions along a surface to hold its weight. Animals from a wide range of genetic lineages use opposed gripping. These animals, including arthropods, birds, and mammals, vary in scale from milligrams to hundreds of kilograms (Fig. 1). Over this range, different animal groups accomplish opposed gripping in different manners. Many insects, for example, have adhesive pads in addition to spines and claws, and they do not have opposable digits [12]. As a result, they may use pure adhesion, and if they do use the opposed

© Springer International Publishing AG 2017
M. Mangan et al. (Eds.): Living Machines 2017, LNAI 10384, pp. 604–610, 2017.
DOI: 10.1007/978-3-319-63537-8_56

Mass (g) 10^{-1} 10^0 10^1 10^2 10^3 10^4 10^5

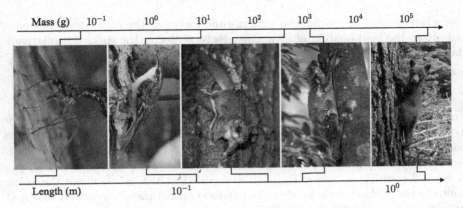

Length (m) 10^{-1} 10^0

Fig. 1. A few of the many species that use opposed claw or spine gripping to attach to vertical arboreal surfaces. From left to right: insect [2], bird (courtesy G. K. Roderick), flying squirrel [4], colugo [1], and bear [3]; masses and lengths approximated from [12].

grip strategy, they require multiple limbs to pull inward [7]. Conversely, birds, which rely on claws for surface attachment, can use the opposed grip strategy with a single foot. Regardless of the body structure, the prevalence of this gripping technique in the animal kingdom over evolutionary time suggests that it may be particularly effective for surface attachment. Previous work on opposed grip techniques for robots on rough surfaces has focused on making functional mechanisms for grasping rough surfaces, force analysis for a fixed mechanism geometry, or the contact dynamics involved in perching [8–11,13,14]. Here, we draw design inspiration from animals and extract insights from the opposed grip technique with a focus on contact geometry and preload.

2   Bio-Inspired Design

2.1   Contact Geometry and Preload

To remain attached to a tree, an animal's contact points with the surface must sustain its weight. From Fig. 1, we observe that aerial and scansorial animals use the opposed grip strategy in conjunction with a support below the center of mass, such as feet or a tail. As a result, the external load on the limbs is a combination of part of the weight and the force needed to counteract the pitch-back moment.

To increase the maximum external load, animals can apply preload forces. Claws and spines rely on friction and interlocking with the surface, which allows them to sustain large shear forces parallel to a surface and small normal forces away from a surface [6]. As long as the contact forces stay within friction limitations, animals can vary the preload forces to keep the overall force vector in bounds. From a geometric standpoint, if we let λ be the characteristic length of a bird foot or insect limb, what matters is the ratio λ/ρ where ρ is the radius of curvature of the surface (e.g. a tree branch or trunk). As $\lambda \to \rho$, the foot or limb can wrap around the surface so that normal forces at the claws or spines contribute to the load bearing ability. Conversely, if $\lambda \ll \rho$, the surface is relatively flat, and large shear forces are required.

2.2 Modeling

To examine the effects of normal and shear forces in more detail, we modify a flat surface spine model from [5,6]. A schematic of the local claw/surface contact is shown in Fig. 2A. A foot with opposing claws engages with a rough, curved surface in the global frame, in which is fixed the orthogonal basis $\hat{n}_{x,y,z}$. A second coordinate frame, $\hat{a}_{x,y,z}$, is embedded at the point where the spine contacts the surface, with $\hat{a}_x$ along the macroscopic surface tangent and rotated by $-\alpha$ (clockwise) with respect to $\hat{n}_x$.

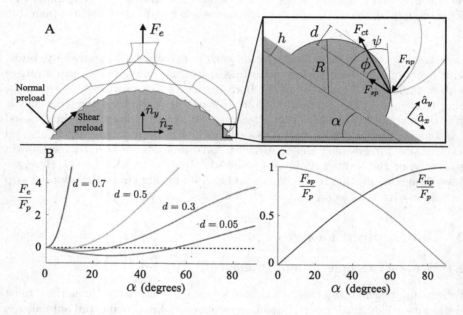

Fig. 2. Claw/asperity interaction: (A) A foot is externally loaded with force F_e. An inset shows the relevant geometry and forces at the claw tip. For the plots, $h = 0$ and $R = 1$ so d is a fraction of R, and $\mu = 0.3$. (B) F_e/F_p as a function of α and d illustrates how the benefit of preload varies with α and d. (C) Plots of F_{sp}/F_p and F_{np}/F_p as a function of α show how the normal and shear components of preload vary with macroscopic surface orientation.

At a finer scale, we consider that the surface is rough. In the example in Fig. 2, the tip of the claw is resting on the surface of a rounded bump, modeled here as a cylindrical surface of radius R. At the microscopic scale, the local surface tangent is further rotated by $-\psi$.

We now conduct a planar force balance on a toe, considering the effects of an initial preload and the forces required to balance an external load, $F_e \hat{n}_y$. The preload, which is equal and opposite at each claw, can be decomposed into shear and normal components (F_{sp}, F_{np}) in the $\hat{a}_{x,y,z}$ coordinate system. The force exerted by the right claw to maintain static equilibrium is then

$$\boldsymbol{F}_{ct} = (F_e/2)\hat{n}_y - F_{int}\hat{n}_x + F_{sp}\hat{a}_x + F_{np}\hat{a}_y \tag{1}$$

where F_{int} is an internal force that can arise when the external load is applied and depends on the bending moment, if any, at the joint at the base of the toe.

Next, we consider the case in which the toe is completely flexible and exerts no bending moments, which would be the case to minimize effort. In this case, the toe can be replaced virtually by a tendon running from the claw tip to the external load, illustrated by the dotted lines in Fig. 2. In this case, $\phi = 0°$. Then the ratio of the maximum possible external force, F_e, to the magnitude of the preload $(F_p = \sqrt{F_{np}^2 + F_{sp}^2})$ is

$$\frac{F_e}{F_p} = \frac{2\sin\alpha}{\cos\psi - \mu\sin\psi}\left(\frac{\sin\psi + \mu\cos\psi}{\sqrt{1 + \cot^2\alpha}} + \frac{\mu\sin\psi - \cos\psi}{\sqrt{1 + \tan^2\alpha}}\right) \tag{2}$$

where $\tan\psi = \sqrt{(R^2 - (R-d)^2)/(R-d)}$ and μ is the coefficient of static friction. This equation is valid only for $\mu < \text{arccot}\psi$. If $\mu \geq \text{arccot}\psi$ the claw will not slip for any magnitude of F_e and the foot and claws function somewhat like lifting tongs, with no preload required.

A biomechanist or mechanical designer can use the ratios F_e/F_p, F_{sp}/F_p, and F_{np}/F_p to study the forces on a foot or mechanism at the highest level. The maximum load to preload ratio, F_e/F_p, plotted in Fig. 2B, is effectively a gain that denotes how effective preload is for increasing the maximum external load. A negative F_e/F_p indicates geometry for which adding preload would cause the foot or mechanism to slip. Figure 2C shows how the normal and tangential components of the preload vary with surface orientation.

The simple model in Fig. 2 illustrates some basic points. First, for a given preload, contact farther around the asperity increases the maximum load for a given loading angle. At the foot or mechanism level, for maximum weight support, it is therefore important that the claws have a large travel in the normal direction to reach deep into the local surface bumps and pits. Second, as α increases for a given contact depth, F_e/F_p is initially 0, decreases to a negative minimum value (where positive F_p is unhelpful) and then increases again.

On a relatively flat surface (small α), the preload is dominated by shear forces. However, if the contact point is not deep along the asperity, shear preload will cause the contact point to slip, which results in a negative F_e/F_p. Thus, shear preload is beneficial on gently curved surfaces as long as the contact point is sufficiently deep. As α increases, the preload is dominated by F_{np} as the foot can squeeze the surface.

In the more general case, one could consider external loads with components in both the $\hat{n}_x$ and $\hat{n}_y$ directions and perhaps even moments about $\hat{n}_z$, producing

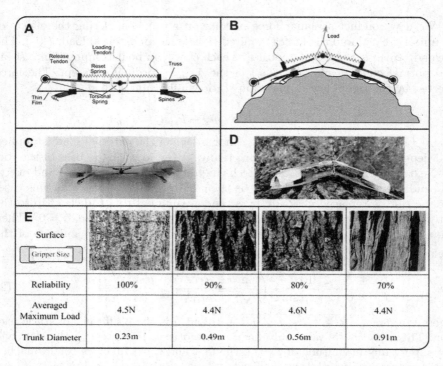

Fig. 3. A−D: Schematics and photos of the opposed-grip spine mechanism. E: Results from load tests on various tree textures and diameters. Reliability (percentage of successful attachments) and averaged maximum load are computed from 10 trials per tree type where the mechanism was placed at a random location on the surface. To avoid damaging the mechanism, pull-off tests were limited to 5 N.

bending moments in the toes. In this case, the forces at the two claws would be unequal, although the general trends concerning normal and tangential preloads would still apply.

2.3   Mechanism Design

To explore the mechanical principles behind rough, curved surface attachment in more depth, we developed a mechanism that takes advantage of opposed spines to increase the maximum allowable load away from the surface.

To enable animals to perform activities including agile locomotion, feeding, and mating, it is advantageous that the hands and feet of aerial animals be small relative to the weight of the animal. Similarly, the weight of the mechanism must be kept low while also being able to sustain the force from an aerial robot. Rather than attempting to copy the complex anatomy and structure of animal limbs, the mechanism draws on a few of the main insights regarding curved surface perching. A schematic and photo of the mechanism are shown in Fig. 3A−D. Just as arboreal animals use multiple digits on their hands and legs, we use arrays of articulated segments. The attachment arises from small spines or hooks with $\approx 10\,\mu m$ tip radius, which are suspended by flexures so that they can catch on bumps and pits (asperities) on a surface [5]. Due to the different sizes of asperities, each spine in a group must have independent compliance in the surface normal direction to reach around an asperity and apply a load as in Fig. 2. To achieve this behavior, we use a simplified design in which groups or tiles of spines are attached to compliant fingers. The tiles are bonded to a flexible film that is pretensioned by a truss and several torsional springs. The truss is bistable. In the open state, because the central pivot is below the members of the truss, it is held open by a reset spring (Fig. 3A and C). Upon contact with the surface, the truss can be pushed to embrace the surface such that the reset spring is overcome by the middle torsional spring. As the spines contact the surface, the mechanism settles into the closed state (Fig. 3B and D). The spines are loaded through the film along the macroscopic surface tangent. This feature keeps the loading angle near the surface, which disproportionately increases the shear force as a load away from the surface is applied. To release, the truss is folded away from the surface into the first stable configuration.

The span of the mechanism is 180 mm, working over a range of $15° < \alpha < 35°$ for tree trunks studied. For large tree trunks, the spines are made to reach up to 1 cm deep into the tree bark valleys and crevices. In the case that $\lambda \to \rho$, the effective α can reach 45°. According to the model, for the smallest α value of 15°, a preload increases the maximum allowable F_e, as long as $\psi > 60°$, which is ensured by the deep reach of the spines. For this mechanism, the total preload F_p is approximately 0.4 N for a relatively small α of 20°.

Pull-off tests were performed to assess the reliability and maximum load capabilities of this gripping mechanism. Figure 3E shows the reliability and averaged maximum load from 10 trials per tree bark type where the mechanism was placed at a random location on the surface. To avoid destroying the mechanism, the pull-off tests were limited to 5 N. With a mass of 16.1 g, this mechanism weighs only 3% of the average maximum load. These results are promising for perching aerial robots.

For integration with an aerial platform, it is desirable that the mechanism be passive. However, to trigger the grasp and detachment, a small amount of energy needs to be put into the mechanism to toggle between the bistable states. The energy from landing can be used to trigger the grasp, and if some of the landing energy is stored, the stored energy can also be used for release. With more stored energy, a jump and takeoff maneuver becomes possible.

3   Conclusion

Perching on rough, vertical curved surfaces depends strongly on multiple interconnected factors including contact geometry in relation to platform size as well as preload. On relatively flat surfaces, increasing the shear preload can substantially increase the maximum outward normal load. As the surface curvature increases, normal preload dominates and facilitates a firm grip. In both cases, it is desirable to provide the spines with a long suspension travel so that they can reach deep into features for maximum loads. Future work will focus on improving claw/surface interaction models, with extensions to multiaxial loading and improved claw and spine designs for arboreal fliers.

Acknowledgments. This work is supported by ARL MAST MCE 14.4; W. Roderick is supported by a NSF GRF (DGE-114747).

References

1. Yap, L.K.: www.flickr.com/photos/64565252@N00/453554996/. cc by 2.0
2. Bartz, R.: Munich aka makro freak. wikimedia commons. cc by-sa 2.5
3. User: Lassennps. www.flickr.com/photos/lassennps/9403813124/. cc by 2.0
4. User: Mimimiaphotography. wikimedia commons. cc by-sa 3.0
5. Asbeck, A.T., Cutkosky, M.R.: Designing compliant spine mechanisms for climbing. J. Mech. Robot. **4**(3), 031007 (2012)
6. Dai, Z., Gorb, S.N., Schwarz, U.: Roughness-dependent friction force of the tarsal claw system in the beetle pachnoda marginata (coleoptera, scarabaeidae). J. Exp. Biol. **205**(16), 2479–2488 (2002)
7. Goldman, D.I., Chen, T.S., Dudek, D.M., Full, R.J.: Dynamics of rapid vertical climbing in cockroaches reveals a template. J. Exp. Biol. **209**(15), 2990–3000 (2006)
8. Jiang, H., Pope, M.T., Hawkes, E.W., Christensen, D.L., Estrada, M.A., Parlier, A., Tran, R., Cutkosky, M.R.: Modeling the dynamics of perching with opposed-grip mechanisms. In: 2014 IEEE International Conference on Robotics and Automation (ICRA), pp. 3102–3108. IEEE (2014)
9. Kovač, M., Germann, J., Hürzeler, C., Siegwart, R.Y., Floreano, D.: A perching mechanism for micro aerial vehicles. J. Micro-Nano Mechatron. **5**(3–4), 77–91 (2009)
10. Lam, T.L., Xu, Y.: A novel tree-climbing robot: treebot. In: Tree Climbing Robot. Spring Tracts in Advanced Robotics, pp. 23–54. Springer, Heidelberg (2012)
11. Parness, A., Frost, M., Thatte, N., King, J.P., Witkoe, K., Nevarez, M., Garrett, M., Aghazarian, H., Kennedy, B.: Gravity-independent rock-climbing robot and a sample acquisition tool with microspine grippers. J. Field Robot. **30**(6), 897–915 (2013)
12. Roderick, W.R., Cutkosky, M.R., Lentink, D.: Touchdown to take-off: at the interface of flight and surface locomotion. Interface Focus **7**(1), 20160094 (2017)
13. Spenko, M., Haynes, G.C., Saunders, J., Cutkosky, M.R., Rizzi, A.A., Full, R.J., Koditschek, D.E.: Biologically inspired climbing with a hexapedal robot. J. Field Robot. **25**(4–5), 223–242 (2008)
14. Xu, F., Wang, B., Shen, J., Hu, J., Jiang, G.: Design and realization of the claw gripper system of a climbing robot. J. Intell. Robot. Syst. 1–17 (2017). doi:10.1007/s10846-017-0552-3

Collisional Diffraction Emerges from Simple Control of Limbless Locomotion

Perrin E. Schiebel$^{(\boxtimes)}$, Jennifer M. Rieser, Alex M. Hubbard,
Lillian Chen, and Daniel I. Goldman

Georgia Institute of Technology, Atlanta, GA 30332, USA
perrin.schiebel@gatech.edu, {jennifer.rieser,
daniel.goldman}@physics.gatech.edu

Abstract. Snakes can utilize obstacles to move through complex terrain, but the development of robots with similar capabilities is hindered by our understanding of how snakes manage the forces arising from interactions with heterogeneities. To discover principles of how and when to use potential obstacles, we studied a desert-dwelling snake, *C. occipitalis*, which uses a serpenoid template to move on homogeneous granular materials. We tested the snake in a model terrestrial terrain—a single row of vertical posts—and compared its performance with a robophysical model. Interaction with the post array resulted in reorientation of trajectories away from the initial heading. Combining trajectories from multiple trials revealed an emergent collisional diffraction pattern in the final heading. The pattern appears in both the living and robot snake. Furthermore, the pattern persisted when we changed the maximum torque output of the robot motors from 1.5 N-m to 0.38 N-m in which case local deformation of the robot from the serpenoid curve appears during interaction with the posts. This suggests the emergent collisional diffraction pattern is a general feature of these systems. We posit that open-loop control of the serpenoid template in sparse terrains is a simple and effective means to progress, but if adherence to a heading is desired more sophisticated control is needed.

Keywords: Locomotion · Snakes · Complex terrain

1 Introduction

Principles governing movement in heterogeneous terrains remain largely undiscovered. During terrestrial locomotion, contacts with the surroundings are often intermittent and can lead to unexpected emergent behavior [1]. Snakes are remarkable in their ability to use a seemingly simple morphology—a limbless, elongate trunk—to navigate many habitats including a wide range of terrestrial environments. Previous research on terrestrial snake locomotion focused on so-called generalist snakes which encounter a variety of terrain (forest, grassland, wetland, etc.) consisting of many different materials. These snakes use posteriorly-propagating body bends to push the trunk laterally against obstacles and generate the forces needed for forward movement [2, 3]. The versatility and simplicity of this scheme makes it an attractive model for robots [4]. However, the challenge of controlling the many degrees of freedom to effectively

© Springer International Publishing AG 2017
M. Mangan et al. (Eds.): Living Machines 2017, LNAI 10384, pp. 611–618, 2017.
DOI: 10.1007/978-3-319-63537-8_57

manage interaction with obstacles can stymie robotic implementation of slithering locomotion.

Locomotor templates [5] can simplify control and aid understanding. We previously found the desert-dwelling sand-specialist Mojave Shovel-nosed snake (*Chionactis occipitalis* Fig. 1a) uses a highly-stereotyped waveform which adheres to a sinusoidal curvature in both time and arclength along the body (a "serpenoid" curve [6], Eq. 1) when moving on the surface of homogeneous sand. Given the observed stereotypy of the waveform both between individuals and trials, we hypothesized that *C. occipitalis* uses open-loop control, where the muscle activation is not modified in response to perturbation from the terrain.

Therefore, to begin a systematic search for principles of slithering movement in terrestrial environments, we studied *C. occipitalis* navigating a model terrestrial terrain —a single row of vertical, rigid posts embedded in a homogeneous substrate (Fig. 1d) —inspired by the omnipresent sand substrate and sparse obstacles in their natural habitat (Fig. 1b). We compared the performance of the living snake to a multi-link snake robot in a similar model terrain (Fig. 1c). The robophysical model provided the benefit of behaving in a controlled way which facilitated understanding both of the control strategy of the animal as well as the benefits and drawbacks of using the simple open-loop serpenoid template scheme in multi-modal terrain.

2   Living Snake Experiments

A schematic of our terrain model is shown in Fig. 1d. A carpet with long fibers mimicked the yielding properties of sand without the experimental challenge of using granular materials, namely hysteresis. We verified that the snake used the sand-swimming serpenoid template to move on the carpet. A single row of six 0.64 cm diameter polyurethane rods was placed perpendicular to the direction of motion of the snake. The open space between the posts was 1.7 cm. The rubber rods would deform slightly ($\sim 5 \times$ less than the body width) at the point of contact of the snake. This deformation was measured at 200 Hz using a high-speed camera (X-PRI, AOS) and used to calculate the force applied. A seventh rigid post was included at the end of the row to act as a fiducial.

Nine *C. occipitalis* were used in our trials. To simplify comparison with the robot, we "blindfolded" the snakes by obscuring the spectacle scale with water-based face paint. Snakes were tested individually and the trials captured at 200 Hz using a high-speed camera (S-Motion, AOS) and digitized using custom MATLAB (R2015b, MathWorks) code.

Because the living and robotic snakes are different sizes (Fig. 1 caption) we chose to use as a unit of measurement of length the average distance travelled in one undulation, v_oT. For the living snake we calculated v_oT as the average center-of-mass speed times the average temporal period of *C. occipitalis* moving freely when no obstacles were present.

The scattering angle, θ, was measured as the angle between a point on a snake's trajectory and the z-axis as the snake passed through a circular arc centered between the 3rd and 4th posts and in line with the row (Fig. 1d). We chose to measure the angle

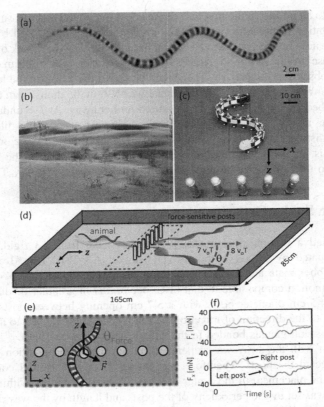

Fig. 1. *C. occipitalis* and robotic snake in heterogeneous terrain. (a) The Mojave Shovel-nosed snake, *Chionactis occipitalis*, is a small desert-dwelling Colubrid specialized to move on and within sand (9 individuals; Mass: 21 ± 3 g; Length: 38.4 ± 2.3 cm) (b) An example of the natural habitat of *C. occipitalis*. The sand substrate is omnipresent. Obstacles are sparse and include small plants, twigs, and rocks. (c) The 12-joint robot snake. (Mass: 1.1 kg; Length: 80 cm) Local joint angles were commanded to vary sinusoidally both along the arclength and in time according to Eq. 1. The array of 5 evenly-spaced posts is seen at the bottom of the image. In order to sample all initial conditions, the robot CoM was placed throughout the initial conditions box indicated by the blue rectangle. (d) Snake experiments were carried out in a 165 cm × 85 cm arena. The substrate was a high-pile carpet. Six force-sensitive posts and one fiducial post were oriented perpendicular to the direction of travel. Red and black traces are example snake trajectories. The scattering angle θ was calculated by averaging the angle between the trajectory and the dashed line indicating the z-axis between the dashed arcs at seven and eight v_oT. (e) Schematic of a snake interaction with a post. The snake applied a force vector $\vec{F} = \langle F_x, F_z \rangle$ to the post. $\theta_{Force} = atan(F_x/F_z)$ is the angle between the positive z-axis and $\vec{F}$. θ_{Force} is always less than 180°. $F_x > 0$ yields positive θ_{Force} and $F_x < 0$ yields negative θ_{Force}. (f) Example time-resolved forces from a single living snake trial. The snake contacted two posts and applied forces in both x and z. (Color figure online)

with respect to all points on the first half of the body. We found this measurement was a fair representative of the path of the snake given the noise introduced by tracking as well as differences between the waveforms of individuals. The drawback of this method was that close to the array the tracks from trajectories which ultimately diverge overlapped, obscuring the pattern. Therefore, we found the peaks in the histogram did not become obvious to the eye until the snake was 5–6 undulations from the array and continued to become more clear as we measured further away. At 7–8 undulations from the array the angles have discernable peaks. These peaks would likely become increasingly distinct further away from the posts, however the tracks begin ending past 8 v_oT. That is, at 7–8 v_oT the number of data points included in the histogram is comparable to the number of points included at, for example, 1 to 2 v_oT.

3   Robotic Snake Experiments

We constructed a robophysical model [7] of the snake from 13 rigid, 3D printed segments actuated by 12 servo motors (Dynamixel AX-12A) (Fig. 1c). The robot moved on rubber mats and LEGO wheels on the underside of the robot facilitated low-slip locomotion comparable to that of *C. occipitalis* in granular media [8]. A row of five rigid 4.5 cm diameter posts with a 5.7 cm opening between posts was placed perpendicular to the direction of travel of the robot. The force applied to the posts was measured via strain gauges bonded to the square Aluminum-rod base.

Interaction with the array was dependent on the phase and position of the robot when it contacted the posts. Therefore, to explore all possible initial conditions, we varied the initial placement of the center-of-mass (CoM) of the robot within a rectangle whose width was set by the periodicity of the posts and length by the wavelength of the waveform (blue rectangle, Fig. 1c). The x and z coordinates of each segment were captured at 120 Hz by a system of four OptiTrack cameras (Flex 13, Natural Point) tracking infrared reflective markers on the robot.

The robot was controlled using a Robotis CM-700 controller and powered using an external supply. The actuator positions were determined by the equation for a serpenoid curve (Eq. 1).

$$\zeta_i = \zeta_{Max}\sin(ks_i + 2\pi ft) \tag{1}$$

ζ_i is the angular position of actuator $i = [1, ..., 12]$ with a set maximum angular excursion $\zeta_{Max} = 0.62$ rad, spatial frequency k = 1, and temporal frequency $f = 0.15$ Hz. The waveform seen on the robot in Fig. 1c is this serpenoid curve at time t = 0. The control signal sent to the robot was open-loop such that these parameters were not changed at any point in any of the trials and the control signals would continue to be sent as a function of time and position on the body regardless of external forces or tracking accuracy of the actuators.

We tested two versions of this control on the robot. The first case was high-torque (HT). In this case the maximum torque each actuator could produce was 1.5 N-m. The HT robot could accurately track the desired waveform in most cases. We verified the tracked robot positions using the OptiTrack data and found the tracking error was < 5%.

For the limited-torque (LT) case we kept all other aspects of the robot and controller the same but limited the torque output of each actuator to 25% of the overall maximum (0.38 N-m). For reference, the largest torque measured in the robot moving on the rubber mats alone was 20% max, or 0.3 N-m. In the LT case the actuator would track the commanded trajectory up until the torque exceeded 0.38 N-m. At this point the motor continued attempting to track the commanded angles but did not exert more than 0.38 N-m of torque. The motor resumed successful tracking of the commanded trajectory once this was possible with ≤ 0.38 N-m of torque. The inability of one actuator to achieve the desired position did not change the commands to it nor to the other actuators. The observed result of the LT condition was that local deformations from the serpenoid curve appeared during interaction with the post array. The distribution of tracking error during interaction with the posts had a similar mean to the HT case but the tails of the distribution were longer and asymmetric with a greater number of large positive errors (maximum tracking error measured was 34%) than seen in the HT case. This reflects the observed local deviations from the commanded angles. We compared the kinematics of the HT to the LT case for the robot moving in a steady state with no pegs present to verify that decreasing the maximum torque available to the actuators did not otherwise change their behavior.

4   Results

The array acts to scatter the snakes. Two example trajectories are shown in the schematic in Fig. 1d (black and red tracks). The action of the array is further illustrated when all trajectories from all trials are combined as in Fig. 2d. The snakes move from bottom to top, in the direction of positive z. The units are normalized by v_0T, the average CoM velocity times the period of the motion, i.e., the average distance travelled in one undulation. The trajectories are colored according to the scattering angle θ. To calculate θ we averaged the polar angle of the trajectory with respect to the z-axis as it passed through a band between seven and eight v_0T from the array (see Fig. 1d for a schematic).

The trajectories of the LT and HT robot trials are shown in Fig. 2e and f, respectively. The trajectories are colored by scattering angle as before, and as in the trials with *C. occipitalis* some trajectories were deflected away from the z-axis by the interaction with the array. The LT robot generally scattered at smaller angles than the HT robot. We note that the largest scattering angles of *C. occipitalis* were greater than those of the robot, but we cannot say whether this is of any significance. During these trials we found that the scattering angle was sensitive to a number of factors related to the various dimensions of the system, and it is as of yet unclear which of these drove the differences between *C. occipitalis* and the robot, or if it was to a greater degree due to differences in the neuromechanical systems (e.g. the use of bilateral muscle versus a single servo to actuate the trunk).

The emergent pattern of the trajectory re-orientations was further illustrated in a histogram of the scattering angle. These histograms are above their corresponding trajectory maps in Fig. 2. It is clear that both the living and robotic snakes are more

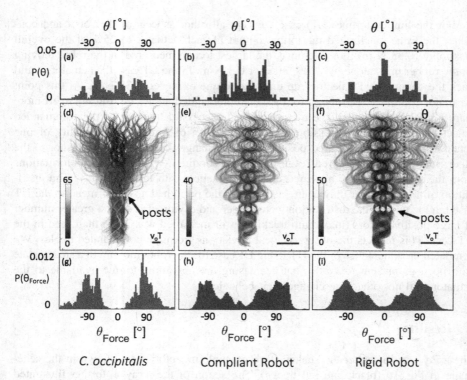

θ [°] θ [°] θ [°]

C. occipitalis   Compliant Robot   Rigid Robot

Fig. 2. Emergent collisional diffraction pattern (a–c) Scattering angle θ for the snake (181 trials), limited-torque (LT) robot (216 trials), and maximum-torque (HT) robot (366 trials), respectively (left to right). Snake scattering angles are calculated as the mean polar angle of a trajectory when it is a distance between 7 and 8 v_oT from the center of the post array (see diagram in Fig. 1d and discussion in Sect. 2). As the robot waveform has less natural variation, scattering angle is calculated by fitting a line to the maxima/minima of the trajectory for all body segments and calculating the angle between these lines and the vertical. θ is taken to be the average of all of these values. (d–f) Trajectories for the snake, LT robot, and HT robot, respectively (left to right). To help differentiate trajectories each run is colored according to the absolute value of its scattering angle. Light gray circles indicate the position of the posts. Prior to the posts the group of trajectories is "collimated" whereas after interacting with the array some trajectories deflect away from the vertical z-axis. (g–i) Force orientation angle θ_{Force} for the snake, LT robot, and HT robot, respectively (left to right). θ_{Force} is calculated for each contacted post throughout each run by finding the angle between the force vector and the z-axis. A value of zero corresponds to the snake pushing directly forward (+z) while 180° is the snake pushing directly opposite the direction of motion (−z). The peaks in the distributions occur around ±90°, meaning the snakes are most likely to push left/right on the posts.

likely to travel in certain directions than others upon exiting the array, and this pattern is qualitatively similar for the three systems tested.

The forces applied to the obstacles revealed a similar emergent pattern. The angle θ_{Force} is the angle between the force vector and the positive z-axis, i.e. atan(F_x/F_z)

(Fig. 1e). The bottom row of Fig. 2 shows histograms of θ_{Force} for all trials. We find that both the living and robotic systems are more likely to push left/right against the array. It seemed reasonable to expect that θ_{Force} and θ would be correlated. However, we did not find any relationship between the two. This may be attributed to the complexity of the interactions acting simultaneously between the posts, the body, and the substrate; perhaps in combination with the highly dissipative nature of the surroundings.

5 Discussion

This study highlights the benefits and repercussions of using an open-loop template during limbless locomotion in multi-modal terrain. Control of the serpenoid-template was easy to implement, and we note that it was exceedingly rare ($\sim 1\%$ of trials) for the rigid robot to become wedged in the array, while the compliant robot and living snake always transited the array. We therefore argue open-loop control of the serpenoid template is an effective strategy for transit of sparse terrain which requires no external sensors, with the caution that the pattern of trajectory reorientation appears to be a general feature of these systems. A more sophisticated control scheme which can correct the heading changes caused by collision with obstacles may be necessary if a specific trajectory is desired.

Acknowledgements. Supported by Army Research Office (ARO) grant W911NF-11-1- 0514; NSF grants PoLS PHY-1150760; National Defense Science and Engineering Graduate (NDSEG) Fellowship. The authors would like to thank Dr. Joseph Mendelson III, director of research at Zoo Atlanta and Adjunct Associate Professor, School of Biological Sciences, Georgia Institute of Technology for facilitating the acquisition of *C. occipitalis*.

References

1. Qian, F., Goldman, D.I.: The dynamics of legged locomotion in heterogeneous terrain: universality in scattering and sensitivity to initial conditions. In: Robotics: Science and Systems (2015)
2. Gray, J., Lissmann, H.: The kinetics of locomotion of the grass-snake. J. Exp. Biol. **26**(4), 354–367 (1950)
3. Kelley, K., Arnold, S., Gladstone, J.: The effects of substrate and vertebral number on locomotion in the garter snake Thamnophis elegans. Funct. Ecol. **11**(2), 189–198 (1997)
4. Murphy, R.R., Tadokoro, S., Nardi, D., Jacoff, A., Fiorini, P., Choset, H., Erkmen, A.M.: Search and rescue robotics. In: Siciliano, B., Khatib, O. (eds.) Springer Handbook of Robotics, pp. 1151–1173. Springer, Heidelberg (2008). doi:10.1007/978-3-540-30301-5_51
5. Full, R.J., Koditschek, D.E.: Templates and anchors: neuromechanical hypotheses of legged locomotion on land. J. Exp. Biol. **202**(23), 3325–3332 (1999)
6. Hirose, S., Morishima, A.: Design and control of a mobile robot with an articulated body. Int. J. Robot. Res. **9**(2), 99–114 (1990)

7. Aguilar, J., Zhang, T., Qian, F., Kingsbury, M., McInroe, B., Mazouchova, N., Li, C., Maladen, R., Gong, C., Travers, M.: A review on locomotion robophysics: the study of movement at the intersection of robotics, soft matter and dynamical systems. Rep. Prog. Phys. **79**(11), 110001 (2016)
8. Sharpe, S.S., Koehler, S.A., Kuckuk, R.M., Serrano, M., Vela, P.A., Mendelson, J., Goldman, D.I.: Locomotor benefits of being a slender and slick sand swimmer. J. Exp. Biol. **218**(3), 440–450 (2015)

Binocular Vision Using Synthetic Nervous Systems

Anna Sedlackova, Nicholas S. Szczecinski[(✉)], and Roger D. Quinn

Case Western Reserve University, Cleveland, OH, USA
{anna,nss36}@case.edu

Abstract. This paper presents a system that allows our 29 degree-of-freedom robot, MantisBot to detect the distance to an object labeled as "prey". The prey is detected using two Pixy cameras with build-in software for object recognition. Each camera outputs analog signals encoding the azimuth and elevation of the prey, which are interpreted by MantisBot's synthetic nervous system. An analytical derivation of the distance of an object from the robot as a function of the azimuth values from each camera produces training data for a neural network. The network learns this relationship, endowing MantisBot with binocular depth perception.

Keywords: Binocular vision · Synthetic nervous systems · Object recognition

1 Introduction

Insects are great models for walking robots. Our 29 degree-of-freedom robot, MantisBot, is a testbed for synthetic nervous systems that controls locomotion, and how descending commands from the head might alter local reflexes to change the speed and direction of locomotion [1]. Praying mantises are a good model for robotic goal-directed locomotion because they use proficient visual systems to hunt prey [2]. Mantises can perceive depth without moving their head which enables them to catch prey using ballistic movements. According to the works of Jenny Reed and her team, the optomotor response is governed by stimuli in the central visual field and not the periphery [7]. We can simulate this reaction using MantisBot by presenting a cue to its central field of vision and triggering a reaction by calculating the distance to the prey in real time.

In our previous work, we used cues from a simplified visual system to identify the azimuthal position of simulated "prey", after which the robot directly stepped in that direction [1]. For the robot to freely walk; however, it must be able to detect the distance of the prey to adjust its speed of movement. Praying mantises have 20° of overlapping vision between their eyes [3], and they can use that for binocular depth perception [4]. A binocular vision system is presented for MantisBot, using two Pixy cameras for object detection. With the collected information, an analytical solution to the depth of an object as a function of the azimuth input from both cameras is computed. A simple neural network is used to approximate this calculation in MantisBot. This paper presents preliminary data from the robot.

© Springer International Publishing AG 2017
M. Mangan et al. (Eds.): Living Machines 2017, LNAI 10384, pp. 619–625, 2017.
DOI: 10.1007/978-3-319-63537-8_58

2   Methods

a. Color Recognition Using Pixy Cameras

MantisBot uses two color recognition Pixy cameras to detect "prey" (Charmed Labs, Austin, TX). These cameras have onboard processors that use hue-based color recognition algorithms to locate blobs. This data is then sent to MantisBot's microcontroller over a serial connection. Figure 1 shows MantisBot's head, with the Pixy cameras.

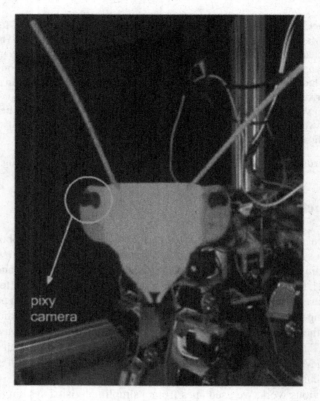

Fig. 1.   MantisBot's face with two Pixy cameras at 45° angle.

Each camera has a wide-angle lens, giving it a 110° field of view. As presented in Fig. 2, the cameras are placed at a 90° angle with respect to each other (i.e. $\beta = 45°$), such that their field of view overlaps by 20°. The azimuthal coordinate of the object's position (γ_1 and γ_2) are streamed to two separate microcontrollers that convert the coordinates to analog values, which are interpreted by the neural network.

b. Computation of Distance from Desired Object Using Analytical Mathematics

The two analog values γ_1 and γ_2, represent the azimuthal coordinate of the object as seen from each camera. Using the set distance between the center of the two cameras, d, and

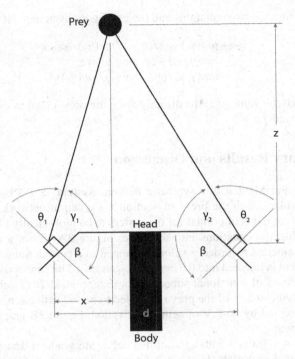

Fig. 2. Explaining the values used in calculating the distance from the head

the two angle values γ_1 and γ_2, the distance from the viewed object can be computed as long as it is in the binocular field of view, that is, where the viewed object can be seen with both cameras at the same time. Figure 2 illustrates these values. The distance to the object, z, can be expressed as a function of γ_1 and γ_2. The calculations are as follows:

$$\gamma_1 = 180° - \theta_1 - \beta$$
$$\gamma_2 = 180° - \theta_2 - \beta$$

where θ_1 and θ_2 are values obtained by the Pixy cameras and β is a constant angle between the frontal plane of the head, and the plane of the camera. The values can also be expressed as follows:

$$tan(\gamma_1) = z/(d-x)$$
$$tan(\gamma_2) = z/x$$

z is the distance from the face of the MantisBot to the viewed object; x can be seen in Fig. 2. Having two equations with two unknowns, x and z, the expression for z can be obtained through arithmetic manipulation:

$$x = z/tan(\gamma_2)$$

x is eliminated from the two equations and the expression is manipulated:

$$tan(\gamma_1) = z/(d - z/tan(\gamma_2))$$
$$tan(\gamma_1)(d - z/tan(\gamma_2)) = z$$
$$z = tan(\gamma_1)d/(1 + tan(\gamma_1)/tan(\gamma_2))$$

It can be seen that the value of z, the distance from the prey, changes directly based on the values of $\gamma1$ and $\gamma2$.

3   Preliminary Results and Discussion

We wish to expand MantisBot's synthetic nervous system (SNS) by enabling it to perform the calculations in the previous section via a neural network. Although this requires an extra step, it is important for the system to be biologically plausible instead of directly parsing the equations into the code. Figure 3A shows a diagram of the network. Each camera's recorded γ value is mapped to a current value between 0 and 20 nA. This current is injected into the neurons U_1 and U_2. The connectivity was chosen based on our method of functional subnetworks (Szczecinski 2017 submitted). Similarly, the perceived distance of the prey is mapped to neural activation, with a distance of 300 cm represented by 20 mV of neuron activation. Figure 3B graphically demonstrates this mapping.

The network was trained with a traditional delta-rule gradient descent method [5]. Individual training data triplets (U_1, U_2, U_{out}) were used to compute the network's error, and the gradient of the error was used to determine in what proportion to change the parameter values. The activation function was the steady-state solution to the neuron's dynamics [6]. Because the head is symmetrical, the parameter values were also assumed to be symmetrical. The parameters tuned were the synaptic conductance, g, and synaptic potential, E, of each layer. Figure 3C shows that the parameter values g_1, E_1, g_2, and E_2 converge and the network's calculation error decreases as the training progresses. Figure 3D shows that the trained network closely reproduces the intended data. A plot of the error between the trained network and the training data reveals a maximum error of 2 mV, which corresponds to about 30 cm.

Figure 4 shows data from a 30 s trial in which "prey" was moved within MantisBot's field of view, and it approximated its distance. MantisBot's approximation, drawn in blue, also includes a shaded area of 30 cm, which is the maximum error of the neural network. This data is quite noisy, because the Pixy cameras detect the prey less reliable at long distances. The peaks in the data, however, follow the distance to the prey. We may be able to improve performance in the future by intelligently filtering the output data, for instance by ignoring input when only one eye can see the prey. When both eyes see the prey, the distance may be saved in memory until the next time the robot can make the measurement.

Simple Network Trained to Compute Depth Perception

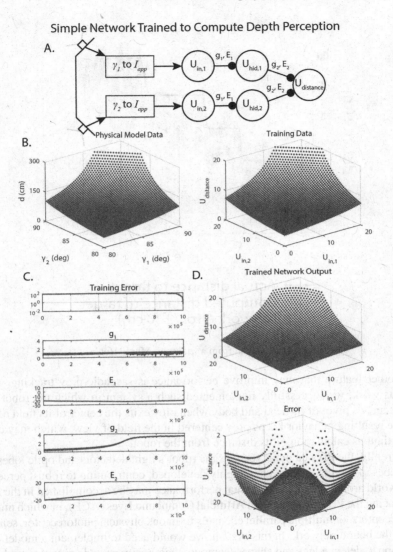

Fig. 3. A – Diagram of the synthetic nervous system, B & D – Graphic demonstration of the data, network output and error, C – Graphs showing parameter value and the network's calculation error

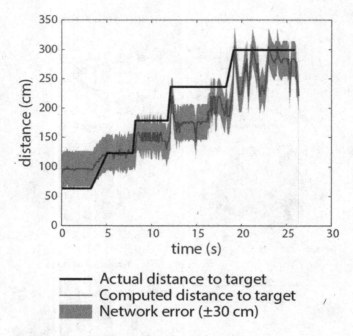

Fig. 4. Data from a trial with MantisBot

Another feature that may improve performance is saccadic prey tracking. In our previous work, we successfully implemented such a system in which the robot plans and executes a pivot of its head and body when prey exits the center of its field of view [1]. The resulting behavior keeps prey centered in the field of view, which may enable the continuous calculation of its distance from the robot.

Currently, the Pixy cameras serve as the robot's light detectors and optic lobes. The optic lobes of insects are large and highly developed, contributing to robust perception of the world around them. Better visual performance may be accomplished in the future by replacing the Pixy cameras with Artificial Compound Eyes (ACE), in which multiple photoreceptors, or multiple parallel channels from one physical photoreceptor, send raw data to the neural network. In this system, we would need to implement a model of the optic lobe to detect edges and shapes, compute their positions and velocities, and direct attention at specific items. Recent advances in Deep Neural Networks (DNN) suggest that such networks can be developed. In fact, the training method used in this paper is based on DNN training methods. Our SNS system, however, adds neuron dynamics, which may prove even more useful for interpreting visual data.

References

1. Szczecinski, N.S., Getsy, A.P., Martin, J.P., et al.: MantisBot is a robotic model of visually guided motion in the praying mantis. Arthropod. Struct. Dev. (2017). doi:10.1016/j.asd.2017.03.001
2. Mittelstaedt, H.: Prey capture in mantids. Recent Adv. Invertebr. Physiol. 51–72 (1957)
3. Rossel, S.: Foveal fixation and tracking in the praying mantis. J. Comput. Physiol. A Neuroethol. Sens. Neural Behav. Physiol. **139**, 307–331 (1980)
4. Rossel, S.: Binocular stereopsis in an insect. Lett. Nat. **302**, 821–822 (1983)
5. Trappenberg, T.: Fundamentals of Computational Neuroscience, 2nd edn. Oxford University Press, Oxford (2009)
6. Szczecinski, N.S., Hunt, A.J., Quinn, R.D.: Design process and tools for dynamic neuromechanical models and robot controllers. Biol. Cybern. (2017). doi:10.1007/s00422-017-0711-4
7. Nityananda, V., Tarawneh, G., Errington, S., Serrano-Pedraza, I., Read, J.: The optomotor response of the praying mantis is driven predominantly by the central visual field. J. Comp. Physiol. A **203**, 77–87 (2017)

Cell Patterning Method by Vibratory Stimuli

Ippei Tashiro[✉], Masahiro Shimizu, and Koh Hosoda

Graduate School of Engineering Science, Osaka University,
1-3 Machikaneyama, Toyonaka, Osaka 560-8531, Japan
tashiro.ippei@arl.sys.es.osaka-u.ac.jp

Abstract. Recently, micro robots driven by muscle cells attract a lot of attention. The shape of the robot can be controlled by modifying the growing of cells. This paper proposes a new cell patterning method by vibratory stimuli. The stimuli can detach redundant cells, and a certain cell pattern can be formed. In the experiment, around 80 Hz vibratory stimulus was applied to skeletal muscle cells via a PDMS scaffold and the cells were observed by camera. As a result, we found that a vibratory stimulus promotes detaching of the cells, especially at the moment of cell division. We also found that the increment of the cells depends on the vibratory energy. These results suggested that we can control the number of cells by vibration stimulus. Actually, we made specific cell pattern by the vibration stimulus.

Keywords: Cell patterning · Vibratory stimuli

1 Introduction

Recently, micro robots driven by muscle cells are attracting a lot of attention [1–3]. We can utilize the characteristic features of the living cells such as self-growth and self-regeneration. Various types of micro cells robots have been developed so far. Shimizu et al. created muscle actuator using mouse myoblast cell line C2C12 driven by photostimulation [1]. Kim et al. developed a crub-like microrobot using rat cardiomyocytes [2] that could walk over ten days. The cells were plated on the grooved surface of the body, resulting in a high concentration of pulsating cells. Park et al. developed a soft-robotic ray using rat cardiomyocytes [3]. The serpentine cell pattern leads to coordinated undulatory swimming. In order to design the bodies of these robots, the cell patterns must be controlled. Cell micropatterning [4,5] is one of conventional methods to make cell patterns in which polydimethylsiloxane (PDMS) stamps with desired microfeatures are utilized to print extracellular matrix proteins onto particular areas of cell culture substrates. Although this method is accurate, it is costly and only can be applied to cells in 2D plane.

This paper proposes another cell patterning method by utilizing vibratory stimuli. We adopt skeletal muscle cells and apply vibratory stimuli to cellular scaffold during multiplication. We expected to detach regional cells by utilizing vibratory stimuli. If only regional cells can be promoted detaching during

© Springer International Publishing AG 2017
M. Mangan et al. (Eds.): Living Machines 2017, LNAI 10384, pp. 626–630, 2017.
DOI: 10.1007/978-3-319-63537-8_59

multiplication, it is considered to be able to control the final cell patterns. Furthermore, the cell patterning method using mechanical stimuli may be able to be applied to the cells cultured in 3D space.

The purpose of this study is to make the patterns of skeletal muscle cells by vibratory stimuli. We cultured mouse myoblast cell line C2C12 on a 2D PDMS film and applied a vibratory stimulus to the cells via the film. We observed cell condition and number of cells by time-lapse imaging. We also observed the distribution of cells after stimulation.

2   Materials and Methods

Cells were cultured in a circular PDMS chamber whose bottom is a thin PDMS film. In the experiment, we used vibration motors to apply vibratory stimuli. By attaching a vibration motor to the center of the film, cells were applied a vibratory stimulus via the film. In order to capture images of cells under stimulation at fixed time intervals, we developed a time-lapse imaging system inside a hand-made incubator (Fig. 1). This system is composed of a phase-contrast microscopy, a PDMS chamber with a vibration motor, CulturePal CO_2 pack, a sealed jar, a glass heater for chamber, a heater for inner air, a fan, a temperature sensor, a camera, and an insulating container. This system allows both cell culture under a microscope and microphotography at fixed time intervals. It is necessary for cell growth to keep the concentration of CO_2 at 5% and temperature at 37 °C. CO_2 in the jar is maintained 5% with CulturePal CO_2 pack. Temperature in

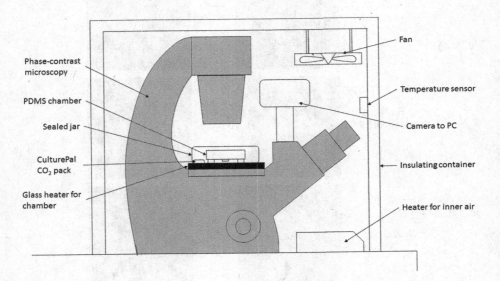

Fig. 1. Time-lapse imaging system inside a hand-made incubator. The PDMS chamber seeded with cells is placed in a sealed jar with CulturePal CO_2 pack. The jar is placed under the microscope.

the insulating container is maintained 37 °C with the heaters. The camera is attached to a phase-contrast microscopy. It is controlled to take photomicrographs at fixed time intervals. Cells were seeded in the chamber and cultured statically for 12 h, then applied a vibratory stimulus for 24 h. We photographed every 5 min. Finally, we observed the distribution of cells after stimulation.

3   Result

We counted number of cells in each time-lapse image. Figure 2 shows comparison of cells before and after the vibratory stimulation. We varied applying vibratory stimulus, voltages applied to the motor were set to 1.0 V, 1.5 V and static culture (no vibration). As a result, (1) number of cells increases under the static culture condition; (2) There is no significant difference in number of cells under the 1.0 V condition; (3) Number of cells decreases under the 1.5 V condition. It is suggested that we can control the number of cells by changing vibratory energy. From observation, we could see that the decrease was caused by detaching of cells. Furthermore, cells detached when its divide.

We tried to perform the specific cell patterning. We used a vibration motor bonded a rectangular plastic plate as a vibration source. After stimulation, the

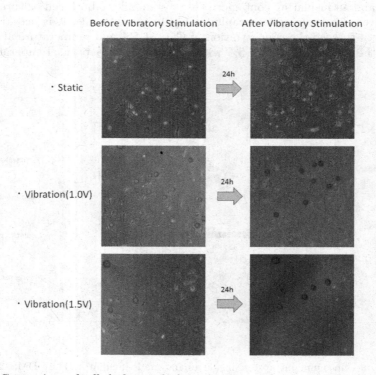

Fig. 2. Comparison of cells before and after vibratory stimulation. Orange round particles are observation markers embedded in the PDMS film. (Color figure online)

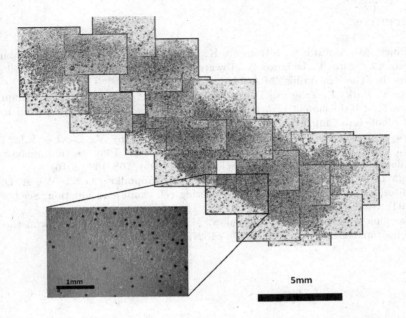

Fig. 3. A rectangular cell pattern. This image is stitched microphotographs of the center of the chamber and performed sharpening and edge detection. Cells are shown in black. The image at the bottom left is a magnification around boundary of cell aggregation. We also observed disc-like cell pattern with vibratory stimuli (not shown in this paper).

motor was removed, then cells were cultured statically for 36 h. Figure 3 shows the image which is stitched microphotographs of the center of the chamber and performed image processing. The black dots in the image represent cells. It can be seen that a rectangular cell pattern was formed.

4 Conclusion

In this report, we investigated the effect of a vibratory stimulus on cells. As a result, cells are promoted to detach by the stimulus, and it is suggested that cell increment depends on vibratory energy. Furthermore, we could find the difference of the local number of cells by observing the cell distribution after stimulus. Based on those results, we tried to perform the specific cell patterning. After this, it is necessary to investigate about cell death. The next challenge is to study a method to precisely control detaching of cells and to form minuter pattern of cells.

References

1. Shimizu, M., Yawata, S., Miyamoto, K., Miyasaka, K., Asano, T., Yoshinobu, T., Yawo, H., Ogura, T., Ishiguro, A.: Toward biorobotic systems with muscle cell actuators. In: The Proceedings of AMAM, pp. 87–88 (2011)
2. Kim, J., Park, J., Yang, S., Baek, J., Kim, B., Lee, S.H., Yoon, E.-S., Chun, K., Park, S.: Establishment of a fabrication method for a long-term actuated hybrid cell robot. Lab Chip **7**(11), 1504–1508 (2007)
3. Park, S.-J., Gazzola, M., Park, K.S., Park, S., Di Santo, V., Blevins, E.L., Lind, J.U., Campbell, P.H., Dauth, S., Capulli, A.K., et al.: Phototactic guidance of a tissue-engineered soft-robotic ray. Science **353**(6295), 158–162 (2016)
4. Singhvi, R., Kumar, A., Lopez, G.P., Stephanopoulos, G.N., Wang, D.I.C., Whitesides, G.M., Ingber, D.E.: Engineering cell shape and function. Science 696 (1994). New York then Washington
5. Yokoyama, S., Matsui, T.S., Deguchi, S.: Microcontact peeling as a new method for cell micropatterning. PloS One **9**(7), e102735 (2014)

Dry Adhesion of Artificial Gecko Setae Fabricated via Direct Laser Lithography

Omar Tricinci[1]([⊠]), Eric V. Eason[2,3], Carlo Filippeschi[1], Alessio Mondini[1], Barbara Mazzolai[1], Nicola M. Pugno[4,5,6], Mark R. Cutkosky[3], Francesco Greco[1,7], and Virgilio Mattoli[1]

[1] Center for Micro-BioRobotics @SSSA, Istituto Italiano di Tecnologia, Viale Rinaldo Piaggio 34, 56025 Pontedera, Italy
omar.tricinci@iit.it

[2] Lockheed Martin Advanced Technology Center, Palo Alto, CA 94304, USA

[3] Biomimetics and Dexterous Manipulation Laboratory, Department of Mechanical Engineering, Center for Design Research, Stanford University, Stanford, CA 94305-2232, USA

[4] Laboratory of Bio-Inspired and Graphene Nanomechanics, Department of Civil Environmental and Mechanical Engineering, University of Trento, 38123 Trento, Italy

[5] School of Engineering and Materials Science, Queen Mary University of London, Mile End Road, E1 4NS London, UK

[6] Italian Space Agency, Via del Politecnico Snc, 00133 Rome, Italy

[7] Department of Life Science and Medical Bio-Science, Graduate School of Advanced Science and Engineering, Waseda University, 2-2 Wakamatsu-Cho Shinjuku-Ku, 169-8480 Tokyo, Japan

Abstract. Biomimetics has introduced a new paradigm: by constructing structures with engineered materials and geometries, innovative devices may be fabricated. According to this paradigm, both shape and material properties are equally important to determine functional performance. This idea has been applied also in the field of the microfabrication of smart surfaces, exploiting properties already worked out by nature, like in the case of self-cleaning, drag reduction, structural coloration, and dry adhesion. Regarding dry adhesive properties, geckos represent a good example from which we take inspiration, since they have the extraordinary ability to climb almost every type of surface, even smooth ones, thanks to the hierarchical conformation of the fibrillary setae in their toe pads. Due to this design, they can increase the area of contact with a surface and thus the amount of attractive van der Waals forces. While reproducing with artificial materials the same functional morphology of gecko's pads is typically not achievable with traditional microfabrication techniques, recently Direct Laser Litography offered new opportunities to fabrication of complex three-dimensional structures in the microscale with nanometric resolution. Using direct laser lithography, we have fabricated artificial gecko setae, reproducing with unprecedented faithfulness the natural morphology in the same dimensional scale. Adhesion force of artificial setae toward different surfaces have been tested in dry condition by means of a dedicated setup and compared with natural ones.

Keywords: Biomimetics · Direct laser lithography · Gecko · Dry adhesion

M. Mangan et al. (Eds.): Living Machines 2017, LNAI 10384, pp. 631–636, 2017.
DOI: 10.1007/978-3-319-63537-8_60

1   Introduction

Biomimetic principles have attracted great interest in the field of microfabrication techniques, since they represent the cutting edge in the production of smart functional surfaces [1]. In particular, several studies have tried to reproduce the dry adhesion capability of gecko foot [2–5]. Geckos are able to climb up surfaces with various roughness, even smooth, supporting their weight thanks to the hierarchical conformal morphology of the pads at the micro- and nano-scale that provides a huge number of short range interactions mainly governed by van der Waals forces [6]. As an example, *Gekko gecko* can produce 10 N of adhesive force since each of the 5000 setae mm^{-2} ensures an average adhesion force of about 20 μN, but this value can even be tenfold higher depending on the preload [7].

Artificial surfaces inspired by the gecko foot could have several technological applications in the realization of films with high and reversible adhesiveness to different substrates. A surface inspired by the gecko foot would have an asymmetric behavior: it provides strong adhesion and requires low force for the detachment.

It has been possible to mimic the Gecko effect by reproducing with traditional microfabrication techniques the adhesive terminal units of the gecko pads. However, the faithful reproduction of three-dimensional hierarchical morphology, and comparable adhesion force, is still far to be achieved. Direct laser lithography recently demonstrated its potentiality for the fabrication of biomimetic surfaces patterned with three-dimensional features in the micro- and nano-scale [8, 9]. Here this microfabrication technique has been used for the replication of complex artificial hairs inspired by the gecko's pads setae. A dedicated setup based on a single-axis force sensor, capable of measuring down to the nN scale, has been implemented for the measurement of the adhesion force of the structures on a flat silicon surface.

2   Microfabrication of Artificial Setae

The gecko foot skin is composed by a multilevel structure of lamellae, setae, branches and spatulae composed of β-keratin. The length of the setae is in the range of 30–130 μm and the diameter is 5–10 μm while the spatulae, the end portions ramified from the branches, have tips 0.5 μm long, 0.2–0.3 μm wide and 0.01 μm thick. Such type of structure is complex since the long stalk has a high aspect ratio and the tips are composed by nanoscale contact units. Direct laser lithography allows to overcome the challenge of replicating such complex design respect to other fabrication techniques.

Two different models have been prepared for the fabrication of artificial gecko setae. The first model tries to reproduce the rough morphology of the gecko setae and it has been used for the calibration of all the parameters of the 3D laser lithography, such as, for example, the laser power and the writing speed. The second model has been realized for the definition of a more realistic morphology that can reproduce the effective shape of the natural one with a higher level of details. The structures realized according to this type of model have been used for the adhesion test.

For the calibration of parameters of the 3D laser lithography a CAD model of the setae has been realized with Solid Works. The setae represent the basic element of the gecko toe hierarchical structures (Fig. 1a). The CAD model of the stalk has been sliced with Nanoslicer (Nanoscribe GmbH), obtaining the coordinates of the points of the sections parallel to the substrate. An external shell has been added to the stalk to cover the internal construction and enhance the structural stability.

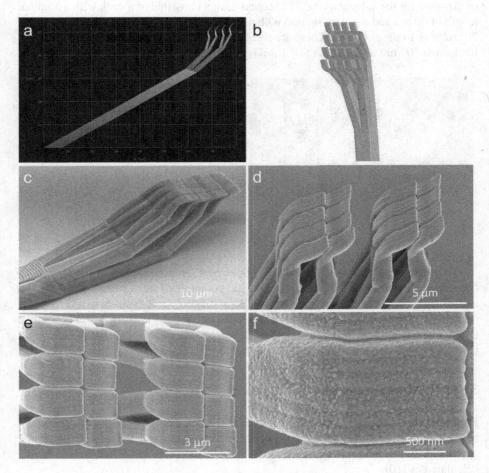

Fig. 1. First design of the setae for the calibration of the lithographic parameters: (a) Design in DeScribe; (b) Simulation of the tips in Matlab; (c-f) Results of the microfabrication.

The tips have been modelized as paths of coordinates in Matlab, and the photo polymerized structures have been simulated by sweeping an ellipsoid along the paths (Fig. 1b). The sorting strategy is based on a breadth-first algorithm, in which all adjacent edges at the same depth from the base are written before moving to the next level of edges that are further from the base. In this way the setae are better supported while being built. The programmed stalk has a nominal length of 68 μm and a square section with nominal sides of 3.4 µm; it forms an angle of 30° relative to the substrate. The branches grow of about 20 µm along the direction perpendicular to the substrate. There are 16 rectangular final tips with sides of 1 µm and 1.2 µm. The microfabrication produced outstanding results (Fig. 1c), even at the level of the tips (Fig. 1d–f).

The lithographic parameters have been defined selecting those providing the shortest fabrication time but with the best resolution of the features and used for the fabrication of the setae for the adhesion test. The second design consisted of a stalk with a nominal length of 60 µm and a square section with nominal sides of 2.8 µm; it forms an angle of 30° relative to the substrate. There are three levels of branches ending with 128 tips in total, each 300 nm in thickness and 1 µm in length (Fig. 2a, b).

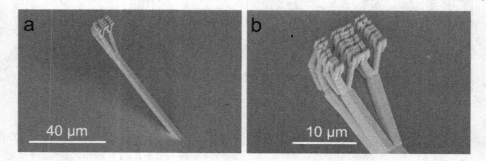

Fig. 2. Second design of the setae: (a) View of the seta; (b) Detail of the tips.

The setae were fabricated in IP-DiLL photoresist (Nanoscribe GmbH) by means of Nanoscribe (Nanoscribe GmbH). A square silicon sample of side of 1 mm was glued on a glass slide; the resist was poured on the silicon substrate and exposed to a laser beam (center wavelength 780 nm), using a writing speed of 100 µm s^{-1} with a power of 5.6 mW (Calman laser source). The sample was developed for 20 min in SU-8 Developer (MicroChem Corp) and rinsed in IPA and deionized water.

3   Adhesion Test of Artificial Setae

Frictional adhesion in gecko mostly relies on relatively weak van der Waals forces which become significant at very small setae-surface distances, that is in the range of an atomic gap distance of 0.3 nm. The normal force between the setae and a flat surface can be calculated as [10]:

$$F_{VDW} = -\frac{AL\sqrt{R}n_f n_s}{8\sqrt{2}D^{5/2}}$$

where n_f and n_s are the number of tips in the spatulae and the number of spatulae in the setae, respectively; A, the material-dependent Hamaker constant, is in the range of 10^{-20}–10^{-19} J, the typical values for condensed phases. According to this model, the adhesion force for this type of artificial setae range from 0.5 to 5 µN. The parallel friction forces, that were investigated in this work, are instead in the range of 20 µN [7].

The measurement of such small forces required a dedicated setup, prepared for the test of adhesion properties of the artificial gecko setae (Fig. 3a). The sample was mounted on the end effector of a nano-manipulator (Kleindiek Nanotechnik) which allows the movement with three degrees of freedom. The single-axis sensor (FemtoTools), able to detect perpendicular force in both directions in the range of ± 100 µN was positioned on a micro-manipulator used to move it in the correct configuration relative to the sample under an optical microscope (Hirox KH-7700). The cycles of movement of the nano-manipulator with the artificial setae against the force sensor was controlled by a dedicated software. Adhesion to the silicon smooth surface of force sensor was tested.

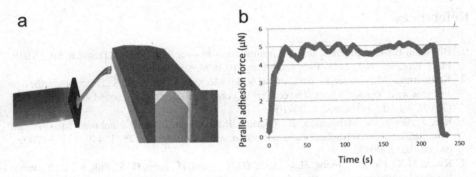

Fig. 3. Experimental results: (a) Experimental setup for the measurement of the adhesion of the setae; (b) Adhesion test result.

For the adhesion test a cyclic procedure was repeated. It consisted of a phase of approaching and perpendicular preloading, a phase of parallel displacement of about 5 µm, in order to ensure the proper conformation of the tips to the substrate, a sliding phase of the setae along the surface of the micro-force sensor during which the parallel adhesive force is measured, and finally a perpendicular detachment. Also a calibration procedure for the estimation of the perpendicular preload as a function of the bending of the setae was required. All the measurement of the adhesion force was carried out at a pushing speed of 35 nm s^{-1} that can be considered quasi-static.

In Fig. 3b the typical behavior of the parallel adhesive force as a function of time is reported. A linear relation between the adhesive force and the perpendicular preload was found. The range of interest of the preload is around 5 µN, since this is the typical value observed in gecko setae, which can provide an average force of 20 µN in this condition. Finally, in order to compare the performances of the artificial setae and the natural ones, it is necessary to normalize their contact area. The area covered by the artificial spatulae is about 50% of the area of the natural ones, so that the normalized adhesion force value is 10 µN. Compared with the performance of the natural gecko, the measured force is lower but in the same range, meaning that there could be the possibility to improve it by tuning the morphology of the setae, by modifying the angles of the branches, the dimensions and the numbers of the adhesive features.

4   Conclusions

Here it has been presented a procedure for the fabrication of gecko foot-like micro-structures by means of direct laser lithography. These structures faithfully reproduced the features at the microscale of the animal model, representing a new goal in the micro-fabrication field respect to standard techniques. The adhesion test of the artificial setae provided encouraging results; in order to achieve higher forces and emulate the natural model the results obtained so far permit to envision as a promising strategy an interplay of morphology, dimensional scaling and materials. Thus artificial setae could be the starting point of a new generation of dry and reversible adhesives able to conform to almost every surface.

References

1. Bhushan, B.: Biomimetics: lessons from nature–an overview. Philos. Trans. R Soc. Math. Phys. Eng. Sci. **367**, 1445–1486 (2009). doi:10.1098/rsta.2009.0011
2. Kamperman, M., Kroner, E., del Campo, A., McMeeking, R.M., Arzt, E.: Functional adhesive surfaces with "Gecko" effect: The concept of contact splitting. Adv. Eng. Mater. **12**, 335–348 (2010). doi:10.1002/adem.201000104
3. Lee, J., Bush, B., Maboudian, R., Fearing, R.S.: Gecko-inspired combined lamellar and nanofibrillar array for adhesion on nonplanar surface. Langmuir **25**, 12449–12453 (2009). doi:10.1021/la9029672
4. Kwak, M.K., Pang, C., Jeong, H.-E., Kim, H.-N., Yoon, H., Jung, H.-S., Suh, K.-Y.: Towards the next level of bioinspired dry adhesives: new designs and applications. Adv. Funct. Mater. **21**, 3606–3616 (2011). doi:10.1002/adfm.201100982
5. Hawkes, E.W., Eason, E.V., Asbeck, A.T., Cutkosky, M.R.: The gecko's toe: scaling directional adhesives for climbing applications. IEEE/ASME Trans. Mechatron. **18**, 518–526 (2013). doi:10.1109/TMECH.2012.2209672
6. Autumn, K., Sitti, M., Liang, Y.A., Peattie, A.M., Hansen, W.R., Sponberg, S., Kenny, T.W., Fearing, R., Israelachvili, J.N., Full, R.J.: Evidence for van der Waals adhesion in gecko setae. Proc. Natl. Acad. Sci. **99**, 12252–12256 (2002). doi:10.1073/pnas.192252799
7. Autumn, K., Liang, Y.A., Hsieh, S.T., Zesch, W., Chan, W.P., Kenny, T.W., Fearing, R., Full, R.J.: Adhesive force of a single gecko foot-hair. Nature **405**, 681–685 (2000). doi: 10.1038/35015073
8. Tricinci, O., Terencio, T., Mazzolai, B., Pugno, N.M., Greco, F., Mattoli, V.: 3D micropatterned surface inspired by salvinia molesta via direct laser lithography. ACS Appl. Mater. Interfaces **7**, 25560–25567 (2015). doi:10.1021/acsami.5b07722
9. Marino, A., Filippeschi, C., Mattoli, V., Mazzolai, B., Ciofani, G.: Biomimicry at the nanoscale: current research and perspectives of two-photon polymerization. Nanoscale **7**, 2841–2850 (2015). doi:10.1039/C4NR06500J
10. Israelachvili, J.N.: Intermolecular and Surface Forces: Revised, 3rd edn. Academic Press, Cambridge (2011)

Gesture Recognition Through Classification of Acoustic Muscle Sensing for Prosthetic Control

(Extended Abstract)

Samuel Wilson$^{(\boxtimes)}$ (iD) and Ravi Vaidyanathan

Imperial College London, London, UK
s.wilson14@imperial.ac.uk

Abstract. In this paper we present the initial evaluation of a new upper limb prosthetic control system to be worn on the residual limb, which is capable of identifying hand gestures through muscle acoustic signatures (mechanomyography, or MMG) measured from the upper arm. We report the development of a complete system consisting of a bespoke inertial measurement unit (IMU) to monitor arm motion and a skin surface sensor capturing acoustic muscle activity associated with digit movement. The system fuses the orientation of the arm with the synchronized output of six MMG sensors, which capture the low frequency vibrations produced during muscle contraction, to determine which hand gesture the user is making. Twelve gestures split into two test categories were examined, achieving a preliminary average accuracy of 89% on the offline examination, and 68% in the real time tests.

Keywords: Mechanomyography · Prosthetic control · Gesture recognition

1 Introduction

Mechanomyography (MMG) refers to the recording of mechanical vibrations produced during muscle contraction, and therefore provides a non-invasive method of deriving control signals from neuromuscular activity. The most commonly examined Human Computer Interface (HCI) used to control powered prosthetics is Surface Electromyography (sEMG). sEMG monitors the electrical signals propagating through the muscle. It is generally accepted that MMG signals and EMG signals are generated through the same process, however the signals from MMG have a much higher signal to noise ratio than EMG, and can be reliably recorded outside the clinical environment [1]. EMG calibration relies on a number of transient factors, (skin impedance, sweat level), which do not affect MMG. This is of particular importance for sensors within an airtight socket.

MMG suffers from an increased level of crosstalk when compared to EMG. While this makes it harder to determine which specific muscle the signal originated from, it was hypothesized by the authors that the interactions of the

© Springer International Publishing AG 2017
M. Mangan et al. (Eds.): Living Machines 2017, LNAI 10384, pp. 637–642, 2017.
DOI: 10.1007/978-3-319-63537-8_61

component signals from complex gestures would produce more complex surface signals, which may allow gesture recognition to take places without the need to identify the activated muscles, which has potential advantages over currently used systems in the practical applications of this work.

We can categorize gestures as either large or small, where large gestures require the activation of an entire muscle group, and small gestures require specific areas of muscle activation. Small gestures include moving the first finger to touch the thumb, and moving the second finger to do the same. Large gestures include opening/closing the hand and rotating the wrist. Previous MMG studies examining large gestures have achieved an accuracy of 79.66% across four gestures [2]. To the best of the authors' knowledge, gesture identification using a combination of large and small gestures has not previously been attempted using MMG.

Previous studies with EMG have used large 12 sensor arrays to gather data, and using this approach, up to 50 movements have been classified with an accuracy of 75% and 46% for healthy subjects and amputees respectively [3]. Other studies have examined up to 15 small finger movements using 6–12 sensors, achieving a classification accuracy of 98% and 90% [4].

We present an MMG based prosthetic control system, where the user's intent is derived and acted upon. The method of interaction outlined in this paper records a combination of arm orientation and muscle activation, and predicts the users' intent before translating that into commands that can be executed.

2   Experimental Protocol

A self-contained portable system that was capable of recording high frequency signal interactions could not be found, therefore the authors designed and manufactured suitable hardware. Two types of sensors were used; MMG sensors for recording muscle activity and an IMU to determine the orientation of the arm.

MMG signals have a maximum useful frequency of 100 Hz [5], however, due to the potential for signal iterations between sensors, the sampling frequency of this hardware was set to 1 kHz. The mechanomyography sensors were built utilizing a MEMS microphone and a conical echo chamber, which has been shown to give the best frequency response for MMG signals in previous studies [6,7].

2.1   Data Collection

The assembled armband consisted of 6 MMGs mounted around the arm, and one IMU. It was positioned with the IMU placed roughly over the extensor carpi radialis brevis, ensuring the MMG sensors were spread evenly over major flexors and extensors in the forearm/residual limb. Five healthy participants and one transradial amputee were instructed through 2 recording sessions, which consisted of them making 100 instances of 7 distinct gestures during the first session, and 5 distinct gestures during the second (The test length was halved for the transradial amputee as that subject suffered from a faster onset of fatigue). Each test involved the subject making a single gesture in response to a visual cue.

A number of common gestures were selected for the Test 1 protocol, with a view to future development to integrate the test output with real-time control of a prosthetic hand. The gestures chosen were open hand, close hand, pinch using first finger, pinch using second finger, thumbs up, point with first finger, and close all fingers sequentially. The subjects began in the rest position with the arm held vertically, and returned to the rest position between each gesture by completely relaxing the hand, so as not to produce unintended MMG signals.

For Test 2, the participants were instructed to allow their hand to hang by their side, and then to twitch each digit on their hand in turn. This test was designed to determine whether it was possible to distinguish between similar, small gestures.

3   Data Processing

Gestures were extracted from the data stream using a variation of the segmentation algorithm outlined in [8], which used a threshold on the energy of the data to indicate whether a gesture had been made. The IMU was used to identify motion induced artefacts within the data and disregard them. Using this technique to find the start of a gesture, a 200 ms dataset was recorded, band pass filtered and passed to the template based classification algorithm as $\tilde{s}$. 200 ms was chosen because visualisation of the data showed that the first contractions to initiate movements were the most consistent part of each gesture.

A template based classification algorithm was chosen due to its low computational expense, allowing for future implementation of the classification process on the hardware. The templates were created by averaging multiple instances of $\tilde{s}$ to form an N row matrix P where each row was a template for each MMG channel. Thus the l_{th} row of P can be expressed as:

$$P_l = \frac{1}{R} \sum_{i=1}^{R} \tilde{s}_i \tag{1}$$

where R is the total amount of repeated actions in the gesture training set and s_i is the segmented vector for each repetition. A dataset for each of the 12 gestures was created with each participant, then randomly split 70/30 into a training set and a test set. The training set was used to create a template for each MMG channel for 12 gestures, thus creating a matrix P for each gesture. This ensured that any features consistent across the majority of the gestures remained, while those that were specific to individual instances were lost.

Each gesture in the test set was then correlated to the templates. First, the data for each candidate gesture was segmented to create a matrix M where each row represented the data from one MMG channel for the candidate action. Each row of M was then correlated to the P matrix for each gesture using

Pearson product-moment correlation coefficient. Thus, the correlation coefficient ρ between the ith row of M and P will be:

$$\rho_{M_i P_i} = \frac{\sum_{j=1}^{b} [M_{i_j} - \bar{M}_{i_j}][P_{i_j} - \bar{P}_{i_j}]}{\sigma_{M_i} \sigma_{P_i}} \tag{2}$$

where σ is the standard deviation for each vector and $\bar{M}_i$ and $\bar{P}_i$ are the average values of the vectors. For each row of M and P, the correlation coefficient will be between -1 and 1. The correlation coefficient for each row of M to every P gesture matrix was generated and summed to produce Q. The systems prediction for the gesture was that with the largest Q value.

4   Results

The classification algorithm was applied to the test data, and used to generate a confusion matrix for each subject. Table 1 shows the combined confusion matrix for all participants. Additionally, a dataset containing 38 gestures was recorded from one subject. These included both fast and slow variations of hand movements, wrist movements and mouse movements, including left and right clicks. The classification accuracy for these 38 gestures was 72.8%.

The results of this extended test appear to indicate that the results from the 12 gesture tests are not dependant on which gestures are chosen, and therefore a user could pick gestures which they are able to comfortably and consistently perform and assign them to specific control outcomes. This is essential in a prosthetic system, as it allows the wearer to pick gestures that are convenient and comfortable for them, enabling more natural control, and therefore increasing the probability that they will be consistent. It may also have significant benefit when applied to subjects who have never had a fully functional limb, and therefore do not have prior experience and muscle memory of the gestures.

Table 1. Confusion matrix for all subjects performing all gestures

		Gesture Recognized											
		1	2	3	4	5	6	7	8	9	10	11	12
Actual Gesture	1	78.9%	3.4%	0.0%	2.0%	1.4%	0.0%	1.4%	0.0%	4.1%	2.7%	3.4%	2.7%
	2	2.8%	89.0%	0.0%	0.0%	0.7%	1.4%	1.4%	1.4%	0.7%	0.0%	2.1%	0.7%
	3	2.0%	2.0%	85.8%	3.4%	0.7%	0.7%	0.0%	0.0%	5.4%	0.0%	0.0%	0.0%
	4	3.5%	1.4%	5.6%	84.6%	0.0%	0.0%	2.1%	0.7%	0.7%	0.7%	0.0%	0.7%
	5	1.4%	1.4%	3.4%	1.4%	81.4%	3.4%	0.0%	0.7%	0.7%	0.7%	0.7%	4.8%
	6	0.0%	2.7%	0.7%	0.0%	4.1%	87.0%	1.4%	0.7%	0.0%	2.1%	1.4%	0.0%
	7	3.1%	3.8%	0.8%	3.0%	5.3%	3.8%	68.7%	3.1%	2.3%	1.5%	1.5%	3.1%
	8	0.8%	2.3%	0.0%	2.3%	1.5%	3.8%	3.0%	73.5%	1.5%	3.8%	5.3%	2.3%
	9	2.2%	3.6%	2.2%	0.0%	2.9%	0.0%	0.7%	0.7%	77.0%	3.6%	5.0%	2.2%
	10	0.0%	2.1%	0.0%	1.4%	0.0%	1.4%	1.4%	2.1%	3.5%	84.7%	2.8%	0.7%
	11	0.0%	4.2%	0.0%	0.0%	1.4%	0.7%	0.0%	1.4%	0.0%	4.9%	80.4%	7.0%
	12	0.7%	0.7%	0.7%	2.8%	5.6%	2.1%	2.1%	2.1%	4.9%	0.7%	11.3%	66.2%
											Total Accuracy		79.9%

Table 2. Real time classification accuracy

Classification

Subject	T1	T2	T3
1	93.2%	74.0%	74.3%
2	85.0%	61.0%	52.7%
3	91.5%	58.0%	64.7%
4	98.3%	79.3%	78.6%
5	95.0%	28.0%	

Table 3. Real time segmentation accuracy

Segmentation

Subject	T1	T2	T3
1	98.3%	97.3%	98.7%
2	100.0%	99.3%	98.6%
3	98.3%	100.0%	100.0%
4	100.0%	100.0%	100.0%
5	100.0%	97.3%	

The offline analysis was seamlessly ported to the real-time system. The recorded data was filtered in real time and written to a rolling buffer with a length of 50 samples (50 ms). Once the signal segmentation threshold had been reached, the remaining 150 ms of the gesture were recorded.

These real time experiments consisted of three tests. For the first (T1), the user trained the system with two gestures (Open and Close). The second test (T2) involved five of the seven gestures from the first recording session (the thumb up and sequential finger close were removed after multiple participants were unable to reliably repeat the gestures). The third test (T3) consisted of the five gestures from the second recording session.

Each user created their templates during the training period, where they recorded 70 instances of each gesture. The subject was then instructed to make 30 sets of each gesture, which were classified in real time. Tables 2 and 3 show the classification and segmentation accuracies for the real time tests.

5 Conclusion

The average accuracy of the real time pattern matching technique was 68% over 7 gestures. Offline, an accuracy of 80% was achieved over 12 gestures. We believe that the experimental protocol led to the lower accuracy for the real time recognition, and using a template that uses positive feedback on successful recognition to emphasize the more recent gestures would account for muscle fatiguing during use. With this protocol in place, we expect the online accuracy to increase by adapting to any changes in the muscle activity signals. The segmentation algorithm achieved an accuracy of 99.2% in the real time tests.

Based on this work, we believe that the complex interaction of mechanical signals within the arm is capable of producing repeatable patterns, and that these patterns offer a comparable alternative to previously examined EMG solutions. This is particularly evident in the 38 gesture data set, classified with an accuracy of 72.8%. In addition, the independence of the MMG sensors to transient local and physiological conditions means that mechanomyography may be better suited to commercial prosthetic control than electromyography, the currently accepted solution. This work will be advanced by examining the effect of removing and replacing the sensor array on the signal consistency.

642   S. Wilson and R. Vaidyanathan

Acknowledgements. We express our gratitude to Dr Enrico Franco and to those tested. The work is supported by the US Office of Naval Research Global (ONR-G) (N62909-14-1-N221) and UK-India Educational Research Initiative (UK-IERI) (IND/CONT/E/14-15/366).

References

1. Woodward, R., Gardner, M., Angeles, P., Shefelbine, S., Vaidyanathan, R.: A novel acoustic interface for bionic hand control. In: Natraj, A., Cameron, S., Melhuish, C., Witkowski, M. (eds.) TAROS 2013. LNCS (LNAI), vol. 8069, pp. 296–297. Springer, Heidelberg (2014). doi:10.1007/978-3-662-43645-5_32
2. Zeng, Y., Yang, Z.Y., Cao, W., Xia, C.M.: Hand-motion patterns recognition based on mechanomyographic signal analysis. In: 2009 International Conference on Future Biomedical Information Engineering (Fbie 2009), pp. 21–24 (2009)
3. Atzori, M., et al.: Electromyography data for non-invasive naturally-controlled robotic hand prostheses. Sci. Data **1**, 140053 (2014). PMC Web. 11 May 2017
4. Al-Timemy, A.H., Bugmann, G., Escudero, J., Outram, N.: Classification of finger movements for the dexterous hand prosthesis control with surface electromyography. IEEE J. Biomed. Health Inf. **17**(3), 608–618 (2013)
5. Silva, J., Chau, T., Goldenberg, A.: MMG-based multisensor data fusion for prosthesis control, pp. 2909–2912 (2003)
6. Posatskiy, A.O., Chau, T.: Design and evaluation of a novel microphone-based mechanomyography sensor with cylindrical and conical acoustic chambers. Med. Eng. Phys. **34**, 1184–1190 (2012)
7. Silva, J., Chau, T., Naumann, S., Heim, W., Goldenberg, A.A.: Optimization of the signal-to-noise ratio of silicon-embedded microphones for mechanomyography. In: Canadian Conference on Electrical and Computer Engineering, vol. 3, pp. 1493–1496 (2003)
8. Mace, M., Abdullah-Al-Mamun, K., Vaidyanathan, R., Wang, S., Gupta, L.: Real-time implementation of a non-invasive tongue-based human-robot interface. In: 2010 IEEE/RSJ International Conference on Intelligent Robots and Systems, Taipei, pp. 5486–5491 (2010)

Erratum to: Bio-inspired Design of a Double-Sided Crawling Robot

Jong-Eun Lee[1], Gwang-Pil Jung[2], and Kyu-Jin Cho[1(✉)]

[1] Seoul National University, Seoul 08826, Republic of Korea
{yhjelee,kjcho}@snu.ac.kr
[2] Seoul National University of Science and Technology,
Seoul 01811, Republic of Korea
gwangpiljung@gmail.com

Erratum to:
**Chapter "Bio-inspired Design of a Double-Sided
Crawling Robot" in: M. Mangan et al. (Eds.):
Biomimetic and Biohybrid Systems, LNAI,
DOI: 10.1007/978-3-319-63537-8_50**

By mistake, the initially published version of chapter 50 omitted the acknowledgements. This has been updated.

The updated online version of this chapter can be found at
http://dx.doi.org/10.1007/978-3-319-63537-8_50

Author Index

Printed in the United States
By Bookmasters